NANOTECHNOLOGY FOR ENVIRONMENTAL POLLUTION DECONTAMINATION

Tools, Methods, and Approaches for Detection and Remediation

NANOTECHNOLOGY FOR ENVIRONMENTAL POLLUTION DECONTAMINATION

Tools, Methods, and Approaches for Detection and Remediation

Edited by

Fernanda Maria Policarpo Tonelli, PhD

Rouf Ahmad Bhat, PhD

Gowhar Hamid Dar, PhD

AAP | APPLE ACADEMIC PRESS

First edition published 2023

Apple Academic Press Inc.
1265 Goldenrod Circle, NE,
Palm Bay, FL 32905 USA

760 Laurentian Drive, Unit 19,
Burlington, ON L7N 0A4, CANADA

CRC Press
6000 Broken Sound Parkway NW,
Suite 300, Boca Raton, FL 33487-2742 USA

4 Park Square, Milton Park,
Abingdon, Oxon, OX14 4RN UK

Library and Archives Canada Cataloguing in Publication

Title: Nanotechnology for environmental pollution decontamination : tools, methods, and approaches for detection and remediation / edited by Fernanda Maria Policarpo Tonelli, PhD, Rouf Ahmad Bhat, PhD, Gowhar Hamid Dar, PhD.
Names: Tonelli, Fernanda Maria Policarpo, editor. | Bhat, Rouf Ahmad, 1981- editor. | Dar, Gowhar Hamid, editor.
Description: First edition. | Includes bibliographical references and index.
Identifiers: Canadiana (print) 20220249563 | Canadiana (ebook) 2022024975X | ISBN 9781774910405 (hardcover) | ISBN 9781774910412 (softcover) | ISBN 9781003279563 (ebook)
Subjects: LCSH: Nanotechnology—Environmental aspects. | LCSH: Pollutants. | LCSH: Detectors.
Classification: LCC T174.7 .N38 2023 | DDC 628.5—dc23

Library of Congress Cataloging-in-Publication Data

CIP data on file with US Library of Congress

ISBN: 978-1-77491-040-5 (hbk)
ISBN: 978-1-77491-041-2 (pbk)
ISBN: 978-1-00327-956-3 (ebk)

About the Editors

Fernanda Maria Policarpo Tonelli, PhD

Fernanda Maria Policarpo Tonelli, PhD, is currently working as associate professor at Federal University of São João del Rei, Divinópolis-MG, Brazil. She was previously with the Department of Morphology, Federal University of Minas Gerais, BH—Minas Gerais, Brazil. She specializes in molecular biology and has been studying biotechnological topics such as gene delivery approaches (using engineered viral particles and nanomaterials) aiming for transgenesis. She has taught topics related to biochemistry and molecular biology and has authored scientific articles and more than 30 book chapters from international publishers. She has also reviewed various articles/book proposals. She has presented and participated in many national and international conferences and has helped organize various scientific events. Dr. Tonelli has also dedicated herself to the promotion of science and technology through cofunding with an oil and gas corporation. Dr. Tonelli is active in scientific advocation groups for women. Her efforts as a researcher have been recognized with various awards, including "For Women in Science Brazil-L'Oreal/UNESCO/ABC" and "Under30 Brazil–Forbes" and with certificates of merit.

Rouf Ahmad Bhat, PhD

Rouf Ahmad Bhat, PhD, presently working at the Department of School Education, Government of Jammu and Kashmir, India. He has been teaching graduate and postgraduate students of environmental sciences for the past three years. He is an author of more than 50 research articles (h-index 20; i-index 28; total citation >980) and 40 book chapters. He has published more than 28 books with international publishers (Springer, Elsevier, CRC Press Taylor and Francis, Apple Academic Press, John Wiley and IGI Global). He specializes in limnology, toxicology, phytochemistry, and phytoremediation. Dr. Bhat has presented and participated in numerous state, national, and international conferences, seminars, workshops, and symposium. In addition, he has worked as an associate environmental expert on the World Bank-funded Flood Recovery Project and also as environmental support staff for Asian Development Bank (ADB)-funded development projects. He has received many awards, appreciations, and recognition for his services to the science of water testing and air and noise analysis. He has served as an editorial board member and reviewer for reputed international journals. Dr. Bhat is still writing and experimenting with diverse capacities of plants for use in aquatic pollution remediation.

He pursued his doctorate at Sher-e-Kashmir University of Agricultural Sciences and Technology Kashmir (Division of Environmental Science), India.

Gowhar Hamid Dar, PhD

Gowhar Hamid Dar, PhD, is currently working as an Assistant Professor in Environmental Science, Sri Pratap College, Cluster University Srinagar, Department of Higher Education (J&K), India. He has a PhD in Environmental Science with specialization in Environmental Microbiology (Fish Microbiology, Fish Pathology, Industrial Microbiology, Taxonomy and Limnology). He has been teaching post graduate and graduate students for the past many years at Post graduate Department of Environmental Science, Sri Pratap College, Cluster University Srinagar. He has more than 50 papers (h-index 12; i-index 14; total citation >360) in international and national journals of repute and a number of books with international publishers (Springer, Elsevier, CRC Press Taylor and Francis, Apple Academic Press, John Wiley, IGI Global) to his credit. Moreover, he is supervising a number of students for the completion of degrees. He has been working on the Isolation, Identification and Characterization of microbes, their pathogenic behavior and impact of pollution on development of diseases in fish fauna for the last several years. He has received many awards and appreciations for his services toward the science and development. Besides, he is also an acting member of various research and academic committees.

Contents

Contributors ... *xi*

Abbreviations ... *xv*

Foreword by ***Mairaj Ud Din Sheikh*** ... *xxi*

Preface ... *xxv*

PART I: ENVIRONMENTAL POLLUTION ... **1**

1. Organic Pollutants Threatening Human Health ... **3**
Cássia Michelle Cabral, Kamila Cabral Mielke, Fábio Ribeiro Pires, and José Barbosa Dos Santos

2. The Risk of Inorganic Environmental Pollution to Humans ... **39**
Zia Ur Rahman Farooqi, Muhammad Moaz Khursheed, Nouman Gulzar, Muhammad Ahmad Akram, Muhammad Mahroz Hussain, Abdul Qadeer, Waqas Mohy-Ud-Din, and Sadia Younas

PART II: NANOTECHNOLOGY AND NANOSENSORS ... **59**

3. Electrochemical, Optical, Magnetic, and Colorimetric Nanosensors as Tools to Detect Environmental Pollution ... **61**
Bambang Kuswandi and Muhammad Afthoni

4. Nanotechnological Revolution: Carbon-Based Nanomaterials Detecting Pollutants ... **101**
Vandana Singh and Shalvi Upadhyay

5. Applying Noncarbon-Based Nanomaterials to Detect Environmental Contaminants ... **135**
Kannan Deepa, Ashish Kapoor, and Prabhakar Sivaraman

PART III: NANOTECHNOLOGY AND NANOBIOSENSORS ... **155**

6. Introduction: Nanozymes—Nouvelle Vague of Artificial Enzymes ... **157**
Manmeet Kaur, Inderpal Kaur, and Gautam Chhabra

7. **Potential of Nanobiosensors for Environmental Pollution Detection: Nanotechnology Combined with Enzymes, Antibodies, and Microorganisms 189**
Tamoghni Mitra, Saurav Kumar Sahoo, Arpita Banerjee, Anjani Kumar Upadhyay, and Kazi Nahid Hasan

8. **Nanobiosensors Containing Noncarbon-Based Nanomaterials to Assess Environmental Pollution Levels 219**
Vinars Dawane, Satish Piplode, Vishnu K. Manam, Prabodh Ranjan, and Abhishek Chandra

9. **Carbon-Based Nanomaterials: Nanobiosensors Detecting Environmental Pollution 263**
Swapnali Jadhav, Ekta B. Jadhav, Swaroop S. Sonone, Mahipal Singh Sankhla, and Rajeev Kumar

PART IV: NANOREMEDIATION 299

10. **Nanomaterials to Remediate Water Pollution 301**
Fernanda Maria Policarpo Tonelli, Flávia Cristina Policarpo Tonelli, Moline Severino Lemos, and Danilo Roberto Carvalho Ferreira

11. **Carbonaceous Materials for Nanoremediation of Polluted and Nutrient-Depleted Soils**
Guilherme Max Dias Ferreira, Gabriel Max Dias Ferreira, José Romão Franca, and Jenaina Ribeiro-Soares

12. **Air Pollution Management by Nanomaterials 409**
Yassine Slimani, Essia Hannachi, and Ghulam Yasin

PART V: NANOBIOREMEDIATION 443

13. **Nano-Phytoremediation: Using Plants and Nanomaterials to Environmental Pollution Remediation 445**
Ahmed Ali Romeh

14. **Mass Production of Arbuscular Mycorrhizal Fungus Inoculum and Its Use for Enhancing Biomass Yield of Crops for Food, for In Situ Nano-Phyto-Mycorrhizo Remediation of Contaminated Soils and Water, and for Sustainable Bioenergy Production 469**
A. G. Khan and A. Mohammad

PART VI: NANOMATERIALS FEASIBILITY 485

15. Hazardous and Safety and Management of Nanomaterials for the Personal Health and Environment 487
J. Immanuel Suresh and A. Judith

16. Economic Impact of Appslied Nanotechnology: An Overview 495
Mir Zahoor Gul and Beedu Sashidhar Rao

17. Sustainability Aspects of Nano-Remediation and Nano-Phytoremediation 509
Misbah Naz, Muhammad Ammar Raza, Sarah Bouzroud, Essa Ali, Syed Asad Hussain Bukhari, Muhammad Tariq, and Xiaorong Fan

Index 527

Contributors

Muhammad Afthoni
Pharmacy Department, Faculty of Science and Technology, Ma Chung University,
Villa Puncak Tidar N-01 65151, Malang, Indonesia

Muhammad Ahmad Akram
Environmental Protection Agency, Faisalabad 38000, Pakistan

Essa Ali
Institute of Plant Genetics and Developmental Biology, College of Chemistry and Life Sciences,
Zhejiang Normal University, Jinhua 321000, China

Arpita Banerjee
School of Biotechnology, Kalinga Institute of Industrial Technology, Bhubaneswar, Odisha, India

Syed Asad Hussain Bukhari
Department of Agronomy, Bahauddin Zakariya University, Multan 60800, Pakistan

Sarah Bouzroud
Laboratoire de Biotechnologieet Physiologie Végétales, Centre de biotechnologievégétale et
microbiennebiodiversité et environnement, Faculté des Sciences, Université Mohammed V de Rabat,
Rabat 1014, Morocco

Cássia Michelle Cabral
Department of Agronomy, Federal University of Vales do Jequitinhonha e Mucuri, Diamantina, MG, Brazil

Abhishek Chandra
Department of Chemistry, HVHP Institute of Post Graduate Studies and Research,
Kadi Sarva Vishwavidhyalaya, Kadi 382715, Gujarat, India

Gautam Chhabra
School of Agricultural Biotechnology, Punjab Agricultural University, Ludhiana 141001, India

Vinars Dawane
School of Environment and Sustainable Development, Central University of Gujarat,
Gandhinagar 382030, Gujarat, India; E-mail: vinars27dawane2009@gmail.com

Kannan Deepa
Department of Chemical Engineering, SRM Institute of Science and Technology, Kattankulathur 603203,
Tamil Nadu, India

Waqas Mohy-Ud-Din
Institute of Soil and Environmental Sciences, University of Agriculture Faisalabad 38040, Pakistan

Xiaorong Fan
Key Laboratory of Plant Nutrition and Fertilization in Lower-Middle Reaches of the Yangtze River,
Ministry of Agriculture, and Nanjing Agricultural University, Nanjing 210095, China;
E-mail: xiaorongfan@njau.edu.cn
State Key Laboratory of Crop Genetics and Germplasm Enhancement, Nanjing Agricultural University,
Nanjing 210095, China

Zia Ur Rahman Farooqi
Institute of Soil and Environmental Sciences, University of Agriculture Faisalabad 38040, Pakistan; E-mail: ziaa2600@gmail.com

Danilo Roberto Carvalho Ferreira
Department of Materials, CDTN/CNEN, Belo Horizonte, Brazil

Gabriel Max Dias Ferreira
Department of Chemistry, Federal University of Ouro Preto, Campus Morro do Cruzeiro, 35400-000, Ouro Preto, Minas Gerais, Brazil

Guilherme Max Dias Ferreira
Department of Chemistry, Institute of Natural Science, Federal University of Lavras, 37200-900, Lavras, Minas Gerais, Brazil

José Romão Franca
Department of Physics, Institute of Natural Science, Federal University of Lavras, 37200-900, Lavras, Minas Gerais, Brazil

Mir Zahoor Gul
Department of Biochemistry, University College of Sciences, Osmania University, Hyderabad, 500007, Telangana, India; E-mail: ziahgul@gmail.com

Nouman Gulzar
Institute of Soil and Environmental Sciences, University of Agriculture Faisalabad 38040, Pakistan

Essia Hannachi
Laboratory of Physics of Materials - Structures and Properties, Department of Physics, Faculty of Sciences of Bizerte, University of Carthage, Zarzouna 7021, Tunisia

Kazi Nahid Hasan
School of Biotechnology, Kalinga Institute of Industrial Technology, Bhubaneswar, Odisha, India

Moline Severino Lemos
Department of Cell Biology, ICB/UFMG, Belo Horizonte, Brazil

Muhammad Mahroz Hussain
Institute of Soil and Environmental Sciences, University of Agriculture Faisalabad 38040, Pakistan

Ekta B. Jadhav
MSc Forensic Science, Government Institute of Forensic Science, Aurangabad, Maharashtra, India

Swapnali Jadhav
M.Sc. Forensic Science, Government Institute of Forensic Science, Aurangabad, Maharashtra, India

A. Judith
PG Department of Microbiology, The American College, Madurai 625002, Tamil Nadu, India

Ashish Kapoor
Department of Chemical Engineering, SRM Institute of Science and Technology, Kattankulathur 603203, Tamil Nadu, India

Inderpal Kaur
Department of Biochemistry, Punjab Agricultural University, Ludhiana141001, India

Manmeet Kaur
Department of Microbiology, Punjab Agricultural University, Ludhiana 141001, India; E-mail: manmeet1-mb@ pau.edu

A. G. Khan
Grad Life Member Western Sydney University, Sydney, Australia; E-mail: lasara@gmail.com

Muhammad Moaz Khursheed
Institute of Soil and Environmental Sciences, University of Agriculture Faisalabad 38040, Pakistan

Rajeev Kumar
Department of Forensic Science, School of Basic and Applied Sciences, Galgotias University, Greater Noida, India

Bambang Kuswandi
Chemo and Biosensors Group, Faculty of Pharmacy, University of Jember, Jl, Kalimantan 37, Jember 68121, Indonesia; E-mail: b_kuswandi.farmasi@unej.ac.id

Vishnu K. Manam
Department of Plant Biology and Biotechnology, Unit of Algal Biotechnology and Nanobiotechnology, Pachaiyappa's College, University of Madras, Chennai 600030, Tamil Nadu, India

Kamila Cabral Mielke
Department of Agronomy, Federal University of Viçosa, Viçosa, MG, Brazil

Tamoghni Mitra
School of Biotechnology, Kalinga Institute of Industrial Technology, Bhubaneswar, Odisha, India

A. Mohammad
W. Booth School of Engineering Practice and Technology, Engineering Technology Building, McMaster University, Hamilton, Ontario, Canada

Misbah Naz
State Key Laboratory of Crop Genetics and Germplasm Enhancement, Nanjing Agricultural University, Nanjing 210095, China; E-mail: raymisbah@ymail.com

Satish Piplode
Department of Chemistry, Sahid Bhagat Singh Govt. P. G. College Pipariya, Hoshangabad 461775, Madhya Pradesh, India

Fábio Ribeiro Pires
Department of Agricultural and Biological Sciences, Federal University of Espírito Santo, São Mateus, ES, Brazil; E-mail: pires.fr@gmail.com

Abdul Qadeer
Institute of Soil and Environmental Sciences, University of Agriculture Faisalabad 38040, Pakistan

Prabodh Ranjan
Department of Chemical Engineering, Indian Institute of Technology Madras, Chennai 600036, Tamil Nadu, India

Beedu Sashidhar Rao
Department of Biochemistry, University College of Sciences, Osmania University, Hyderabad, 500007, Telangana, India

Muhammad Ammar Raza
College of Food Science and Biotechnology, Key Laboratory of Fruits and Vegetables Postharvest and Processing Technology Research of Zhejiang Province, Zhejiang Gongshang University, Hangzhou 310018, China

Ahmed Ali Romeh
Plant Production Department, Faculty of Technology and Development, Zagazig University, Zagazig, Egypt; E-mail: ahmedromeh2006@yahoo.com

Saurav Kumar Sahoo
MITS School of Biotechnology, Bhubaneswar, Odisha, India

Mahipal Singh Sankhla
Department of Forensic Science, School of Basic and Applied Sciences, Galgotias University, Greater Noida, India; E-mail: mahipal4n6@gmail.com

José Barbosa Dos Santos
Department of Agronomy, Federal University of Vales do Jequitinhonha e Mucuri, Diamantina, MG, Brazil

Yassine Slimani
Department of Biophysics, Institute for Research and Medical Consultations (IRMC), Imam Abdulrahman Bin Faisal University, Dammam 31441, Saudi Arabia; E-mail: yaslimani@iau.edu.sa; slimaniyassine18@gmail.com

Vandana Singh
Department of microbiology, School of Allied Health Science, Sharda University, Greater Noida, 201306, Uttar Pradesh, India; E-mail: vandana.singh@sharda.ac.in

Prabhakar Sivaraman
Department of Chemical Engineering, SRM Institute of Science and Technology, Kattankulathur 603203, Tamil Nadu, India; E-mail: prabhaks@srmist.edu.in

Jenaina Ribeiro Soares
Department of Physics, Institute of Natural Science, Federal University of Lavras, 37200-900, Lavras, Minas Gerais, Brazil; E-mail: jenaina.soares@ufla.br

Swaroop S. Sonone
MSc Forensic Science, Government Institute of Forensic Science, Aurangabad, Maharashtra, India

J. Immanuel Suresh
PG Department of Microbiology, The American College, Madurai 625002, Tamil Nadu, India; E-mail: immanuelsuresh1978@gmail.com

Muhammad Tariq
Faculty of Pharmacy and Alternative Medicine, The Islamia University Bahawalpur 6300, Pakistan

Flávia Cristina Policarpo Tonelli
Department of Pharmacy, UFSJ/CCO, Divinópolis, Brazil; E-mail: flacristinaptonelli@gmail.com

Fernanda Maria Policarpo Tonelli
Department of Cell Biology, ICB/UFMG, Belo Horizonte, Brazil; E-mail: tonellibioquimica@gmail.com

Anjani Kumar Upadhyay
School of Biotechnology, Kalinga Institute of Industrial Technology, Bhubaneswar, Odisha, India; E-mail: upadhyayanjanikumar6@gmail.com

Shalvi Upadhyay
Department of Forensic Sciences, School of Allied Health Science, Sharda University, Greater Noida, 201306, Uttar Pradesh, India

Ghulam Yasin
State Key Laboratory of Chemical Resource Engineering, College of Materials Science and Engineering, Beijing University of Chemical Technology, Beijing 100029, China

Sadia Younas
Department of Chemistry, The Government Sadiq College Women University, Bahawalpur 63100, Pakistan

Abbreviations

AAS	absorption spectroscopy
ACs	activated carbons
AChE	acetylcholinesterase
AdS-DPV	adsorptive stripping-DPV
AFM	atomic force microscopy
AFS	fluorescence spectroscopy
AMF	arbuscular mycorrhizal fungus
ANN	artificial neural networks
AOB	antimony oxide bromide
AOPs	advanced oxidative processes
APAP	acetaminophen
APIs	active pharmaceutical ingredients
APTES	aminopropyltriethoxysilan
APTS	3-aminopropyltriethoxysilane
AQ-SPEC	Air Quality Sensor Performance Evaluation Center
ASV	anodic stripping voltammetry
ATP	adenosine-tri-phosphate
BBF	biochar-based fertilizer
BC	biochar
BCF	bioconcentration factor
BDD	boron-doped diamond
BFR	brominated compounds
BOD	biochemical oxygen demand
BPA	bisphenol-A
BSI	British Standard Institute
CAP	chloramphenicol
CBM	carbamate
CBNs	carbon black nanoparticles
CBNs	carbon-based nanomaterials
CD	circular dichroism
CDOM	colored dissolved organic matters
CEC	cation exchange capacity
CFCs	chlorofluorocarbons
CNM	carbon nanomaterials

CNTs	carbon nanotubes
CNT-FET	carbon-based field-effect transistors
COD	chemical oxygen demand
COPD	chronic obstructive pulmonary disease
CPX	antimycotin cycloperoxolamine
CQDs	carbon quantum dots
CSPE	carbon screen-printed electrode
CTC	cation exchange capacity
CV	cyclic voltammetry
CVD	chemical vapor deposition
CWA	chemical warfare agents
c-DecaBDE	decabromodiphenyl ether
DAP	diammonium phosphate
DDT	dichlorodiphenyltrichloroethane
DFT	density functional theory
DLCs	dioxin-like compounds
DLS	dynamic light scattering
DNB	dinitrobenzene
DNT	dinitrotoluene
DO	dissolved oxygen
DPV	differential pulse voltammetry
ECN	engineered carbonaceous nanomaterials
EDA	ethylene diamine
EDCs	endocrine disrupting compounds
EDX	energy-dispersive X-ray
EELS	electron energy loss spectroscopy
EFSA	European Food Protection Authority
EIA	US Energy Information Administration
EIS	electrochemical impedance spectroscopy
ELISA	enzyme-linked immunosorbent assay
ENR	enrofloxacin
EPA	environmental protection agency
EPO	European Patent Office
ERMN	extraradical mycelium network
EWNS	Engineered Water Nanostructures
FAM	fluorescein amidite
FETs	field-effect transistors
FRET	fluorescence resonance energy transfer
FTIR	fourier transform infrared spectroscopy

GC	gas chromatography
GCEs	glassy carbon electrodes
GC-MS	gas chromatography-mass spectrometry
GCNTs	graphene coated CNTs
GLDH	glutamate dehydrogenase
GN	graphene
GO	graphene oxide
GOx	glucoseoxidase
GQDs	graphene quantum dots
GR	graphene
GST	glutathione S-transferase
HBCDD	hexabromocyclododecane
HCB	hexachlorobenzene
HCH	hexachlorocyclohexane
HOC	hydrophobic organic compounds
HPLC	high-performance liquid chromatography
HPNPP	hydroxypropyl p-nitrophenyl phosphate
HQ	hydroquinone
HRP	horseradish peroxidase
HTA	high throughput analysis
ICT	intramolecular charge transfer
IFE	inner filter effect
IgG	immunoglobulin G
IL	ionic liquid
IL-CCE	ionic liquid-based carbon ceramic electrode
IPA	adsorption of Isopropyl alcohol
IPCC	Intergovernmental Panel on Climate Change
ITO	indium-tin-oxide
LMCT	ligand to metal charge transfer
LOC	lab-on-a-chip
LOD	limit of detection
LR	linear regression
LRET	luminescence resonance energy transfer
LSPR	localized surface plasmon resonance
LSV	linear sweep voltammetry
MAP	monoammonium phosphate
MCP	monocrotophos
ME	mechanical exfoliation
MFC	microbial fuel cell

MHB	mycorrhiza-helping bacteria
MIP	molecularly imprinted polymer
MLCT	metal to ligand charge transfer
MLR	multivariate linear regression
MMP	magnetic microparticle
MNPs	magnetic nanoparticles
MOFs	metal–organic frameworks
MOX	metallic-oxide-semiconductor
MP	methyl parathion
MRI	magnetic resonance imaging
MRSA	methicillin-resistant Staphylococcus aureus
MSDS	material safety data sheets
MSOs	mercury-specific-oligonucleotides
MWCNT	multiwall carbon nanotube
NBs	nanobelts
ND	nano-diamonds
NEP	evaluation of norepinephrine
NFs	nanofibers
NMs	nanomaterials
NMBs	nanomembranes
NMPR	nano-mycorrhizo-phytoremediation
NOM	natural organic matter
NPs	nanoparticles
NPGF	nanoporous gold film
NRBs	nanoribbons
NRs	nanorods
NTs	nanotubes
NWS	nanowires
OCP	organochlorinated pesticides
OPs	organophosphorus
PAHs	polynuclear aromatic hydrocarbons
PB	Prussian blue
PBDEs	pentabromodiphenyl ether
PCBs	polychlorinated biphenyls
PCDDs	polychlorodibenzo-p-dioxins
PCDFs	polychlorodibenzofurans
PCNs	polychlorinated naphthalenes
PCP	pentachlorophenol
PCR	polymerase chain reaction

PDMS	polydimethylsiloxane
PEC	photo-electro-chemical
PeCB	pentachlorobenzene
PECVD	plasma-enhanced chemical vapor deposition
PEI	polyethyleneimine
PET	photo-induced electron transfer
PFAS	poly-fluoroalkyl substances
PFCs	perfluorinated compounds
PGPR	plant-growth promoting rhizobacteria
P-HAP	p-hydroxyamenophen
PL	photo luminance
PNS	per-oxy-nano-sensor
POPs	persistent organic pollutants
PPA	phenylpropanolamine
PPCPs	pharmaceuticals, personal care products
PSi	porous Si
PTEs	potentially toxic elements
QCM	quartz crystal microbalance
QDs	quantum dots
RBH	rhodamine b hydrazine
rGO	reduced graphene oxide
RISC	RNA-induced silencing complex
ROS	reactive oxygen species
RRS	Rayleigh scattering
SAW	surface acoustic wave
SBS	styrene-butadiene-styrene
SDD	Secchi disk depth
SEM	scanning electron microscope
SERS	surface-enhanced Raman scattering
SPR	surface plasmon resonance
SSS	sea surface salinity
STEM	scanning transmission electron microscopy
STM	scanning tunnel microscopy
SWCNT	single-wall carbon nanotube
SWV	square wave voltammetry
TCDD	tetrachlorodibenzodioxin
TCE	trichloroethylene
TCh	thiocholine
TCPP	tetrakis (4-carboxyphenyl) porphyrin

TEM	transmission electron microscopy
TEOS	tetraethoxysilane
TLC	thin-layer chromatography
TNB	trinitrobenzene
TNT	trinitrotoluene
TP	total phosphorus
TPI	Terras Pretas de Índios
TPRS	two-photon Rayleigh scattering
TSS	total suspended sediments
VOC	volatile-organic compounds
VOMs	volatile organic materials
WB	wheat bran
WT	water temperature
XPS	x-ray fluorescence spectroscopy
XPS	x-ray photoelectron spectra

Foreword

Knowledge is an ever-expanding domain, and scientific enquiry forms an important part of it. Production of scientific literature, even though at times repetitive, serves its purpose of outreaching to the stakeholders. An important task of science, besides the theoretical construct it builds up, is to give solutions to the problems and help humankind to live in harmony with the environment. The ever-increasing population and the fight for finite resources has built up a strained relationship between humans and the environment. Since resource exploitation cannot be halted, the next best thing to do is to explore the scientific methods of limiting or eliminating damage to the environment so as to provide the Earth with sustainable use of resources. This book is one such attempt to bring together techniques (using nanotechnology) to obtain the objective of reducing harm to the environment, particularly in Indian conditions.

Rapid urbanization/industrialization, while not being given necessary attention to sustainability, has converted a large array of organic and inorganic substances into a serious risk to living beings. Environmental pollution has become a worldwide threat that severely affects ecosystems in a negative way. Air, soil, and water polluted by persistent contaminants such as pesticides and heavy metals require urgent attention, and efficient strategies to deal with pollutants aiming for remediation are extremely necessary.

The nanotechnology field offers interesting tools to deal with pollution, and this book is dedicated to exploring approaches and methods involving nanomaterials, both to detect contaminants and also to clean and restore polluted spots.

The book *Nanotechnology for Environmental Pollution Decontamination: Tools, Methods, and Approaches for Detection and Remediation* includes 17 chapters and is divided into six parts to allow readers to have access to a comprehensive view on the theme.

The introductory chapter, *Nanoenzymes: Nouvelle Vague of Artificial Enzymes,* is conceptualized by Indian authors to address a topic of nanoparticles containing enzyme-like features capable of catalyzing reactions efficiently to achieve a specific goal, such as nanoremediaton, which has been receiving increasing attention over recent years.

The first part of the book, *Environmental Pollution*, contains two chapters dedicated to exploring threats to human beings arising out of organic and inorganic environmental pollution. The first one, *Organic Pollutants Threatening Human Health*, contributed by Brazilian authors, reviews organic pollutants' negative impacts, and Chapter 2, which is written by authors from Pakistan, reviews the harmful consequences related to inorganic contaminants.

The second part, *Nanotechnology and Nanosensors*, presents three chapters and addresses the different types of nanosensors (nanomaterials used in devices to detect analytes) to detect environmental pollution. Chapter 3 has been written by authors from Indonesia and reviews these different types of nanosensors, emphasizing the technology advances on their performance for environmental pollutant detections. Chapter 4, produced by Indian authors is focused on nanosensors that contain nanomaterials from which chemical constitution is based on carbon atoms. Chapter 5, also by Indian authors, addresses nanosensors containing nanomaterials from which chemical constitution is not based on carbon atoms.

The third part, containing Chapters 6, 7, and 8, drafted by Indian authors, explores nanosensors containing biological material (enzymes, antibodies, whole microorganisms): the nanobiosensors to monitor environmental contamination. Chapter 6 discusses nanomaterials as strategic tools to promote an extremely responsive, active, and high-frequency detection of environmental contaminants. Chapter 7 reviews the importance, recent advancements, and key future challenges associated with noncarbon-based nanobiosensors regarding detecting pollutants, and Chapter 8 has a similar approach regarding carbon-based nanobiosensors.

The fourth part also presents three chapters and is dedicated to the use of nanomaterials to remediate polluted areas. Chapters 9 and 10 were produced by Brazilian authors, and Chapter 11 by authors from Saudi Arabia, Tunisia, and China. Chapter 9 is dedicated to discussing the use of materials in nanoscale to treat contaminated water, Chapter 10 addresses the use of these materials to remediate polluted soil, and Chapter 11 is focused on nanoremediating air pollution.

The fifth part contains Chapters 12 and 13 and discusses bioremediation using nanomaterials. Chapter 12, produced by an Egyptian author, focuses on the joint use of plants and nanomaterials, nanophytoremediation, to remediate environmental pollution. Chapter 13, written by an Australian author, addresses the use of microorganisms and nanomaterials to remediate pollution.

The sixth part explores the feasibility of nanotechnologies and contains two chapters. Chapter 14, produced by an Indian author, is dedicated to briefly reviewing the main aspects regarding safety of the use of these nanoscale materials. Chapter 15, on its turns, written by Indian authors, is dedicated to discussing the impacts of nanotechnology on the world's economy. Chapter 16, written by authors from China and Morocco, discusses strategies of remediation using nanomaterials that should be biocompatible and environment friendly to ecosystems.

The structure adopted to present the book chapters was designed to cover the most important aspects regarding the nanotechnology field as a very relevant contributor to offer strategic and efficient tools to remediate environmental pollutants, taking important steps on the path toward sustainable development. Therefore, the book can act as a source of attraction to the global scientific community. The editors must be highly complimented for their sincere hard work in bringing this volume.

—**Prof. Mairaj Ud Din Sheikh**
Department of Geography
Sri Pratap College
Srinagar, Jammu and Kashmir, India

The sixth part explores the feasibility of nanotechnologies and contains two chapters. Chapter 14 (produced by an Indian author) is dedicated to briefly reviewing the main aspects regarding safety in the use of these nanoscale materials. Chapter 15, on the other, written by Indian authors, is dedicated to discussing the impacts of nanotechnologies on the world economy. Chapter 16, written by authors from [illegible] [illegible]

Preface

Human actions have been ignoring sustainability principles to promote rapid industrialization/urbanization, which has severely impacted ecosystems worldwide in a negative way.

Organic and inorganic environmental pollutants generated at high levels have caused contamination of water, soil, and air, and have been threatening the lives of living beings. Strategies to reduce and eliminate the deleterious effects of environmental pollutant are, therefore, necessary in an urgent basis.

Diverse solutions are being developed in this regard; however, the nanotechnology field possesses enormous potential to provide materials in nanoscale that are designed specifically to perform efficiently a desirable task and at a low cost.

When it comes to environmental pollution, nanomaterials are useful not only to remediate the polluted areas but also to detect pollutants by means of sensors and also monitor the quality of environmental restoration performed by nanotechnologies or nanobiotechnologies.

This book is an attempt to offer undergraduate students and researchers an extensive and comprehensive knowledge on nanotechnology that can be applied to pollution detection and remediation, assisted or otherwise, by biological strategies from a biotechnological point of view.

There are 17 chapters in this book that address themes such as nanozymes; organic and inorganic pollutants threatening human health; different types of carbon-based and noncarbon-based nanomaterials in nanosensors and nanobiosensors to detect environmental pollution; nanomaterials to specially deal with water, soil, or air pollution; and assisted nanoremediation promoted by plants (nanophytoremediation) or microorganism (e.g., mycorrhizal fungus promoting in situ nano-phyto-mycorrhizo-remediation; and also to address; aspects related to a macroperspective of nanoremediation highlighting economy aspects related to nanotechnology, safety aspects on the use of nanomaterials, and also sustainability aspects related to the use of nanomaterials for strategies of environmental restoration.

We are extremely grateful to all the authors who have contributed various chapters in this book and to AAP for their generous cooperation in publishing this book.

—Dr. Fernanda Maria Policarpo Tonelli
Dr. Rouf Ahmad Bhat
Dr. Gowhar Hamid Dar

PART I
Environmental Pollution

CHAPTER 1

Organic Pollutants Threatening Human Health

CÁSSIA MICHELLE CABRAL[1], KAMILA CABRAL MIELKE[2], FÁBIO RIBEIRO PIRES[3*], and JOSÉ BARBOSA DOS SANTOS[1]

[1]*Department of Agronomy, Federal University of Vales do Jequitinhonha e Mucuri, Diamantina, MG, Brazil*

[2]*Department of Agronomy, Federal University of Viçosa, Viçosa, MG, Brazil*

[3]*Department of Agricultural and Biological Sciences, Federal University of Espírito Santo, São Mateus, ES, Brazil*

[*]*Corresponding author. E-mail: pires.fr@gmail.com*

ABSTRACT

Persistent organic pollutants (POPs) are a worldwide concern due to their bioaccumulative nature and persistence for a long period of time, resulting in large-scale environmental contamination. These pollutants are extremely resistant to biodegradation and subject to transfer over long distances via the atmosphere. They may be present even in regions with no historical use. POPs can bioaccumulate in adipose tissue due to their lipophilicity and can seriously affect the nervous, hepatic, reproductive, and hormonal systems of the contaminated organisms, including plants, animals, and humans. Therefore, they are responsible for various lethal diseases and environmental problems. The diversity of diseases due to POPs includes diabetes, endocrine disorders, cancer, and cardiovascular, and reproductive problems. As to achieve sustainability and safety in relation to the integrity

Nanotechnology for Environmental Pollution Decontamination: Tools, Methods, and Approaches for Detection and Remediation. Fernanda Maria Policarpo Tonelli, Rouf Ahmad Bhat, & Gowhar Hamid Dar (Eds.)

of public and environmental health, monitoring the effects of POPs and taking measures to mitigate the negative consequences is imperative.

1.1 INTRODUCTION

Technological and economic advances run parallel to the extraction of natural resources. The nature of the impacts from this relationship on the natural world and human health is not fully understood. Over time, natural resources are used indiscriminately for the success of economic activities. A primary consequence of this is environmental pollution, which is a serious concern that is discussed by researchers, government entities, and society in general (Alharbi et al., 2018; Rauert et al., 2018; Encarnação et al., 2019).

In recent decades, there has been an increase in the concentration of hazardous substances released into the environment from human activities, such as from the agricultural and industrial sectors (Guo et al., 2019). It is estimated that approximately 394,000 chemical substances are commercially available, and since the last decade, it has been found that around 2000 new substances enter the market each year (CAS, 2020), with spending on such chemicals projected to reach 14 billion dollars by 2050 (OPAS, 2018). The chemical landscape is constantly changing with new products and substances replacing the old ones, and the quantities produced and used vary according to their effectiveness and demand.

A diversity of pollutants can be found in the majority of environments. Among these, those resistant to degradation, whether through chemical, biological, or photolytic reactions are of the most important in the context of human and environmental health. These products are called persistent organic pollutants (POPs) and are composed of industrial chemicals (PCBs, polychlorinated biphenyls, PBDEs, PFOS, etc.), pesticides, and residues from industrial processes (dioxin and furano) (Alharbi et al., 2018; Guo et al., 2019). POPs are a class of carbon-based organic chemicals that are highly stable, bioaccumulative, and with progressive accumulation capability along the food chain (Guo et al., 2019).

Such molecules can be transported over long distances crossing over international borders, and their presence has been detected in places without any historical use (Alharbi et al., 2018). A large number of synthetic chemicals became commercially available in large quantities after the Second World War (Encarnação et al., 2019). Though synthetic chemical production rate is not directly causative of POP release into the environment and human exposure. Some chemicals are degradation products, such as hexachlorocyclohexane (HCH), resulting from the degradation of lindane, and p, p′-DDE (dichlorodiphenyldichlorethylene), a DDT metabolite; however, it is an indicator of the potential impact on human health and the environment (Encarnação et al., 2019).

An increasing number of reports on the impact of POPs on the endocrine system underline the concerns regarding POP exposure and human health (Encarnação et al., 2019). Such concerns are highlighted by the increasing number of studies that have reported environmental and ecological occurrences (Durante et al., 2016; Gaur et al., 2018; Fernandes et al., 2019).

1.2 WHY ARE POPS DANGEROUS?

POPs are synthetic organic chemical substances, distinguished by their unique chemical and physical characteristics, such as semivolatility, persistence, bioaccumulation, and toxicity (CETESB, 2020). Such physicochemical properties determine the environmental fate of POPs (Torre et al., 2016).

POPs are semivolatile and evaporate slowly (Durante et al., 2016). In this sense, they are carried by the air currents, reaching long distances in the atmosphere until they reach lower temperatures. When this occurs, POPs condense and precipitate to reach the earth's surface and accumulate in the ecosystems in soil and water, including in regions where there is no historical use. (Alharbi et al., 2018). Therefore, polar regions are the most affected regions by POP contamination. Thus, although POPs can be produced and applied in tropical and temperate regions, high concentrations of these substances, such as PCBs used in transformers and pesticides (such as DDT and toxafene), have been identified in ecosystems and humans in the polar regions where POPs have never been used (CETESB, 2020). As POPs have different volatilization capacities, they migrate through environments at different speeds, and contaminate watercourses and migratory species, thereby increasing the global reach of these compounds (Durante et al., 2016). The propagation of POPs around the world by air and ocean currents symbolizes that they are not only a local toxic threat but also a global issue (Yilmaz et al., 2020).

Some substances can remain in the environment for long periods without any change in their composition, demonstrating a resistance to biological and chemical degradation (Alharbi et al., 2018; Guo et al., 2019). Persistence is based on the measure of half-life, that is, the time required for half of the substance to be degraded, whether in hours, days, months, or even years (CETESB, 2020). Thus, adopting an average half-life of 60 days as an example, we will have 50% of the initial value measured after that period, falling by another 50% after another 60 days, and so on until the residue is no longer considered dangerous. So, persistence is defined from the time of emission to the point at which the contaminant is no longer dangerous. Furthermore, according to the Stockholm Convention, a contaminant is persistent if its half-life is greater than 2 months in water, more than 6 months in sediment and soil, and more than 2 days in the air. This persistence favors the dispersion of pollutants by volatilization, leaching or carryover, which contaminates air, water, and consequently soil and plants, and then animals and humans, which is the mechanism characterizing biomagnification or bioaccumulation (Li et al., 2021).

The persistent nature of POPs combined with their lipophilic characteristics means they have a high degree of toxicity to humans (Yilmaz et al., 2020). These compounds can bioaccumulate along the food chain, through diet or through the airway and lodge in adipose tissue, allowing these compounds to persist in biota that have a low metabolism rate (Rolle-Kampczyk et al., 2020). Consequently, the elimination of contaminants through excretion or biotransformation is also low. Consequently, they can reach concentrations that have harmful effects on human health and the environment. Evidence of bioaccumulation is given in some regions, where the presence of POPs is detected (regardless of whether there is local production or use), highlighting that these substances enter the food chains and accumulate in fish, birds, marine mammals, and humans (CETESB, 2020).

Exposure to these pollutants causes multiple health problems, such as hormonal disorders, cancer, cardiovascular diseases, obesity, reproductive and neurological diseases, learning difficulties, immune system dysfunction, susceptibility to disease, and diabetes (Alharbi et al., 2018). In addition, these pollutants are teratogenic (Guo et al., 2019). Many studies have described the mechanisms of POPs that lead to the development of various diseases and health problems, thus establishing through laboratory studies that POPs are toxic agents and represent an evident danger to human health (Guo et al., 2019; Kuang et al., 2020; Yilmaz et al., 2020) (Table 1.1).

1.3 ORIGIN OF POPS

POPs are largely of an anthropogenic origin, such as from pesticides used in agriculture to industrial production of chemical compounds. Some compounds, such as dioxins and furans can have a natural origin, they can also be released into the environment through volcanic activities or forest fires (Jacob and Cherian, 2013) (Fig. 1.1).

TABLE 1.1 Health Problems Resulting from Pollution Caused by Persistent Organic Pollutants (POPs).

POPs	Health problems	References
Dichlorodiphenyltrichloroethane (DDT), hexachlorobenzene (HCB) and polychlorinated biphenyls (PCBs)	842 cases associated with type 2 Diabetes	Wu et al. (2013)
DDT	Association between p, p' DDT level and body mass index and waist circumference.	Elobeid et al. (2010)
DDT and chlordane	May adversely affect survival after breast cancer diagnosis.	Parada et al. (2016)
Chlordane and PCBs	Serum concentrations are associated with the risk of lung cancer in the general population.	Park et al. (2020)
Pentabromodiphenyl ether (PBDEs)	Risks associated with neurotoxicity.	Martin et al. (2017)
HCB, HCHS, DDT, Mirex, and PCBs	Low doses similar to current exposure levels can increase the risk of type 2 diabetes, possibly through endocrine disruption	Lee et al. (2010)
Dioxins	Dioxin compounds can cause disruption and differentiation of tissue, cellular and biochemical processes	Haffner and Schecter (2014)
Endosulfan	Can induce oxidative stress and mitochondrial injury, activate autophagy, induce inflammatory response, and lead to endothelial dysfunction. This indicates that exposure to endosulfan is a potential risk factor for cardiovascular disease	Zhang et al. (2017)
Endosulfan	Reduces fertility levels in male animals, induces DNA damage leading to undesirable processing of broken DNA ends	Sebastian and Raghavan (2017)
HCB	Exposure can contribute to the development of endometriosis, affecting the parameters of inflammation and invasion of human endometrial cells	Chiappini et al. (2016)
Mirex	Among women aged 45–55 years, serum mirex was positively associated with menopause	Grindler et al. (2015)

TABLE 1.1 *(Continued)*

POPs	Health problems	References
Chlordecone	Risk of prostate cancer	Multigner et al. (2010)
Chlordane, heptachlor and mirex	Potential to disrupt fertilization, blastocyst implantation or even the duration of pregnancy of animals.	Wrobel and Mlynarczuk (2017)
Furans	Potentially carcinogenic to humans	Kanan and Samara (2018)
Aldrin	Potential to cause damage in the early stages of spermatogenesis in animals	Wrobel et al. (2015); Das Neves et al. (2018)

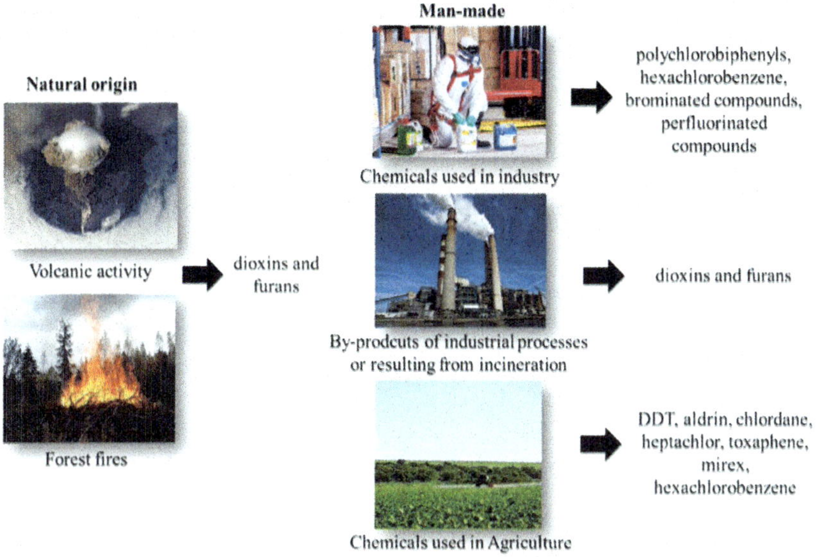

FIGURE 1.1 Classification of POPs according to origin.
Source: Adapted from Curtean-Bănăduc (2016).

POPs are introduced to ecosystems through direct pesticide applications, industrial waste, and as by-products of industrial processes of industries such as agriculture, forestry, horticulture, industries, and medicine, (Alharbi et al., 2018; Guo et al., 2019). Therefore, they are routinely detected in human systems and the environment (Yilmaz et al., 2020).

The source of the most representative POPs can also be compiled into intentional and unintentional substances. Synthetic substances produced as by-products of the chemical manufacturing industry like HCB, HCH, or as flame-retardants (PCBs, hexabromobiphenyl) are classified as intentional. Substances classified as unintentional are the result of industrial processes or the combustion of materials (dioxins and furans) (Liu et al., 2008; Zhang et al., 2010; Ferreira, 2013). The main sources of anthropogenic POP pollution (Curtean-Bănăduc, 2016) are summarized in Figure 1.2.

1.4 PRIORITY POLLUTANTS

POPs are toxic components that interact with the body. This contamination occurs by ingestion of food and drinking contaminated water (Liu et al., 2016), and can occur by inhalation or contact with the skin (Ye et al., 2015; Wrobel and Mlynarczuk, 2017). Inside the organism, it is absorbed, distributed to tissues, stored, biotransformed, and excreted (Rolle-Kampczyk et al., 2020; Bhuvaneswari et al., 2021).

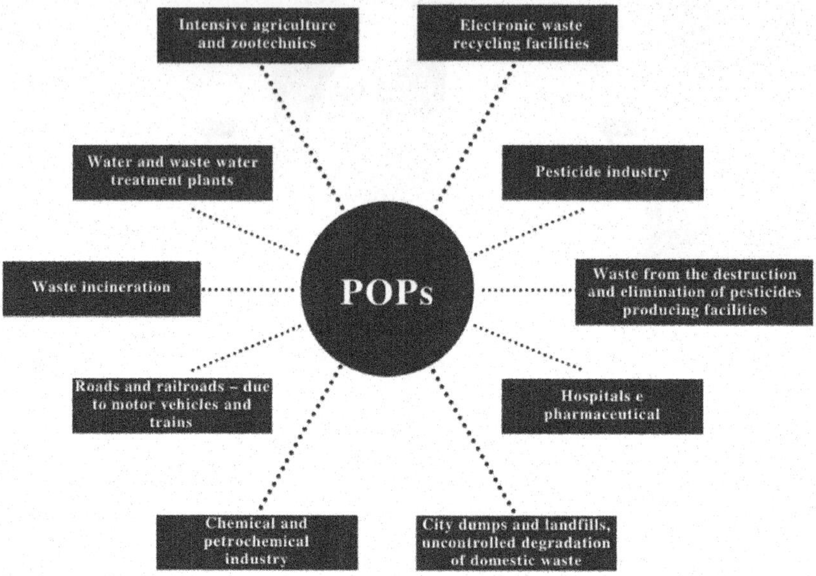

FIGURE 1.2 Main sources of anthropogenic pollution of POPs.

POPs include the following classes of substances: organochlorinated pesticides (OCP), PCBs, perfluorinated compounds (PFCs), brominated compounds (BFR), dioxins and furans (Curtean-Bănăduc, 2016). Most of these components are limited and restricted globally based on the Recommendations of the Stockholm Convention on POPs, an international treaty aimed at deliberating on and/or safe disposal of these pollutants as well as controlling their production and use, to which Brazil is a signatory. Thus, the Stockholm Convention implements measures to prevent the release of POPs into the environment in order to mitigate contamination. The main POPs are presented in Table 1.2.

1.4.1 ALDRIN

Aldrin is a broad-spectrum insecticide belonging to the cyclodiene class of organochlorinated pesticides. Very toxic in nature, it is volatile, bioaccumulative due to its high lipophilic properties and it persists in the environment

TABLE 1.2 Main POPs Classified at the Stockholm Convention.

Chemical	Annex	Use		Molecular formula	Chemical structure
Aldrin	A	Pesticide	Persistent organochlorine insecticide (DT50~365 days) that was mainly used to control soil-dwelling insects	$C_{12}H_8Cl_6$	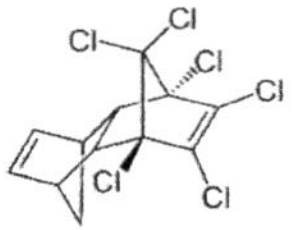
Chlordane	A	Pesticide	Organochlorine insecticide once commonly used to control a range of pests	$C_{10}H_6Cl_8$	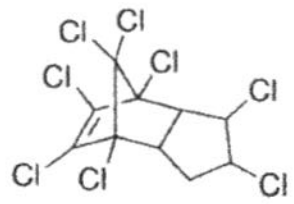
Chlordecone	A	Pesticide	A banned insecticide that is effective against leaf-cutting insects but less effective on sucking pests	$C_{10}Cl_{10}O$	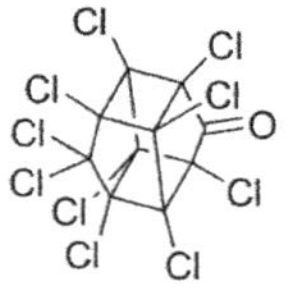
DDT	B	Pesticide	A banned insecticide, highly persistent in soil (4–30 years), that was used to control insect vectors of disease, especially malaria	$C_{12}H_9C_{15}$	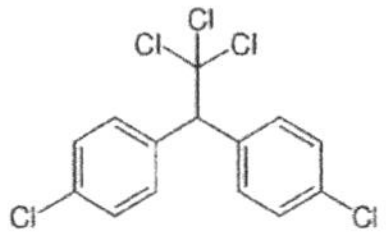
Decabromodiphenyl ether (c-DecaBDE)	A	By-product	It is one of the commercial formulations of PBDEs. It is widely used as brominated flame-retardants in furniture	$C_{12}Br_{10}O$	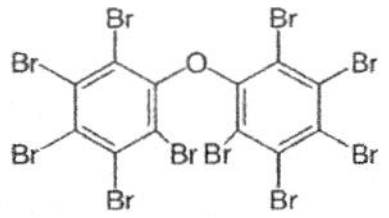
Dioxins	C	By-product	Burning processes, such as commercial or municipal waste incineration produce dioxins. Chlorinated dioxins have 75 different forms	$C_{12}H_4Cl_4O_2$	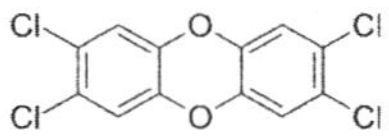

TABLE 1.2 *(Continued)*

Chemical	Annex	Use		Molecular formula	Chemical structure
Furans	C	By-product	Found in cigarette smoke, and is used in the production of resins and lacquers, agrochemicals, and pharmaceuticals. There are 135 different chlorinated furans	C_4H_4O	
Endosulfan	A	Pesticide	An isomer mixture of alpha- and beta-endosulfan which is an insecticide and acaricide, used to control sucking, chewing and boring insects	$C_9H_6Cl_6O_3S$	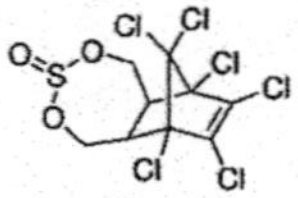
Heptachlor	A	Pesticide	Insecticide once used to kill termites, ants, and other insects in agricultural and domestic situations	$C_{10}H_5Cl_7$	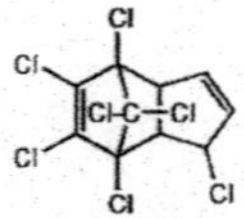
Hexabromobiphenyl	A	Industrial chemical	An industrial chemical that has been used as a flame-retardant mainly in the 1970s	$C_{12}H_4Br_6$	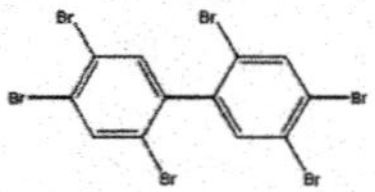
Hexabromo-cyclododecane (HBCDD)	A	Industrial chemical	Used as flame-retardant additive, providing fire protection during the service life of vehicles, buildings or articles, as well as protection while stored	$C_{12}H_{18}Br_6$	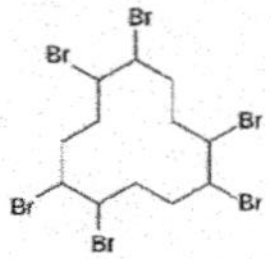
HCB	A and C	Pesticide/ industrial chemical	A chlorinated hydrocarbon fungicide used to control bunt and a pesticide transformation product	C_6Cl_6	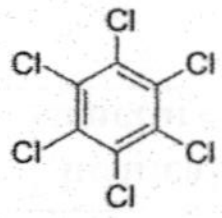

TABLE 1.2 *(Continued)*

Chemical	Annex	Use		Molecular formula	Chemical structure
HCH	A	Pesticide/ by-product	A broad-spectrum insecticide used mainly during seed treatment to control phytophagous and soil-inhabiting insects	$C_6H_6Cl_6$	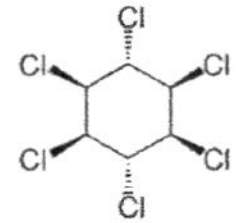
Lindane	A	Pesticide	Lindane has been used as a broad-spectrum insecticide for seed and soil treatment, foliar applications, tree and wood treatment, and against ectoparasites in both veterinary and human applications	$C_6H_6Cl_6$	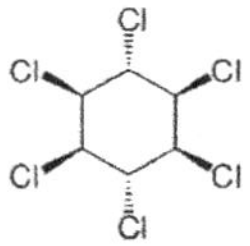
Mirex	A	Pesticide	This insecticide is used mainly to combat fire ants, and it has been used against other types of ants and termites. It has also been used as a flame-retardant in plastics, rubber, and electrical goods	$C_{10}Cl_{12}$	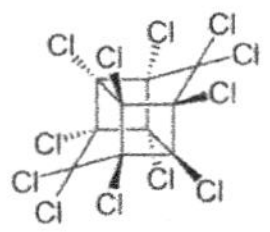
PBDEs	A	Industrial chemical	Additive flame-retardant. Commercial mixture of pentaBDE is highly persistent in the environment, bioaccumulative and has a high potential for long-range environmental transport	$C_{12}H_5Br_5O$	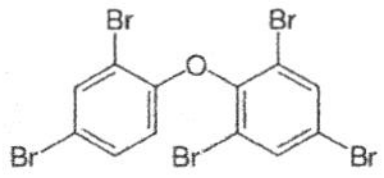
Pentachlorobenzene (PeCB)	A	Pesticide/ industrial chemical	PeCB was used in PCB products, in dyestuff carriers, as a fungicide, a flame-retardant and as a chemical intermediate, for example, previously for the production of quintozene	C_6HCl_5	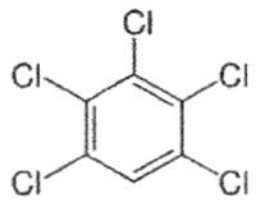
Pentachlorophenol (PCP) and its salts and esters	A	Pesticide	PCP has been used as herbicide, insecticide, fungicide, algaecide, disinfectant and as an ingredient in antifouling paint	C_6HCl_5O	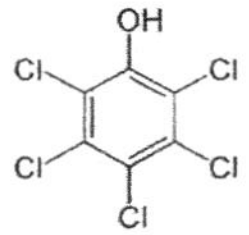

TABLE 1.2 *(Continued)*

Chemical	Annex	Use		Molecular formula	Chemical structure
PCB	A and C	Industrial chemical	These compounds are used in industry as heat exchange fluids, in electric transformers and capacitors, and as additives in paint, carbonless copy paper, and plastics	$C_{12}H_{10-n}Cl_n$	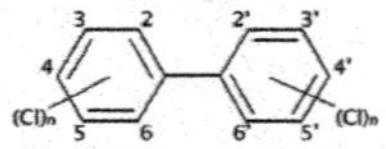
Polychlorinated naphthalenes (PCNs)	A and C	Industrial chemical	PCNs make effective insulating coatings for electrical wires. Others have been used as wood preservatives, as rubber and plastic additives, for capacitor dielectrics and in lubricants	$C_{10}H_8Cl_4$	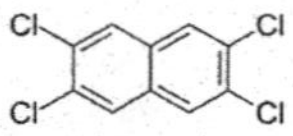

Source: Stockholm Convention (2020), PPDB (2020).

Annex A: measures are necessary to eliminate the production and use of these POP

Annex B: restriction measures are necessary in the production and use of these POP

Annex C: measures are needed to reduce the emission of such unintentional POP for up to three decades after application (Jhamtani et al., 2018). Aldrin undergoes biotransformation in the environment, turning into dieldrin (Najam and Alam, 2015). It is banned and restricted in several countries through the Stockholm Convention in the 1970s. It is still detected in the environment at high concentrations (Jhamtani et al., 2018).

Aldrin acts as an endocrine disruptor, interfering with the release and synthesis of hormones (Yilmaz et al., 2020). Aldrin inhibits hormonal action throughout the chorion, increasing the risk of abortions or premature births in cattle, necessitating the use of chorionic villus hormones to support pregnancy in animals (Mlynarczuk et al., 2020). Studies conducted in Jordan in the Middle East have verified pesticide residues in the yolk and white of 95% of the eggs tested, with mean concentrations of 0.067 and 0.058 mg/kg, respectively (0.02 mg/kg maximum egg residue limit according to the European Commission, 2016), demonstrating the inappropriate use of pesticides in home gardens, thus exposing the environment to pollution, representing a risk to human health (Alaboudi et al., 2019). Aldrin is highly toxic to humans and animals, causing seizures, instability, and excitation of the central nervous system (Jhamtani et al., 2017).

1.4.2 CHLORDANE

Chlordane is an organochloric pesticide that is persistent in the environment and is volatile, bioaccumulative, and highly toxic (Xiong, 2017). Chlordane was used in agriculture for termite control and was banned in 1988 in the USA (Parada et al., 2016).

Zebrafish larvae exposed to chlordane have low survival rates, delayed development and hatching times, reduced embryonic productivity, and abnormal heart rate and blood flow. These results suggest that exposure to chlordane in ecosystems causes direct morphological effects and phenotypic changes related to development and reproduction (Xiong, 2017). Chlordane has been associated with the incidence of breast cancer and subsequent mortality in women. Organochlorinated compounds are lipophilic, and in humans, breast tissue is rich in adipose tissue, causing bioaccumulation in this region (Parada et al., 2016).

1.4.3 CHLORDECONE

Chlordecone is an organochloric and lipophilic insecticide and is highly persistent in soil (Fournier et al., 2017). It was widely applied in banana plantations in the French Caribbean, Guadeloupe, and Martinique between 1970s and 1990s. With extremely slow degradation, chlordecone is still detected in soils in these regions (Benoit et al., 2020). In addition to soil, this pollutant has been found in surface and groundwater, coastal ecosystems, terrestrial ecosystems, and aquatic food chains. (Benoit et al., 2020). This persistence in the soil can lead to contamination of animals, mainly ruminants, since during grazing, soil injection may occur (Fournier et al., 2017).

It is classified as a carcinogenic and can interfere with the development of babies by affecting cognitive and motor development when exposed during pregnancy or breastfeeding (Fournier et al., 2017). Chlordecone is toxic to the reproductive system, neurotoxic, and disrupts the endocrine system due to its estrogenic properties *in vitro* and *in vivo* (Multigner et al., 2016). This pollutant was found in the umbilical cord blood of neonatal children and was associated with growth irregularities in boys up to 3 months and girls from 8 to 18 months, mainly in relation to height (Costet et al., 2015).

1.4.4 DICHLORODIPHENYLTRICHLOROETHANE

Dichlorodiphenyltrichloroethane (DDT) is an organochlorinated pesticide that is persistent in the environment, has high solubility in lipids and therefore has a bioaccumulation capacity, and can biomagnify in food chains (Mahugija et al., 2018). It was the first synthetic insecticide, and was used in agriculture and in the fight against insect-borne diseases, subsequently, negative impacts on the environment and human health have been verified (Thi Thuy, 2015).

This pesticide gained negative notoriety with the book "Silent Spring" by Rachel Carson. Carson associates DDT with reduced eggshell thickness, resulting in reproductive problems and bird death. In this sense, recent studies have reported that exposure can result in neurotoxic, carcinogenic, and immunologic effects and negative effects on reproductive function in free-ranging chickens in the northern area of KwaZulu-Natal (Thompson et al., 2017). Even in past life exposures, DDT is still detectable among Canadians and has been associated with dysfunction of the respiratory system (Ye et al., 2015). South Africa is one of the few countries that still actively sprays DDT to control the malaria vector (Gerber et al., 2016; Mahugija et al., 2018). Assessments in the muscle tissue of tiger fish, collected in the Kruger National Park region of South Africa, showed organochlorine bioaccumulation, and among them DDT was found at very high levels, surpassing those already registered in South Africa's freshwater systems. This is of serious concern to local populations that consume and depend economically on fishing (Gerber et al., 2016).

1.4.5 DECABROMODIPHENYL ETHER

Decabromodiphenyl ether (c-deca-BDE) is a commercial formulation of PBDEs. C-deca-BDE is mainly used in textiles and invariably in electrical and electronic products (Andrade et al., 2017). Due to its physicochemical properties, including hydrophobicity and lipophilicity, it has high persistence and sorption. C-deca-BDE is bioaccumulative in the environment, resulting in adverse effects on human health and the ecosystem (Kim et al., 2017). BDE-209 is the main analog of c-deca-BDE and is listed as a chemical of concern. C-deca-BDE has been under review in the USA since 2010, and some states have restricted its manufacture and use (Andrade et al., 2017).

c-DecaBDE tends to accumulate in household dust, and thus, can pose serious health risks, as they find their way into the human body through possible emissions (Maddela et al., 2020). BDE-209 (analogous to c-deca-BDE) has thyroid toxicity, endocrine, reproductive, behavioral, and neurological disorders as well as pulmonary disorders (Maddela et al., 2020). In a recent study, a link between neurobehavioral toxicity and BDE-209-induced visual dysfunction was observed in zebrafish larvae (Zhang et al., 2020).

1.4.6 DIOXIN

Dioxin-like compounds (DLCs) are composed of two benzene rings connected by two oxygen atoms and consist of four to eight chlorine atoms. 2,3,7,8-tetrachlorodibenzo-p-dioxin (TCDD), one of the most toxic dioxins, is persistent and has wide dispersion and potential for bioaccumulation in the food chain (Tavakoly Sany et al., 2015).

TCDD has serious toxicological effects, such as immune system disorders, teratogenesis, and induction of tumors (Tavakoly Sany et al., 2015). Evidence of reduced reproductive performance with adverse effects on subsequent generations in rodents and zebrafish exposed to TCDD was observed. Such changes are characterized by epigenetic changes in the placenta and/or sperm, suggesting that this effect can occur with human exposure (Viluksela and Pohjanvirta, 2019).

1.4.7 FURAN

Isomer 2,3,7,8 or tetrachlorodibenzo furan and dibenzofurans (PDCFs) are structurally similar to dioxins and share many of their toxic effects (Kanan and Samara, 2018). They are persistent and are the by-products of the synthesis or combustion of chlorine-based compounds (Rawn et al., 2017).

The main route of exposure is ingestion. A study aimed at establishing the Canadian national estimates of maternal and child exposure to environmental contaminants found that in 1017 human milk samples collected, 298 had traces of PCDFs (average of 2.1 pg/g lipid) accompanied by other POPs (Rawn et al., 2017). PCDFs are classified as toxic and carcinogenic (Kanan and Samara, 2018).

1.4.8 ENDOSULFAN

Endosulfan is a synthetic organochlorinated insecticide that is persistent in the environment, with a bioaccumulation capacity, volatility, and toxicity (Kumar et al., 2016). This insecticide has been widely used for crops in the last three decades. It degrades microbially and endosulfan sulfate is the primary metabolite, which presents toxicity analogous to the original compound and is harmful to several organisms (Supreeth and Raju, 2017).

It is an acute neurotoxic compound for insects, mammals, and humans (Lakroun et al., 2015). In addition, there are reports of serious poisonings and fatal cases of human contamination by ingestion (provoked), for doses above 260 mL of endosulfan (Patočka et al., 2016). It has been associated with cases of environmental contamination, soil and water pollution, and metabolic dysfunctions in living organisms (Mudhoo et al., 2019). In a study that investigated the physiological and molecular aspects of endosulfan, it was found that in addition to reducing fertility levels in male animals, the endosulfan induced DNA damage (Sebastian and Raghavan, 2017). In addition, there is no antidote for endosulfan poisoning (Menezes et al., 2017). Furthermore, the potential neurotoxic effects of endosulfan were investigated in mice by Lee et al. (2015). After a single neonatal exposure (0.1 or 0.5 mg/kg body weight) during a critical period of brain development, the researchers found that the pesticide could induce changes in neuroprotein levels, which is important for normal brain development, in addition to causing neurobehavioral abnormalities that persist in adulthood (Lee et al., 2015).

Cardiovascular problems may also be linked to endosulfan. This pesticide can induce oxidative stress and mitochondrial lesions, activating autophagy, leading to an inflammatory response that causes endothelial dysfunction via the AMPK/mTOR pathway. This indicates that exposure to endosulfan poses a potential risk for cardiovascular disease (Zhang et al., 2017). In the environment, it can cause an imbalance in food chains due to its toxicity to animals. Tests carried out on tilapia revealed that endosulfan negatively affected the liver function, including liver enzymes and plasma proteins, in addition to reducing the number of red blood cells and white blood cells. Endosulfan decreased the transcription levels of glutathione S-transferase (GST) mRNA and altered the normal histological structure of the liver, gills, and spleen of affected fish (Hussein et al., 2019).

1.4.9 HEPTACHLOR

Heptachlor is a synthetic organochloric insecticide used against insects, ants, and termites (Bhuvaneswari et al., 2021). It is persistent in the environment, nonsoluble in water, bioaccumulative, volatile, and toxic (Martínez-Ibarra et al., 2017).

Heptachlor residues or their biotransformation products have been found in environmental and biological matrices, such as breast milk, semen, serum, follicular fluids, liver, muscle tissue, adipose tissue, kidney, and human blood after exposure (Bhuvaneswari et al., 2021). Research with rats that were exposed to the compound through oral consumption during pregnancy and lactation indicated that heptachlor induced female reproductive changes, such as increased female anogenital distance, delayed vaginal opening, and reduction in progesterone production. Consequently, exposure to heptachlor may pose a risk to the reproductive health of humans (Martínez-Ibarra et al., 2017). This pesticide causes cellular oxidative damage, which interferes with the formation of the human proximal renal tubule. This oxidative stress may be responsible for the activation of growth factor signaling that leads to kidney damage through epithelial transition to the mesenchyme (Singh et al., 2019). Researchers suggest that a biotransformation product of heptachlor, heptachlor epoxide, is linked to inducing Lewy pathology. This disease is characterized by the progressive loss of mental function by the development of Lewy bodies, which culminates in Parkinson's disease (Zhang et al., 2020a). Studies on human aging in Honolulu-Asia revealed that the prevalence of Lewy pathology almost doubled in the presence of heptachlor epoxide in the samples of occipital and temporal lobes from frozen human brains compared with samples where the pesticide was not applied (30.1% vs. 16.3%, $P<0.001$) (Ross et al., 2019).

1.4.10 HEXABROMOBIPHENYL

Hexabromobiphenyl is lipophilic, persistent, and highly stable and can travel long distances in the environment and can bioaccumulate, resulting in the potential risk of contamination of the environment and food chains (Liu et al., 2016). It belongs to a class of additive type brominated flame-retardants. It is used in several products, including paints, plastics, textiles, and electronics to reduce its flammability (Rantakokko et al., 2019).

Every day, we are exposed to this type of pollutant by means of fine particles in suspension (Rantakokko et al., 2019). Research conducted in Beijing, Tianjin-Hebei region, China, verified the presence of these pollutants in atmospheric particle samples, denoting that human exposure by inhalation can cause adverse reactions, and that children are more likely to be contaminated than adults (Zhang et al., 2019). Liu et al. (2016) measured hexabromobiphenyl in human milk in China and found a positive correlation between human milk and consumption of foods of animal origin (Liu et al., 2016), highlighting the biomagnification capacity of these pollutants.

1.4.11 HEXABROMOCYCLODODECANE

Hexabromocyclododecane (HBCDD) is a persistent compound that occurs in different environments, and therefore in various bioaccumulative matrices (Koch et al., 2015). It can leach or volatilize, enter the environment, and consequently contaminate humans (Drage et al., 2017). It is a brominated flame-retardants, highly efficient in reducing flammability, and is used in products, such as textile upholstery, rigidly shaped plastics, polystyrene foam, and polypropylene resin in household appliances, and as a textile coating additive (Drage et al., 2017).

From an environmental point of view, it is a ubiquitous compound, the occurrence of which has been observed in fish and seafood species of the Mediterranean Sea (Chessa et al., 2019). Toxicity studies conducted with *Caenorhabditis elegans* revealed that chronic exposure to HBCDD at concentrations greater than 20 nM influenced growth, locomotion, lipofuscin accumulation, and cellular apoptosis of nematodes, with increased levels of stress-related gene expression (Wang et al., 2018). A positive linear correlation was found between HBCDD (lipid weight concentrations) and trophic levels based on hydrogen isotopes in aquatic organisms (Liu et al., 2020).

1.4.12 HEXACHLOROBENZENE

Hexachlorobenzene (HCB) is an organochlorine pesticide that is similar to dioxins. it is stable and therefore persists in the environment. It accumulates in the food chain, and has been found in adipose tissue, breast milk, blood, and umbilical blood (Chiappini et al., 2016). For long, it has been used as a fungicide, it is still released into the environment as a waste by-product of various industrial processes (Miret et al., 2019).

HCB is an endocrine disruptor that mainly induces toxic effects on the reproductive system (Specht et al., 2015). Experimental studies indicate that exposure to organochlorines can interfere with hormonal regulation and immune function, and therefore can cause endometriosis (Chiappini et al., 2019). HCB acts as an endocrine disruptor in the thyroid, uterus, and mammary gland, and has been classified as possibly carcinogenic to humans. It acts by stimulating the proliferation of preneoplastic cells, migration, invasion, and metastasis, in addition to angiogenesis in breast cancer (Miret et al., 2019). Contaminant exposure can induce porphyria, nephrotoxicity, and oxidative stress hepatotoxicity both in laboratory animals (female Sprague-Dawley rats and adult male Wistar rats) and in humans (Khan et al., 2017; Starek-świechowicz et al., 2017).

1.4.13 HEXACHLOROCYCLOHEXANE

Hexachlorocyclohexane (HCH) isomers are organochlorinated compounds that pose potential risks to humans and the ecosystem. They are persistent in the environment, have a capacity for bioaccumulation and biomagnification, and can be transported over long distances (Di et al., 2018). HCHs were used between 1940s and 1980s in the production of insecticides (Kuang et al., 2020).

Studies carried out with samples of human milk in the city of Jinhua in China revealed the presence of β-hexachlorocyclohexane (β-HCH) in 36.4% of the analyzed samples. A positive correlation was detected between the presence of HCB residues and the decrease in birth weight of children (Kuang et al., 2020). Assessments to determine the risk of food exposure for adults and children to pollutants were carried out in five cities in India (Bangalore, Bhubaneswar, Guwahati, Ludhiana, and Udaipur). HCBs were detected in some of the bovine milk samples, indicating contamination which is likely due to agricultural and animal husbandry practices (Gill et al., 2020). In this study, it was verified that children are comparatively more exposed to pesticides than adults are, when body weight is taken into account, meaning that the risk of cancer was also higher for children.

1.4.14 LINDANE

Lindane (γ-hexachlorocyclohexane, γ-HCH) is an isomer of HCH with insecticidal properties and is persistent, bioaccumulative, and toxic (Zhang et al., 2020b). It is very stable in freshwater and saltwater environments (Yuksel et al., 2016). For each ton of lindane, 8–12 tons of HCH isomer residue is generated. A considerable amount of this compound is currently disposed of as waste in unmanaged landfills, constituting one of the most common and easily detectable pesticides in the environment (Wacławek et al., 2019).

Lindane is highly toxic to fish, bees, and aquatic invertebrates. This pollutant acts as an endocrine disruptor, causing hormonal imbalance in exposed fish, causing the breakdown of steroidogenesis in ovaries and causing a reduction in egg production (Yuksel et al., 2016). Lindane is also a risk for the central, reproductive, and endocrine nervous system in the fish and human being (Yuksel et al., 2016; Wacławek et al., 2019). Research conducted in northeastern South Africa, investigating the presence of pesticides in the muscle tissue of catfish (*Clarias gariepinus*), found a variety of POPs, including γ-HCH, highlighting the risk to human health when consuming contaminated fish (Barnhoorn et al., 2015).

1.4.15 MIREX

Mirex is a synthetic organochlorinated pesticide that is volatile, toxic, and is highly persistence in the environment, and it has a bioaccumulation capacity (Gandhi et al., 2015). Initially, it was used as an insecticide to control ants, termites, and other insects. It was later used as a flame-retardant for plastics, rubbers, and electrical materials (Wohlers et al., 2019). Mirex is absorbed by the skin of animals and because of its high lipophilicity, it has a tendency for biomagnification (Wrobel and Mlynarczuk, 2017). Photodegradation of Mirex results mainly in photomirex, which is equally stable in the environment (Gandhi et al., 2015).

The most common adverse effect attributed to mirex is endocrine disruption. This xenobiotic can inhibit the secretion of testosterone, estradiol, and progesterone, and stimulates the production of oxytocin in cattle. In this sense, it unbalances the hormonal system by altering the secretory function of ovarian cells, indicating that there is potential for this insecticide to hinder fertilization, blastocyst implantation, or even gestation time (Wrobel and Mlynarczuk, 2017). Mirex is toxic to a variety of aquatic biota animals and is considered a carcinogen to humans (Gandhi et al., 2015). This pesticide is also associated with maternal and infant thyroid hormone disorders, which can alter child development (Yamazaki et al., 2020).

1.4.16 PENTABROMODIPHENYL ETHER

Polybrominateddiphenyl ethers (PBDEs) are a large group that includes tetra-, penta-, hexa-, hepta-, and octa-bromodiphenyl ethers. PBDEs have high persistence in the environment, a bioaccumulation capacity, and toxicological potential (Chou et al., 2019). Brominated flame-retardants are used in households and electronic furniture (Parry et al., 2018). Studies reveal that these pollutants can leach and reach greater depths in the soil due to the persistence of these compounds. In addition, they can be absorbed by cultures that will later be consumed by humans (Chou et al., 2019).

Exposure to PBDE during fetal development is associated with deficiencies in the performance of executive functions and attention deficits. therefore, they are considered neurotoxic. Prenatal and postnatal exposure to PBDE negatively affects externalization behavior (e.g., hyperactivity and conduct problems) (Vuong et al., 2018). Biomonitoring data indicate that in recent years, concentrations of PBDEs have increased in wild animals and humans. The body load of PBDEs is three to nine times higher in babies and children than in adults due to exposure via breast milk and household dust. Tetra-, penta-, and hexa-BDEs are the most commonly found isomers in humans (Linares et al., 2015). Studies show that PBDEs can cause changes in homeostasis, delay in neurodevelopment, and can cause changes in the reproductive system, and cause cancer (Linares et al., 2015; Chou et al., 2019).

1.4.17 PENTACHLOROBENZENE

Pentachlorobenzene (PeCB) is part of the chlorobenzene group and participates in the composition of the HCB of technical grade. It is persistent, has bioaccumulation potential, and is very toxic (Wiltschka et al., 2020). In synthetic dyes, PeCB is discharged into watercourses during the treatment of textile dyeing effluents (Yuan et al., 2020).

PeCB, among other organochlorines, was observed in the maternal serum collected during delivery and in the umbilical cord blood of newborns in Tarragona (Spain). In all these cases, the highest concentrations were observed for women with higher levels of schooling and social class (Junqué et al., 2020). This greater exposure may be linked to different eating habits, since these groups tend to have a higher proportion of fish and seafood in their diet (Junqué et al., 2018). PeCB and others are invariably present in textile dyeing effluents, and represent an ecological risk to the aquatic ecosystem and impacts human health and the environment (Wiltschka et al., 2020; Yuan et al., 2020).

1.4.18 PENTACHLOROPHENOL, ITS SALTS, AND ESTERS

Pentachlorophenol (PCP) (2,3,4,5,6-PCP) and its sodium salt (Na-PCP) are hydrocarbons of the chlorophenol family (Cheng et al., 2014; Lopez-Echartea et al., 2016). Volatile, persistent, stable metabolites, with a bioaccumulative, and biomagnification capacity (Verbrugge et al., 2018). They have been used as a preservative for wood, herbicide, insecticide, biocide, disinfectant, defoliant, anti-sap, and as antimicrobial agents since 1930 (Cheng et al., 2014; Lopez-Echartea et al., 2016).

PCP alters oxidative phosphorylation, which interferes with cellular respiration and results in increased metabolism (Verbrugge et al., 2018). High levels of this pollutant have been found in free-range eggs. The source of contamination in the poultry farm was from wood treated with PCP, which was used as a structural component in the 1940s. This highlights the persistence, bioaccumulation, and biomagnification characteristics inherent to chlorophenenols (Piskorska-Pliszczynska et al., 2016). Widespread use of PeCP in areas endemic to schistosomes enabled occupational exposure to PeCP, which increased the incidence of cancer in the Chinese population targeted by biomonitoring (Cheng et al., 2014).

1.4.19 POLYCHLORINATED BIPHENYLS

Polychlorinated biphenyls (PCBs) are artificial chlorinated aromatic hydrocarbons that are persistent, lipophilic, can bioaccumulate, and therefore undergo biomagnification (Fu et al., 2018). The polychlorinated biphenyl family consists of 209 individual PCBs or “congeners,” although in reality, only 130 congeners have been found in commercial chemical formulations (CETESB, 2020). They are used as dielectric fluids in transformers, condensers, hydraulic lubricants, paints, and adhesives, among others (Dhakal et al., 2018). PCBs can undergo biotransformation in animals and plants, resulting in hydroxylated PCBs (OH-PCBs), which are less persistent (Dhakal et al., 2018). In mammalian metabolism, PCBs are converted into hydroxyl, methylsulfonyl, and sulfated metabolites, which are persistent in human blood (Grimm et al., 2015).

A study of the spatial distribution of PCBs was carried out in rivers in Northeast China through the evaluation of fishing products. PCBs, among others, were the main pollutants with detection rates of 100%. The health risk assessment showed that the consumption of fish from these areas would pose a carcinogenic risk to humans (Fu et al., 2018). Biomonitoring performed on serum samples from 46 individuals from rural (Columbus Junction area in East Iowa) and urban (East Chicago) communities to characterize the population's exposure levels to PCBs, verified the presence of OH-PCBs in human serum. Such fact can lead to bioactivation of electrophilic metabolites, interference in the hormone signaling process, inhibition and influence on hormone transport, classifying this PCB biotransformation product as an endocrine disruptor (Dhakal et al., 2018; Grimm et al., 2017). Underlining the risks to the ecological system caused by PCBs, research conducted in Iran found that at least five commercially important fish species are contaminated with this xenobiotic. From this perspective, there are human health risks attributed to the dietary incorporation of locally caught fish (Ranjbar Jafarabadi et al., 2019). Epidemiological evidence indicates that exposure to POPs, such as PCBs, increases the risk of developing diabetes, hypertension, and obesity. Furthermore, such comorbidities are important for the onset and progression of cardiovascular diseases (Perkins et al., 2016).

1.4.20 POLYCHLORINATED NAPHTHALENES

Polychlorinated naphthalenes (PCNs) are classified as chlorinated polycyclic aromatic hydrocarbons (Fernandes et al., 2017). They are a global concern because of their persistence, toxicity, and tendency for bioaccumulation (Wu et al., 2018). PCNs are used as stabilizers, dosolants, and flame-retardants (Gu et al., 2019). Polychlorinated naphthalenes may be produced accidentally concomitantly with dibenzo-p-dioxins and dibenzofurans during industrial processes and released in the form of gases, which can pose risks to human health and the environment (Yang et al., 2017).

Cod liver oil is a very popular food supplement; however, researchers have found high levels of PCN contamination in samples from the North Atlantic.

Although fish oils currently produced are subject to rigorous purification processes, in the Baltic region, cod liver oil harvested and consumed locally continues to contribute considerably to the food intake of these contaminants (Fernandes et al., 2017). In the analysis of soil samples collected from an industrial area and surrounding residential area in Shandong Province, China, extremely high concentrations of PCNs were observed in the two soil samples. This finding was related to the carcinogenic risk for workers and residents (Wu et al., 2018). Reports of serious or even fatal effects on humans and animals, according to the degree of exposure, are warnings of PCN intoxication manifestations of PC. Among the earliest records of PCN, poisoning is an incident that occurred at the end of the Second World War (1939–1945). A group of six individuals inadvertently consumed a technical product containing PCNs (referred to as "chlorinated paraffin"), which was mistakenly replaced for butter in a meal. Contamination can cause disruption of the gastrointestinal system with abdominal pain, neuropathy, and depression, followed by chloracne (Fernandes et al., 2017).

1.5 EMERGING CONTAMINANTS AND ACTIVE PHARMACEUTICAL INGREDIENTS

There are several emerging contaminants that have aroused global concern, including new brominated flame-retardants, organophosphate flame-retardants, polar pesticides, triclosan, synthetic musks, bisphenol-A, perchlorate and polycyclic siloxanes, PFCs, and illicit drugs (Covaci et al., 2012).

Pharmaceutical products or active pharmaceutical ingredients (APIs), such as antibiotics, are a group of emerging contaminants that have received remarkable attention in the last decade (Kar et al., 2018). Although in small concentrations, the increasing and widespread use of APIs in hospitals and as prescribed by doctors has resulted in the continuous introduction of APIs and their metabolites into the environment. This new scenario is due to the continuous development of advanced instruments and improved analytical methodologies that have enabled the detection of these compounds as micro-contaminants, at low levels, in different environmental matrices. Traces of APIs have been found in groundwater and surface water that are used to supply drinking water. Therefore, there has been concern about the potential risk to human health due to exposure to pharmaceutical waste through drinking water (Kar et al., 2018). However, there is no evidence that any serious risk may arise from low concentrations of pharmaceuticals found in drinking water. Nonetheless, the impacts of long-term and low-level exposure to a blend of APIs and their metabolites on ecosystems and human health are not understood and requires considerable research attention (Jelić et al., 2012). As a result, the environmental risk assessment of medicines should be carefully evaluated, especially in hospital effluents and urban sewage.

1.6 INDEXES/OBTAINING FEES

In order to adequately assess the risk of adverse effects that POPs may have on human health, it is necessary to obtain data in both acute and chronic toxic effects as well as the levels of exposure through diet studied with regards to the general population, those that consume considerable quantities, and sensitive subjects (Oancea, 2016).

Humans can be exposed to POPs both directly and indirectly. Direct exposure can be obtained as follows (Oancea, 2016):

- **Inhalation:** for volatile and semivolatile compounds.
  The exposure factor CE (mg/m^3) can be calculated according to the formula:
  $$CE = (CA \times TE \times FE \times DE)/(TM \times 365),$$
  where CA = air concentration (mg/m^3), TE = time of exposure (h/day), FE = frequency of exposure (days/year), DE = exposure duration of (years), TM = averaging time (years).

- **Ingestion from soil:** considered to be the main POP exposure path.
  The exposure through ingestion can be calculated according to the following formula:
  $$Exping = (CS \times 10^{-6} \times FE \times VIS)/(BW \times 365),$$
  where CS = soil concentration (mg/kg), FE = frequency of exposure (days/year), VIS = rate of ingestion from soil (mg/day), GC = bodyweight (kg).

- **Skin absorption from soil**
  Exposure through skin absorption can be calculated according to the formula:
  $$Expderm = (CS \times 10^{-6} \times FA \times FAD \times FE \times AS)/(GC \times 365),$$
  where CS = soil concentration (mg/kg), FA = soil adherence factor (mg/cm), FAD = skin absorption ratio (nonunit), AS = surface area (cm^2/day), GC = bodyweight (kg).

Indirect exposure can be achieved as follows:

- **Diet**
  The dietary exposure can be calculated with the following formula:

$$\text{Expdiet} = \text{VIA} \times \text{CA/GC}$$

  where VIA = rate of ingestion from food (kg/day), CA = concentration in food (mg/kg), GC = bodyweight (kg).

- **Contaminated drinking water**
  The drinking water exposure can be calculated according to the following formula:

$$\text{Expwater} = \sum \text{VIAp} \times \text{CAp/GC}$$

  where VIAp = rate of ingestion from water (L/day), Cap = concentration in water (mg/L), GC = bodyweight (kg).

The acute or chronic toxic effects determined by humans upon exposure to POP depend on several factors, such as exposure dose, absorption and distribution mechanism, metabolism, excretion capacity, and health condition of the body.

Chronic exposure to chemicals in the diet occurs daily, for a long period, or throughout life. The estimation of chronic intake will depend on the availability and quality of the data involved in its calculation; the closer to reality of the data, the more significant the result will be (Jardim and Caldas, 2009).

While chronic risk assessment estimates the average food intake over a long period, acute risk assessment assesses exposure through consumption of a single meal or for 24 h (Jardim and Caldas, 2009). The importance of acute exposure to pesticides was recognized in the early 1990s in California after reports of poisoning from the consumption of highly contaminated food due to inadequate application of POPs in the field (Goldman et al., 1990; Maff, 1993). More recently, human intoxication from organophosphate insecticides and carbamates present in the diet has been reported (Tsai et al., 2003; Mendes et al., 2005).

1.7 FINAL CONSIDERATIONS

We have made great progress in the use of toxic molecules and have achieved effective applicability for products in terms of reducing the impact on nontarget organisms as well as recommendations for dwindler doses to achieve the proposed objectives. However, we have a wide variety of POPs that concern society due their environmental dispersal.

Factors such as climate change can alter the persistence rates of these compounds in nature, a phenomenon which should be carefully evaluated. Similarly, new eating habits direct society to consuming supplements beneficial to health, but which, even of "natural" sources, can be sources of contamination by POPs.

Because POPs are synthetic chemicals that are semi-volatile/volatile, persistent, toxic and have a tendendy for bioaccumulation, they are found ubiquoutsly in the environment. Consequently, POPs pose a serious problem to human health through environmental exposure.

With a global population of 8 billion people, that continues to increase, and current and future demands for water and energy growing, topics such as food security, public health risk, water scarcity, and mass migrations put pressure on the sustainability of all sectors of the society. It is essential that strategies are established to monitor the effects of POP on the health of the population and a safe policy for waste disposal is determined. It is evident that measures should be adopted in the short, medium, and long terms to decrease the use and coexistence with POPs residues to avoid the potential effects on society and the environment, both in abiotic environments as well as for microorganisms, fauna, and flora.

KEYWORDS

- **persistent organic pollutants**
- **human health**
- **atmospheric pollutants**
- **human exposure**

REFERENCES

Alaboudi, A. R.; Osaili, T. M.; Alrwashdeh, A. Pesticides (Hexachlorocyclohexane, Aldrin, and Malathion) Residues in Home-Grown Eggs: Prevalence, Distribution, and Effect of Storage and Heat Treatments. *J. Food Sci.* **2019,** *84*, 3383–3390.

Alharbi, O. M. L.; Basheer, A. A.; Khattab, R. A.; Ali, I. Health and Environmental Effects of Persistent Organic Pollutants. *J. Mol. Liq.* **2018,** *263*, 442–453.

Andrade, N. A.; McConnell, L. L.; Anderson, M. O.; Torrents, A.; Ramirez, M. Polybrominated Diphenyl Ethers: Residence Time in Soils Receiving Biosolids Application. *Environ. Pollut.* **2017,** *222*, 412–422.

Barnhoorn, I. E. J.; Van Dyk, J. C.; Genthe, B.; Harding, W. R.; Wagenaar, G. M.; Bornman, M. S. Chemosphere Organochlorine Pesticide Levels in Clarias gariepinus from Polluted Freshwater Impoundments in South Africa and Associated Human Health Risks. *Chemosphere* **2015,** *120*, 391–397.

Benoit, P.; Cravedi, J. P.; Desenclos, J. C.; Mouvet, C.; Rychen, G.; Samson, M. Environmental and Human Health Issues Related to Long-Term Contamination by Chlordecone in the French Caribbean. *Environ. Sci. Pollut. Res.* **2020,** *27*, 40949–40952.

Bhuvaneswari, R.; Nagarajan, V.; Chandiramouli, R. Chemisorption of Heptachlor and Mirex Molecules on Beta Arsenene Nanotubes—A First-Principles Analysis. *Appl. Surf. Sci.* **2021,** *537*, 147835.

CAS- Scientific Information, Designed For Your Needs | CAS. https://www.cas.org/pt-br/products (accessed Nov 26, 2020).

CETESB -Poluentes Orgânicos Persistentes (POPs) | Centro Regional. https://cetesb.sp.gov.br/centroregional/a-convencao/poluentes-organicos-persistentes-pops/ (accessed Nov 25, 2020).

Cheng, P.; Zhang, Q.; Shan, X.; Shen, D.; Wang, B.; Tang, Z.; Jin, Y.; Zhang, C.; Huang, F. Cancer Risks and Long-Term Community-Level Exposure to Pentachlorophenol in Contaminated Areas, China. *Environ. Sci. Pollut. Res.* **2014,** *22*, 1309–1317.

Chessa, G.; Cossu, M.; Fiori, G.; Ledda, G.; Piras, P.; Sanna, A.; Brambilla, G. Occurrence of Hexabromocyclododecanes and Tetrabromobisphenol A in Fish and Seafood from the Sea of Sardinia—FAO 37.1.3 Area: Their Impact on Human Health within the European Union Marine Framework Strategy Directive. *Chemosphere* **2019,** *228*, 249–257.

Chiappini, F.; Bastón, J. I.; Vaccarezza, A.; Singla, J. J.; Pontillo, C.; Miret, N.; Farina, M.; Meresman, G.; Randi, A. Enhanced Cyclooxygenase-2 Expression Levels and Metalloproteinase 2 and 9 Activation by Hexachlorobenzene in Human Endometrial Stromal Cells. *Biochem. Pharmacol.* **2016,** *109*, 91–104.

Chiappini, F.; Sánchez, M.; Miret, N.; Cocca, C.; Zotta, E.; Ceballos, L.; Pontillo, C.; Bilotas, M.; Randi, A. Exposure to Environmental Concentrations of Hexachlorobenzene Induces Alterations Associated with Endometriosis Progression in a Rat Model. *Food Chem. Toxicol.* **2019,** *123*, 151–161.

Chou, T. H.; Ou, M. H.; Wu, T. Y.; Chen, D. Y.; Shih, Y. H. Temporal and Spatial Surveys of Polybromodiphenyl Ethers (PBDEs) Contamination of Soil Near a Factory Using PBDEs in Northern Taiwan. *Chemosphere* **2019,** *236*, 124117.

Costet, N.; Pelé, F.; Comets, E.; Rouget, F.; Monfort, C.; Bodeau-Livinec, F.; Linganiza, E. M.; Bataille, H.; Kadhel, P.; Multigner, L. Perinatal Exposure to Chlordecone and Infant Growth. *Environ. Res.* **2015,** *142*, 123–134.

Covaci, A.; Geens, T.; Roosens, L.; Ali, N.; Van Den Eede, N.; Ionas, A. C.; Malarvannan, G.; Dirtu, A. C. Human Exposure and Health Risks to Emerging Organic Contaminants. In *Emerging Organic Contaminants and Human Health*; Barceló, D., Ed.; Springer: Londres, 2012; pp 243–305.

Curtean-Bănăduc, A., Ed. *The Impact of Persistent Organic Pollutants on Freshwater Ecosystems and Human Health*; Editura Universităţii "Lucian Blaga" din Sibiu: Sibiu, 2016.

Das Neves, J.; Barnhoorn, I. E. J.; Wagenaar, G. M. The Effects of Environmentally Relevant Concentrations of Aldrin and Methoxychlor on the Testes and Sperm of Male *Clarias gariepinus* (Burchell, 1822) after Short-Term Exposure. *Fish Physiol. Biochem.* **2018,** *44*, 1421–1434.

Dhakal, K.; Gadupudi, G. S.; Lehmler, H. J.; Ludewig, G.; Duffel, M. W.; Robertson, L. W. Sources and Toxicities of Phenolic Polychlorinated Biphenyls (OH-PCBs). *Environ. Sci. Pollut. Res.* **2018,** *25*, 16277–16290.

Di, S.; Liu, R.; Chen, L.; Diao, J.; Zhou, Z. Selective Bioaccumulation, Biomagnification, and Dissipation of Hexachlorocyclohexane Isomers in a Freshwater Food Chain. *Environ. Sci. Pollut. Res.* **2018** *25*, 18752–18761.

Drage, D. S.; Mueller, J. F.; Hobson, P.; Harden, F. A.; Toms, L. M. L. Demographic and Temporal Trends of Hexabromocyclododecanes (HBCDD) in an Australian Population. *Environ. Res.* **2017,** *152*, 192–198.

Durante, C. A.; Santos-Neto, E. B.; Azevedo, A.; Crespo, E. A.; Lailson-Brito, J. POPs in the South Latin America: Bioaccumulation of DDT, PCB, HCB, HCH and Mirex in Blubber of Common Dolphin (*Delphinus delphis*) and Fraser's Dolphin (*Lagenodelphis hosei*) from Argentina. *Sci. Total Environ.* **2016,** *572*, 352–360.

El-Shahawi, M. S.; Hamza, A.; Bashammakh, A. S.; Al-Saggaf, W. T. An Overview on the Accumulation, Distribution, Transformations, Toxicity and Analytical Methods for the Monitoring of Persistent Organic Pollutants. *Talanta* **2010,** *80*, 1587–1597.

Elobeid, M. A.; Padilla, M. A.; Brock, D. W.; Ruden, D. M.; Allison, D. B. Endocrine Disruptors and Obesity: An Examination of Selected Persistent Organic Pollutants in the Nhanes 1999–2002 Data. *Int. J. Environ. Res. Public Health.* **2010,** *7*, 2988–3005.

Encarnação, T.; Pais, A. A. C. C.; Campos, M. G.; Burrows, H. D. Endocrine Disrupting Chemicals: Impact on Human Health, Wildlife and the Environment. *Sci. Prog.* **2019,** *102*, 3–42.

Fernandes, A.; Rose, M.; Falandysz, J. Polychlorinated Naphthalenes (PCNs) in Food and Humans. *Environ. Int.* **2017,***104*, 1–13.

Fernandes, A. R.; Mortimer, D.; Rose, M.; Smith, F.; Steel, Z.; Panton, S. Recently Listed Stockholm Convention POPs: Analytical Methodology, Occurrence in Food and Dietary Exposure. *Sci. Total Environ.* **2019,** *678*, 793–800.

Ferreira, A. P. Polychlorinated Biphenyl (PCB) Congener Concentrations in Aquatic Birds. Case Study: Ilha Grande Bay, Rio de Janeiro, Brazil. *An Acad Bras Cienc.* **2013,** *85*, 1379–1388.

Fournier, A.; Feidt, C.; Lastel, M. L.; Archimede, H.; Thome, J. P.; Mahieu, M.; Rychen, G. Toxicokinetics of Chlordecone in Goats: Implications for Risk Management in French West Indies. *Chemosphere* **2017,** *171*, 564–570.

Fu, L.; Lu, X.; Tan, J.; Zhang, H.; Zhang, Y.; Wang, S.; Chen, J. Bioaccumulation and Human Health Risks of OCPs and PCBs in Freshwater Products of Northeast China. *Environ. Pollut.* **2018,** *242*, 1527–1534.

Gandhi, N.; Tang, R. W. K.; Bhavsar, S. P.; Reiner, E. J.; Morse, D.; Arhonditsis, G. B.; Drouillard, K.; Chen, T. Is Mirex Still a Contaminant of Concern for the North American Great Lakes? *J. Great Lakes Res.* **2015,** *41*, 1114–1122.

Gaur, N.; Narasimhulu, K.; PydiSetty, Y. Recent Advances in the Bio-Remediation of Persistent Organic Pollutants and Its Effect on Environment. *J. Clean. Prod.* **2018,** *198*, 1602–1631.

Gerber, R.; Smit, N. J.; Van Vuren, J. H. J.; Nakayama, S. M. M.; Yohannes, Y. B.; Ikenaka, Y.; Ishizuka, M.; Wepener, V. Bioaccumulation and Human Health Risk Assessment of DDT and Other Organochlorine Pesticides in an Apex Aquatic Predator from a Premier Conservation Area. *Sci. Total Environ.* **2016,** *550*, 522–533.

Gill, J. P. S.; Bedi, J. S.; Singh, R.; Fairoze, M. N.; Hazarika, R. A.; Gaurav, A.; Satpathy, S. K.; Chauhan, A. S.; Lindahl, J.; Grace, D. Pesticide Residues in Peri-Urban Bovine Milk from India and Risk Assessment: A Multicenter Study. *Sci. Rep.* **2020,** *10*.

Goldman, L. R.; Beller, M.; Jackson, R. J. Aldicarb Food Poisonings in California, 1985–1988: Toxicity Estimates for Humans. *Arch. Environ. Health.* **1990,** *45*, 141–147.

Grimm, F. A.; Hu, D.; Kania-Korwel, I.; Lehmler, H. J.; Ludewig, G.; Hornbuckle, K. C.; Duffel, M. W.; Bergman, A.; Robertson, L. W. Metabolism and Metabolites of Polychlorinated Biphenyls. *Crit. Rev. Toxicol.* **2015,** *45*, 245–272.

Grimm, F. A.; Lehmler, H. J.; Koh, W. X.; DeWall, J.; Teesch, L. M.; Hornbuckle, K. C.; Thorne, P. S.; Robertson, L. W.; Duffel, M. W. Identification of a Sulfate Metabolite of PCB 11 in Human Serum. *Environ Int.* **2017,** *98*, 120–128.

Grindler, N. M.; Allsworth, J. E.; Macones, G. A.; Kannan, K.; Roehl, K. A.; Cooper, A. R. Persistent Organic Pollutants and Early Menopause in U.S. Women. *PLoS One*. **2015,** *10*, e0116057.

Gu, W.; Li, Q.; Li, Y. Fuzzy Risk Assessment of Modified Polychlorinated Naphthalenes for Enhanced Degradation. *Environ. Sci. Pollut. Res.* **2019,** *26*, 25142–25153.

Guo, W.; Pan, B.; Sakkiah, S.; Yavas, G.; Ge, W.; Zou, W.; Tong, W.; Hong, H. Persistent Organic Pollutants in Food: Contamination Sources, Health Effects and Detection Methods. *Int. J. Environ. Res. Public Health.* **2019,** *16*, 108828.

Haffner, D.; Schecter, A. Persistent Organic Pollutants (POPs): A Primer for Practicing Clinicians. *Curr. Environ. Heal. Rep.* **2014,** *1*, 123–131.

Hussein, M. M. A.; Elsadaawy, H. A.; El-Murr, A.; Ahmed, M. M.; Bedawy, A. M.; Tukur, H.A.; Swelum, A. A. A.; Saadeldin, I. M. Endosulfan Toxicity in Nile tilapia (*Oreochromis niloticus*) and the Use of Lycopene as an Ameliorative Agent. *Comp. Biochem. Physiol. Part—C Toxicol. Pharmacol.* **2019,** *224*, 108573.

Jacob, J.; Cherian, J. Review of environmental and human exposure to persistent organic pollutants. *Asian Soc. Sci.* **2013,** *9*, 107–120.

Jardim, A. N. O.; Caldas, E. D. Chemical Dietary Exposure and the Risks to Human Health. *Quim. Nova.* **2009,** *32*, 1898–1909.

Jelić, A.; Petrović, M.; Barceló, D. Pharmaceuticals in Drinking Water. In *Emerging Organic Contaminants and Human Health*; Barceló, D., Ed.; Springer: Londres, 2012; pp 47–70.

Jhamtani, R. C.; Dahiya, M. S.; Agarwal, R. Forensic Toxicology Research to Investigate Environmental Hazard. *J. Forensic Sci. Crim. Investig.* **2017,** *3*, 1–4.

Jhamtani, R. C.; Shukla, S.; Sivaperumal, P.; Dahiya, M. S.; Agarwal, R. Impact of Co-exposure of Aldrin and Titanium Dioxide Nanoparticles at Biochemical and Molecular Levels in Zebrafish. *Environ. Toxicol. Pharmacol.* **2018,** *58*, 141–155.

Junqué, E.; Garí, M.; Llull, R. M.; Grimalt, J. O. Drivers of the Accumulation of Mercury and Organochlorine Pollutants in Mediterranean Lean Fish and Dietary Significance. *Balear. Islands Sci. Total Environ.* **2018,** *634*, 170–180.

Junqué, E.; Garcia, S.; Martínez, M. Á.; Rovira, J.; Schuhmacher, M.; Grimalt, J. O. Changes of Organochlorine Compound Concentrations in Maternal Serum During Pregnancy and Comparison to Serum Cord Blood Composition. *Environ. Res.* **2020,** *182*, 108994.

Kanan, S.; Samara, F. Dioxins and Furans: A Review from Chemical and Environmental Perspectives. *Trends Environ. Anal. Chem.* **2018,** *17*, 1–13.

Kar, S.; Roy, K.; Leszczynski, J. Impact of Pharmaceuticals on the Environment: Risk Assessment Using QSAR Modeling Approach. In *Methods in Molecular Biology*; Humana Press Inc., 2018; pp 395–443.

Khan, S.; Priyamvada, S.; Khan, S.A.; Khan, W.; Yusufi, A. Studies on Hexachlorobenzene (HCB) Induced Toxicity and Oxidative Damage in the Kidney and Other Rat Tissues. *Int. J. Drug Metab. Toxicol.* **2017,** *1*, 1–9.

Kim, M.; Li, L. Y.; Gorgy, T.; Grace, J. R. Review of Contamination of Sewage Sludge and Amended Soils by Polybrominated Diphenyl Ethers Based on Meta-Analysis. *Environ. Pollut.* **2017,** *220*, 753–765.

Koch, C.; Schmidt-Kötters, T.; Rupp, R.; Sures, B. Review of Hexabromocyclododecane (HBCD) with a Focus on Legislation and Recent Publications Concerning Toxicokinetics and -Dynamics. *Environ. Pollut.* **2015,** *199*, 26–34.

Kuang, L.; Hou, Y.; Huang, F.; Guo, A.; Deng, W.; Sun, H.; Shen, L.; Lin, H.; Hong, H. Pesticides in Human Milk Collected from Jinhua, China: Levels, Influencing Factors and Health Risk Assessment. *Ecotoxicol. Environ. Saf.* **2020,** *205*, 111331.

Kumar, N.; Ambasankar, K.; Krishnani, K. K.; Gupta, S. K.; Bhushan, S.; Minhas, P. S. Acute Toxicity, Biochemical and Histopathological Responses of endosulfan in Chanos chanos. *Ecotoxicol. Environ. Saf.* **2016,** *131*, 79–88.

Lakroun, Z.; Kebieche, M.; Lahouel, A.; Zama, D.; Desor, F.; Soulimani, R. Oxidative Stress and Brain Mitochondria Swelling Induced by Endosulfan and Protective Role of Quercetin in Rat. *Environ. Sci. Pollut. Res.* **2015,** *22*, 7776–7781.

Lee, D. H.; Steffes, M. W.; Sjödin, A.; Jones, R. S.; Needham, L. L.; Jacobs, D. R. Low Dose of Some Persistent Organic Pollutants Predicts Type 2 Diabetes: A Nested Case-Control Study. *Environ. Health Perspect.* **2010,** *118*, 1235–1242.

Lee, I.; Eriksson, P.; Fredriksson, A.; Buratovic, S.; Viberg, H. Developmental Neurotoxic Effects of Two Pesticides: Behavior and Neuroprotein Studies on Endosulfan and Cypermethrin. *Toxicology* **2015,** *335*, 1–10.

Li, H.; Zhang, Z.; Sun, Y.; Wang, W.; Xie, J.; Xie, C.; Hu, Y.; Gao, Y.; Xu, X.; Luo, X. Tetrabromobisphenol A and Hexabromocyclododecanes in Sediments and Biota from Two Typical Mangrove Wetlands of South China: Distribution, Bioaccumulation and Biomagnification. *Sci. Total Environ.* **2021,** *750*, 141695.

Linares, V.; Bellés, M.; Domingo, J. L. Human exposure to PBDE and critical evaluation of health hazards. *Arch. Toxicol.* **2015,** *89*, 335–356.

Liu, W. X.; Chen, J. L.; Hu, J.; Ling, X.; Tao, S. Multi-Residues of Organic Pollutants In Surface Sediments from Littoral Areas of the Yellow Sea, China. *Mar. Pollut. Bull.* **2008,** *56*, 1091–1103.

Liu, X.; Wen, S.; Li, J.; Zhang, L.; Zhao, Y.; Wu, Y. A Study on the Levels of a Polybrominated Biphenyl in Chinese Human Milk Samples Collected in 2007 and 2011. *Environ. Monit. Assess.* **2016,** *188*.

Liu, Y.; Luo, X.; Zeng, Y.; Deng, M.; Tu, W.; Wu, Y.; Mai, B. Bioaccumulation and Biomagnification of Hexabromocyclododecane (HBCDD) in Insect-Dominated Food Webs from a Former E-Waste Recycling Site in South China. *Chemosphere* **2020,** *240*, 124813.

Lopez-Echartea, E.; Macek, T.; Demnerova, K.; Uhlik, O. Bacterial Biotransformation of Pentachlorophenol and Micropollutants Formed during Its Production Process. *Int. J. Environ. Res. Public Health* **2016,** *13*.

Maddela, N. R.; Venkateswarlu, K.; Kakarla, D.; Megharaj, M. Inevitable Human Exposure to Emissions of Polybrominated Diphenyl Ethers: A Perspective on Potential Health Risks. *Environ. Pollut.* **2020,** *266*, 115240.

MAFF. *Annual Report of the Working Party on Pesticide Residues: 1992, Supplement to the Pesticides Register 1993*; HMSO: London, UK, 1993.

Mahugija, J. A. M.; Nambela, L.; Mmochi, A. J. Determination of Dichlorodiphenyltrichloroethane (DDT) and Metabolites Residues in Fish Species from Eastern Lake Tanganyika. *South African J. Chem.* **2018,** *71*, 86–93.

Martin, O. V.; Evans, R. M.; Faust, M.; Kortenkamp, A. A Human Mixture Risk Assessment for Neurodevelopmental Toxicity Associated with Polybrominated Diphenyl Ethers Used as Flame Retardants. *Environ. Health Perspect.* **2017,** *125*, 087016.

Martínez-Ibarra, A.; Morimoto, S.; Cerbón, M.; Prado-Flores, G. Effects on the Reproductive Parameters of Two Generations of *Rattus norvegicus* Offspring from Dams Exposed to Heptachlor during Gestation and Lactation. *Environ. Toxicol.* **2017,** *32*, 856–868.

Mendes, C. A.; Mendes, G. E.; Cipullo, J. P.; Burdmann, E. A. Acute Intoxication Due to Ingestion of Vegetables Contaminated with Aldicarb. *Clin.Toxicol. (Phila)* **2005,** *43*, 117–118.

Menezes, R. G.; Qadir, T. F.; Moin, A.; Fatima, H.; Hussain, S. A.; Madadin, M.; Pasha, S. B.; Al Rubaish, F. A.; Senthilkumaran, S. Endosulfan Poisoning: An overview. *J. Forensic Leg. Med.* **2017,** *51*, 27–33.

Miret, N. V.; Pontillo, C. A.; Zárate, L. V.; Kleiman, D.; Pisarev, D.; Cocca, C.; Randi, A. S. Impact of Endocrine Disruptor Hexachlorobenzene on the Mammary Gland and Breast Cancer: The Story Thus Far. *Environ. Res.* **2019,** *173*, 330–341.

Mlynarczuk, J.; Górska, M.; Wrobel, M. H. Effects of DDT, DDE, Aldrin and Dieldrin on Prostaglandin, Oxytocin and Steroid Hormone Release from Smooth Chorion Explants of Cattle. *Anim. Reprod. Sci.* **2020,** *223*, 106623.

Mudhoo, A.; Bhatnagar, A.; Rantalankila, M.; Srivastava, V.; Sillanpää, M. Endosulfan Removal through Bioremediation, Photocatalytic Degradation, Adsorption and Membrane Separation Processes: A Review. *Chem. Eng. J.* **2019,** *360*, 912–928.

Multigner, L.; Ndong, J. R.; Giusti, A.; Romana, M.; Delacroix-Maillard, H.; Cordier, S.; Jégou, B.; Thome, J.P.; Blanchet, P. Chlordecone Exposure and Risk of Prostate Cancer. *J. Clin. Oncol.* **2010,** *28*, 3457–3462.

Multigner, L.; Kadhel, P.; Rouget, F.; Blanchet, P.; Cordier, S. Chlordecone Exposure and Adverse Effects in French West Indies Populations. *Environ. Sci. Pollut. Res.* **2016,** *23*, 3–8.

Najam, L.; Alam, T. Levels and distribution of OCPs, (specially HCH, Aldrin, Dieldrin, DDT, Endosulfan) in Karhera Drain and Surface Water of Hindon River and Their Adverse Effects. *Orient. J. Chem.* **2015,** *31*, 2025–2030.

OPAS -Organização Pan-Americana da Saúde. *O impacto de substâncias químicas sobre a saúde pública: Fatores conhecidos e desconhecidos*; DF: Brasília, 2018.

Oancea, S. The Impact of Persistent Organic Pollutants on Human Health. In *The Impact of Persistent Organic Pollutants on Freshwater Ecosystems and Human Health*; Curtean-Bănăduc, A., Ed.; Editora Universităţii: Sibiu, 2016; pp 57–83.

Parada, H.; Wolff, M. S.; Engel, L. S.; White, A. J.; Eng, S. M.; Cleveland, R. J.; Khankari, N. K.; Teitelbaum, S. L.; Neugut, A. I.; Gammon, M. D. Organochlorine Insecticides DDT and Chlordane in Relation to Survival Following Breast Cancer. *Int. J. Cancer.* **2016,** *138*, 565–575.

Park, E. Y.; Park, E.; Kim, J.; Oh, J. K.; Kim, B.; Hong, Y. C.; Lim, M. K. Impact of Environmental Exposure to Persistent Organic Pollutants on Lung Cancer Risk. *Environ. Int.* **2020,** *143*, 105925.

Parry, E.; Zota, A. R.; Park, J. S.; Woodruff, T. J. Polybrominated Diphenyl Ethers (PBDEs) and Hydroxylated PBDE Metabolites (OH-PBDEs): A Six-Year Temporal Trend in Northern California Pregnant Women. *Chemosphere* **2018,** *195*, 777–783.

Patočka, J.; Wu, Q.; França, T. C. C.; Ramalho, T. C.; Pita, R.; Kuča, K. Clinical Aspects of the Poisoning by the Pesticide Endosulfan. *Quim. Nov.* **2016,** *39*, 987–994.

Perkins, J. T.; Petriello, M. C.; Newsome, B. J.; Hennig, B. Polychlorinated Biphenyls and Links to Cardiovascular Disease. *Environ. Sci. Pollut. Res.* **2016,** *23*, 2160–2172.

Piskorska-Pliszczynska, J.; Strucinski, P.; Mikolajczyk, S.; Maszewski, S.; Rachubik, J.; Pajurek, M. Pentachlorophenol from an Old Henhouse as a Dioxin Source in Eggs and Related Human Exposure. *Environ. Pollut.* **2016,** *208*, 404–412.

PPDB–Pesticide Properties Database. Footprint: Creating Tools for Pesticide Risk Assessment and Management in Europe. Developed by the Agriculture & Environment Research Unit (AERU), University of Hertfordshire, Funded by UK National Sources and the EU-funded FOOTPRINT project (FP6-SSP-022704). https://sitem.herts.ac.uk/aeru/ppdb/en/index.htm (accessed Nov 24, 2020).

Ranjbar Jafarabadi, A.; Riyahi Bakhtiari, A.; Mitra, S.; Maisano, M.; Cappello, T.; Jadot, C. First Polychlorinated Biphenyls (PCBs) Monitoring in Seawater, Surface Sediments and Marine Fish Communities of the Persian Gulf: Distribution, Levels, Congener Profile and Health Risk Assessment. *Environ. Pollut.* **2019,** *253*, 78–88.

Rantakokko, P.; Kumar, E.; Braber, J.; Huang, T.; Kiviranta, H.; Cequier, E.; Thomsen, C. Concentrations of Brominated and Phosphorous Flame Retardants in Finnish House Dust and Insights Into Children's Exposure. *Chemosphere* **2019,** *223*, 99–107.

Rauert, C.; Harner, T.; Schuster, J. K.; Eng, A.; Fillmann, G.; Castillo, L. E.; Fentanes, O.; Ibarra, M. V.; Miglioranza, K. S. B.; Rivadeneira, I. M. Air Monitoring of New and Legacy POPs in the Group of Latin America and Caribbean (GRULAC) Region. *Environ. Pollut.* **2018,** *243*, 1252–1262.

Rawn, D. F. K.; Sadler, A. R.; Casey, V. A.; Breton, F.; Sun, W. F.; Arbuckle, T. E.; Fraser, W. D. Dioxins/Furans and PCBs in Canadian Human Milk: 2008–2011. *Sci. Total Environ.* **2017,** *595*, 269–278.

Rolle-Kampczyk, U.; Gebauer, S.; Haange, S.-B.; Schubert, K.; Kern, M.; Moulla, Y.; Dietrich, A.; Schön, M. R.; Klöting, N.; Von Bergen, M. Accumulation of Distinct Persistent Organic Pollutants Is Associated with Adipose Tissue Inflammation. *Quim. Nova.* **2020,** *39*, 987-994

Ross, G. W.; Abbott, R. D.; Petrovitch, H.; Duda, J. E.; Tanner, C. M.; Zarow, C.; Uyehara-Lock, J. H.; Masaki, K. H.; Launer, L. J.; Studabaker, W. B. Association of Brain Heptachlor Epoxide and Other Organochlorine Compounds with Lewy Pathology. *Mov. Disord.* **2019,** *34*, 228–235.

Sebastian, R.; Raghavan, S. C. Molecular Mechanism of Endosulfan Action in Mammals. *J. Biosci.* **2017,** *42*, 149–153.

Singh, N.; Siddarth, M.; Ghosh, R.; Tripathi, A. K.; Banerjee, B. D. Heptachlor-Induced Epithelial to Mesenchymal Transition in HK-2 Cells Mediated via TGF-β1/Smad Signalling. Hum. *Exp. Toxicol.* **2019,** *38*, 567–577.

Specht, I. O.; Bonde, J. P. E.; Toft, G.; Giwercman, A.; Spanò, M.; Bizzaro, D.; Manicardi, G. C.; Jönsson, B. A. G.; Robbins, W. A. Environmental Hexachlorobenzene Exposure and Human Male Reproductive Function. *Reprod. Toxicol.* **2015,** *58*, 8–14.

Starek-świechowicz, B.; Budziszewska, B.; Starek, A. Hexachlorobenzene as a Persistent Organic Pollutant: Toxicity and Molecular Mechanism of Action. *Pharmacol. Rep.* **2017,** *6*, 1232–1239.

Supreeth, M.; Raju, N. Biotransformation of Chlorpyrifos and Endosulfan by Bacteria and Fungi. *Appl. Microbiol. Biotechnol.* **2017,** *101*, 5961–5971.

Stockholm Convention. The Listing of POPs in the Stockholm Convention. http://chm.pops.int/TheConvention/ThePOPs/ListingofPOPs/tabid/2509/Default.aspx (accessed Nov 24, 2020).

Tang, B.; Zeng, Y. H.; Luo, X. J.; Zheng, X. B.; Mai, B. X. Bioaccumulative Characteristics of Tetrabromobisphenol A and Hexabromocyclododecanes in Multi-Tissues of Prey and Predator Fish from an E-Waste Site, South China. *Environ. Sci. Pollut. Res.* **2015,** *22*, 12011–12017.

Tavakoly Sany, S. B.; Hashim, R.; Salleh, A.; Rezayi, M.; Karlen, D. J.; Razavizadeh, B. B. M.; Abouzari-lotf, E. Dioxin Risk Assessment: Mechanisms of Action and Possible Toxicity in Human Health. *Environ. Sci. Pollut. Res.* **2015,** *22*, 19434–19450.

Thi Thuy, T. Effects of DDT on Environment and Human Health. *J. Educ. Soc. Sci.* **2015,** *2*, 108–114.

Thompson, L. A.; Ikenaka, Y.; Yohannes, Y. B.; Van Vuren, J. J.; Wepener, V.; Smit, N. J.; Darwish, W. S.; Nakayama, S. M. M.; Mizukawa, H.; Ishizuka, M. Concentrations and Human Health Risk Assessment of DDT and Its Metabolites in Free-Range and Commercial Chicken Products from KwaZulu-Natal, South Africa. *Food Addit. Contam.—Part A Chem. Anal. Control. Expo. Risk Assess.* **2017,** *34*, 1959–1969.

Torre, A. de L.; Sanz, P.; Navarro, I.; Martínez, M. A. Time Trends of Persistent Organic Pollutants in Spanish Air. *Environ. Pollut.* **2016,** *217*, 26–32.

Tsai, M. J.; Wu, S. N.; Cheng, H. A.; Wang, S. H.; Chiang, H. T. An Outbreak of Food-Borne Illness Due to Methomyl Contamination. *J. Toxicol. Clin. Toxicol.* **2003,** *41*, 969–973.

Verbrugge, L. A.; Kahn, L.; Morton, J. M. Pentachlorophenol, Polychlorinated Dibenzo-P-Dioxins and Polychlorinated Dibenzo Furans in Surface Soil Surrounding Pentachlorophenol-Treated Utility Poles on the Kenai National Wildlife Refuge, Alaska USA. *Environ. Sci. Pollut. Res.* **2018,** *25*, 19187–19195.

Vijgen, J.; Borst, B. De.; Weber, R.; Stobiecki, T.; Forter, M. HCH and Lindane Contaminated Sites : European and Global Need for a Permanent Solution for a Long-Time Neglected Issue. *Environ. Pollut.* **2019,** *248*, 696–705.

Viluksela, M.; Pohjanvirta R. Multigenerational and Transgenerational Effects of Dioxins. *Int. J. Mol. Sci.* **2019,** 20, 2947.

Vuong, A. M.; Yolton, K.; Dietrich, K. N.; Braun, J. M.; Lanphear, B. P.; Chen, A. Exposure to Polybrominated Diphenyl Ethers (PBDEs) and Child Behavior: Current Findings and Future Directions. *Horm. Behav.* **2018,** *101*, 94–104.

Wacławek, S.; Silvestri, D.; Hrabák, P.; Padil, V. V. T.; Torres-Mendieta, R.; Wacławek, M.; Černík, M.; Dionysiou, D. D. Chemical Oxidation and Reduction ofHexachlorocyclohexanes: A Review. *Water Res.* **2019,** *162*, 302–319.

Wang, X.; Yang, J.; Li, H.; Guo, S.; Tariq, M.; Chen, H.; Wang, C.; Liu, Y. Chronic Toxicity of Hexabromocyclododecane (HBCD) Induced by Oxidative Stress and Cell Apoptosis on Nematode Caenorhabditis Elegans. *Chemosphere* **2018,** *208*, 31–39.

Wiltschka, K.; Neumann, L.; Werheid, M.; Bunge, M.; Düring, R. A.; Mackenzie, K.; Böhm, L. Hydrodechlorination of Hexachlorobenzene in a Miniaturized Nano-Pd(0) Reaction System Combined with the Simultaneous Extraction of all Dechlorination Products. *Appl. Catal. B Environ.* **2020,** *275*, 119100.

Wohlers, D. W.; Ingerman, L.; Mcllroy, L. *Toxicological Profile for Mirex and Chlordecone—Draft for Public Comment*; Atlanta, Georgia, 2019.

Wrobel, M. H.; Grzeszczyk, M.; Mlynarczuk, J.; Kotwica, J. The Adverse Effects of Aldrin and Dieldrin on Both Myometrial Contractions and the Secretory Functions of Bovine Ovaries and Uterus In Vitro. *Toxicol. Appl. Pharmacol.* **2015,** *285*, 23–31.

Wrobel, M. H.; Mlynarczuk, J. Secretory Function of Ovarian Cells and Myometrial Contractions in Cow Are Affected by Chlorinated Insecticides (Chlordane, Heptachlor, Mirex) In Vitro. *Toxicol. Appl. Pharmacol.* **2017,** *314*, 63–71.

Wu, H.; Bertrand, K. A.; Choi, A. L.; Hu, F. B.; Laden, F.; Grandjean, P.; Sun, Q. Persistent Organic Pollutants and Type 2 Diabetes: A Prospective Analysis in the Nurses' Health Study and Meta-Analysis. *Environ. Health Perspect.* **2013,** *121*, 153–161.

Wu, J.; Hu, J. C.; Wang, S. J.; Jin, J. X.; Wang, R.; Wang, Y.; Jin, J. Levels, Sources, and Potential Human Health Risks of PCNs, PCDD/Fs, and PCBs in an Industrial Area of Shandong Province, China. *Chemosphere* **2018,** *199*, 382–389.

Xiong, J. Toxicity of Chlordane At Early Developmental Stage of Zebrafish. *BioRxiv* **2017,** 119248.

Yamazaki, K.; Itoh, S.; Araki, A.; Miyashita, C.; Minatoya, M.; Ikeno, T.; Kato, S.; Fujikura, K.; Mizutani, F.; Chisaki, Y. Associations between Prenatal Exposure to Organochlorine Pesticides and Thyroid Hormone Levels in Mothers and Infants: The Hokkaido Study on Environment and Children's Health. *Environ. Res.* **2020,** *189*, 109840.

Yang, L.; Liu, G.; Zheng, M.; Jin, R.; Zhu, Q.; Zhao, Y.; Zhang, X.; Xu, Y. Atmospheric Occurrence and Health Risks of PCDD/Fs, Polychlorinated Biphenyls, and Polychlorinated Naphthalenes by Air Inhalation in Metallurgical Plants. *Sci. Total Environ.* **2017,** *580*, 1146–1154.

Ye, M.; Beach, J.; Martin, J. W.; Senthilselvan, A. Association between Lung Function in Adults and Plasma DDT and DDE Levels: Results from the Canadian Health Measures Survey. *Environ. Health Perspect.* **2015,** *123*, 422–427.

Yilmaz, B.; Terekeci, H.; Sandal, S.; Kelestimur, F. Endocrine Disrupting Chemicals: Exposure, Effects on Human Health, Mechanism of Action, Models for Testing and Strategies for Prevention. *Rev. Endocr. Metab. Disord.* **2020,** *21*, 127–147.

Yuan, Y.; Ning, X. An.; Zhang, Y.; Lai, X.; Li, D.; He, Z.; Chen, X. Chlorobenzene Levels, Component Distribution, and Ambient Severity in Wastewater from Five Textile Dyeing Wastewater Treatment Plants. *Ecotoxicol. Environ. Saf.* **2020,** *193*, 110257.

Yuksel, H.; Ispir, Ü.; Ulucan, A.; Turk, C.; Taysı, M. R. Effects of Hexachlorocyclohexane (HCH- γ -Isomer, Lindane) on the Reproductive System of Zebrafish (*Danio rerio*). *Fish. Aquat. Sci.* **2016,** *16*, 917–921.

Zhang, B.; Xu, T.; Yin, D.; Wei, S. The Potential Relationship between Neurobehavioral Toxicity and Visual Dysfunction of BDE-209 on Zebrafish Larvae: A Pilot Study. *Environ. Sci. Eur.* **2020,** *32*.

Zhang, L.; Wei, J.; Ren, L.; Zhang, J.; Wang, J.; Jing, L.; Yang, M.; Yu, Y.; Sun, Z.; Zhou, X. Endosulfan Induces Autophagy and Endothelial Dysfunction via the AMPK/mTOR Signaling Pathway Triggered by Oxidative Stress. *Environ. Pollut.* **2017,** *220*, 843–852.

Zhang, W.; Wang, P.; Zhu, Y.; Yang, R.; Li, Y.; Wang, D.; Matsiko, J.; Han, X.; Zhao, J.; Zhang, Q. Brominated Flame Retardants in Atmospheric Fine Particles in the Beijing-Tianjin-Hebei Region, China: Spatial and Temporal Distribution and Human Exposure Assessment. *Ecotoxicol. Environ. Saf.* **2019,** *171*, 181–189.

Zhang, W.; Zhang, Q.; Yang, Q.; Liu, P.; Sun, T.; Xu, Y.; Qian, X.; Qiu, W.; Ma, C. Contribution of Alzheimer's Disease Neuropathologic Change to the Cognitive Dysfunction in Human Brains with Lewy Body–Related Pathology. *Neurobiol. Aging.* **2020a,** *91*, 56–65.

Zhang, W.; Lin, Z.; Pang, S.; Bhatt, P.; Chen, S. Insights into the Biodegradation of Lindane (γ -Hexachlorocyclohexane) Using a Microbial System. *Front. Microbiol.* **2020b,** *11*, 1–12.

Zhang, Y.; Luo, X. J.; Wu, J. P.; Liu, J.; Wang, J.; Chen, S. J.; Mai, B. X. Contaminant Pattern and Bioaccumulation of Legacy and Emerging Organhalogen Pollutants in the Aquatic Biota from an E-Waste Recycling Region in South China. *Environ. Toxicol. Chem.* **2010,** *29*, 852–859.

[illegible] Wang, [illegible] Y.; [illegible] Levels, Sources, and [illegible] PCBs, PCDD/Fs, and [illegible] in an Industrial Area of [illegible] Chemosphere 2018, 192, [illegible] 359.

Krung, [illegible] Toxicity of [illegible] Developmental Stage of Zebrafish [illegible] 2015, [illegible]

Yamazaki, [illegible] M.; Beng, [illegible] [illegible]

CHAPTER 2

The Risk of Inorganic Environmental Pollution to Humans

ZIA UR RAHMAN FAROOQI[1*], MUHAMMAD MOAZ KHURSHEED[1], NOUMAN GULZAR[1], MUHAMMAD AHMAD AKRAM[2], MUHAMMAD MAHROZ HUSSAIN[1], ABDUL QADEER[1], WAQAS MOHY-UD-DIN[1], and SADIA YOUNAS[3]

[1]Institute of Soil and Environmental Sciences, University of Agriculture Faisalabad 38040, Pakistan

[2]Environemtnal Protection Agency, Faisalabad 38000, Pakistan

[3]Department of Chemistry, The Government Sadiq College Women University, Bahawalpur 63100, Pakistan

[*]Corresponding author. E-mail: ziaa2600@gmail.com

ABSTRACT

This chapter aims to convey the risk levels of major inorganic pollutants (heavy metals or potentially toxic elements) to human lives. The mobilization of these potentially toxic elements (PTEs) through numerous anthropogenic activities have badly contaminated the environment. Since these PTEs are nonbiodegradable, they are accumulated into the human food chain and results in negative effects on human health. This phenomenon exerts some serious environmental health implications with concerns on the agricultural produce and quality as well as health of organisms living in the environment. Some pollutants are mutagenic, endocrine disruptors, carcinogenic, and teratogenic, while some trigger neurological changes in organism's behavior.

Nanotechnology for Environmental Pollution Decontamination: Tools, Methods, and Approaches for Detection and Remediation. Fernanda Maria Policarpo Tonelli, Rouf Ahmad Bhat, & Gowhar Hamid Dar (Eds.)

Because of their negative effects, due consideration is also required for the remediation of PTE-contaminated sites. Many limitations such as higher initial costs, expensive labor, resource modification, and disturbance in native soil microflora are the negative points while using various available chemical and physical methods for this purpose. Phytoremediation (use of plants species for chemical removal from contaminated sites) is a cheaper, economical, and safe approach to be practiced in PTE-contaminated soils when compared with physical and chemical methods of remediation. This chapter allows readers to get cost-effective ideas about the remediation of PTEs removal. This chapter also discusses the occurrence of PTEs, their sources, and their removal techniques with future recommendations.

2.1 INTRODUCTION

Unusual pollution is a natural phenomenon, but pollution levels due to human activities of goods production has emerged as a main cause of pollution in the environment. Heavy metals or potentially toxic substances currently contain high environmental toxic substances (PTEs). These exist naturally on the surface of the earth, with anthropogenic practices, such as smelting, refining, electrification, power processing, energy delivery, sewage disposal, melting efficiency, and intensive agriculture (Tepanosyan et al., 2017). Due to agricultural practices, farming, and environmental processes, PTEs are released in the environment. They could be found in lakes, ponds, and dams, which are actually the most likely to be the sources of PTEs in the world. Certain nutrients remain in the atmosphere for a long time, and due to their high availability in water, can concentrate on the entire food chain and have a detrimental effect on birds that feed on aquatic habitats (Kumar et al., 2019). The unrestricted use of inorganic fertilizers, sewage, animal waste, and fertilizer in agricultural systems also increases the number of PTEs in agricultural land. Studying the presence of PTEs on agricultural soil is important because of the mutation and accumulation in the human body through the use of plants, animals and the use of their products. The increase in natural threats and the effects of chemical pollution on water and soil have been identified over the past 100 years as a result of intensified crop production, as well as international conservation practices around the world. In addition, due to the excessive use of agrochemicals and PTE-loaded fertilizer and amendments such as sewage sludge, cultivated concentrations have contributed to the introduction of PTEs into the soil (Zhang et al., 2018). Therefore, it is important in policy formulation and could assist in developing strategies to reduce the imposition of PTEs on agricultural land. The agribusiness is a simple example of agriculture, and in its production phase, it is one of the industries that process agricultural products as raw materials. This was built based on intensive farming. Through the incorporation of large amounts of organic chemicals based fertilizer and pesticides, there has been increased soil contamination caused by PTEs, such as cadmium (Cd), copper (Cu), zinc (Zn), nickel (Ni), chromium (Cr), lead (Pb), and arsenic (As) (Yuan et al., 2021). When PTEs join the human food chain, soil and water pollution by PTEs contributes to crop losses and adverse human health effects. High levels of PTEs in the soil can adversely affect plant growth. However, certain types of plants, without showing signs of stress, can accumulate PTEs and endanger human health. In addition to large quantities of plants, plants can have a combination of both PTEs and nutrients (El-Meihy et al., 2019). Their contamination could be the cause by irrigation with contaminated water, improper use of organic (natural and unconventional) pesticide-based pesticides. There are some research indicators for the assessment of the extent and intensity of soil and carcass contamination in environmental studies. These indicators are used to compare, measure, monitor, and control the effects of pathogens (He et al., 2019).

Biological variants are used as bioindicators in many trophic phases to provide evidence of potential adverse effects of exposure to contaminants. Monitoring of metals in bird species is important in assessing their well-being, and at the same time, in assessing how polluted their habitats are. Various psychological, metabolic, reproductive, growth, and behavioral disorders of various types are associated with exposure to high levels of iron (Fe). In addition, it is responsible for oxidative damage, as the most expensive dosage processes that contribute to oxidative stress are toxins from the body and excretion from the body. High levels of PTEs can lead to shrinkage of the egg shell, infertility and self-harm, adverse developmental outcomes, and embryonic malformations and bird deaths, both of which contribute to human loss. For example, by reducing their growth or body weight, exposure to Cd, mercury (Hg), and selenium (Se) negatively affects bird health and ultimately harms longevity and reproductive success. Mercury implanted in the body tissues will adversely affect reproduction, especially in species with a high trophic stage, such as fish-eating birds (Grúz et al., 2018; Bada and Omotoriogun, 2019).

Of particular importance in Europe are the colonial herds; their conservation status has been greatly improved over the past 30 years and now several herdsmen have been protected. However, a variety of species are at risk of conditions that overwhelm local and European authorities, such as drought and loss of habitat in winter migrants in Africa, improvements in food supply and quality of water and food in Southern Europe, and pollution and deforestation of adequate breeding grounds in Italy. Since almost all animals feed on aquatic animals and are unable to count and classify toxic toxins, water pollution can be a significant risk factor for adults and chicks. Many bird tissues like feathers, liver, bones and eggs have been used to track and expose birds (Egwumah et al., 2017). Among these, feathers have many advantages because, without damaging their well-being, they can be easily and repeatedly collected from the same person, and their storage does not require refrigeration. In contrast, feathers represent local pollutants than other tissues in the case of wolves, because the chicks eat deer collected near the colony, and large amounts of material are deposited in the feathers during the breeding season (Liang et al., 2016).

It is well known that people are exposed to PTEs through food (food and drinking water) and respiration (air pollution). For such chemicals, however, skin sensitivity should not be reduced. In terms of skin contamination, while most of the chemicals used during textile processing are washed, the remaining amounts of these compounds may and may not be released during consumer use. Nitrogen (N) is one of the most important biogeochemical elements, as well as trace elements (Co, Zn, Fe, Cu, etc.), and represents important inputs for agricultural sustainability. In trees, plants, wildlife, human waste, such as sewage sludge, ammonium nitrate fertilizers, and industrial-grade N-compounds, and car compounds released after rain, the N-nitrate form (NO_3^-) is found. It usually exists in groundwater and surface water as a natural resource, but as N levels rise, the causes are primarily associated with anthropogenic influences. In fact, while there are many N-factors that can contribute to the fight against groundwater pollution, human actions are the cause of NO_3^- level rise to a dangerous level (Motevalli et al., 2019; Rahmati et al., 2019). The groundwater contamination is an environmental problem because it is a low-toxic compound, but it is harmful to human health as it is reduced to nitrite (NO_2^-), which causes diseases, such as methemoglobinemia in children and adult stomach cancer. Nitrate in groundwater can be found in a variety of agricultural sectors. Due to the volume of this mineral fertilizer, which is greater than the needs of the crop, NO_3^- leaching from cultivated fields to water occurs (groundwater and surface water). Therefore, the main source of global water pollution is the use of farmland. Therefore, to protect water from NO_3^- contamination from agricultural products, the European Union's Directive 91/676/EEC defines the appropriate level of NO_3^- pollution in groundwater with a limit of 50 mg/L. Therefore, it has been established. ways to monitor and track groundwater and enforce measures to promote groundwater safety.

2.2 ENVIRONMENTAL POLLUTANTS

A common concern with the environment is air pollution from natural and uncommon pollution. Several of these pollutants are extremely persistent and replicate legal levels beyond the atmosphere. They are particularly toxic and can be obtained by contaminating food chains to the highest trophic areas. Pollution involves some development in the environment, but the use is limited to reflecting any degradation, chemicals and environmental degradation. Human lives are affected by these types of releases either by direct or indirect mechanisms. Pollution is a subset of xenobiotic chemicals released by anthropogenic processes entering the atmosphere and are present at a higher concentration of "natural levels." Pollutants are often classified as perishable and nonperishable: decaying organic matter consists of compost and organic matter that are easily separated under normal conditions. Those that are not exposed to microorganisms, for example, PTEs, plastics, and soaps are nonperishable compounds. The exponential population growth, industrial expansion, and city sprawl has contributed to the massive release in the xenobiotic chemical atmosphere. A significant amount of toxic chemicals produced by factories that are widely used to improve agricultural production in developed countries (Hashemi et al., 2017; Zhang et al., 2019).

2.2.1 ORGANIC POLLUTANTS

Excessive use of unconventional materials has led to widespread and long-term environmental issues over the past few decades. In fact, the widespread use of pesticides in the agricultural and public health sectors around the world has led to many cases of food poisoning, agricultural land, and water shortages. And several years after its introduction, fossils of pesticides known to be extinct still survive. Organic pollutants include some pesticides, aromatic compounds and phenyls based compounds (Alharbi et al., 2018).

2.2.2 INORGANIC POLLUTANTS

A number of intentional or unintentional chemicals are released into the soil or bodies of water on a daily basis. Some of them are important for having a purpose while most of them are challenging because of their long-term persistence and resistance to corruption. This category of pollutants include PTEs including metal and metalloid and radionuclides. Unusual pollutants are more likely to be present in the ground and in groundwater from the leakage of a polluted resource field, such as waste management and mining facilities. Pollutants from these sources will flow through the groundwater flow, resulting in a phase-polluted water farm. Among the rare toxins, heavy metals and metalloids are in close proximity to PTEs and PTEs are a major concern due to their high toxicity in low-concentration areas. They may be attached to solid soil layers or easily accessible by organic matter for extraction (Mishra and Maiti, 2017).

2.3 INORGANIC POLLUTANTS: ORIGIN AND TYPES

Inorganic emissions from various sources of air, water, and soil. Standing stores include fossil fuel emissions, processing plants and factories, as well as sheds and other types of fuel-fired heating systems. Traditional biomass burning is a major cause of air pollution in developed and poor countries. Standard biomass includes wood, plant waste, and manure. Mobile services include cars, boats, and ships. Controlled agricultural fire helps in good forest management activities. Controlled or determined temperatures are a common practice used for forest conservation, planting, regeneration of plains, or removal of greenhouse gases. Fire is a common phenomenon of nature for both trees and grasslands, and a controlled fire can be a forest tool. Controlled temperatures encourage appropriate forest trees to sprout, thus rejuvenating the forest. Sources from nonfire processes occur. They are all available in oil-based oils, hair spray, varnish, aerosol sprays, and other solvent. Waste collection in landfills, which produce methane. Methane (CH_4) is highly flammable and can cause air to explode. It also acts as an asphyxiant, and in a closed space, can remove oxygen. If oxygen concentration is reduced by migration to less than 19.5%, asphyxia or congestion may occur. Atomic resources, such as nuclear bombs, toxic agents, rocket, and biological warfare result in radioactive pollutants. An important source of nitrogen oxides can be from agriculture due to inadequate agricultural management (Dubovina et al., 2018; Salvo et al., 2018).

As water is important for health, it poses health risks due to its flow and improved access to health. The climate of natural rocks and soil primarily adds to the inanimate pollution of water. Disposal of industrial waste, construction and sewage containing minerals, PTEs, and trace elements (Cd, Pb, As, Cr etc.) have also reduced the water supply over the past few years and has adversely affected the human health.

2.4 TYPES OF INORGANIC POLLUTANTS

2.4.1 POTENTIALLY TOXIC ELEMENTS AND ALIKE ELEMENTS (METALLOIDS)

Any chemical substance that has a relatively high density and toxicity in low concentrations is called PTEs. Hg, Cd, As, Cr, thallium (Tl), and Pb are examples of heavy metals. There are common parts on the surface of the earth. There is no need to weaken or kill them. PTEs reach our bodies only in limited amounts through food, drinking water, and air. Any PTE (e.g., Cu, Se, Zn) is important as criteria for maintaining the body's metabolism. However, they can contribute to tackling toxicity at higher altitudes. For example, their toxicity can be caused by contamination of drinking water, high concentration of air near the outlets, or ingestion by food chain. As they appear to be compacted, PTEs are toxic. Compared with the concentration of atmospheric chemicals, the bioaccumulation shows an increase in the concentration of chemicals in the body of organisms over time. Whenever they are taken and processed faster than demolished (made into bodies) or extracted, compounds combine into living organisms (Rehman et al., 2018).

Potentially hazardous substances can reach the water system through industrial and consumer waste, or by acid rain that breaks down the soil and deposits heavy metals in streams, wetlands, water, and groundwater. Another PTE types is the substance having properties in-between PTEs and semi-liquid PTEs (metalloids). Examples of these are boron (B), As, Hg, and Antimony (Sb). Table 2.1 describes the inorganic pollutants and their sources.

2.4.1.1 HUMAN HEALTH RISKS

There has long been concerns about the pollution of PTEs due to their toxicity to plants, animals, and humans as they lack biodegradability. They have a remarkable effect on the colors and texture of aquatic animals that through biomagnification enters the human food chain and eventually affect humans. Toxicity levels of the pollutant depends on the type of pollutant, its role in organism's body, and the organ it affects or toxin it produces. PTEs in drinking water are often linked to human toxins by Pb, Fe, Cd, Cu, Zn, and Cr. They are needed in the body in small amounts but can also be toxic in large doses (Fig. 2.1). They form one important group of environmental hazards if any. Some PTEs like Cu are important in tracking devices, but they

TABLE 2.1 Inorganic Pollutants and Their Sources.

Contaminants	Sources	References
Inorganic fertilizers (e.g., nitrates and phosphates etc.)	Used in agriculture	Farmer (2018), Garzon-Vidueira et al. (2020)
Sulfides	Mined minerals produce and sulfides when combined with water and action of microorganism	Quevedo et al. (2020)
Heavy metals (PTEs)	Motor vehicles, acid mine drainage	Tepanosyan et al. (2017)
Arsenic	Pesticides, chemical wastes, mining bi-products	Zhang et al. (2017)
Cadmium	Industrial discharge, metal plating, Ni–Cd batteries, mining waste	Kubier et al. (2019)
Chromium	Metal plating industries, tanning process	Hausladen et al. (2018)
Lead	Plumbing, mining, coal, gasoline	Li et al. (2020)
Mercury	Mining, industrial waste, coal	Sundseth et al. (2017)
Zinc	Metal plating industries, industrial waste	Olukosi et al. (2018)
Nickel	Sewage, industrial waste or mine washing	Meshram and Pandey (2018)

show toxicity when there are too many amounts in drinking water. Cr, Cu, and Zn can cause noncarcinogenic risks, such as neurologic involvement, headaches, and liver disease when they exceed their safety standards. There is also evidence that exposure to low-dose of toxins can cause cancer to different body organs. The scientist found an increased risk of dying from lung cancer due to exposure to dust and fog containing Cr. Cd in diets due to the consumption of contaminated rice and other foods results in kidney damage and different types of cancers which are serious and long-lasting, as exposure can also cause many health problems. These include skin, respiratory, cardiovascular, vascular, abdominal, hematological, hepatic, renal, neurological, growth, reproductive, immunological, genotoxic, mutagenetic, and carcinogenic effects (as liver cancer) (Ali and Khan, 2018; Liu et al., 2018; Amqam et al., 2020).

Arsenic: Vomiting, abdominal pain, diarrhea, muscle cramping, skin pigmentation, and cancers.

Cadmium: Chills, fever and muscle pain, lungs and kidney damage, and cancers.

Chromium: Asthma, cough, breathing difficulty, anemia, damage to male reproductive system, and cancers

Lead: Weakness, anemia, kidney and brain damage, damage to developing baby's central nervous system and cancer

Mercury: Damage to skin, eye and digestive tract, tremors, insomnia, memory loss, headaches, and cancers.

Nickel: Chronic bronchitis, impaired lung function, lung and nasal sinus, and cancers.

Zinc: Nausea, vomiting, loss of appetite, abdominal cramps, diarrhea, and headache.

Nutrients: Nitrates cause blue baby syndrome to babies and anemia to adults, phosphates cause osteoporosis and kidney damage, while potassium cause eye, nose, throat and lungs irritations.

FIGURE 2.1 Potentially toxic elements and their risks to human health.

2.4.2 RADIOACTIVE SUBSTANCES

Radioactive substances produce rays and known as physical pollution sources. They decrease the quality of living due to the release of radioactive material during the explosion and exploration of nuclear weapons, nuclear production, and termination of operations, radioactive mines, waste management and waste disposal, and risks at nuclear power plants. Naturally occurring radioactive materials and technologically advanced radioactive materials contain materials which are usually industrial waste or products derived from organic matter, such as uranium, thorium, and plutonium and any of their decay products, such as radium and radon. Naturally occurring radioactive material is found in the lower concentrations in the earth's crust and pollute the environment when humans do mining and some natural processes, such as radon gas leaks into the atmosphere or by groundwater depletion (Steffan et al., 2018; Fadlallah et al., 2019).

2.4.2.1 HUMAN HEALTH RISKS

Radioactive pollution is accompanied with the ionizing radiation emissions of high-energy particles or gamma radiation that has a high frequency and therefore high energy. Particles or high-energy electromagnetic radiation penetrates into the human body and causes ionization of molecules present in the body. This ionizing radiation is a serious threat to human health. These free radicals that are formed react with the components of the living organism, causing the destruction of proteins, membranes, and nucleic acids (Vieira et al., 2020). Depending on the intensity and duration of exposure to ionizing radiation, its effects on a living organism may be accompanied by mild, moderate, or even fatal consequences. Low-level exposure may induce only superficial effect and mild skin irritation. The short-range exposure which occurs for few days provokes the loss of hair or nails, subcutaneous bleeding, and impairment of cells. The long but low-intensity exposure leads to nausea, vomiting, diarrhea, and bruises. The acute exposure to ionizing radiation is characterized by the damage of DNA cells that results in cancer, genetic defects, and even death (Li et al., 2018).

2.4.3 NUTRIENT POLLUTION

Pollution is a common thread that connects many problems along the coast of nations, including eutrophication, harmful algal flowers, dead areas, fish stocks, certain shellfish toxins, loss of seaweed and kelp beds, the destruction of certain corals, and even other animals, marine mammals, and seabirds. More than 60% of our rivers and coastal areas in all the coastal states of the United States have been moderately polluted. This degradation is worse in the mid-Atlantic, southeast and in the Gulf of Mexico. Over the past 40 years, pollution control laws have significantly reduced the release of toxic substances in our coastal waters. It is known that (1) benthic chlorophyll distribution is strongly associated with the total N and phosphorus (P) in water, (2) nutrient uptake in many rivers and streams, (3) bioassays typically show N alone or concert responses with P autotrophic (primary production and chlorophyll) and heterotrophic (respiratory) responses, (4) both heterotrophic and autotrophic processes are influenced by the presence of N and P, and (5) Transmission of cyanobacteria generally appears to be unable to fully satisfy the N limit in rivers and streams where P is present above the N level. This suggests that the control of N and P should be considered in the management of streams (Glibert et al., 2018; Sarma et al., 2020).

The livestock from agriculture sector provide up to one-third human protein requirements and a major employment provider in developing countries. While it offers such great benefits to people, livestock mistreatment can have serious environmental consequences that have not been addressed adequately in many emerging economies. About 60% of the world's biomass is harvested annually to support all human activities consumed by the livestock industry, making it impossible to distribute such large-scale resources to the industry. Nitrogen from animal waste that falls directly into the surface water or nitrogen is regenerated in the air as ammonia can be one major source of N from agricultural activity to coastal waters (Shortle and Horan, 2017).

2.5 MANAGEMENT OF INORGANIC ENVIRONMENTAL POLLUTANTS

Air, land, and water pollution caused less than 9 million deaths in 2016, or 16% of all deaths worldwide. About 92% of all deaths related to pollution are found in developing countries, with the poor, the marginalized, and the very young are affected by the health effects of pollution. The economic burden is enormous in 2016, air pollution occurring alone costing the global economy US$ 5.7 trillion or 4.8% of global GDP (Bennett et al., 2018).

The deterioration of environmental health due to rapid industrialization, urbanization, and increasing human pressure, is a major concern for developing countries, including India. Environmental pollution (soil, water, and air) is a toxic waste from various natural and anthropogenic activities and its adverse effects on biodiversity require further research to reduce and solve these problems effectively. Compared with conventional remediation methods, modern methods are much higher and more effective in removing large amounts of natural and uncommon pollution from contaminated sources (Fig. 2.2).

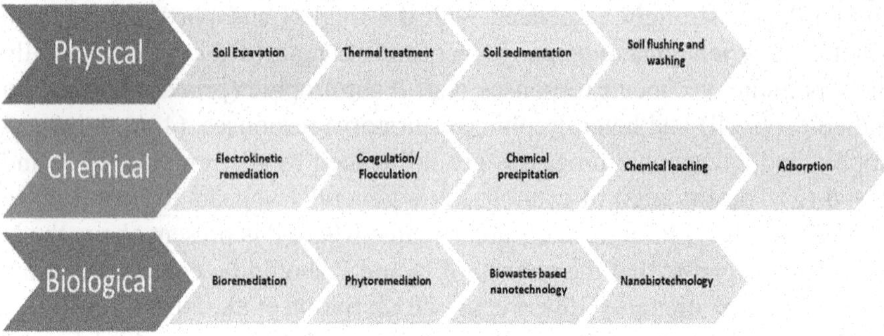

FIGURE 2.2 Remediation approaches for contaminants.

2.5.1 CHEMICAL MANAGEMENT

The identification and removal of hazardous environmental pollutants is indeed a major challenge in this age of industrial development. Changing the old way of managing progressively in the study of nanoparticles will go a long way in overcoming this problem. Because of the unique features of this process, such as surface area, size, photoelectronic, photocatalytic, and absorptivity properties, nanoparticles are considered to be the most useful and important source for the detection and repair of hazardous pollutants.

The preparation of inorganic ions and metals of natural clay particles and metal oxides has been tested. PTEs such as Cu, Cd, Ni, and Pb are better absorbed by oxidized hydroxylated carbon nanotubes (CNTs). The best adsorber for organometallic elements has been developed in the form of various pure CNTs. CNTs act as adsorbent for a wide variety of organic compounds, such as dioxin and its metabolites, polynuclear aromatic hydrocarbons (PAHs), chlorophenols, chlorobenzenes, and bisphenol A (Xue et al., 2017).

Advertising material will be of great help and comfort in use to solve this problem. While the adsorption capacity of any nanomaterial depends on the height of the structure and the surface characteristics. For this, the most advanced carbonaceous nanoparticles are graphene oxide and graphene. Theoretically, a certain area in graphene is high and it contains active groups, which show their potential in adsorption. In the past, air and water pollution controls have been based on graphene or composite compounds. The different active groups of graphene make it more acidic, making it more exposed to alkaline chemicals and cations. Graphene has a hydrophobic surface and due to its high oku-π interaction, it has a greater tendency to chemical reactions. Different nanocomposites can be made by converting graphene oxide or graphene with iron or natural oxides, which increases the adsorptive tendency and isolation efficiency (Ali et al., 2019).

The use of different sorbents and adsorbents such as biochar for environmental management has been the cause of the world's recent interest in advancing biochar production technology and its management. The use of biochar in the soil has great potential for enhancing extended and long-lasting carbon stability due to the fact that biochar carbon is very restorative and has a fragrant structure. The recurring biochar condition persists in the soil from 1000 to 10,000 years and helps to increase soil storage and pollution regulation. Recently, biochar application to contaminated soil has been considered as a tool to rectify soil contaminated by PTEs. It has more surface and has a porous structure. Biochar, a carbon-rich natural substance, has been widely used in the formation of particles for carbon recovery, soil fertility, and environmental remediation. Synergistic methods of biochar in composite membranes have been proposed to determine the superior membrane efficacy in the treatment of pollutants. Multifunctional biochar composite membranes not only effectively prevent pollution problems caused by using the biochar particle as a sorbent but can also be produced on a large scale, demonstrating the great potential for practical use. The promising ability to extract biochars to eliminate pollution depends on its connected structures, surface chemistry, surface area, and porous structure, which differs significantly from the immature material and the proposed conditions for pyrolysis (Wu et al., 2017; Zhu et al., 2017; Wang et al., 2019).

2.5.2 BIOLOGICAL APPROACHES

The most harmful factors affecting aquatic and human lives are hazardous metals and organic color dyes. The biological technique can be used for their remediation is known as microbial fuel cell (MFC), the most hopeful, environment friendly, and an emergent technology. In this biological remediation method, which involves microbial role, an electric current is generated along with treatment of aquatic pollutants. While discussing about the MFC technology, it is clear that this technique provides an alternative with high potency in the area of wastewater management. Long-term useable green energy sources have been introduced by this technology, and the researchers are actively focusing on this methodology as biodegradable matter and waste material are used in MFC devices as fuel to operate the device. Some important parts of MFC device include electrodes, variety of microbes, and MFC design. These parts have an important participation in the removal of pollutants caused by dyes from water and reduction of heavy metals from highly toxic area making that less toxic. To check the working efficiency of the MFCs devices for water treatment different researches have been conducted (Yaqoob et al., 2020).

Many living organisms could be used in the degradation and detoxification of major aquatic pollutants. The green algae (*Chlorella*) has been used for metal biosorption. Bioremediation of toxins through plant is also considered the most important way of bioremediation of the biosphere. Moreover, phytochelatins and metallothionein can detoxify the PTEs. It shows that plants have the ability in the reclamation of air, soil, and water (Shrestha et al., 2019).

2.5.2.1 TYPES OF BIOREMEDIATION

Plants and microbes use a variety of methods to repair contaminated areas. Bioremediation, specifically phytoremediation technology can be categorized on the basis of the process and its functionality as follows:

2.5.2.1.1 Mycoremediation

Recently, algae have become a material for the natural purification of polluted water and polluted areas. They are able to accumulate/remove toxins from the body nutrients, PTEs, pesticides, and other nontoxic substances and inorganic matter, and radioactive material. Wastewater treatment programs with low algae have become increasingly important over the past 50 years, and it is now widely accepted that wastewater treatment methods work as standard treatments. These specific factors have made algal wastewater treatment systems less costly and inexpensive treatment systems, especially for municipal sanitation. In this way, 70% of oxygen demand, 66% of chemical requirements, 71% of total N, 67% of P, 54% of dry solids, and 51% of soluble solids are reduced (Prasad, 2017).

2.5.2.1.2 Phytoextraction

Phytoextraction, a common phytoremediation process, involves the removal of pollutants by plant roots by subsequent accumulation in the airborne plant parts. Parts of the aerial plants are harvested and removed, thus ensuring permanent removal of metals, such as Pb, Cd, Ni, Cu, and Cr from sites they are unclean. However, it only applies to those sites that contain low to moderate levels of iron pollution because plant growth does not reside in highly polluted areas (Chin, 2020).

2.5.2.1.3 Rhizofiltration

This approach relies on the ability of plant roots to absorb and destroy metals or excess nutrients from aqueous plants (wastewater streams, nutrient regeneration systems). Rhizofiltration fixes metals, such as Pb, Cd, Ni, Cu, and Cr radioactive material (U, Cs, Sr). Efficient plants should produce a large amount of biomass root or root zone, be able to collect and tolerate a significant amount of target metals, include easy handling and low maintenance costs, and should have secondary waste that needs to be disposed of. Earth plants are more suitable for rhizofiltration because they produce longer, larger, and more common root systems with larger metal spells (Benavides et al., 2018).

2.5.2.1.4 Phytostabilization

In this way, plants are used to convert toxic soil compounds into less toxic species, which can be removed from the soil. Phytostabilization stops debris, prevents air and water erosion exposure, provides water control that reduces direct migration of contaminants into groundwater and disables physical and chemical pollution through root sorption and chemical rehabilitation through various soil amendments. Phytostabilization requires plants that can grow in contaminated soil and their roots must grow in a polluted environment, and to convert biological, chemical, or physical conditions in the soil that convert toxic metals into less toxic ones. Plant toxicity by plants can be improved by supplementing the soil with increasing soil organisms and pH (using lime), or by binding certain nutrients with phosphate or carbonate without the use of soil amendments (Visconti et al., 2020).

2.5.2.1.5 Phytovolatization

The use of plants to absorb PTEs and to convert them into flexible, nontoxic chemical forms for respiration is called phytovolatization. Some PTEs, such as As, Hg, and Se may exist as natural gas types. Other plants that are naturally occurring or genetically modified, such as *Chara canescens* (musk grass), *Brassica juncea* (Indian mustard), and *Arabidopsis thaliana* have reportedly been able to absorb heavy metals and convert them into oxygen-rich species inside the plant and release them into space. Phytovolatization is mainly used for the extraction of mercury when mercuric ions are converted to nontoxic mercury elemental gas. However, unlike other remedial techniques, once impurities are removed by volatilization, one cannot control their migration (Guarino et al., 2020).

2.5.2.1.6 Phytodegradation

In phytodegradation, plants reduce pollution through metabolic processes and utilize rhizospheric associations between plants and soil insects. Plant enzymes that digest waste can be released into the rhizosphere, where they can play a major role in the conversion of pollutants. Enzymes, such as dehalogenase, nitroreductase, peroxidase, laccase, and nitrilase have been found in vessels and soil. In addition, natural compounds, such as blasting machines, chlorine-containing solvents, herbicides and insecticides and inorganic nutrients can also be undermined by this technology (Chlebek and Hupert-Kocurek, 2019).

2.6 CONCLUSIONS

In the human body, inorganic compounds play a very significant function. Some metals, including zinc, copper, and iron are very important to human survival. But they are considered highly hazardous compounds like other metals as they reach their allowable levels in the human body. There is no doubt, though, that these potentially dangerous chemicals have given comfort to human life, even though this calmness has resulted in human destruction. They affect the human body from soil malfunction to malfunction. Both living organisms are changed by their detrimental consequences in every way. It is not enough for humanity to escape natural catastrophes, but man should somehow get rid of his own troubles. When we begin to stick to the ideals of sustainable development, we will prevent the impact of humanitarian crises on the environment. In order to avoid environmental contamination at the start, certain steps need to be taken. The degradation of potentially hazardous contaminants can be avoided by premature mitigation, but once they enter the atmosphere, they must be resolved by intelligent action, and such steps may be to rid the environment of metals in an affordable, simple, and permanent manner. Physical, chemical, and biological remediation steps are included.

KEYWORDS

- **heavy metals**
- **nutrient pollution**
- **human health risks**

REFERENCES

Alharbi, O.M.; Khattab, R.A.; Ali, I. Health and Environmental Effects of Persistent Organic Pollutants. *J. Mol. Liquids* **2018,** *263*, 442–453.

Ali, H.; Khan, E. Bioaccumulation of non-Essential Hazardous Heavy Metals and Metalloids in Freshwater Fish. Risk to Human Health. *Environ. Chem. Lett.* **2018,** *16*, 903–917.

Ali, I.; Mbianda, X.; Burakov, A.; Galunin, E.; Burakova, I.; Mkrtchyan, E.; Tkachev, A.; Grachev, V. Graphene Based Adsorbents for Remediation of Noxious Pollutants from Wastewater. *Environ. Int.* **2019,** *127*, 160–180.

Amqam, H.; Thalib, D.; Anwar, D.; Sirajuddin, D.; Mallongi, A. Human Health Risk Assessment of Heavy Metals via Consumption of Fish from Kao Bay. *Rev. Environ. Heal.* **2020,** *1*, 257–263.

Bada, A. A..; Omotoriogun, T. C. Incidence of Heavy Metals in Feathers of Birds in a Human-Impacted Forest, South-West Nigeria. **2019.**

Benavides, L. C. L.; Pinilla, L. A. C.; Serrezuela, R.; Serrezuela, W. F. R. Extraction in Laboratory of Heavy Metals through Rhizofiltration Using the Plant Zea Mays (Maize). *Int. J. Appl. Environ. Sci.* **2018,** *13*, 9–26.

Bennett, J.E.; Stevens, G.A.; Mathers, C.D.; Bonita, R.; Rehm, J.; Kruk, M.E.; Riley, L.M.; Dain, K.; Kengne, A.P.; Chalkidou, K. Ncd Countdown 2030: Worldwide Trends in Non-Communicable Disease Mortality and Progress towards Sustainable Development Goal Target 3.4. *Lancet* **2018,** *392*, 1072–1088.

Chin, M. 2020. Improving Metal Hyperaccumulator Wild Plants to Develop Commercial Phytoextraction Systems: Approaches and Progress. In *Phytoremediation of Contaminated Soil and Water*; CRC Press, 2020; pp 129–158.

Chlebek, D.; Hupert-Kocurek, K. Endophytic Bacteria in the Phytodegradation of Persistent Organic Pollutants. *Adv. Microbiol.* **2019,** *58*, 70–79.

Dubovina, M.; Krčmar, D.; Grba, N.; Watson, M.A.; Rađenović, D.; Tomašević-Pilipović, D.; Dalmacija, B. Distribution and Ecological Risk Assessment of Organic and Inorganic

Pollutants in the Sediments of the Transnational Begej Canal (Serbia-Romania). *Environ. Poll.* **2018,** *236*, 773–784.

Egwumah, F.; Egwumah, P.; Edet, D. Paramount Roles of Wild Birds as Bioindicators of Contamination. *Int. J. Avian Wildlife Biol.* **2017,** *2*, 00041.

El-Meihy, R.M.; Abou-Aly, H.E.; Youssef, A.M.; Tewfike, T.A.; El-Alkshar, E. A. Efficiency of Heavy Metals-Tolerant Plant Growth Promoting Bacteria for Alleviating Heavy Metals Toxicity on Sorghum. *Environ. Exp. Bot.* **2019,** *162*, 295–301.

Fadlallah, H.; Taher, Y.; Haque, R.; Jaber, A. Oradiex: A Big Data Driven Smart Framework for Real-Time Surveillance and Analysis of Individual Exposure to Radioactive Pollution, 2019.

Farmer, A. *Phosphate Pollution: A Global Overview of the Problem. Phosphorus: Polluter and Resource of the Future: Motivations, Technologies and Assessment of the Elimination and Recovery of Phosphorus from Wastewater*. IWA Publishing, UK, 2018.

Garzon-Vidueira, R.; Rial-Otero, R.; Garcia-Nocelo, M.L.; Rivas-Gonzalez, E.; Moure-Gonzalez, D.; Fompedriña-Roca, D.; Vadillo-Santos, I.; Simal-Gandara, J. Identification of Nitrates Origin in Limia River Basin and Pollution-Determinant Factors. *Agric. Ecosys. Environ.* **2020,** *290*, 106775.

Glibert, P.M.; Beusen, A.H.; Harrison, J.A.; Dürr, H.H.; Bouwman, A.F.; Laruelle, G. G. Changing Land-, Sea-, and Airscapes: Sources of Nutrient Pollution Affecting Habitat Suitability for Harmful Algae. In *Global Ecology and Oceanography of Harmful Algal Blooms*; Springer, 2018; pp 53–76.

Grúz, A.; Déri, J.; Szemerédy, G.; Szabó, K.; Kormos, É.; Bartha, A.; Lehel, J.; Budai, P. Monitoring of Heavy Metal Burden in Wild Birds at Eastern/North-Eastern Part of Hungary. *Environ. Sci. Pollu. Res.* **2018,** *25*, 6378–6386.

Guarino, F.; Miranda, A.; Castiglione, S. Cicatelli, A. Arsenic Phytovolatilization and Epigenetic Modifications in Arundo Donax l. Assisted by a PGPR Consortium. *Chemosphere* **2020,** *251*, 126310.

Hashemi, B.; Zohrabi, P.; Kim, K.-H.; Shamsipur, M.; Deep, A.; Hong. J. Recent Advances in Liquid-Phase Microextraction Techniques for the Analysis of Environmental Pollutants. *TrAC Trends Anal. Chem.* **2017,** *97*, 83–95.

Hausladen, D.M.; Alexander-Ozinskas, A.; McClain, C.; Fendorf, S. Hexavalent Chromium Sources and Distribution in California Groundwater. *Environ. Sci. Technol.* **2018,** *52*, 8242–8251.

He, J.; Yang, Y.; Christakos, G.; Liu, Y.; Yang, X. Assessment of Soil Heavy Metal Pollution Using Stochastic Site Indicators. *Geoderma.* **2019,** *337*, 359–367.

Kubier, A.; Wilkin, R.T.; Pichler, T. Cadmium in Soils and GROUNDWATEr: A Review. *Appl. Geochem.* **2019,** *108*, 104388.

Kumar, S.; Prasad, S.; Yadav, K.K.; Shrivastava, M.; Gupta, N.; Nagar, S.; Bach, Q.-V.; Kamyab, H.; Khan, S.A.; Yadav, S. Hazardous Heavy Metals Contamination of Vegetables and Food Chain: Role of Sustainable Remediation Approaches-a Review. *Environ. Res.* **2019,** *179*, 108792.

Li, M.; Weis, D.; Smith, K.E.; Shiel, A.E.; Smith, W.D.; Hunt, B.P.; Torchinsky, A.; Pakhomov, E. A. Assessing Lead Sources in Fishes of the Northeast Pacific Ocean. *Anthropocene* **2020,** *29*, 100234.

Li, P.; Zhang, R.; Zheng, G. Genetic and Physiological Effects of the Natural Radioactive Gas Radon on the Epiphytic Plant Tillandsia Brachycaulos. *Plant Physiol. Biochem.* **2018,** *132*, 385–390.

Liang, J.; Liu, J.; Yuan, X.; Zeng, G.; Yuan, Y.; Wu, H.; Li, F. A Method for Heavy Metal Exposure Risk Assessment to Migratory Herbivorous Birds and Identification of Priority Pollutants/Areas in Wetlands. *Environ. Sci. Pollu. Res.* **2016,** *23*, 11806–11813.

Liu, J.; Liu, Y.J.; Liu, Y.; Liu, Z.; Zhang, A. N. Quantitative Contributions of the Major Sources of Heavy Metals in Soils to Ecosystem and Human Health Risks: A Case Study of Yulin, China. *Ecotoxicol. Environ. Saf.* **2018,** *164*, 261–269.

Meshram, P.; Pandey, B. D. Advanced Review on Extraction of Nickel from Primary and Secondary Sources. *Mineral Proces. Extract. Metall. Rev.* **2018.**

Mishra, S.; Maiti, A. The Efficiency of Eichhornia Crassipes in the Removal of Organic and Inorganic Pollutants from Wastewater: A Review. *Environ. Sci. Pollu. Res.* **2017,** *24*, 7921–7937.

Motevalli, A.; Naghibi, S.A.; Hashemi, H.; Berndtsson, R.; Pradhan, B.; Gholami, V. Inverse Method Using Boosted Regression Tree and k-Nearest Neighbor to Quantify Effects of Point and Non-Point Source Nitrate Pollution in Groundwater. *J. Cleaner Prod.* **2019,** *228*, 1248–1263.

Olukosi, O.A.; van Kuijk,S.; Han, Y. Copper and Zinc Sources and Levels of Zinc Inclusion Influence Growth Performance, Tissue Trace Mineral Content, and Carcass Yield of Broiler Chickens. *Poultry Sci.* **2018,** *97*, 3891–3898.

Prasad, R. *Mycoremediation and Environmental Sustainability*; Springer, 2017.

Quevedo, C.P.; Jiménez-Millán, J.; Cifuentes, G.R.; Jiménez-Espinosa, R. Electron Microscopy Evidence of Zn Bioauthigenic Sulfides Formation in Polluted Organic Matter-Rich Sediments from The Chicamocha River (Boyacá-Colombia). *Minerals* **2020,** *10*, 673.

Rahmati, O.; Choubin, B.; Fathabadi, A.; Coulon, F.; Soltani, E.; Shahabi, H.; Mollaefar, E.; Tiefenbacher, J.; Cipullo, S.; Ahmad, B. B. Predicting Uncertainty of Machine Learning Models for Modelling Nitrate Pollution of Groundwater Using Quantile Regression and Uneec Methods. *Sci. Total Environ.* **2019,** *688*, 855–866.

Rehman, K.; Fatima, F.; Waheed, I.; Akash, M. S. H. Prevalence of Exposure of Heavy Metals and Their Impact on Health Consequences. *J. Cell. Biochem.* **2018,** *119*, 157–184.

Salvo, A.; La Torre, G.L.; Mangano, V.; Casale, K.E.; Bartolomeo, G.; Santini, A.; Granata, T.; Dugo, A. Toxic Inorganic Pollutants in Foods from Agricultural Producing Areas of Southern Italy: Level and Risk Assessment. *Ecotoxicol. Environ. Saf.* **2018,** *148*, 114–124.

Sarma, V.; Krishna, M.; Srinivas, T. Sources of Organic Matter and Tracing of Nutrient Pollution in the Coastal Bay of Bengal. *Marine Pollu. Bull.* **2020,** *159*, 111477.

Shortle, J.; Horan, R. D. Nutrient Pollution: A Wicked Challenge for Economic Instruments. Water *Econ. Policy*. **2017,** *3*, 1650033.

Shrestha, P.; Bellitürk, K.; Görres, J. H. Phytoremediation of Heavy Metal-Contaminated Soil by Switchgrass: A Comparative Study Utilizing Different Composts and Coir Fiber on Pollution Remediation, Plant Productivity, and Nutrient Leaching. *Int. J. Environ. Res. Public Heal.* **2019,** *16*, 1261.

Steffan, J.; Brevik, E.; Burgess, L.; Cerdà, A. The Effect of Soil on Human Health: An Overview. *Eur. J. Soil Sci.* **2018,** *69*, 159–171.

Sundseth, K.; Pacyna, J.M.; Pacyna, E.G.; Pirrone, N.; Thorne, R. J. Global Sources and Pathways of Mercury in the Context of Human Health. *Int. J. Environ. Res. Public Heal.* **2017,** *14*, 105.

Tepanosyan, G.; Maghakyan, N.; Sahakyan, L.; Saghatelyan, A. Heavy Metals Pollution Levels and Children Health Risk Assessment of Yerevan Kindergartens Soils. *Ecotoxicol. Environ. Saf.* **2017,** *142*, 257–265.

Vieira, C.L.; Garshick, E.; Alvares, D.; Schwartz, J.; Huang, J.; Vokonas, P.; Gold, D.R.; Koutrakis, P. Association between Ambient Beta Particle Radioactivity and Lower Hemoglobin Concentrations in a Cohort of Elderly Men. *Environ. Int.* **2020,** *139*, 105735.
Visconti, D.; Álvarez-Robles, M.J.; Fiorentino, N.; Fagnano, M.; Clemente, R. Use of Brassica Juncea and Dactylis Glomerata for the Phytostabilization of Mine Soils Amended with Compost or Biochar. *Chemosphere* **2020,** *260*, 127661.
Wang, R.-Z.; Huang, D.-L.; Liu, Y.-G.; Zhang, C.; Lai, C.; Wang, X.; Zeng, G.-M.; Gong, X.-M.; Duan, A.; Zhang, Q. Recent Advances in Biochar-Based Catalysts: Properties, Applications and Mechanisms for Pollution Remediation. *Chem. Engg. J.* **2019,** *371*, 380–403.
Wu, H.; Lai, C.; Zeng, G.; Liang, J.; Chen, J.; Xu, J.; Dai, J.; Li, X.; Liu, J.; Chen, M. The Interactions of Composting and Biochar and Their Implications for Soil Amendment and Pollution Remediation: A Review. *Cri. Rev. Biotechnol.* **2017,** *37*, 754–764.
Xue, X.-Y.; Cheng, r.; Shi, l.; Ma, Z.; Zheng, X. Nanomaterials for Water Pollution Monitoring and Remediation. *Environ. Chem. Lett.* **2017,** *15*, 23–27.
Yaqoob, A.A.; Khatoon, A.; Mohd Setapar, S.H.; Umar, K.; Parveen, T.; Mohamad Ibrahim, M.N.; Ahmad, A.; Rafatullah, M. Outlook on the Role of Microbial Fuel Cells in Remediation of Environmental Pollutants with Electricity Generation. *Catalysts*. **2020,** *10*, 819.
Yuan, X.; Xue, N.; Han, Z. A Meta-Analysis of Heavy Metals Pollution in Farmland and Urban Soils in China over the Past 20 Years. *J. Environ. Sci.* **2021,** *101*, 217–226.
Zhang, F.; Li, Y.-H.; Li, J.-Y.; Tang, Z.-R.; Xu, Y.-J. 3D Graphene-Based Gel Photocatalysts for Environmental Pollutants Degradation. *Environ. Pollut.* **2019,** *253*, 365–376.
Zhang, L..; Qin, X..; Tang, J..; Liu, W..; Yang, W. Review of Arsenic Geochemical Characteristics and Its Significance on Arsenic Pollution Studies in Karst Groundwater, Southwest China. *Appl. Geochem.* **2017,** *77*, 80–88.
Zhang, L..; Zhu, G..; Ge, X..; Xu, G..; Guan, Y. Novel Insights Into Heavy Metal Pollution of Farmland Based on Reactive Heavy Metals (RHMS): Pollution Characteristics, Predictive Models, and Quantitative Source Apportionment. *J. Hazard. Mater.* **2018,** *360*, 32–42.
Zhu, X..; Chen, B..; Zhu, L..; Xing, B. Effects and Mechanisms of Biochar-Microbe Interactions in Soil Improvement and Pollution Remediation: A Review. *Environ. Pollu.* **2017,** *227*, 98–115.

PART II

Nanotechnology and Nanosensors

CHAPTER 3

Electrochemical, Optical, Magnetic, and Colorimetric Nanosensors as Tools to Detect Environmental Pollution

BAMBANG KUSWANDI[1*] and MUHAMMAD AFTHONI[2]

[1]*Chemo and Biosensors Group, Faculty of Pharmacy, University of Jember, Jl, Kalimantan 37, Jember 68121, Indonesia*

[2]*Pharmacy Department, Faculty of Science and Technology, Ma Chung University, Villa Puncak Tidar N-01 65151, Malang, Indonesia*

[*]*Corresponding author. E-mail: b_kuswandi.farmasi@unej.ac.id*

ABSTRACT

The integration between nanomaterials and sensor devices allows for a various number of nanosensor construction and applications, including monitoring of environmental pollutions. Nanosensor is a smart alternative to the classical analytical instrument that allows highly sensitive, selective, and online detection of environmental pollutants with simple or less preparation. This chapter describes technology advances in nanosensors that utilize electrochemical, optical, colorimetric, and magnetic properties to improve the sensor analytical performance for environmental pollutant detections. Gold nanoparticles, magnetic nanoparticles, carbon nanotubes, quantum dots, etc., as nanomaterials have been actively investigated for nanosensor developments based on various transducer modes. The nanosensor technology will eventually create tiny sensing devices for rapid detection and screening of a wide range of environmental pollutants with high selectivity

Nanotechnology for Environmental Pollution Decontamination: Tools, Methods, and Approaches for Detection and Remediation. Fernanda Maria Policarpo Tonelli, Rouf Ahmad Bhat, & Gowhar Hamid Dar (Eds.)

and sensitivity at an affordable price that have become a smart analytical tool in environmental monitoring in the near future.

3.1 INTRODUCTION

Environmental pollutions from human activities often cause a severe effect globally, and many pollutants are harmful to humans, animals, and plants (Appannagari, 2017). Especially for humans, pollution causes morbidity and mortality (Manisalidis et al., 2020). In 2015, it was predicted that 9 million premature deaths, and 16% of all deaths in the world were due to diseases associated with environmental pollutions (Landrigan et al., 2018). Pollution is classified as human-made pollution and natural pollution. Human-made pollution is caused by human activity, where human-made pollution can be a trigger for biological pollution, such as our flood, etc. On the other hand, environmental pollution can also be categorized as water pollution, air pollution, land pollution, food pollution, etc. While natural pollution is often caused by natural phenomena, such as flood, drought, earthquake, etc. (Appannagari, 2017).

Air pollution is defined as the presence of chemical compounds in atmospheric air, with toxicity at a high concentration that may be hazardous to animal vegetation, human. Air pollution decreases the quality of air. Common gases causing air pollution include sulfur oxide (e.g., SO_2), nitrogen oxide (e.g., NO, NO_2), and carbon monoxide (CO). For human, SO_2 gase may damage skin. CO gas has higher affinity for hemoglobin compares with oxygen causing carboxyhemoglobin. In the case of environmental pollutants causing health problems, polyaromatic hydrocarbon (PAH), persistent organic pollutants (POPs) (Ukaogo et al., 2020), incomplete combustion of organic material, etc. had caused cancer (Zamora-León and Delgado-López, 2020).

Water pollution is a critical global public policy issue. Sometimes, pollutants at levels that could harm human has caused drinking water pollution (Ewuzie et al., 2020). The parameter of quality water include turbidity, total suspended sediments (TSS), dissolved oxygen (DO), biochemical oxygen demand (BOD), chemical oxygen demand (COD), water temperature (WT), secchi disk depth (SDD), colored dissolved organic matters (CDOM), total phosphorus (TP), and sea surface salinity (SSS) (Gholizadeh et al., 2016). The effect of water pollutant can cause acute toxicity and genotoxicity for human (Bashar and Fung, 2020).

Classically, the determination of environmental pollution was done using chromatography technique, such as gas chromatography (GC) for air pollution, thin-layer chromatography (TLC), and high-performance liquid chromatography (HPLC) for water pollution (Parihar et al., 2020). However, these analytical instruments required sample preparation, lengthy procedures, relatively expensive, skilled personnel, and cannot be used in the field (Reis et al., 2009). Other approaches that have been developed currently are chemical sensors and biosensors employing nanomaterials that are often called nanosensors.

A suitable nanosensor for analyzing the environmental pollution should be inexpensive, small, easy-to-operate, and portable that can be deployed in the field. During the last decades, nanosensors that can be employed in situ analysis have gained increasing concern in environmental pollution monitoring. In this chapter, we presented various nanosensors based on their transducers, such as electrochemical, optical, colorimetric, and magnetic/acoustic, for the detection of environmental pollutants as target analytes.

3.2 NANOSENSORS

Nanosensors measure physicochemical change and convert it to signals that proportional to the analyte detected. They are nanoscale sensors that are constructed to about 10 nm in dimension and can detect a few attogram scales of masses. Today, there are various approaches proposed to create nanosensors. These include bottom-up assembly, top-down lithography, and molecular self-assembly (Peng et al., 2009). They are created to imitate the natural nanomaterials, such as DNA, membranes, proteins, and other biomolecules that could detect physicochemical changes in a minute via various detection mechanisms (Kuswandi, 2019).

Nanosensors are used in different nanomaterials that could enhance detection at a very low level of harmful contaminants in foods associated with environmental pollution that could occur during processing, distribution, or handling (Jianrong et al., 2004). Currently, nanosensors have widely been considered as analytical tools in various fields, such as medicine, agriculture, the food industry, and the environment (Akimov et al., 2008). Nanosensors share the same basic principle of sensor, that is, a selective reaction with an analyte, signal generation from the physicochemical reaction with the recognized element, and the signal processing into measurable quantities as shown in Figure 3.1. In pollutant detection, the recognized element contains a reagent or bioreagent that could detect the target analyte of pollutants from environmental samples. In nanosensor development, nanomaterials are used in the recognition element reagent in order to enhance the sensor analytical performance, such as sensitivity, selectivity, etc.

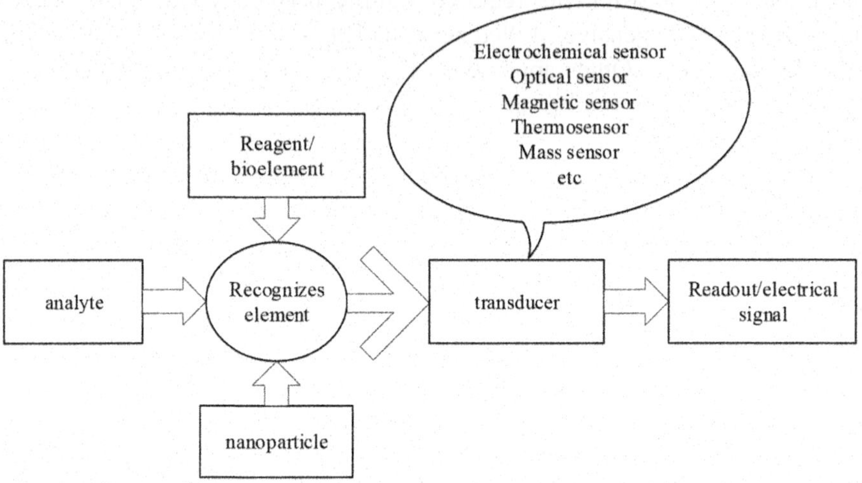

FIGURE 3.1 Mechanism principles of nanosensor.

Nanosensors have advantages in terms of specificity and sensitivity compared with sensor-based bulk materials, since their nanoscale properties of nanomaterial are not present in classical materials (Sadik et al., 2009). Nanosensors could improve specificity because they work at a scale similar to biological processes, capable of functionalization with reagents and biomolecules, so that the recognition processes could be improved, which in turn, increasing detectable signal. Increasing in sensitivity due to nanomaterials have a high surface-to-volume ratio, as well as new physical features, such as nanophotonics or localized surface plasmon resonance.

Nanosensors based on their applications can be classified into four classes, that is, chemical nanosensors, nanobiosensors, electrometers, and deployable nanosensors, as illustrated in Figure 3.2 (Agrawal and Prajapati, 2012). Chemical nanosensor generally uses cantilevers and electronics with a capacitive signal to analyze the target analytes. This nanosensor is very sensitive to detect a target analyte as a single chemical species or biomolecule. Nanobiosensor is a biosensor based on nanomaterials which contains various types of biorecognition elements. Diverse kinds of nanobiosensors are developed, such as enzyme biosensor, cell-based biosensor, immunosensors and DNA biosensors (Kuswandi, 2019). Electrometer is a mechanical electrometer in a nanometer scale that consists of a torsional resonator, a gate electrode, and a detection electrode that are employed to couple charge to the mechanical element. Deployable nanosensor is a chemical detection system that comprises a nanomaterial for sample collection and concentration with a microelectromechanical-based chemical lab-on-a-chip detector. This nanodevice can be employed in homeland security that allows chemical agent detection in the air without human life risking by sending it up in the air.

In the case of nanobiosensors, by using nanomaterials in the biosensor construction, some features, such as sensitivity, specificity, and rapid response are improved significantly (Jin et al., 2003). These features allow a wide range of the nanobiosensor applications in daily environmental monitoring activities, for example, raw materials inspection, on-line process control, storage conditions monitoring, etc. In addition, the nanobiosensors could allow cost-effective improvements in environmental detection and monitoring.

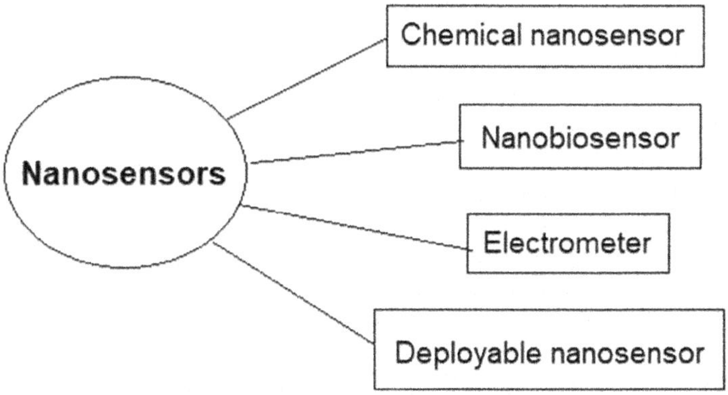

FIGURE 3.2 The nanosensors classification based on their applications can be grouped into four classes of applications, that is, chemical nanosensors, nanobiosensors, electrometers, and deployable nanosensors.

3.3 NANOSENSOR MATERIALS

Nanomaterials are important tools with physicochemical features that are specific because of their quantum-size phenomena when compared with traditional bulk materials (Willner and Vikesland, 2018). In addition, they also have very large surface area due to their nanosize (Buzea et al., 2007). For example, nanoporous carbon's surface areas can reach up to 2000 sqm/g

(Promphet et al., 2015). Carbon nanotubes (CNTs) are content of unique variations of common graphite, with molecular tubes made up of hexagonally bonded sp2 carbon atoms structure, with dimensions in the range 1–50 nm, that is, single-walled carbon nanotube (SWCNT) or multiwalled carbon nanotube (MWCNT) (Mueller and Nowack, 2008). The surface area of SWCNT lies in the range of 400–900 sqm/g (Kurniawan et al., 2018). Zeolites, naturally occurring materials, have nanoscale pores, while anodic etching of monocrystalline silica using HF buffered with ethanol creates nanoporous silica. These nanomaterials have a high surface area-to-volume ratio (hundreds of sqm/cc) (Xiao and Li, 2008).

The study of these different features allows the possibility to increase the sensitivity and selectivity of nanosensors. Interesting techniques have been developed on the improvement of electronic characteristics using metallic nanostructures (Agüí et al., 2008). These include the utilization of the nanostructured materials with specific shapes and dimensions, such as nanoparticles and quantum dots (0D), carbon nanotubes and nanowires (1D), and graphene sheets or metallic platelets (2D) orientations that show their features. Table 3.1 describes the different structure of nanomaterials based on their sizes, shapes, dimensions, and their applications. These nanomaterials provide increasing sensitivities in nanosensor, since their high surface-to-volume ratios that allow target analyte molecules to bond significantly that affect the bulk electrical characteristics of the structure. Because of their nanosize, they may be taken up by cells (Soares et al., 2018), and thus these nanomaterials are promising candidates for in vivo sensing applications. In many cases, the nanosensors with inherent electrical features such as CNTs are particularly excellent and allow to improve the sensitivity of detection by themselves. Several sensing platforms have been constructed with nanomaterials that extremely change in sensor signal output (Jain, 2003).

In the nanobiosensor technology, one of the most important advances has led to the employment of various nanoparticles, such as metal nanoparticles (He et al., 2008), carbon materials (Tortorich et al., 2018), quantum dots (Keçili and Hussain, 2018), oxide nanoparticles (Bhateria and Singh, 2019), and magnetic nanomaterials (Haun et al., 2010). These are employed to improve the electrochemical signals of biocatalytic reactions that occur at the electrode/electrolyte interface. In addition, functional nanomaterials that are bound to biomolecules, such as proteins, nucleic acids, antibodies, aptamer etc., have been designed to detect and amplify transducer signals in nanobiosensor systems (Kuswandi, 2019).

TABLE 3.1 Structure of Nanomaterials Based on Their Size, Shape, Dimensions, and Their Applications.

Dimension	Direction of confinement	Example	Shape	Application
0D	x, y, z	Nanoparticles		Photocatalytic activity
1D 	x, y	Rods, Wires, Tubes		Wave guides
2D 	x	Nanofilms Nanocoatings		Components for PC, mobile phone
3D Bulk	Nil	Nanocomposite		

In the last decade, the use of nanomaterials, such as nanoparticles, nanowire, nanotube, nanoneedle, nanosheet, nanocomposites, nanorod, and nanobelt, for nanosensor constructions have seen increased growth and development. Nanoparticles with surface modifications, such as gold nanoparticles (AuNPs), silver nanoparticles (AgNPs), Quantum dots (QDs), magnetic nanoparticles (MNPs), and CNTs can have specific target-binding features that provide high selectivity and sensitivity toward pollutants as target analytes. Various types of nanoparticles show different optical, fluorescence, and magnetic features, and interactions between these features allow nanoparticles as a great promising candidate for pollutants detection as given in Figure 3.3 (Peng et al., 2009).

Table 3.2 describes the main types of nanomaterials that have been used for more improvement in the sensing mechanisms in nanosensor technology, either using a top-down or bottom-up approach (Kuswandi, 2019). Table 3.2 highlights the advantage features of several nanomaterials used and some examples of applications (shown by corresponding references). The details of various nanobiosensors reported by the use of various nanomaterials, especially toward environmental pollutants, are discussed in the applications section, such as heavy metal ions, organic contaminants, and others.

MNPs made up of iron and its oxides have been employed for specific and excellent magnetism detection events and interactions, such as magnetic resonance imaging (MRI) and other magnetism detection techniques. These MNPs can be incorporated with fluorescent molecules or can be created to deliver specific responses by coupling with externally applied magnetic fields. Similarly, nanostructures-based zinc and zinc oxide have been extensively employed for biochemical phenomenon detection to be a more precise and sensitive scheme. These magnetic nanosensors have been used in the optimized detection of organic pollutants and other target analytes (Koh and Josephson, 2009). Similarly, CNTs have also been employed and optimized to detect the biosensing processes that allow for rapid and sensitive detection, since much better interactions between the target analyte and the bioreceptor molecule have occurred. CNTs-based biosensors have been actively employed in body fluids, such as for glucose (Yogeswaran and Chen, 2008) and insulin detections (Qu et al., 2006). In addition, their application for pollutants detection have been reported in the literature (Badihi-Mossberg et al., 2007; Hussain and Keçili, 2020).

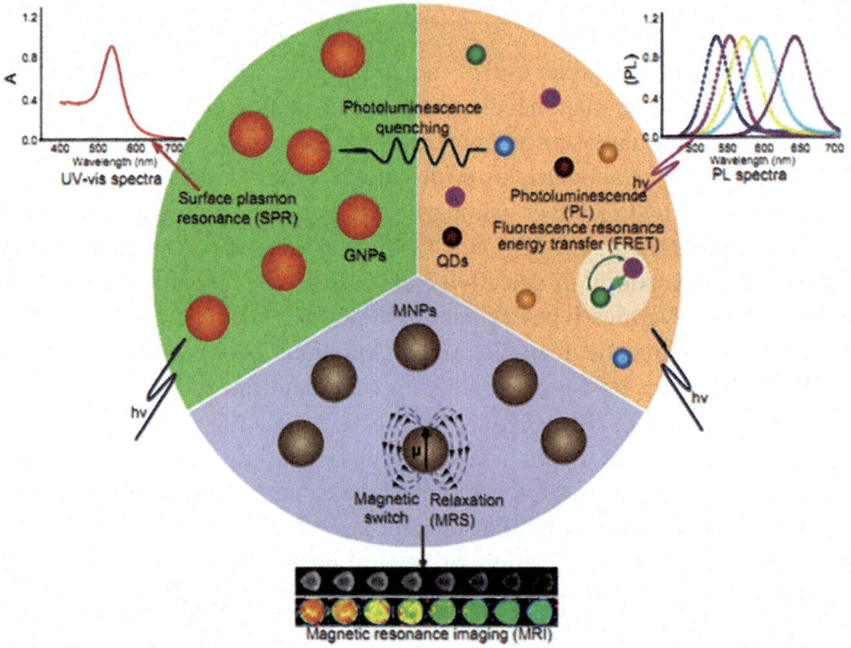

FIGURE 3.3 A schematic features of AuNPs/GNPs, QDs, and MNPs for nanosensor applications toward environmental pollutants.

Source: Reprinted with permission from Peng et al., 2009. © 2009 American Chemical Society.

3.4 SENSING TYPES

A nanosensor generally consists of recognition elements that can be referred to as chemical or biochemical recognition elements or receptors attached to the transducer. The receptor interaction toward the target analyte is designed to create physicochemical events that can be converted proportionally into a quantifiable and processable signal, such as an electrical, optical, or magnetic signal (Agrawal and Prajapati, 2012). In this regard, nanosensors are new generations of sensor types that allow sensing or detection of various target analytes, such as chemicals species, compounds, or microorganisms at absolute low levels with rapid response and high accuracy (Asmatulu and Khan, 2019).

TABLE 3.2 Some Examples of Nanomaterials Used for Nanosensor Technology.

No.	Nanomaterial type	Advantages	References
1	Nanoparticles	Possess good catalytic properties, aid in immobilization, and better loading of biomolecule as target analyte,	Luo et al. (2006)
2	Nanowires	Good electrical and sensing features for biochemical sensing, better charge conduction, and high versatility	Pal et al. (2007)
3	Nanorods	Can be coupled with MEMS, good plasmonic materials for coupling with sensing phenomenon and size-tunable energy regulation, and induce specific field responses	Sawant and Sawant (2020)
4	Carbon nanotubes	Higher aspect ratios, better electrical transducer, ability to be functionalized, and improved enzyme loading	Pumera et al. (2007)
5	Quantum dots	Excellent in fluorescence, including size-tunable band energy and quantum confinement of charge carriers	Huang et al. (2005)

Nanosensors are classified on the basis of recognition element types or the mode of physicochemical transduction. Therefore, the type of nanosensors depend on the system of transducer mechanism, such as electrochemical, optical, colorimetric, and magnetic. Thus, nanosensor can be classified as electrochemical nanosensor, optical nanosensor, colorimetric nanosensor, and magnetic nanosensor (Agrawal and Prajapati, 2012) that are discussed in this chapter.

3.4.1 ELECTROCHEMICAL TECHNIQUES

In classical interfacial electrochemical techniques, these techniques can be classified into different subtypes on the basis of signal types, such as potentiometry, amperometry, voltammetry, and colorimetry, as given in Figure 3.4. This electrochemical reaction occurs between electrode and target analyte that measures in terms of the transduced signal, for example, a potential or current (Sabahudin Hrapovic et al., 2003). This signal is quantified directly and is proportional to the concentration of the target analyte in the sample. Therefore, in electrochemical nanosensor, the integration of recognition elements employing nanomaterials with electrochemical transducer occurs, other developing technique such as electrochemical impedance spectroscopy is also involved. Numerous electrochemical sensing devices, such as amperometric, voltammetric, conductometric, and electrochemical luminescence are employed in electrochemical nanosensors (Zhu et al., 2015).

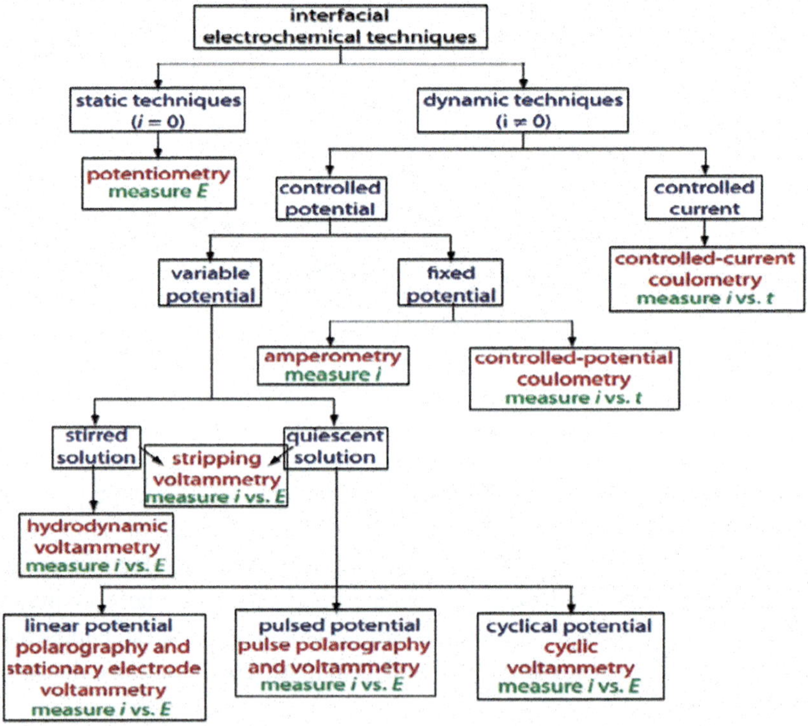

FIGURE 3.4 Schematic diagram of classical interfacial electrochemical techniques.

In the electrochemical sensing mechanism, the chemical reaction takes place in which electron moves between two electrodes in an electrochemical cell containing an analyte. This reaction is called oxidation–reduction reaction (redox). The electrons flow between the two atoms in the redox reaction produce an electrical signal. Here on one side, electrons keep moving into the electrode, and on the other side, it is moving out of the electrode. The electrochemical nanosensors utilized in the detection of pollutants as target analytes have been used for heavy metals, textile dye compounds, pesticides, and herbicides detections (Hussain and Keçili, 2020)

Nanotechnology has great potential to increase selectivity and sensitivity in electrochemical sensor development as it synergies the catalytic activity effects among biocompatibility and conductivity to enhance the transducer amplification and read-out signal (Zhu et al., 2015). The voltammetric sensor is a versatile electroanalytical approach for the determination of pollutant substances. The mechanism of the voltammetric sensor measurement is based on a potential waveform function that was applied to the electrode. This method was successfully determined for various pollutants, such as herbicide, heavy metal, a pharmaceutical compound, etc. (Kuswandi, 2019).

The amperometric sensor is a specific electrochemical technique involving oxidized or reduced reaction (redox reactions) of certain chemical species as target analytes are driven at a constant applied potential on the inert metal electrodes. The target analytes are called electroactive species. Generally, an amperometric cell consists of two or three electrodes, that is, a working electrode, a counter electrode, and a reference electrode. In this cell type, a metal, such as platinum or gold, is usually used as the working electrode, where it works as the recognition element in amperometric nanosensor (D'Orazio, 2003).

A conductometric sensor measures the current produced through the two electrodes in a conductometric cell. This sensor detects and measures the resistance changes so that the target analyte is allowed to restrict the charge that flows through it. Generally, in an electrochemical setup, a target analyte allows the current to pass through between two electrodes, that is, working and reference electrodes. Since the target analyte has high conductance and low resistance, it will enable more current to flow through, and the other way around, due to the presence of two electrodes the analyte restricts the flow of charge differently (Riordan and Barry, 2016).

Besides interfacial electrochemical techniques, solid-state electrochemical approaches are also employed in electrochemical nanosensors, such as field-effect transistors (FETs), as the first developed nanosensor of this type. These FETs have tunable characteristics and can be associated with the simple and quantitative detection system. They are using rod-shaped nanomaterials, such as nanorods, nanowires, nanotubes, nanoribbons, and nanotowers. In the case of pollutants as targeted analytes, the targeted analytes cling to the FET active area and this causes the impedance change and creates a signal (Kuila et al., 2011). Nanomaterial based on this shape provides enhanced sensitivity and current flow active area compared with the activities of other nanoscale morphologies on the detector surface. In this case, the nanomaterial that is most commonly employed is silicon nanowires, since they have high sensitivity and are simple to be chemically modified or functionalized on their surface (Peng et al. 2009).

Nanowires, for the first time, were used in the solid-state-based nanosensor system for sensitive and direct detection of chemical and the biochemical analyte in solution (Cui et al., 2001). This shows that electronic solid-state sensors using nanowire could detect 10 pM concentrations of a macromolecule, while further improvement to reach the detection limit of fM is also possible. In addition, nanosensors can be constructed and optimized for organic pollutant detection based on their conformational change. Here, the sensor sensitivity depends on the electrical resistance that originates from the silicon nanowire tips, producing quicker transfer of electron on the tip, compared with the sidewall. The nanosensor mechanism is based on permittivity and electrical resistance modification on the surface of the nanomaterials resulting from the attachment of the macromolecule as a target analyte.

In addition, graphene and its derivatives are another popular nanomaterials for a solid-state electrical nanosensor. Graphene oxide was utilized for the low detection of hydrogen peroxide in a composite with horseradish peroxidase (HRP) and DNA (Zhang et al., 2010). Here, the composite was prepared on a glassy carbon electrode surface where the HRP reduces hydrogen peroxide, while the graphene oxide and DNA stabilize the HRP in order to allow transport of electrons to the electrode. By using this construction, this nanosensor was able to detect a low concentration of hydrogen peroxide (below 1 mM).

3.4.2 OPTICAL

Commonly, an optical nanosensor consists a recognition element where nanomaterials employed could react to the desired target pollutants, and a transducer element is used for signaling the interacting event with the target analyte. Here, the recognition elements, such as enzyme, antibody, DNA, aptamer, molecularly-imprinted polymers, and host-guest recognizer, attract more attention of many researchers to enhance the sensor's analytical performance. The optical nanosensor produces a facile, rapid, and low-cost approach for sensitive and selective detection of pollutants. In direct optical nanosensors the analytes are detected directly via some intrinsic optical properties which include absorption or luminescence as applied in conventional optical sensors. Figure 3.5 shows various signaling for optical sensors (Yan et al., 2018) that are also employed in optical nanosensor.

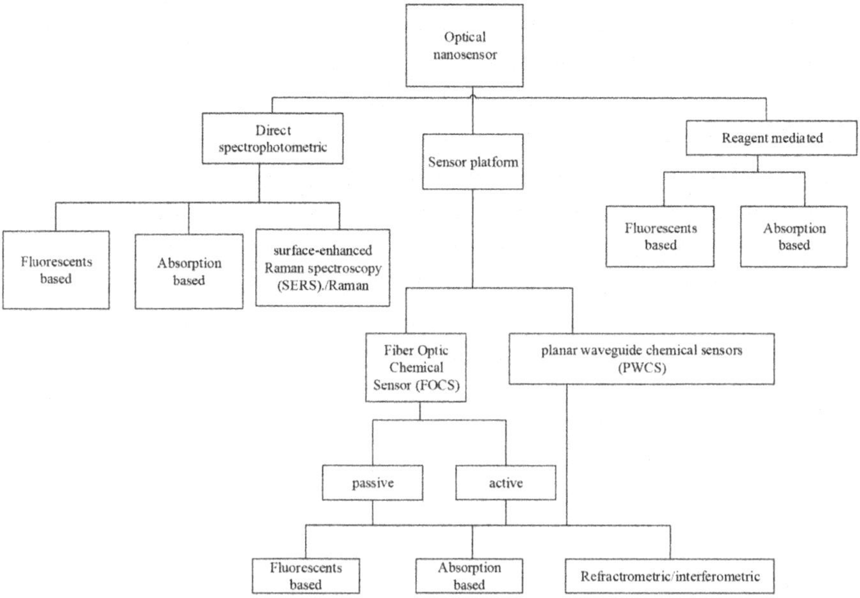

FIGURE 3.5 Classification of optical nanosensor.
Source: Adapted from McDonagh et al. (2008).

Fiber optic nanosensor contains an optoelectric component such as LED, LD, photovoltaic cell, etc. Optoelectronic components produce light that travels through optical fiber, after that the light passes through modulating element that is measurand, then the light enters in the detector for signal processing and finally produces read out. The optical fiber consists of the core and cladding which have a different refractive index (Yadav and Srivastava, 2018). Some of the optical fiber nanosensor head probes consist a nanometer-size Fizeau interferometer doped with the vapochromic material, created onto the cleaved end of a multimode fiber optic pigtail (Cabanillas-Galán et al., 2008). The optical fiber nanosensor function is working using the bright red powders of the vapochromic material [Au2Ag2(C6F5)4(C6H5N)2]. Here, the vapocromic material changes in its optical characteristics, for example, color or refractive index upon exposure to organic vapors, such as air pollutants. The original optical signal recovery is reached as the analyte vapors are disappeared.

An inactive and bioacceptable matrix has been used as a sensing element and an optical reporter in the fluorescence nanosensor developments (Chen et al., 2013). The sensing element of these nanosensors is a fluorophore, a material that emits fluorescence, which is selective and sensitive to the target analyte such as pollutant. Several fluorescence nanosensors employing a polyacrylamide matrix have been developed to target analytes such as zinc (Sumner et al., 2002). Here, microemulsion polymerization techniques are used to prepare uniform sized nanoparticles.

The organic polymerase silica sol–gels have been used as another important matrix in optical nanosensors. They are optically transparent, photostable, porous, high purity, thermally stable, and chemically inert material. For example, an oxygen-sensitive ratiometric sol–gel nanosensor has been developed by employing the probes encapsulated using biologically localized embedding (Xu et al., 2001). Here, the quenching of luminescence byoxygen has been used as the sensing mechanism in the sol–gel nanosensor response, it is determined from the ratio of fluorescence intensities ([Ru(dpp)3]2+/Oregon Green 488-dextran). This nanosensor approach was further developed as ion-selective optode based on fluorescent indicator dyes for adequate sensitivity or selectivity to detect ionic species concentrations of the target analyte (Chan et al., 2001; Kuswandi et al., 2007). Here, an optical ionophore with a flanking optically visible agent acted as a reporter dye was developed for a high degree of selectivity for the particular target analyte. As one of the most commonly used sensing candidates, fluorescence-based nanosensors have been widely used in a broad range of applications (Yan et al., 2018). In fluorescence-based nanosensors, a spectrofluorophotometer is used for the analyte detection or the analyte is detected by the naked eye. This sensor has been employed for the signal change collection via on-site in field applications (Paterson and De La Rica, 2015).

The basic principle of surface plasmon resonance (SPR) nanosensor construction layers rely on a strong electromagnetic field oscillation resonance, where the resonance is created at the interface between nanometal film and p-polarized incident light of a dielectric medium. These phenomena are creating a dark band profile in the light reflectivity at an incident angle and a specific wavelength. Revolutionary research on the blue LED based on III-V compound semiconductor construction allows the LED devices to emit all primary colors (RGB, red, green, and blue) (Prabowo et al., 2018).

In nanosensor type, besides normal SPR, many nanosensors are working based on localized surface plasmon resonance (LSPR) function via variations in transducing of local refractive index of the LSPR extinction band maximum by wavelength modifications. Here, the extinction band is a direct excitation effect of the LSPR. For instance, LSPR nanosensors for Concanavalin A, a carbohydrate-binding protein, where the LSPR spectrum of the mannose-functionalized Ag nanosensor had an λmax of 662.4 nm (Yonzon et al., 2005). Here, Concanavalin A (19.8 μM) was introduced into the flow cell and later followed by complete binding of the Ag nanosensor incubated at room temperature for 20 min. Then, the sample was rinsed in buffer solution and followed by a 6.7 nm red-shift of the LSPR λmax, where in this case, the LSPR λmax of silver nanotriangles was found at 669.1 nm.

Besides LSPR, in optical nanosensors, Raman spectroscopy is often used. This type of optical nanosensor is based on a vibrational spectroscopic approach. This nanosensor allows discriminating between molecules with close resemblance, for example, glucose and fructose, as the structural isomers. Here, surface-enhanced Raman scattering (SERS) has been employed for rapid detection of *Bacillus subtilis* spores as a biological contaminant and as a harmless simulant for *B. anthracis* (Yonzon et al., 2005).

3.4.3 COLORIMETRIC

Colorimetric nanosensor is another type of optical nanosensor that works via a colorimetric basis or color change. Here, a chemical reaction or physicochemical change in the presence of the target analyte causes a color change to occur that can be detected optically or via naked eye. In the case of colorimetric nanosensor applications, AuNPs are employed for the detection of environmental pollutants, for example, heavy metals (Willner and Vikesland, 2018; Ullah et al., 2018; Kuswandi, 2019), in this case, a color change indicates the presence of heavy metals.

The development of colorimetric nanosensors have been increased in the last decades, these nanosensors are versatile, low cost, and can be printed on paper surfaces. In addition, the main advantages of these nanosensors include the color changes can be detected by naked eye or color changes could be measured easily using cameras, smartphones, scanner or other image capturing systems (Chen et al., 2014; Hidayat et al., 2019). Classically, the most common approaches to colorimetric nanosensors were proposed by (1) linking a chromophore with a receptor element based on nanomaterial by means of a covalent bond, (2) the use of competition assays between a dye bonded to a receptor and a certain analyte, and (3) the use of new molecular systems-based nanomaterials that undergo guest-induced chemical reactions coupled to a color change. Therefore, the integration of nanomaterials in the sensing systems has allowed to explore novel properties such as localized surface plasmon resonance. These colorimetric approaches include incorporating nanomaterials in the recognition moiety as hybrid supramolecular chemistry, in the signaling group, such as AuNPs and QDs, or as vehicles for more complex events, such as aggregation/agglomeration processes, etc.

The most common nanoparticles used in the colorimetric method is AuNPs, since AuNPs-based colorimetric approach is a powerful assay for the colorimetric detection using naked eye detection for various target analytes, including pollutants, since the molar extinction coefficient are high (Cao et al., 2011). The AuNPs–LSPR is highly susceptible to their particle size and aggregation degree. When the target analyte is present, the color change occurs due to a change in the wavelength of AuNPs surface plasmon resonance as a result of dispersion change to the aggregation of AuNPs or the disassembly of aggregated AuNPs, as depicted in Figure 3.6 (Chen et al., 2018). Based on the aggregation and anti-aggregation strategies as the sensing mechanism toward analyte, colorimetric nanosensors based on AuNPs have been developed for pollutants detection, such as pesticides, where these colorimetric nanosensors have advantages, such as simple, rapid, user-friendly, low cost, and easily detected by naked eyes (Tseng et al., 2020).

Many harmful gases as air pollutants could be detected by a color change such as by using the commercially available Dräger Tube. This portable device provides an alternative to the bulky and lab-scale systems since the device can be miniaturized to be employed for point-of-sample devices in field applications. For instance, many chemical pollutants are regulated by the EPA (Environmental Protection Agency), they need extensive monitoring to assay that the pollutant levels are within the allowed limits. In this case, the colorimetric nanosensors allow a tool for on-site and on-line monitoring of many environmental pollutants (El-Safty et al., 2008; Ullah et al., 2018; Kuswandi, 2019).

3.4.4 MAGNETIC NANOSENSORS

Magnetic nanosensors are based on the binding of MNPs to a sensor surface, where the nanoparticle magnetic fields alter the magnetic fields of the sensor, resulting in electrical current changes within the sensor, as shown in Figure 3.7. The mechanisms of magnetic nanosensor, where magnetic nanoparticles attach to the sensor surface, can be classified as direct labeling and indirect labeling. Direct labeling means that magnetic probes bind functionality to the sensor surface directly, while indirect labeling uses the principle of the sandwich mode. Magnetic nanosensors are employed via applications of labeled supports to the nanosensors, inserted inside the transducers. The MNPs are distributed into the material by attracting them using an external magnetic field on the sensor active surface (Slimani and Hannachi, 2020). However, MNPs labeled with different sensing mechanisms and instrumentations can be categorized as magnetoresistive nanosensors and magnetic relaxation switches.

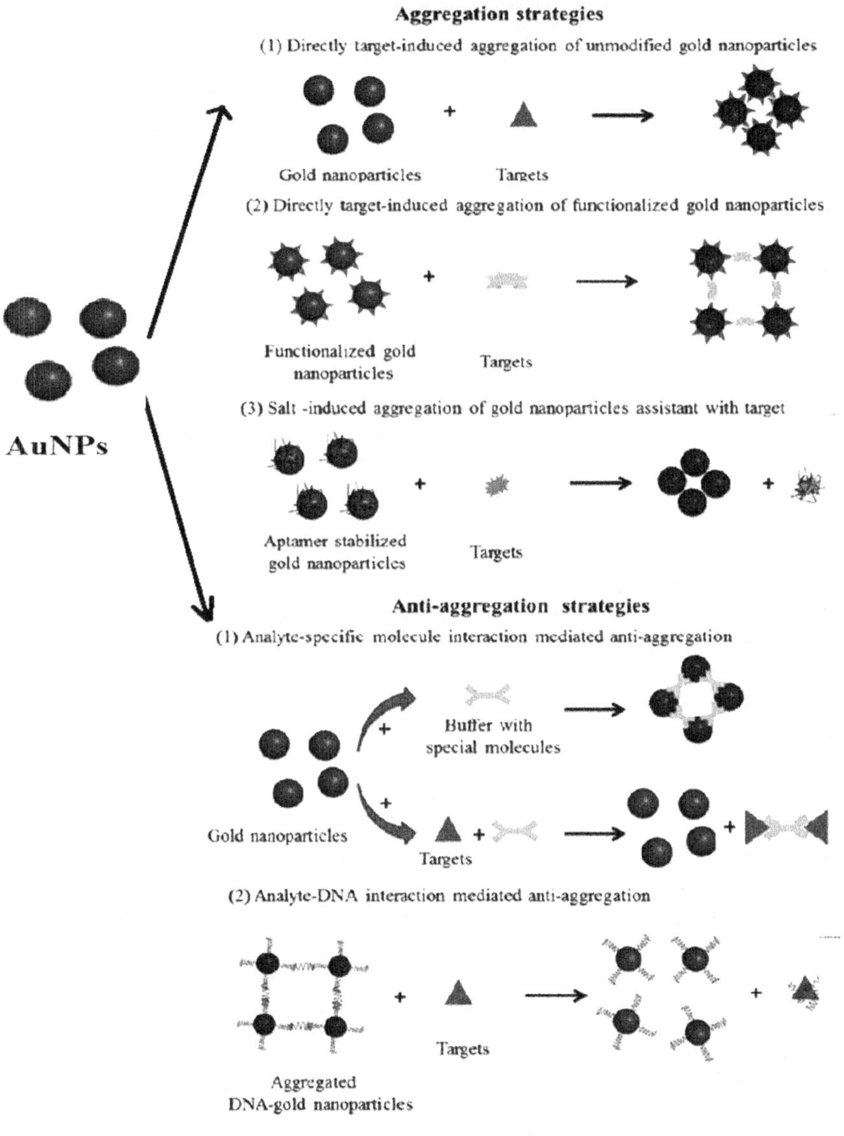

FIGURE 3.6 Commonly used sensing strategies in the colorimetric nanosensor based on AuNPs.

Source: Reprinted with permission from Hai Chen et al. (2018). © 2018 Elsevier.

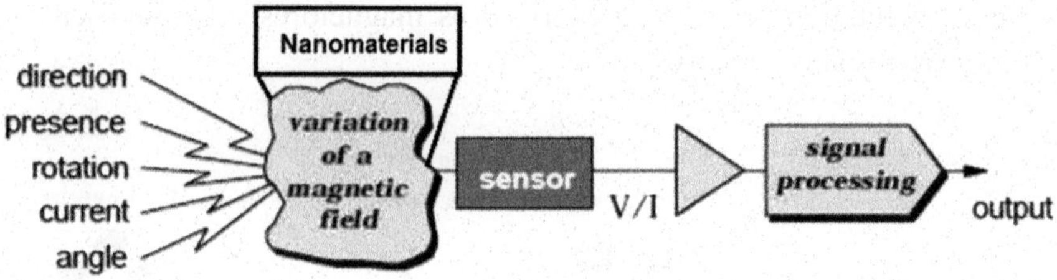

FIGURE 3.7 Schematic diagram of magnetic nanosensor.

Magnetic relaxation switches made from iron oxide and polymeric coating as marker agents were proven to be magnetic resonance agents and are widely used for various target molecules as analytes. When used as a targeted marker, surface-modified nanoparticles bind specifically to the target molecules that produce local inhomogeneities in the applied magnetic field in tissues where molecular targets are present. Thus, these inhomogeneities were causing decreases or increases in the relaxation time, which in turn leads to changes in the contrast of magnetic resonance images in response to the target analyte.

Magnetic nanosensors used magnetic nanoparticles, such as iron nanoparticles, cobalt nanoparticles, and many other elemental nanoparticles that have ferromagnetic properties. Magnetic nanosensors were developed to detect molecular interactions in aqueous solution. Based on target binding. these nanosensors allow changes in their spin-spin relaxation times of neighboring water molecules, which can be detected by magnetic resonance techniques, such as magnetic resonance imaging or nuclear magnetic resonance (Perez et al., 2004; Koh and Josephson, 2009). Magnetic nanosensors have been applied to many target analytes, that is, (1) DNA detection, (2) specific protein detection in solution via antibody-mediated interactions, (3) many enzymatic activities detection, such as proteases, methylases, and restriction endonucleases, And (4) a more complex target detection such as viral particles. (Perez et al., 2004).

3.5 APPLICATIONS

3.5.1 HEAVY METAL POLLUTANTS

Heavy metal means any metallic element that possess high density, and even in trace concentration, it can be toxic for plants and animals (Rascio and Navari-Izzo, 2011). Furthermore, heavy metals, such as mercury (Hg), arsenic (As), copper (Cu), cadmium (Cd), chromium (Cr), lead (Pb), nickel (Ni), and zinc (Zn) are mostly found as pollutant in the environmental samples (Li et al., 2019), the toxicity of heavy metals, such as mercury cause blindness, deafness, attention deficit, ataxia, and decrease the rate of fertility (Ayangbenro and Babalola, 2017). Minamata disease is one of the diseases caused by mercury, and also the mercury neurological syndrome is caused by methylmercury contamination. Clinical symptoms for Minamata disease include sensory disturbance in the distal extremities, ataxia, impairment of gait, loss of balance, speech and hearing, concentric constriction of the visual fields, muscle weakness, tremor, and nystagmus (Semionov, 2018). Therefore, the determination of heavy metal as environmental pollutant is critical for investigating and preventive risk of heavy metal pollution, especially for a human. Tabel 3.3 shows some examples of nanosensor applications for the determination of heavy metal as environmental pollutant.

Besides chemical nanosensors, as given in Table 3.1, various nanomaterials have been employed and integrated as nanosensors to improve their sensitivity and specificity to certain metal detection, for example, heavy metals. The nanomaterials that have been used are MWCNTs (Keçili and Hussain 2018; Jianrong et al., 2004), gold nanoparticles (Zhang et al. 2014; Vijitvarasan et al., 2015), silver nanoparticles (Roto et al., 2019; Chatterjee et al., 2009), silica mesoporous/nanospheres (Zhai et al., 2012), and graphene oxide (Promphet et al., 2015; Zhu et al., 2015). Therefore, in this section, the nanosensors for heavy metal ions are presented.

MWCNT, modified with 2-amino thiophenol has been developed to detect Pb(II) ion in water samples using a potentiometric approach (Guo et al., n.d.). The MWCNT nanosensor functionalized with aminothiophemol showed excellent analytical performance for Pb(I) due to their strong covalent bond between the carbonyl group of MWCNTs and the amino group of aminothiophenol to create the solid framework binding with Pb(II). In another study, an electrochemical nanosensor for the detection of Pb(II), Hg(II), and Cd(II) ions has been developed based on the MWCNT functionalized with triphenylphosphine (Bagheri et al., 2013). Here, the carbon screen-printed electrode (CSPE)-on the MWCNT modified with phosphorus group of triphenylphosphine monolayer binds to the heavy metal ions. In this case, the nanosensor was successfully employed to detect trace metals with a low detection limit at the sub-pM level. A glassy carbon electrode modified with bismuth film/carboxylic acid-functionalized MWCNT-β-cyclodextrin-Nafion nanocomposite (Bi/CMWCNTs-β-CD-Nafion/GCE) has been developed for the low detection of Pb(II) and Cd(II) ions (Zhao et al., 2016). This nanosensor has been successful for detecting

TABLE 3.3 Some Examples of Nanosensor Applications for Heavy Metal Ions Detection.

Analyte	Nanomaterial	Type of nanosensor	Limit of detection	References
Cu^{2+}	Carbon dots-based dual-emission silica nanoparticles	Optical nanosensor	35.2 nM	Liu et al. (2014)
Hg^{2+}	Dithizone nanoloaded membrane	Optical nanoensor	0.057 ppb	Danwittayakul et al. (2008)
Pd^{2+} Hg^{2+}	Carbon nanoparticle probes (CNPs)	Optical nanosensor	58 nM 100 nM	Sharma et al. (2016)
Hg^{2+}	Eu(III)-pyridine-2,4,6-tricarboxylic acid (PTA) on mobil composition of matter No. 41 (Eu(PTA)@MCM-41)	Optical nanosensor	1×10^{-6} M	Zhai et al. (2012)
Co^{2+}	Mesoporous silica monoliths nanoparticle on N,N`di(3carboxysalicylidene)-3,4diamino-5- hydroxypyrazole (MSNPs@DSDH)	Optical nanosensor	0.24 ppb	Shahat et al. (2015)
Cu^{2+}	β-cyclodextrin-modified Iron (II,III) oxide on silica oxide (Fe_3O_4@SiO_2_	Magnetic nanosensor	5.99×10^{-6} M	Zhang et al. (2015)
Mn^{2+} Co^{2+} Cu^{2+} Zn^{2+} Pb^{2+}	Magnetic nanoparticles (MNPs) modified with polyacrylic acid (MNPs- PAA)	Magnetic nanosensor	0.06 μg/L 0.04 μg/L 0.6 μg/L 0.6 μg/L 0.06 μg/L	Lee et al. (2009)
Fe3+	L-cysteine (L-Cys) capped Fe_3O_4@ZnO core-shell nanoparticles (Fe_3O_4@ ZnO@L-Cys)	Magnetic nanosensor	3 nmol/L	Li et al. (2016)
Hg^{2+}	Carbon paste electrode modified with substituted thiourea-functionalized highly ordered nanoporous silica	Electrochemical nanosensor	7.0 × 10−8 M	Javanbakht et al. (2009)
As^{3+}	Sodium 3-mercapto-1-propanesulfonate capped gold nanoparticles with two different cationic polyelectrolytes, the poly (diallyldimethylammonium chloride) (Au/MPS-(PDDA- AuNPs)	Electrochemical nanosensor	0.48 μM	Ottakam Thotiyl et al. (2012)

TABLE 3.3 *(Continued)*

Analyte	Nanomaterial	Type of nanosensor	Limit of detection	References
Hg^{2+}	Carbon paste electrode modified with natural smectite-type clays grafted	Electrochemical nanosensor	60 nm	Tonle et al. (2005)
Cd^{2+}	Mesporous mangan cobalt spinel oxide nanoparticles ($MnCo_2O_4$ NPs)	Electrochemical nanosensor	37.02 nmol/dm	Antunović et al. (2019)
Pb^2			8.063 nmol/dm	

trace levels of Pb(II) and Cd(II) ions in soil samples with excellent recovery that gives a high potential for its use in environmental pollutants detection.

Gold nanoparticles (AuNPs) have been widely used for Hg(II) ion detection in water sample via simple optical nanosensor based on colorimetry approach (Li et al., 2014; Ding et al., 2012; Zhou et al., 2014; Hongming Chen et al., 2015) with the detection level from nM to sub-nM level. This detection limit was much lower compared with the WHO detection limit for Hg(II) in a real water sample (Ding et al., 2012). The nanosensor colorimetric response was associated with the color change (commonly from ruby red to royal purple) of the AuNPs solution as a result of the AuNPs aggregation when Hg(II) was absent. When the Hg(II) ions present, the color of AuNPs solution could have appeared. In addition, AuNPs can also be used with electrochemical nanosensors for the determination of Pb(II) and Cu(II) ions (Wan et al., 2015). Here, AuNPs modified screen-printed electrode was developed for this purpose, where they have a good detection limit at ppb level for Pb(II) and Cu(II) ions.

Another electrochemical nanosensor using graphene oxide–glassy carbon electrode with a monolayer of AuNPs functionalized with the chitosan was developed successfully for rapid and specific detection of Hg(II) (Gong et al., 2010). Here, the nanosensor employing square wave anodic stripping voltammetry with a very low detection limit for Hg(II) ions (<1 ppb). Furthermore, the urine sample has also been used as functional groups in AuNPs as recognition elements of nanosensing for Hg(II) ions. Here, in urine sample, the creatinine and uric acid can bind synergistically to AuNPs as well as can adsorb Hg(II) ions selectively. In this case, a low detection limit of Hg(II) ion (50 nM) was achieved with the low-cost nanosensor fabrication based on AuNPs (Du et al., 2013). It has been proved that by using agglomeration of AuNPs (20 nm chitosan-capped AuNPs), metal ions, that is, Zn(II) and Cu(II) can also be detected based on the suspension color change in the presence of these metal ions (Sugunan et al., 2005). So far, chitosan is a commonly known heavy metal chelating agent. In the case of Zn(II) and Cu(II), their presence is detected by the resulting loose agglomeration and colloidal instability of AuNPs. These effects result in a rapid change in color which is directly associated with the heavy metal ion concentration. Pb(II) ions have also been detected based on an aggregation–dissociation procedure using the DNAzyme-directed assembly of AuNPs. Here DNAzyme-directed assembly of AuNPs cleaves in the presence of Pb(II). In this approach, the color change results in a blue to red color change (Fig. 3.8) with a tunable detection limit of 100 nM–200 μM (Liu and Lu, 2003).

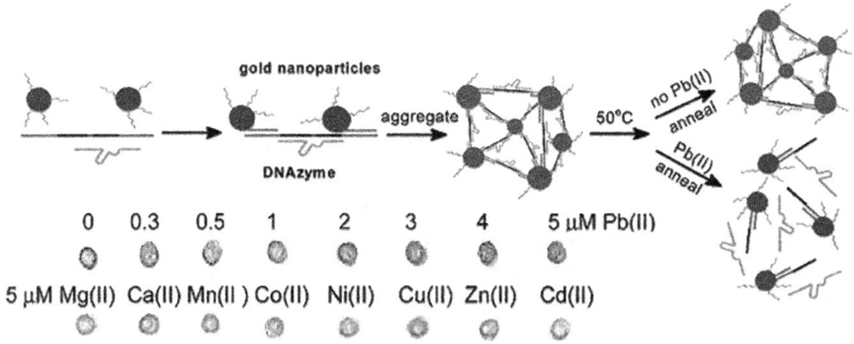

FIGURE 3.8 DNAzyme-directed assembly formation and cleavage of gold nanoparticles in a Pb^+ colorimetric nanosensor.

Source: Reprinted with permission from Liu and Lu (2003). © 2003 American Chemical Society.

Besides AuNPs, silver nanoparticles (AgNPs) have also been employed for optical nanosensor development for detecting Hg(II), Pb(II), and Cr(VI) metal ions in water samples (Vinod Kumar and Anthony, 2014; Farhadi et al., 2012; Ravindran et al., 2012). For example, citrate ion-coated AgNPs have been developed as an optical nanosensor for a low detection limit of Cr(VI) ion (nM) (Ravindran et al., 2012). The graphene oxide-modified carbon/gold screen-printed electrode has been developed for the detection of heavy metals, such as Pb(II) and Cu(II) (Promphet et al., 2015; Wei et al., 2012). In this case, the nanosensor is a simple and rapid analytical tool that demonstrated a wide linear response range with a low detection limit (<10 nM). In addition, the tin oxide (SnO_2) modified with graphene–glassy carbon electrode has also been designed as an integrated heavy metal sensing platform (Wei et al., 2012).

Other strategies have been proposed to develop on-field colorimetric devices for the detection of heavy metal ions based on functionalized AuNPs that can also be applied for field applications. Some works on colorimetric nanosensors developed the smart phone-coupled systems that make the devices very small, compact, and user-friendly as well as preventing the need of bulky UV-Vis spectrophotometers. For instance, a portable device has been constructed for Hg(II) ions detection in water samples employing a smartphone for direct colorimetric data collection and analysis (Wei et al., 2014). The device inn this approach employs LEDs (green and red) to analyze the change in color of the solution. Then, the colorimetric nanosensor system automatically calculates Hg(II) ion concentration by a customized app. This nanosensor has been utilized for preparing contamination map of Hg(II) ions by analyzing samples collected from over 50 locations in California. Furthermore, another work, a paper-based nanosensor, has been developed for contamination detection of Hg(II) ions in real water samples (Chen et al., 2014). They fabricated the sensor array based paper by using a printer, then read and analyze it by a smart phone with a customized app. (Fig. 3.9). They involved cloud computing, where the detection limit of Hg(II) ions was achieved at 50 nM.

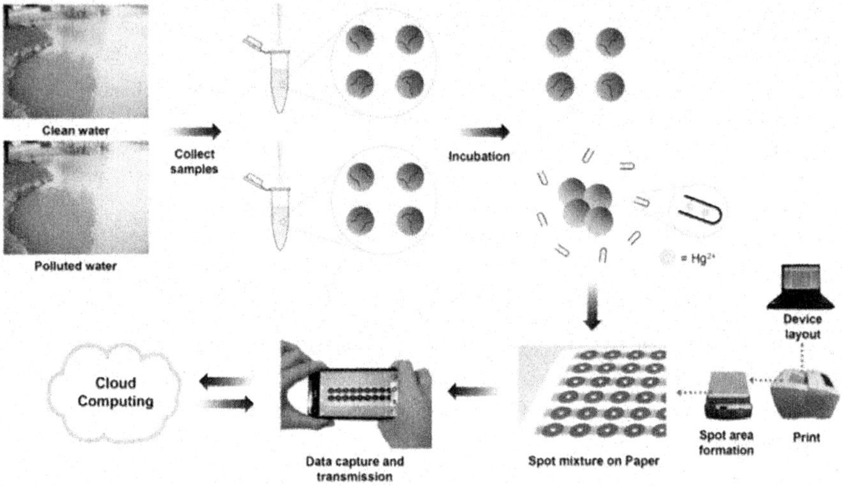

FIGURE 3.9 Schematic digram shows the heavy metal ion sensing based on the paper-based colorimetric nanosensor.

Source: Adapted from Chen et al. (2014).

3.5.2 ORGANIC POLLUTANTS

The organic pollutants that are commonly found in the water environment are pesticides, detergents, organic solvents, hydrocarbons, etc. For example, pesticides. Many of the nanobiosensor approaches designed for pesticide detection are based on the biocatalytic action of acetylcholinesterase (AChE) (Kuswandi, 2019). Generally, pesticides (both organophosphates and carbamates) are attached to a serine moiety within the enzyme active site and thus avoide the deacetylation of acetylcholine (Kuswandi et al., 2008). The disadvantages with this method are that other compounds, that is, heavy metals, detergents, and other pollutants as inhibitors, could also inhibit the enzyme active site and they make the enzyme unstable outside its natural environment (Nagatani et al., 2007; Kuswandi and Mascini, 2005).

An amperometric biosensor using AuNPs and MWCNTs have been developed in AChE-based nanobiosensor to promote electron transfer and catalyze the thiocholine electrooxidation for the detection of monocrotophos (LOD 10 nM) (Norouzi et al., 2010). In this amperometric method, the flow-system has used glassy carbon electrodes modified with AuNPs and MWCNTs. The CNTs impregnated chitosan to improve the immobilization level and to increase the AChE stability. In other work, the amperometric biosensor has been developed using polyaniline (PANI) film as a mediator. (Somerset et al., 2007; Somerset et al., 2009). Here, the amperometric nanobiosensor was designed to employ its dual role as immobilization matrix for AChE and was allowed to utilize its electrocatalytic activity toward thiocholine (TCh). The biosensing mechanism for the Au/MBT/PANI/AChE/PVAc nanobiosensor is given in Figure 3.10. Furthermore, it shows that as acetylthiocholine is catalyzed by AChE to form thiocholine and acetic acid. Then, the electroactive thiocholine is oxidized in the reaction, in turn, conducting PANI film reacts with thiocholine, accepts an electron from the oxidized mercaptobenzothiazole via interaction with the Au electrode.

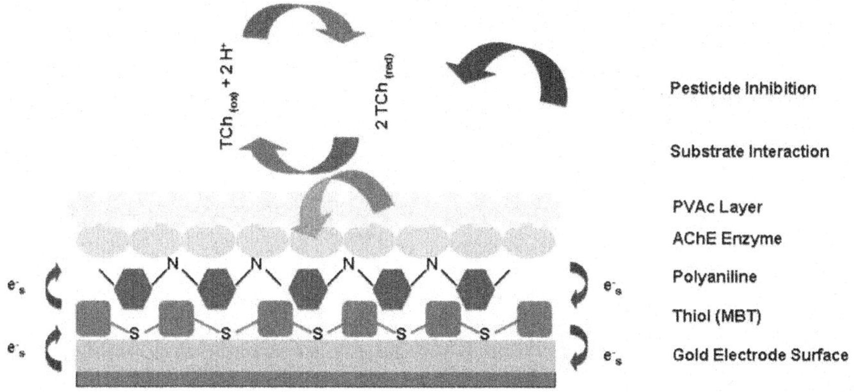

FIGURE 3.10 The Au/MBT/PANI/AChE/PVAc biosensor sensing scheme shows reaction at the gold SAM modified electrode.

Source: Reprinted with permission from Somerset et al. (2009). © 2009 Taylor & Francis.

The nanobiosensors have also been developed for the detection of other pesticides, such as an antiseptic where 2,4-Dinitrophenol (DNP) was used as the active agent. For example, the chromogenic effect of latex microspheres hybridized with AuNPs has been used as a colorimetric nanobiosensor fro DNP detection as a toxin model (Ko et al., 2010). Here, DNP–bovine serum albumin was bound with AuNPs that made the DNP hybridization with an anti-DNP antibody on latex microspheres, which in turn causes the pinkish-red color formation. A DNP–glycine as a model toxin was determined via a competition approach that allows binding between the toxin molecules and the analog-conjugated-AuNPs in the anti-DNP antibody binding pocket. In the presence of the toxin molecules, the AuNPs were displaced from host latex microspheres, a pinkish-red to white color change occurred that is visible to the naked eye.

Organochlorine pesticides are one of the most widely used pesticides in the world, for example, DDT (dichlorodiphenyltrichloroethane). Chemically, it is also known as 1,1,1-trichloro-2,2-bis (4-chlorophenyl) ethane. It has also been detected using nanobiosensors. For example, an immunosensor in dipstick format was designed for DDT detection at the nanogram level, In this case, the biosensing mechanism was based on AuNPs with competitive assay (Lisa et al., 2009). Herein, AuNPs with defined sizes were synthesized specifically and conjugated to anti-DDT antibodies which served as the detecting reagent in the dipstick immunosensor. In this case, the antigen used was DDA (1,1,1-trichloro-2,2-bis(chlorophenyl)acetic acid)-BSA conjugate, it was immobilized on nitrocellulose of the strips. An immunocomplex was formed from AuNPs-conjugated anti-DDT antibodies when treated with free DDT. This immunocomplex was then bound to the DDA-BSA conjugate. Depending on the free DDT concentration in the sample, the AuNPs conjugated anti-DDT antibodies binding to the immobilized DDA-BSA varied, and that was detected by the red color change due to the AuNPs, and the color intensity change was inversely proportional to the concentration of DDT with maximum intensity was in the absence of DDT (i.e., zero concentration).

The nanosensors were also developed for PAHs (polycyclic aromatic hydrocarbons) detection, another well-known organic pollutant besides pesticides. These nanosensors for PAHs detection have been designed using AuNPs (Mailu et al., 2010), SWCNT (Carrillo-Carrión et al., 2009), graphene oxide (Shen et al., 2012), etc. The optical nanobiosensor based on SWCNT-doped cadmium selenium quantum dots fluorescence has been developed for PAHs detection, such as pyrene, benzo(a)pyrene, perylene, and benzo(a) anthracene (Carrillo-Carrión et al., 2009). In this work, the fluorescence nanosensor showed excellent analytical characteristics to pyrenes, such as high sensitivity with a low detection limit (<1 nM), wide linear range, good recovery values, and rapid response time. Moreover, another fluorescence nanosensor was developed for PAHs detection in mineral water, such as anthracene, phenanthrene, and pyrene (Duong et al., 2011). Here, the sensing mechanism was based on CdSe/ZnS QDs functionalized with sol–gel. In this case, PAHs in trace level could increase the fluorescence intensity of the CdSe/ZnS QDs entrapped membranes and reached a low detection limit of PAHs (5 nM).

3.5.3 OTHERS

Other organic pollutants to cause endocrine disruption are also increasing attention today. These substances could be a wide and varied range of organic chemical substances. These chemicals are well-known as endocrine-disrupting compounds besides pesticides, such as diethylstilbestrol, dioxin, and dioxin-like compounds, polychlorinated biphenyls (PCBs). PCDDs (polychlorodibenzo-*p*-dioxins) and PCDFs (polychlorodibenzofurans) are dioxins and dioxin-like compounds that are very toxic environmental pollutants and carcinogens because of their high stability and they are extremely resistance to degradation. A nanosensor for rapid and sensitive detection of dioxin has been developed based on oligopeptide–cysteine–quartz crystal microbalance (QCM) (Mascini et al., 2004, 2005). In this work, the synthesized polypeptide was functionalized with two terminal cysteine residues and structured, including (1) [N]Asn-Phe–Gln–Gly–Ile[C], (2) [N]Asn–Phe-Gln–Gly–Gln[C], and (3) [N]Asn–Phe–Gln–Gly–Phe[C]. The QCM functionalized with polypeptide monolayer on the cysteine could allow directly electrostatic interactions between the amino acids and the dioxins, a low detection limit was reached (1 ppb). Another nanosensor based on QCM for dioxins detection was developed based on immunoassays, such as QCM employing monoclonal antibodies-modified bovine serum albumin (Park et al., 2006), surface plasmon resonance (SPR) using antibody functionalized polypeptide–gold thin layer (Soh et al., 2003), and SPR employing bovine serum albumin conjugated with CM5 sensor chip (Tsutsumi et al., 2008). Herein, the immunosensor has been used for the dioxin determination with good reproducibility and a low detection limit (<1 ppb).

PCBs, as other endocrine disruptors, are generally used in industrial applications such as capacitors, transformers, and condensers. These hazardous synthetic substances are suspected of having a harmful effect on the endogenous hormone that could promote cell malignancies. Traditional techniques for the PCBs determination often used high-performance liquid chromatography and gas chromatography–mass spectrometers (Concejero et al., 2001). However, nanosensor could be an alternative method for simple and rapid PCBs detection. For example, nanosensors for rapid and direct detection of PCBs have been developed based on immunosensor using electrochemical

(Laschi et al., 2003), fluorescence (Endo et al., 2005), piezoelectric (Přibyl et al., 2006), and SPR techniques (Tsutsumi et al., 2008).

An SPR nanobiosensor has been constructed for the detection of PCBs based on gold film modified with cytochrome c as shown in Figure 3.11 (Hong et al., 2008). Herein, the sensing principle is based on the conformational changesof the immobilized cytochrome, then detected by using SPR spectroscopy. The nanobiosensor exhibited a high sensitivity with a low detection limit (<1 ppb). A PCBs biosensor based on *Pseudomonas fluorescence* bacteria was also constructed using a spectrofluorimetric technique. This optical biosensor was immobilized in alginate beads for high sensitivity toward PCBs (Liu et al., 2007, 2010). The nanobiosensor was used successfully to detect the low detection limit of PCBs in samples (<1 ppb). In other work, a microflow immunosensor chip has been constructed based on the monoclonal antibody grafted on polystyrene beads to detect PCBs (Endo et al., 2005). The biosensing mechanism was based on antigen–antibody binding, where the antigen was conjugated to HRP. When hydrogen peroxide and a fluorogenic substrate are present, then fluorescence can be produced as a rapid response time and low detection limit (<1 ppt) toward PCBs.

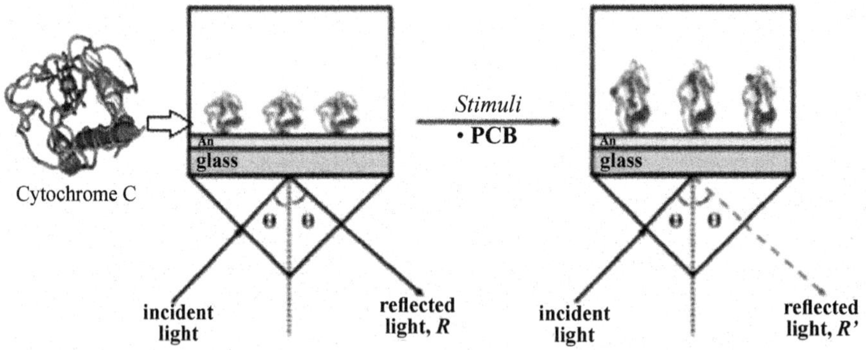

FIGURE 3.11 Schematic of SPR for PCB detection, where the R to R' reflectance change was induced by the PCB solution injection and detected using the Au substrate saturated with Cytochrome c molecule.

Source: Reprinted with permission from Hong et al. (2008). © 2008 Elsevier.

Other approaches using flexible substrates, such as paper-based SERS was also developed for simple and low-cost devices that can be employed for curvy surfaces (Lee et al., 2010). A paper-based SERS swab was constructed by simply immersing a filter paper in gold nanorods suspension. The gold nanorods were adsorbed efficiently onto the surface of filter paper since the electrostatic attraction between the positively charged CTAB-coated gold nanorods and the negatively charged cellulose occurred. The most important benefit of this paper-based SERS is that it allows for simple use for the trace samples collected from a solid surface. By swabbing a 140 pg 1,4-benzene-dithiol (1,4-BDT) residue contaminated on a glass surface, the pollutants were adsorbed easily on the paper surface, and their Raman spectrum was easily produced (Lee et al., 2010).

3.6 CONCLUSIONS

Currently, the integration of nanomaterials in the sensor technology has revolutionized sensor devices by creating the nanosensors in a tiny size, with high sensitivity and selectivity. Nanosensors used wide range of nano-materials, which allow using of biological or chemical reagents to bind or react with a target analyte and transduce it into measurable signals for rapid response of environmental pollutants, ranging from heavy metals to organic pollutants, including other pollutants, such as PAHs, etc. These devices can play an important role in monitoring environmental pollutants rapidly so that preventive actions could be taken if needed. Nanosensor development focuses on creating innovative and novel technologies that allow creating significant change in our better life. These can be achieved by the creation of nanoengineered or active surface modifications, etc., to improve nanosensor performances.

The impact of nanotechnology in nanosensor development has created new horizons for construction sensors with a tiny size suitable for intra-cellular detections, which is also important in environmental pollution monitoring. New nanostructures and novel nanomaterials need to be more investigated to be employed as nanosensors, in particular, to be integrated within tiny chips size in the lab-on-a-chip that consisted of all required functions, such as sample handling and analysis, including on-board electronics. This approach will greatly improve their performances and functionality by developing analytical tools that are very small, disposable, user-friendly, low cost, portable, and highly versatile. Today, a various range of nanosensors have been designed. However, the smart goal of high throughput and inexpensive devices is yet to be really produced in our routine. Therefore, interdisciplinary research and well-structured work that involve chemist, engineers, bioscience, and physics have to be performed to creat reliable and low-cost nanosensors for our daily life in the near future.

ACKNOWLEDGMENT

The author gratefully thank the DRPM, Ministry of Research and Technology (BRIN), the Republic of Indonesia for supporting this work via World-Class Research Grant 2021.

KEYWORDS

- **nanosensor**
- **nanobiosensors**
- **nanotechnology**
- **nanomaterials**
- **transducer**
- **pollutant**

REFERENCES

Ahloowalia, B. S.; Maluszynski, M.; Nichterlein, K. Global Impact of Mutation-Derived Varieties. *Euphytica* **2004,** *135*, 187–204.

Agrawal, S. S.; Prajapati, R. Nanosensors and Their Pharmaceutical Applications: A Review. *Int. J. Pharm. Sci. Nanotechnol.* **2012,** *4* (4), 1528–1535.

Agüí, L.; Yáñez-Sedeño, P.; Pingarrón, J. M. Role of Carbon Nanotubes in Electroanalytical Chemistry. *Analy. Chim. Acta* **2008,** *622* (1–2), 11–47.

Akimov, V.; Alfinito, E.; Bausells, J.; Benilova, I.; Paramo, I. C.; Errachid, A.; Ferrari, G.; Fumagalli, L.; Gomila, G.; Grosclaude, J.; Hou, Y.; Jaffrezic-Renault, N.; Martelet, C.; Pajot-Augy, E.; Pennetta, C.; Persuy, M.A.; Pla-Roca, M.; Reggiani, L.; Rodriguez-Segui, S.; Ruiz, O.; Salesse, R.; Samitier, J.; Sampietro, M.; Soldatkin, A.P.; Vidic, J.; Villanueva, G. Nanobiosensors Based on Individual Olfactory Receptors. *Analog Integr. Circ. Sign. Process.* **2008,** *57* (3), 197–203.

Liu, J.; Lu, Y. A Colorimetric Lead Biosensor Using DNAzyme-Directed Assembly of Gold Nanoparticlesm 2993. https://doi.org/10.1021/JA034775U.

Antunović, V.; Ilić, M.; Baošić, R.; Jelić, D.; Lolić, A. Synthesis of MnCo 2 O 4 Nanoparticles as Modifiers for Simultaneous Determination of Pb(II) and Cd(II). *PLoS ONE* **2019,** *14* (2), 1–19. https://doi.org/10.1371/journal.pone.0210904.

Appannagari, R. R. North Asian International Research Journal of Environmental Pollution Causes and Consequences: A Study. *North Asian Int. Res. J. Soc. Sci. Human.* Aug **2017,** *3*, 2454–9827.

Asmatulu, R.; Khan, W. S. Electrospun Nanofibers for Nanosensor and Biosensor Applications. In *Synthesis and Applications of Electrospun Nanofibers*, 2019; pp 175–96. https://doi.org/10.1016/b978-0-12-813914-1.00009-2.

Ayangbenro, A. S.; Babalola, O. O. A New Strategy for Heavy Metal Polluted Environments: A Review of Microbial Biosorbents. *Int. J. Environ. Res. Public Health* **2017,** *14* (1). https://doi.org/10.3390/ijerph14010094.

Badihi-Mossberg, M.; Buchner, V.; Rishpon, J. Electrochemical Biosensors for Pollutants in the Environment. *Electroanalysis* **2007,** *19* (19–20), 2015–2028. https://doi.org/10.1002/elan.200703946.

Bagheri, H.; Afkhami, A.; Khoshsafar, H.; Rezaei, M.; Shirzadmehr, A. Simultaneous Electrochemical Determination of Heavy Metals Using a Triphenylphosphine/MWCNTs Composite Carbon Ionic Liquid Electrode. *Sens. Actuat. B Chem.* **2013,** *186* (September), 451–460. https://doi.org/10.1016/j.snb.2013.06.051.

Bashar, T.; Fung, I. W. H. Water Pollution in a Densely Populated Megapolis, Dhaka. *Water (Switzerland)* **2020,** *12* (8), 1–13. https://doi.org/10.3390/W12082124.

Bhateria, R.; Singh, R. A Review on Nanotechnological Application of Magnetic Iron Oxides for Heavy Metal Removal. *J. Water Process Eng.* **2019,** *31* (April), 100845. https://doi.org/10.1016/j.jwpe.2019.100845.

Buzea, C.; Pacheco, I. I.; Robbie, K. Nanomaterials and Nanoparticles: Sources and Toxicity. *Biointerphases* **2007,** *2* (4), MR17–MR71. https://doi.org/10.1116/1.2815690.

Cabanillas-Galán, P.; Farmer, L.; Hagan, T.; Nieuwenhuyzen, M.; James, S. L.; Lagunas, M. C. A New Approach for the Detection of Ethylene Using Silica-Supported Palladium Complexes. *Inorg. Chem.* **2008,** *47* (19), 9035–9041. https://doi.org/10.1021/ic800986t.

Cao, X.; Ye, Y.; Liu, S. Gold Nanoparticle-Based Signal Amplification for Biosensing. *Analy. Biochem.* **2011,** *417* (1), 1–16. https://doi.org/10.1016/J.AB.2011.05.027.

Carrillo-Carrión, C.; Simonet, B. M.; Valcárcel, M. Carbon Nanotube-Quantum Dot Nanocomposites as New Fluorescence Nanoparticles for the Determination of Trace Levels of PAHs in Water. *Analy. Chim. Acta* **2009,** *652* (1–2), 278–284. https://doi.org/10.1016/j.aca.2009.08.015.

Chan, W. H.; Yang, R. H.; Wang, K. W. Development of a Mercury Ion-Selective Optical Sensor Based on Fluorescence Quenching of 5,10,15,20-Tetraphenylporphyrin. *Analy. Chim. Acta* **2001,** *444* (2), 261–269. https://doi.org/10.1016/S0003-2670(01)01106-0.

Chatterjee, A.; Santra, M.; Won, N.; Kim, S.; Kim, J. K.; Kim, S. B.; Ahn, K. H. Selective Fluorogenic and Chromogenic Probe for Detection of Silver Ions and Silver Nanoparticles in Aqueous Media. *J. Am. Chem. Soc.* **2009,** *131* (6), 2040–2041. https://doi.org/10.1021/ja807230c.

Chen, B.; Wang, Z.; Hu, D.; Ma, Q.; Huang, L.; Xv, C.; Guo, Z.; Jiang, X. Scanometric Nanomolar Lead (II) Detection Using DNA-Functionalized Gold Nanoparticles and Silver Stain Enhancement. *Sens. Actuat. B Chem.* **2014,** *200*, 310–316. https://doi.org/10.1016/j.snb.2014.04.066.

Chen, G.; Song, F.; Xiong, X.; Peng, X. Fluorescent Nanosensors Based on Fluorescence Resonance Energy Transfer (FRET). *Indust. Eng. Chem. Res.* **2013,** *52* (33), 11228–11245. https://doi.org/10.1021/ie303485n.

Chen, G. H.; Chen, W. Y.; Yen, Y. C.; Wang, C. W.; Chang, H. T.; Chen, C. F. Detection of Mercury(II) Ions Using Colorimetric Gold Nanoparticles on Paper-Based Analytical Devices. *Analy. Chem.* **2014,** *86* (14), 6843–6849. https://doi.org/10.1021/ac5008688.

Chen, H.; Zhou, K.; Zhao, G. Gold Nanoparticles: From Synthesis, Properties to Their Potential Application as Colorimetric Sensors in Food Safety Screening. *Trends Food Sci. Technol.* https://doi.org/10.1016/j.tifs.2018.05.027.

Chen, H.; Hu, W.; Li, C. M. Colorimetric Detection of Mercury(II) Based on 2,2′-Bipyridyl Induced Quasi-Linear Aggregation of Gold Nanoparticles. *Sens. Actuat. B Chem.* Aug **2015,** *215*, 421–427. https://doi.org/10.1016/j.snb.2015.03.083.

Concejero, M. A., Roger, G.; Herradón, B.; Gonzãlez, M. J.; De Frutos, M. Feasibility of High-Performance Immunochromatography as an Isolation Method for Pcbs and Other Dioxin. Like Compounds. *Analy. Chem.* **2001,** *73* (13), 3119–3125. https://doi.org/10.1021/ac001387r.

Cui, Y., Wei, Q.; Park, H.; Lieber, C. M. Nanowire Nanosensors for Highly Sensitive and Selective Detection of Biological and Chemical Species. *Science* **2001,** *293* (5533), 1289–1292. https://doi.org/10.1126/science.1062711.

D'Orazio, P. Biosensors in Clinical Chemistry. *Clin. Chim. Acta* **2003,** *334* (1–2), 41–69. https://doi.org/10.1016/S0009-8981(03)00241-9.

Danwittayakul, S.; Takahashi, Y.; Suzuki, T.; Thanaboonsombut, A. Simple Detection of Mercury Ion Using Dithizone Nanoloaded Membrane. *J. Metals Mater. Miner.* **2008,** *18* (2), 37–40.

Ding, N.; Zhao, H.; Peng, W.; He, Y.; Zhou, Y.; Yuan, L.; Zhang, Y. A Simple Colorimetric Sensor Based on Anti-Aggregation of Gold Nanoparticles for Hg2+ Detection. *Colloids Surf. A Physicochem. Eng. Aspects* Feb **2012,** *395*, 161–167. https://doi.org/10.1016/j.colsurfa.2011.12.024.

Du, J.; Zhu, B.; Chen, X. Urine for Plasmonic Nanoparticle-Based Colorimetric Detection of Mercury Ion. *Small* **2013,** *9* (24), 4104–4111. https://doi.org/10.1002/smll.201300593.

El-Safty, S. A., Ismail, A. A.; Matsunaga, H.; Hanaoka, T.; Mizukami, F. Optical Nanoscale Pool-on-Surface Design for Control Sensing Recognition of Multiple Cations. *Adv. Funct. Mater.* **2008,** *18* (10), 1485–1500. https://doi.org/10.1002/adfm.200701059.

Endo, T.; Okuyama, A.; Matsubara, Y.; Nishi, K.; Kobayashi, M.; Yamamura, S.; Morita, Y.; Takamura, Y.; Mizukami, H.; Tamiya, E. Fluorescence-Based Assay with Enzyme Amplification on a Micro-Flow Immunosensor Chip for Monitoring Coplanar Polychlorinated Biphenyls. *Analy. Chim. Acta* **2005,** *531* (1), 7–13. https://doi.org/10.1016/j.aca.2004.08.077.

Ewuzie, U.; Nnorom, I. C.; Eze, S. O. Lithium in Drinking Water Sources in Rural and Urban Communities in Southeastern Nigeria. *Chemosphere* Apr **2020,** *245*, 125593. https://doi.org/10.1016/j.chemosphere.2019.125593.

Farhadi, K.; Forough, M.; Molaei, R.; Hajizadeh, S.; Rafipour, A. Highly Selective Hg 2+ Colorimetric Sensor Using Green Synthesized and Unmodified Silver Nanoparticles. *Sens. Actuat. B Chem.* **2012,** *161* (1), 880–885. https://doi.org/10.1016/j.snb.2011.11.052.

Gholizadeh, M. H.; Melesse, A. M.; Reddi, L. A Comprehensive Review on Water Quality Parameters Estimation Using Remote Sensing Techniques. *Sensors (Switzerland)* **2016,** *16* (8). https://doi.org/10.3390/s16081298.

Gong, J.; Zhou, T.; Song, D.; Zhang, L. Monodispersed Au Nanoparticles Decorated Graphene as an Enhanced Sensing Platform for Ultrasensitive Stripping Voltammetric Detection of Mercury(II). *Sens. Actuat. B Chem.* **2010,** *150* (2), 491–497. https://doi.org/10.1016/j.snb.2010.09.014.

Guo, J.; Chai, Y.; Yuan, R.; Song, Z.; Zou, Z.;—Sensors and Actuators B: Chemical, and Undefined 2011. Lead (II) Carbon Paste Electrode Based on Derivatized Multi-Walled Carbon Nanotubes: Application to Lead Content Determination in Environmental Samples (n.d.). Elsevier. https://www.sciencedirect.com/science/article/pii/S0925400511000402 (accessed Nov 25, 2018).

Haun, J. B.; Yoon, T-J.; Lee, H.; Weissleder, R. Magnetic Nanoparticle Biosensors. *Wiley Interdiscipl. Rev. Nanomed. Nanobiotechnol.* **2010,** *2* (3), 291–304. https://doi.org/10.1002/wnan.84.

He, B.; Morrow, T. J.; Keating, C. D. Nanowire Sensors for Multiplexed Detection of Biomolecules. *Curr. Opin. Chem. Biol.* **2008,** *12* (5), 522–528. https://doi.org/10.1016/J.CBPA.2008.08.027.

He, H.; Li, H.; Uray, G.; Wolfbeis, O. S. Non-Enzymatic Optical Sensor for Penicillins. *Talanta* **1993,** *40* (3), 453–457. https://doi.org/10.1016/0039-9140(93)80258-S.

Hidayat, M. A.; Illa, R.; Indah, C.; Ningsih, Y.; Yuwono, M. The CUPRAC—Paper Microzone Plates as a Simple and Rapid Method for Total Antioxidant Capacity Determination of Plant Extract. *Eur. Food Res. Technol.* **2019,** no. 0123456789. https://doi.org/10.1007/s00217-019-03312-1.

Hong, S.; Kang, T.; Oh, S.; Moon, J.; Choi, I.; Choi, K.; Yi, J. Label-Free Sensitive Optical Detection of Polychlorinated Biphenyl (PCB) in an Aqueous Solution Based on Surface Plasmon Resonance Measurements. *Sens. Actuat. B Chem.* **2008,** *134* (1), 300–306. https://doi.org/10.1016/J.SNB.2008.05.006.

Huang, Y.; Zhang, W.; Xiao, H.; Li, G. An Electrochemical Investigation of Glucose Oxidase at a CdS Nanoparticles Modified Electrode. *Biosens. Bioelectr.* **2005,** *21* (5), 817–821. https://doi.org/10.1016/J.BIOS.2005.01.012.

Hussain, C. M.; Keçili, R. Electrochemical Techniques for Environmental Analysis. *Modern Environ. Analy. Techniques Pollut.* **2020,** 199–222. https://doi.org/10.1016/b978-0-12-816934-6.00008-4.

Jain, K. K. Nanodiagnostics: Application of Nanotechnology in Molecular Diagnostics. *Expert Rev. Mol. Diagnos.* **2003,** *3* (2), 153–161. https://doi.org/10.1586/14737159.3.2.153.

Javanbakht, M.; Divsar, F.; Badiei, A.; Ganjali, M. R.; Norouzi, P.; Ziarani, G. M.; Chaloosi, M.; Jahangir, A. A. Potentiometric Detection of Mercury(II) Ions Using a Carbon Paste Electrode Modified with Substituted Thiourea-Functionalized Highly Ordered Nanoporous Silica. *Analy. Sci.* **2009,** *25* (6), 789–794. https://doi.org/10.2116/analsci.25.789.

Jianrong, C.; Yuqing, M.; Nongyue, H.; Xiaohua, W.; Sijiao, L. Nanotechnology and Biosensors. *Biotechnol. Adv.* **2004,** *22* (7), 505–518. https://doi.org/10.1016/J.BIOTECHADV.2004.03.004.

Keçili, R.; Hussain, C. M. Recent Progress of Imprinted Nanomaterials in Analytical Chemistry. *Int. J. Analy. Chem.* **2018**. https://doi.org/10.1155/2018/8503853.

Ko, S.; Gunasekaran, S.; Yu, J. Self-Indicating Nanobiosensor for Detection of 2,4-Dinitrophenol. *Food Control* **2010,** *21* (2), 155–161. https://doi.org/10.1016/J.FOODCONT.2009.05.006.

Koh, I.; Josephson, L. Magnetic Nanoparticle Sensors. *Sensors* **2009,** *9* (10), 8130–8145. https://doi.org/10.3390/s91008130.

Kuila, T.; Bose, S.; Khanra, P.; Mishra, A. K.; Kim, N. H.; Lee, J. H. Recent Advances in Graphene-Based Biosensors. *Biosens Bioelectron* **2011**. https://doi.org/10.1016/j.bios.2011.05.039.

Kurniawan, F.; Al Kiswiyah, N. S.; Madurani, K. A.; Tominaga, M. Electrochemical Sensor Based on Single-Walled Carbon Nanotubes-Modified Gold Electrode for Uric Acid Detection. *J. Electrochem. Soc.* **2018,** *165* (11): B515–B522. https://doi.org/10.1149/2.0991811jes.

Kuswandi, B. Nanobiosensor Approaches for Pollutant Monitoring. *Environ. Chem. Lett.* **2019,** *17* (2). https://doi.org/10.1007/s10311-018-00853-x.

Kuswandi, B.; Mascini, M. Enzyme Inhibition Based Biosensors for Environmental Monitoring. *Curr. Enzyme Inhibit.* **2005,** *1* (1), 11–17.

Kuswandi, B.; Nuriman; Dam, H. H.; Reinhoudt, D. N.; Verboom, W. Development of a Disposable Mercury Ion-Selective Optode Based on Trityl-Picolinamide as Ionophore. *Analy. Chim. Acta* **2007,** *591* (2), 208–213. https://doi.org/10.1016/j.aca.2007.03.064.

Kuswandi, B.; Fikriyah, C. I.; Gani, A. A. An Optical Fiber Biosensor for Chlorpyrifos Using a Single Sol–Gel Film Containing Acetylcholinesterase and Bromothymol Blue. *Talanta* **2008,** *74* (4), 613–618. https://doi.org/10.1016/J.TALANTA.2007.06.042.

Landrigan, P. J.; Fuller, R.; Acosta, N. J. R.; Adeyi, O.; Arnold, R.; Basu, N. (N).; Baldé, A. B. et al. The Lancet Commission on Pollution and Health. *Lancet* **2018,** *391* (10119), 462–512. https://doi.org/10.1016/S0140-6736(17)32345-0.

Lee, C. H.; Tian, L.; Singamaneni, S. Paper-Based SERS Swab for Rapid Trace Detection on Real-World Surfaces. *ACS Appl. Mater. Interf.* **2010,** *2* (12), 3429–3435. https://doi.org/10.1021/am1009875.

Lee, P. L.; Sun, Y. S.; Ling, Y. C. Magnetic Nano-Adsorbent Integrated with Lab-on-Valve System for Trace Analysis of Multiple Heavy Metals. *J. Analy. Atom. Spectr.* **2009,** *24* (3), 320–327. https://doi.org/10.1039/b814164a.

Li, C.;; Zhou, K.; Qin, W.; Tian, C.; Qi, M.; Yan, X.; Han, W. A Review on Heavy Metals Contamination in Soil: Effects, Sources, and Remediation Techniques. *Soil Sediment Contam.* **2019,** *28* (4), 380–394. https://doi.org/10.1080/15320383.2019.1592108.

Li, Y-L.; Leng, Y-M.; Zhang, Y-J.; Li, T-H.; Shen, Z-Y.; Wu, A-G. A New Simple and Reliable Hg2+ Detection System Based on Anti-Aggregation of Unmodified Gold Nanoparticles in the Presence of O-Phenylenediamine. *Sens. Actuat. B Chem.* Sept **2014,** *200*, 140–146. https://doi.org/10.1016/J.SNB.2014.04.039.

Lisa, M., Chouhan, R. S.; Vinayaka, A. C.; Manonmani, H. K.; Thakur, M. S. Gold Nanoparticles Based Dipstick Immunoassay for the Rapid Detection of Dichlorodiphenyltrichloroethane: An Organochlorine Pesticide. *Biosens. Bioelectr.* **2009,** *25* (1), 224–227. https://doi.org/10.1016/J.BIOS.2009.05.006.

Liu, X.; Zhang, N.; Bing, T.; Shangguan, D. Carbon Dots Based Dual-Emission Silica Nanoparticles as a Ratiometric Nanosensor for Cu2+. *Analy. Chem.* **2014,** *86* (5), 2289–2296. https://doi.org/10.1021/ac404236y.

Liu, X.; Germaine, K. J.; Ryan, D.; Dowling, D. N. Genetically Modified Pseudomonas Biosensing Biodegraders to Detect PCB and Chlorobenzoate Bioavailability and Biodegradation in Contaminated Soils. *Bioeng. Bugs* **2010,** *1* (3), 198–206. https://doi.org/10.4161/bbug.1.3.12443.

Liu, Y.; Chakrabartty, S.; Alocilja, E. C. Fundamental Building Blocks for Molecular Biowire Based Forward Error-Correcting Biosensors. *Nanotechnology* **2007,** *18* (42), 424017. https://doi.org/10.1088/0957-4484/18/42/424017.

Luo, X.; Morrin, A.; Killard, A. J.; Smyth, M. R. Application of Nanoparticles in Electrochemical Sensors and Biosensors. *Electroanalysis* **2006,** *18* (4), 319–326. https://doi.org/10.1002/elan.200503415.

Mailu, S. N.; Waryo, T. T.; Ndangili, P. M.; Ngece, F. R.; Baleg, A. A.; Baker, P. G.; Iwuoha, E. I. et al. Determination of Anthracene on Ag-Au Alloy Nanoparticles/Overoxidized-Polypyrrole Composite Modified Glassy Carbon Electrodes. *Sensors* **2010,** *10* (10), 9449–9465. https://doi.org/10.3390/s101009449.

Manisalidis, I.; Stavropoulou, E.; Stavropoulos, A.; Bezirtzoglou, E. Environmental and Health Impacts of Air Pollution: A Review. *Front. Public Health* Feb **2020,** *8*, 1–13. https://doi.org/10.3389/fpubh.2020.00014.

Mascini, M., Macagnano, A.; Monti, D.; Del Carlo, M.; Paolesse, R.; Chen, B.; Warner, P.; D'Amico, A.; Di Natale, C.; Compagnone, D. Piezoelectric Sensors for Dioxins: A Biomimetic Approach. *Biosens. Bioelectr.* **2004,** *20* (6), 1203–1210. https://doi.org/10.1016/J.BIOS.2004.06.048.

Mascini, M.; MacAgnano, A.; Scortichini, G.; Del Carlo, M.; Diletti, G.; D'Amico, A.; Di Natale, C.; Compagnone, D. Biomimetic Sensors for Dioxins Detection in Food Samples. *Sens. Actuat. B Chem.* **2005,** *111–112*, 376–384. https://doi.org/10.1016/j.snb.2005.03.054.

McDonagh, C.; Burke, C. S.; MacCraith, B. D. Optical Chemical Sensors. *Chem. Rev.* **2008,** *108* (2), 400–422. https://doi.org/10.1021/cr068102g.

Mueller, N. C.; Nowack, B. 2008. Exposure Modeling of Engineered Nanoparticles in the Environment. *Environ. Sci. Technol.* **2008,** *42* (12), 4447–4453. https://doi.org/10.1021/es7029637.

Nagatani, N.; Takeuchi, A.; Hossain, M. A.; Yuhi, T.; Endo, T.; Kerman, K.; Takamura, Y.; Tamiya, E. Rapid and Sensitive Visual Detection of Residual Pesticides in Food Using Acetylcholinesterase-Based Disposable Membrane Chips. *Food Control* **2007,** *18* (8), 914–920. https://doi.org/10.1016/J.FOODCONT.2006.05.011.

Norouzi, P.; Pirali-Hamedani, M.; Ganjali, M. R.; Faridbod, F. Electrochemical Science a Novel Acetylcholinesterase Biosensor Based on Chitosan-Gold Nanoparticles Film for Determination of Monocrotophos Using FFT Continuous Cyclic Voltammetry. *Int. J. Electrochem. Sci.* **2010,** *5*. www.electrochemsci.org.

Ottakam Thotiyl, M. M.; Basit, H.; Sánchez, J. A.; Goyer, C.; Coche-Guerente, L.; Dumy, P.; Sampath, S.; Labbé, P.; Moutet, J. C. Multilayer Assemblies of Polyelectrolyte-Gold Nanoparticles for the Electrocatalytic Oxidation and Detection of Arsenic(III). *J. Colloid Interf. Sci.* **2012,** *383* (1), 130–139. https://doi.org/10.1016/j.jcis.2012.06.033.

Pal, S.; Alocilja, E. C.; Downes, F. P. Nanowire Labeled Direct-Charge Transfer Biosensor for Detecting Bacillus Species. *Biosens. Bioelectr.* **2007,** *22* (9–10), 2329–2336. https://doi.org/10.1016/j.bios.2007.01.013.

Parihar, S.; Manoj, K.; Pratap, S. Qualitative Estimation of Pesticides in Biological Matrices Using Thin Layer Chromatography Method of Analysis **2020,** 1–5.

Park, J-W.; Kurosawa, S.; Aizawa, H.; Hamano, H.; Harada, Y.; Asano, S.; Mizushima, Y.; Higaki, M. Dioxin Immunosensor Using Anti-2,3,7,8-TCDD Antibody Which Was Produced with Mono 6-(2,3,6,7-Tetrachloroxanthene-9-Ylidene) Hexyl Succinate as a Hapten. *Biosens. Bioelectr.* **2006,** *22* (3), 409–414. https://doi.org/10.1016/J.BIOS.2006.05.002.

Paterson, S.; De La Rica, R. Solution-Based Nanosensors for in-Field Detection with the Naked Eye. *Analyst* **2015,** *140* (10), 3308–3317. https://doi.org/10.1039/c4an02297a.

Peng, C.; Li, Z.; Zhu, Y.; Chen, W.; Yuan, Y.; Liu, L.; Li, Q. et al. Simultaneous and Sensitive Determination of Multiplex Chemical Residues Based on Multicolor Quantum Dot Probes. *Biosens. Bioelectr.* **2009,** *24*, 3657–3662.

Peng, Y.; Cullis, T.; Inkson, B. Bottom-up Nanoconstruction by the Welding of Lndividual Metallic Nanoobjects Using Nanoscale Solder. *Nano Lett.* **2009,** *9* (1), 91–96. https://doi.org/10.1021/nl8025339.

Perez, J. M.; Josephson, L.; Weissleder, R. Use of Magnetic Nanoparticles as Nanosensors to Probe for Molecular Interactions. *ChemBioChem* **2004**. https://doi.org/10.1002/cbic.200300730.

Prabowo, B. A.; Purwidyantri, A.; Liu, K. C. Surface Plasmon Resonance Optical Sensor: A Review on Light Source Technology. *Biosensors* **2018,** *8* (3). https://doi.org/10.3390/bios8030080.

Přibyl, J.; Hepel, M.; Skládal, P. Piezoelectric Immunosensors for Polychlorinated Biphenyls Operating in Aqueous and Organic Phases. *Sens. Actuat. B Chem.* **2006,** *113* (2), 900–910. https://doi.org/10.1016/J.SNB.2005.03.077.

Promphet, N.; Rattanarat, P.; Rangkupan, R.; Chailapakul, O.; Rodthongkum, N. An Electrochemical Sensor Based on Graphene/Polyaniline/Polystyrene Nanoporous Fibers Modified Electrode for Simultaneous Determination of Lead and Cadmium. *Sens. Actuat. B Chem.* Feb **2015,** *207*, 526–534. https://doi.org/10.1016/J.SNB.2014.10.126.

Pumera, M.; Sánchez, S.; Ichinose, I.; Tang, J. Electrochemical Nanobiosensors. *Sens. Actuat. B Chem.* **2007,** *123* (2), 1195–1205. https://doi.org/10.1016/j.snb.2006.11.016.

Qu, F.; Yang, M.; Lu, Y.; Shen, G.; Yu, R. Amperometric Determination of Bovine Insulin Based on Synergic Action of Carbon Nanotubes and Cobalt Hexacyanoferrate Nanoparticles Stabilized by EDTA. *Analy. Bioanaly. Chem.* **2006,** *386* (2), 228–234. https://doi.org/10.1007/s00216-006-0642-8.

Rascio, N.; Navari-Izzo, F. Heavy Metal Hyperaccumulating Plants: How and Why Do They Do It? And What Makes Them so Interesting? *Plant Sci.* **2011,** *180* (2), 169–181. https://doi.org/10.1016/j.plantsci.2010.08.016.

Ravindran, A.; Elavarasi, M.; Prathna, T. C.; Raichur, A. M.; Chandrasekaran, N.; Mukherjee, A. Selective Colorimetric Detection of Nanomolar Cr (VI) in Aqueous Solutions Using Unmodified Silver Nanoparticles. *Sens. Actuat. B Chem.* May **2012,** *166–167*, 365–371. https://doi.org/10.1016/j.snb.2012.02.073.

Reis, P. B.; Ramos, R. M.; Souza, L. F.; Cancado, S. de V.;. Validation of Spectrophotometric Method to Detect and Quantify Nitrite in Ham Pate. *BPJS* **2009,** *45*.

O'Riordan, A.; Barry, S. Electrochemical Nanosensors: Advances and Applications. *Rep. Electrochem.* **2016,** *2016*, *1*. https://doi.org/10.2147/rie.s80550.

Jin, R.; Wu, G.; Li, Z.; Mirkin, C. A.; Schatz, G. C. What Controls the Melting Properties of DNA-Linked Gold Nanoparticle Assemblies? 2003. https://doi.org/10.1021/JA021096V.

Roto, R.; Mellisani, B.; Kuncaka, A.; Mudasir, M.; Suratman, A. Colorimetric Sensing of Pb2+ Ion by Using Ag Nanoparticles in the Presence of Dithizone. *Chemosensors* **2019,** *7* (3). https://doi.org/10.3390/CHEMOSENSORS7030028.

Hrapovic, S.; Liu, Y.; Male, K. B.; Luong, J. H. T. Electrochemical Biosensing Platforms Using Platinum Nanoparticles and Carbon Nanotubes, 2003. https://doi.org/10.1021/AC035143T.

Sadik, O. a.; Zhou, a. L.; Kikandi, S.; Du, N.; Wang, Q.; Varner, K. Sensors as Tools for Quantitation, Nanotoxicity and Nanomonitoring Assessment of Engineered Nanomaterials. *J. Environ. Monit JEM* **2009,** *11* (10), 1782–1800. https://doi.org/10.1039/b912860c.

Sawant, Vijay J.; Sawant, V. J. Biogenic Capped Selenium Nano Rods as Naked Eye and Selective Hydrogen Peroxide Spectrometric Sensor. *Sens. BioSens. Res.* Feb **2020,** *27*. https://doi.org/10.1016/j.sbsr.2019.100314.

Semionov, A. Minamata Disease—Review. *World J. Neurosci.* **2018,** *08* (02), 178–184. https://doi.org/10.4236/wjns.2018.82016.

Laschi, S.; Mascini, M.; Scortichini, G.; Fránek, M.; Mascini, M. Polychlorinated Biphenyls (PCBs) Detection in Food Samples Using an Electrochemical Immunosensor, 2003. https://doi.org/10.1021/JF0208637.

Shahat, A.; Awual, Md. R.; Naushad, M. Functional Ligand Anchored Nanomaterial Based Facial Adsorbent for Cobalt(II) Detection and Removal from Water Samples. *Chem. Eng. J.* **2015,** *271* (Ii), 155–163. https://doi.org/10.1016/j.cej.2015.02.097.

Sharma, V.; Saini, A. K.; Mobin, S. M. Multicolour Fluorescent Carbon Nanoparticle Probes for Live Cell Imaging and Dual Palladium and Mercury Sensors. *J. Mater. Chem. B* **2016,** *4* (14), 2466–2476. https://doi.org/10.1039/c6tb00238b.

Shen, X.; Cui, Y.; Pang, Y.; Qian, H. Graphene Oxide Nanoribbon and Polyhedral Oligomeric Silsesquioxane Assembled Composite Frameworks for Pre-Concentrating and Electrochemical Sensing of 1-Hydroxypyrene. *Electrochim. Acta* Jan **2012,** *59*, 91–99. https://doi.org/10.1016/j.electacta.2011.10.037.

Slimani, Y.; Hannachi, E. *Magnetic Nanosensors and Their Potential Applications. Nanosensors for Smart Cities*. INC, 2020. https://doi.org/10.1016/b978-0-12-819870-4.00009-8.

Soares, S.; Sousa, J.; Pais, A.; Vitorino, C. Nanomedicine: Principles, Properties, and Regulatory Issues. *Front. Chem.* Aug **2018,** *6*, 1–15. https://doi.org/10.3389/fchem.2018.00360.

Soh, N.; Tokuda, T.; Watanabe, T.; Mishima, K.; Imato, T.; Masadome, T.; Asano, Y.; Okutani, S.; Niwa, O.; Brown, S. A Surface Plasmon Resonance Immunosensor for Detecting a Dioxin Precursor Using a Gold Binding Polypeptide. *Talanta* **2003,** *60* (4), 733–745. https://doi.org/10.1016/S0039-9140(03)00139-5.

Somerset, V.; Baker, P.; Iwuoha, E. Mercaptobenzothiazole-on-Gold Organic Phase Biosensor Systems: 1. Enhanced Organosphosphate Pesticide Determination. *J. Environ. Sci. HealthB Pesticides Food Contam. Agric. Wastes* **2009,** *44* (2), 164–178. https://doi.org/10.1080/03601230802599092.

Somerset, V. S.; Klink, M. J.; Baker, P. G. L.; Iwuoha, E. I. Acetylcholinesterase-Polyaniline Biosensor Investigation of Organophosphate Pesticides in Selected Organic Solvents. *J. Environ. Sci. Health B Pesticides Food Contam. Agric. Wastes* **2007,** *42* (3), 297–304. https://doi.org/10.1080/03601230701229288.

Sugunan, A.; Thanachayanont, C.; Dutta, J.; Hilborn, J. G. Heavy-Metal Ion Sensors Using Chitosan-Capped Gold Nanoparticles. *Sci. Technol. Adv. Mater.* **2005,** *6*, 335–340. No longer published by Elsevier. https://doi.org/10.1016/j.stam.2005.03.007.

Sumner, J. P.; Aylott, J. W.; Monson, E.; Kopelman, R. A Fluorescent PEBBLE Nanosensor for Intracellular Free Zinc. *Analyst* **2002,** *127* (1), 11–16. https://doi.org/10.1039/b108568a.

Tonle, I. K.; Ngameni, E.; Walcarius, A. Preconcentration and Voltammetric Analysis of Mercury(II) at a Carbon Paste Electrode Modified with Natural Smectite-Type Clays Grafted with Organic Chelating Groups. *Sens. Actuat. B Chem.* **2005,** *110* (2), 195–203. https://doi.org/10.1016/j.snb.2005.01.027.

Tortorich, R.; Shamkhalichenar, H.; Choi, J-W. Inkjet-Printed and Paper-Based Electrochemical Sensors. *Appl. Sci.* **2018,** *8* (2), 288. https://doi.org/10.3390/app8020288.

Tseng, W-B.; Hsieh, M-M.; Chen, C-H.; Chiu, T-C.; Tseng, W-T. Functionalized Gold Nanoparticles for Sensing of Pesticides: A Review. *J. Food Drug Analy.* **2020,** *28* (4), 522–539. https://doi.org/10.38212/2224-6614.1092.

Tsutsumi, T.; Miyoshi, N.; Sasaki, K.; Maitani, T. Biosensor Immunoassay for the Screening of Dioxin-like Polychlorinated Biphenyls in Retail Fish. *Analy. Chim. Acta* **2008,** *617* (1–2), 177–183. https://doi.org/10.1016/j.aca.2008.02.003.

Ukaogo, P. O.; Ewuzie, U.; Onwuka, C. V. *Environmental Pollution: Causes, Effects, and the Remedies. Microorganisms for Sustainable Environment and Health.* INC, 2020. https://doi.org/10.1016/b978-0-12-819001-2.00021-8.

Ullah, N.; Mansha, M.; Khan, I.; Qurashi, A. Nanomaterial-Based Optical Chemical Sensors for the Detection of Heavy Metals in Water: Recent Advances and Challenges. *TrAC—Trends Analy. Chem.* **2018,** *100*, 155–166. https://doi.org/10.1016/j.trac.2018.01.002.

Vijitvarasan, P.; Oaew, S.; Surareungchai, W. Paper-Based Scanometric Assay for Lead Ion Detection Using DNAzyme. *Analy. Chim. Acta* **2015,** *896*, 152–159. https://doi.org/10.1016/j.aca.2015.09.011.

Vinod Kumar, V.; Anthony. S. P. Silver Nanoparticles Based Selective Colorimetric Sensor for Cd 2+, Hg2+ and Pb2+ Ions: Tuning Sensitivity and Selectivity Using Co-Stabilizing Agents. *Sens. Actuat. B Chem.* Feb **2014,** *191*, 31–36. https://doi.org/10.1016/j.snb.2013.09.089.

Wan, H.; Sun, Q.; Li, H.; Sun, F.; Hu, N.; Wang, P. Screen-Printed Gold Electrode with Gold Nanoparticles Modification for Simultaneous Electrochemical Determination of Lead and Copper. *Sens. Actuat. B Chem.* Mar **2015,** *209*, 336–342. https://doi.org/10.1016/j.snb.2014.11.127.

Wei, Q.; Nagi, R.; Sadeghi, K.; Feng, S.; Yan, E.; Ki, S. J.; Caire, R.; Tseng, D.; Ozcan, A. Detection and Spatial Mapping of Mercury Contamination in Water Samples Using a Smart-Phone. *ACS Nano* **2014,** *8* (2), 1121–1129. https://doi.org/10.1021/nn406571t.

Wei, Y.; Gao, C.; Meng, F. L.; Li, H. H.; Wang, L.; Liu, J. H; Huang, X. J. SnO 2/Reduced Graphene Oxide Nanocomposite for the Simultaneous Electrochemical Detection of Cadmium(II), Lead(II), Copper(II), and Mercury(II): An Interesting Favorable Mutual Interference. *J. Phys. Chem. C* **2012,** *116* (1), 1034–1041. https://doi.org/10.1021/jp209805c.

Willner, M. R.; Vikesland, P. J. Nanomaterial Enabled Sensors for Environmental Contaminants Prof Ueli Aebi, Prof Peter Gehr. *J. Nanobiotechnol.* **2018,** *16* (1), 1–16. https://doi.org/10.1186/s12951-018-0419-1.

Xiao, Y.; Li, C. M. Nanocomposites: From Fabrications to Electrochemical Bioapplications. *Electroanalysis* **2008,** *20* (6), 648–662. https://doi.org/10.1002/elan.200704125.

Xu, H.; Aylott, J. W.; Kopelman, R.; Miller, T. J.; Philbert, M. A. A Real-Time Ratiometric Method for the Determination of Molecular Oxygen inside Living Cells Using Sol-Gel-Based Spherical Optical Nanosensors with Applications to Rat C6 Glioma. *Analy. Chem.* **2001,** *73* (17), 4124–4133. http://www.ncbi.nlm.nih.gov/pubmed/11569801.

Yadav, P. K.; Srivastava, R. Optical Fiber Sensor: Review and Applications Mar **2018,** 0–5.

Yan, X.; Li, H.; Su, X. Review of Optical Sensors for Pesticides. *TrAC–Trends Analy. Chem.* **2018,** *103*, 1–20. https://doi.org/10.1016/j.trac.2018.03.004.

Yogeswaran, U.; Chen, S-M. A Review on the Electrochemical Sensors and Biosensors Composed of Nanowires as Sensing Material. *Sensors* **2008,** *8* (1), 290–313. https://doi.org/10.3390/s8010290.

Yonzon, C. R.; Stuart, D. A.; Zhang, X.; McFarland, A. D.; Haynes, C. L.; Van Duyne, R. P. Towards Advanced Chemical and Biological Nanosensors—An Overview. *Talanta* **2005,** *67* (3), 438–448. https://doi.org/10.1016/j.talanta.2005.06.039.

Zamora-León, S. P.; Delgado-López, F. Polycyclic Aromatic Hydrocarbons and Their Association with Breast Cancer. *Bangladesh J. Med. Sci.* **2020,** *19* (2), 194–199. https://doi.org/10.3329/bjms.v19i2.44995.

Zhai, D.; Zhang, K.; Zhang, Y.; Sun, H.; Fan, G. Mesoporous Silica Equipped with Europium-Based Chemosensor for Mercury Ion Detection: Synthesis, Characterization, and Sensing Performance. *Inorg. Chim. Acta* **2012,** *387*, 396–400. https://doi.org/10.1016/j.ica.2012.02.035.

Zhang, Q.; Qiao, Y.; Hao, F.; Zhang, L.; Wu, S.; Li, Y.; Li, J.; Song, X. M. Fabrication of a Biocompatible and Conductive Platform Based on a Singlestranded DNA/Graphene Nanocomposite for Direct Electrochemistry and Electrocatalysis. *Chem. Eur. J.* **2010,** *16* (27), 8133–8139. https://doi.org/10.1002/chem.201000684.

Zhang, X.; Xu, L-P.; Zhou, S-F.; Liu, G.; Gao, X. Recent Advances in Nanoparticles-Based Lateral Flow Biosensors. *Am. J. Biomed. Sci.* **2014,** *6* (1), 41–57. https://doi.org/10.5099/aj140100041.

Zhang, Y.; Wang, W.; Li, Q.; Yang, Q.; Li, Y.; Du, J. 2015. Colorimetric Magnetic Microspheres as Chemosensor for Cu2+ Prepared from Adamantane-Modified Rhodamine and β-Cyclodextrin-Modified Fe3O4@SiO2 via Host-Guest Interaction. *Talanta* **2015,** *141*, 33–40. https://doi.org/10.1016/j.talanta.2015.03.015.

Zhao, G.; Wang, H.; Liu, G.; Wang, Z. Simultaneous Determination of Cd(II) and Pb(II) Based on Bismuth Film/Carboxylic Acid Functionalized Multi-Walled Carbon Nanotubes-β-Cyclodextrin-Nafion Nanocomposite Modified Electrode. *Int. J. Electrochem. Sci.* **2016,** *11* (10), 8109–8122. https://doi.org/10.20964/2016.10.07.

Zhou, C.; Zhang, X.; Huang, X.; Guo, X.; Cai, Q.; Zhu, S. Rapid Detection of Chloramphenicol Residues in Aquatic Products Using Colloidal Gold Immunochromatographic Assay. *Sensors* **2014,** *14* (11), 21872–21888. https://doi.org/10.3390/s141121872.

Zhu, C.; Yang, G.; Li, H.; Du, D.; Lin, Y. Electrochemical Sensors and Biosensors Based on Nanomaterials and Nanostructures. *Anal. Chem.* **2015,** *87* (1), 230–249. https://doi.org/10.1021/ac5039863.

CHAPTER 4

Nanotechnological Revolution: Carbon-Based Nanomaterials Detecting Pollutants

VANDANA SINGH[1*] and SHALVI UPADHYAY[2]

[1]*Department of microbiology, School of Allied Health Science, Sharda University, Greater Noida, 201306, Uttar Pradesh, India*

[2]*Depatment of Forensic Sciences, School of Allied Health Science, Sharda University, Greater Noida, 201306, Uttar Pradesh, India*

[*]*Corresponding author. E-mail: vandana.singh@sharda.ac.in*

ABSTRACT

Pollution of soil, water, and air are becoming serious issue worldwide due to all the produced toxic chemicals from the different industrial and personal activities. This comportment may give rise to several issues related to the health of humans and environment which increases the application challenges of conventional treatment methodologies. These noxious volatile compounds, inorganic gases, heavy metals, day-to-day personal hygiene and upkeep products, endocrine-disrupting compounds, chemical dyes, pharma products, etc. are disturbing the equilibrium of the globe and enhancing the ecological contamination at a high-alarming frequency. Consequently, their recognition, adsorption, and elimination are of great need. Therefore, this chapter mainly emphasizes on the novel evolutions of nanotechnology and its role in detection of the growing hazardous wastes with high efficiency, less energy along with lower cost.

Nanotechnology for Environmental Pollution Decontamination: Tools, Methods, and Approaches for Detection and Remediation. Fernanda Maria Policarpo Tonelli, Rouf Ahmad Bhat, & Gowhar Hamid Dar (Eds.)

This chapter briefly explains the application of carbon-based nanomaterial for the elimination of various volatile organic compounds (VOC), aerosols, greenhouse gases, photocatalytic, etc., soil pollutants (pesticide, insecticides, drugs, etc.), and water pollutants (biological impurities, microorganisms, heavy metals, pathogenic microbes by adsorption, photocatalysis membrane processes, photocatalysis, and fumigation as well as decontamination processes), all conventional methods used for detection (Zhang et al., 2019).

4.1 INTRODUCTION

The constant development of urbanization and industrialization including transportation, production, manufacturing, refining, mining, etc., depletes naturally present resources and also produces huge amount of hazardous wastes which lead to the pollution of air, soil, and water and consequently are threatening to the human health as well as ecological safety (Kreyling et al., 2010). All the waste generated by humans as well as by industries are released in the environment in various categories, such as toxic gases (nitrogen oxide, ozone, Sulfur oxide, carbon monooxide, etc.), volatile inorganic compounds, etc. in air, whereas water and soil pollutants may consist of pesticides, hydrocarbons, heavy metals (cadmium, mercury, lead, etc.) pathogenic microbial flora, insecticides Zhang et al. (2019), Fereidoun et al. (2007), these pollutants may cause damages to human either directly or indirectly by means of absorption, assimilation, or inhalation. Furthermore to this, few toxicants have the capacity to enter the foodchain which may be very dangerous for human and wild life. Some examples are bioaccumulation of heavy metals in fishes and in aquatic biota (Patil et al., 2016; Kumar et al., 2011; Thamri et al., 2016). So, to remove these pollutants, nanotechnology came in loop.

Nanotechnology mainly deals with nanomaterials which range from 1 to 100 ηm in size. Due to their extraordinary potential, their applications are explored in every field of research, such as in medicine (Ibrahim et al., 2016), in food industry (Shanthilal and Bhattacharya, 2014; Duncan, 2011), in energy (Zang 2011), in bioremediation or pollution treatment (Brame et al., 2011; Ibrahim et al., 2016; Karn et al., 2009) in providing improved methodologies for the development and alteration of present remediation machineries.

Environmentalist are aggressively evaluating nanomaterials as useful tools for contaminants of emerging concerns. The chemical, physical, and electrical properties of carbonaceous nanomaterials motivate pioneering resolution to tenacious ecological problems (Loos, 2015). Recently, carbon-based nanomaterials and their derivatives are explored to enhance the transportation of medicines within compressed tissues, precisely in the case of tumorous cells or tissues and use specially designed carbon nanotubes as artificial trans-membrane apertures. Correspondingly, ecological significance also comprises known-targeted remediation, planned exclusion of harmful pollutants, membrane assemblies for water filtration, etc. (Balamurugan et al., 2011). These carbon-based-nanomaterials empower ecological importance with less consequences.

4.2 CARBON-BASED NANOMATERIALS

The molecular level alterations at wide subclass of elements to endeavor precise characteristics on the nanoscale are very much in trends. The carbon-based nanotubes, MoS_2, chrysolite, boron, WS_2, TiS_2, NbS_2, kaolinite, chrysotile, and other predecessors validate excellent chemical, mechanical, physical, and electrical characteristics in contrast to their bulk and counter variants. The carbon's exceptional hybrid forming (hybridization) characteristic, also compassion of its morphology to agitations in formation conditions, permit it for engineered alterations to that extent that it is not so far matched by any other inorganic nanostructures (Mauter and Elimelech, 2008). The electrical, chemical, and physical, properties of nanocarbon and its hybrid forms are same and steady as it was in the carbon's structures. The arrangement of ground state orbital of its electrons is 1s2, 2s2, 2p2. The less energy gap amid 2s and 2p electron shells enables the raise of one "s" orbital electron to the higher energy "p" orbital that is unfilled in the ground state. The bonding relationships among the adjacent atoms, permits carbon's hybridization to sp, sp2, or sp3 configuration. The energy released due to covalent bond formation with neighboring atoms provides higher energy state to its electronic configuration. This return remains almost equal for the sp2 and sp3 hybridization states (Mauter and Elimelech, 2008; Hu et al., 2007).

Due to the raise in pressure and temperature, carbon adopts the thermodynamically stable trigonometric sp3 shape of diamond. On decreasing the temperature, carbon goes for planar sp2 configuration and forms monolayer sheets with one "π" bond three sigma covalent bonds. The bond configuration of carbonaceous nanomaterials and macroscopic carbon are same, the difference lies only in their size and other properties. Nanocarbons are much stable than the crystalline forms (Hu et al., 2007; Falcao and Wudl, 2007; Mauter and Elimelech, 2008).

4.3 CLASSIFICATION OF CARBON AND CARBON-BASED NANOMATERIAL

Due to its unique electronic construction, carbon has the property to form a polymer with large molecular weight compounds along with elongated chains. As its valency is four, it can readily undergo covalent bonding with both nonmetals and metals and because of this property, carbon-based compounds can occur in varied molecular forms. Carbon also exhibit the property of allotropy which means same kind of atoms can be re-organized in various shapes, and have changed properties (Falcao and Wudl, 2007). Diamond and graphite are the two known examples of naturally occurring allotropes of carbon that exist in the environment. On the basis of their shape and geometrical structures, the carbon nanomaterials are further classified (Fig. 4.1) as tube-shaped as carbon nanotube (CNT), horn-shaped as nano-horns, the zero-dimensional carbon dots known asthe carbon quantum dots (CQDs), and ellipsoidal-shaped carbon nanospheres known as Fullerene.

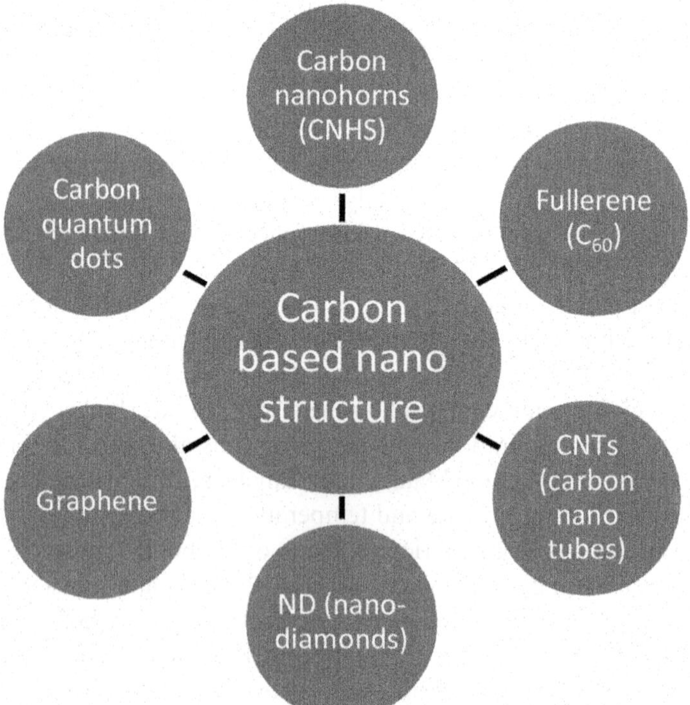

FIGURE 4.1 Types of carbon-based nanomaterial.

4.3.1 CARBON NANOTUBE

CNTs are carbon allotropes that have exceptional electrical and mechanical properties. CNTs are strong, lightweighted, possess high strength compared with other nanomaterials. Because of their decent graphical nature and high specific surface area, their applications are explored in numerous of areas, such as in engineering fieldfor the detection of pollutants, that is, in the field of remediation of pollution, battery formation, used as grout due to their electrochemical property for making nanocarbon–polymer composites etc. CNTs mainly occur in two shapes, that is, either cylindrical, which is designed by systematic rolling of graphene sheets or the other one is called controlled cap fullerene assembly which appeared in half-shape (Meagan et al., 2008; Zhang et al., 2019). As carbon nanotube exist in different geometrical configurations, they can be visualized by using electron microscope which provides their clear image.. Based on the arrangement, CNTs are further classified into two types (Fig. 4.2)

1. Single-walled carbon nanotube (SWCNT).
2. Multiwalled carbon nanotube (MWCNT).

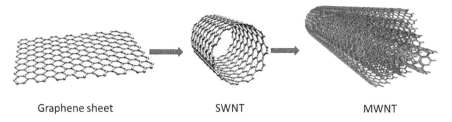

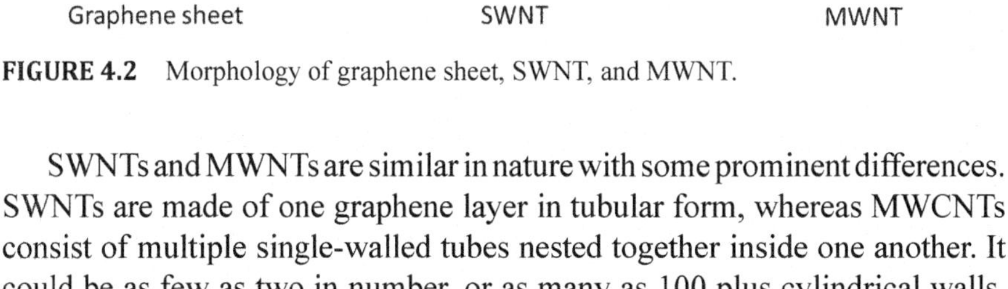

FIGURE 4.2 Morphology of graphene sheet, SWNT, and MWNT.

SWNTs and MWNTs are similar in nature with some prominent differences. SWNTs are made of one graphene layer in tubular form, whereas MWCNTs consist of multiple single-walled tubes nested together inside one another. It could be as few as two in number, or as many as 100 plus cylindrical walls. Carbon nanotubes are shaped by rolling graphene sheets (single layer of carbon atoms) and forming cylindrical molecules. The size of SWCNTs ranges 1–3 ηm in diameter with the length of few micrometers, whereas in MWCNTs, the graphene sheets have the size 0.34 ηm in diameter (interspace distance) and are fixed in similar way as concentric layers in a tubular form and the MWCNTs are approximately 5–40 ηm in diameter. The properties of CNTs are mainly based on their morphology, size, and diameter. When MWCNTs are used in composites, only the outer wall usually contributes to the mechanical and electrical properties. Among these two types of carbonaceous nanotubes, the SWCNTs are more eminent due to their numerous flexible characteristics. SWCNTs possess high electrical property due to their hexagonal alignment or chirality in tubular axis. Numerous approaches have been made by researchers for the preparation of CNTs, such as exfoliation of graphite, chemical vapor deposition (CVD), arc discharge, thermal decomposition of SiC, plasma CVD, and laser ablation. Among all the enumerated procedures discussed above, arc-discharge method was the first method to be used for preparing CNTs, whereas laser ablation was first method used for the preparation the SWCNTs. In contrast to the SWCNTs, the MWCNTs exhibit high thermal and mechanical satiability due to the occurrence of multiple layers. Further, on the basis of their distinct geometries, the SWCNTs are further classified into three subcategories: (1) zig-zag type (good semiconductor) (2) armchair type (high electrical conductivity), and (3) chiral type (semiconductive) (Meagan et al., 2008; Zhang et al., 2019).

4.3.2 FULLERENES

In 1985, the Kroto et al. introduced a new class of carbon family known as fullerene (Fig. 4.3) (represented as buckminsterfullerene (C60)). A fullerene is a carbon allotrope in which the molecules of carbon atoms are connected by single and double bonds in a manner to form a closed or partially closed net, with fused rings of 5–7 atoms. In defect-free form, fullerenes are net-like encircled assemblies composed of 12 (twelve), five-member rings, and an undetermined numeral of six-member rings. Morphological constructions with a smaller number of hexagons show better sp3 bonding property, more reactive sites, and higher straining energies. Isomers with adjacent pentagons show lesser stability than isomers with isolated pentagons due to the delocalized π bonds (Meagan et al., 2008; Zhang et al., 2019). Fullerene is a carbon-based transitional allotrope of diamond and graphite, shaped from the cluster of reiterating units ($n > 20$) of atomic Cn, which are conjointly attached to make a hollow-cored sphere-shaped surface molecule. The carbon atoms are characteristically positioned on the sphere's surface at the apexes of pentagons and hexagons form and allied together by the formation of sp2-hybridization covalent bonds. Usually, this C60 form, two bonding dimensions/lengths, but in the two lengths, the bonds of 6:5 ring are longer than the double bonds of 6:6 type. The main distinct characteristic of C60 is its pentagonal rings which enhances its poor electron delocalization, and consequently, C60 acts like an electron lacking alkene, and therefore, tries to rapidly react with electron-rich classes. The C60 fullerene is odorless because of its tendency of not forming double bonds in the pentagonal ring forms. The C60 form is nonvolatile, odorless, black solid with boiling point of 800 K (Troshin and Lyubovskaya, 2008; Meagan et al., 2008; Rao et al., 1995). Due to all the above characteristics, fullerenes are very much explored in medical therapies, in fuel cell, solar cell, electronics arena, cosmetic products (low-odor fullerenes, such as, C26, C28, etc.), they act as catalyst in chemical reactions, and other relevant fields. C60 fullerenes have exceptional properties, such as superconductivity, extra durability, and they are radical scavengers which enables them to be easily altered as nanocomposites.

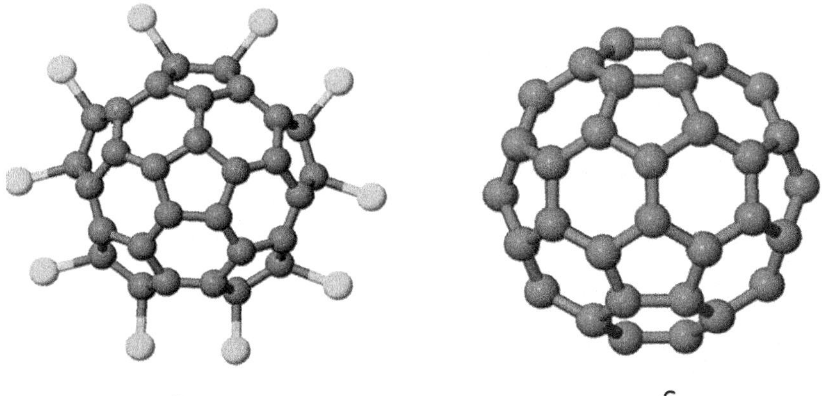

C_{50} C_{60}

FIGURE 4.3 Morphology of fullerene.

4.3.3 GRAPHENE

Graphene (Fig. 4.4) is another allotrope of carbon, composed of one/single layer of carbon atoms organized in a two-dimensional honeycomb mesh. The name Graphene is a combination of two words, that is, graphite and ene, which clearly reflects the meaning that the graphite (an allotrope of carbon) is composed of stacked layers of graphene. Graphene is composed of carbon atoms with sp2 hybridization, secured in a firm hexagonal frame to look like a flat plane. Graphene was first introduced by Canadian physicist Wallace in 2004. Graphene is also a principal structural element of CNTs and fullerenes. The bond length between adjoining atoms of graphene's carbon is about 0.142 nm. The interplanar space between the layers of graphene is about 0.335 nm, which basically keeps them stacked together. Graphene has exceptionally good physical characteristics, such as thermal stability, mechanical rigidity, and electrical conductivity in comparison to their three-dimensional materials (Meijiao et al., 2013; Zhang et al., 2019).

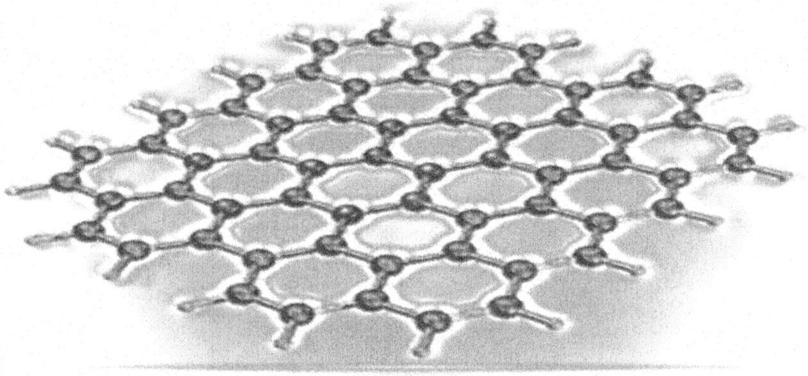

FIGURE 4.4 Morphology of graphene-based nanostructure.

4.3.4 CARBON QUANTUM DOTS

Xu et al. (2004) accidentally discovered a novel class of carbon nanoparticle with unique fluorescent property during the purification of SWCNTs. Later on, Sun et al., in 2006, had renamed these glowing materials as CQDs, and the size of these particles was measured to be less than 10 nm (Fig. 4.5). Due to their diverse properties, such as gathering optical light and emitting multi-color adjusted emission, CQDs were potentially explored in the fields of bioimaging, catalysis applications, photodegradation, etc. (Grobert, 2007). CQDs can be synthesized by using numerous chemical precursors, such as thiourea, ammonium citrate, citric acid, benzene, phytic acid phenylenediamine, and ethylene glycol. Researchers reported many low-energy utilizing techniques for the preparation of CDQs, such as hydrothermal, solvothermal, ultrasonication, electrochemical, microwave-assisted pyrolysis, and chemical oxidation. (Zuo et al., 2015; Georgakilas et al., 2015). The CDQs are often biocompatible also exhibit other valuable properties, such as high aqueous solubility, large surface area, affinity for biomaterial, low toxicity, inexpensive. They contain abundant surface functional groups. CDQs are also explored as gas biosensors and find their application in the detection of heavy metals by using sensors (Niu et al., 2018; Bezzon et al., 2019).

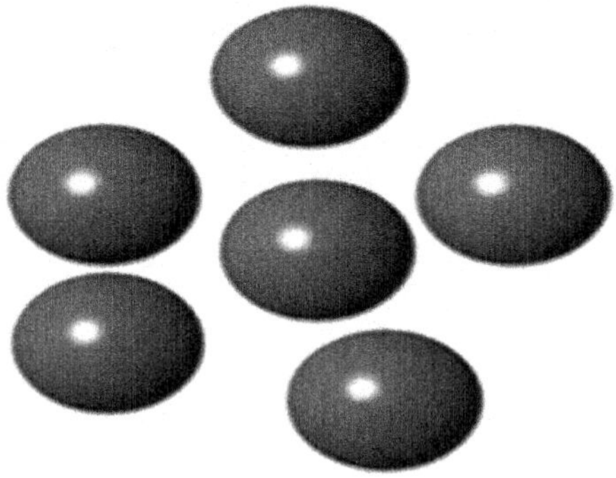

FIGURE 4.5 Structure of carbon quantum dots (CDQ).

4.3.5 NANO HORNS

Nanohorns were first discovered in 1999 by Iijima while forming CNT. Nanohorns are tubule-like or cone-like structures composed of single-layered graphene sheets. They generally occur in large globular shape, and their sizes range from 80 to 100 nm in diameter and very much resembe the dahlia flowers. The cone of single nanohorns is about 1–2 nm (dia) from tips and 4–5 nm (dia) from the base (Karousis et al., 2016) (Fig. 4.6).

4.4 PROPERTIES OF CARBON-BASED NANOMATERIALS

The properties of carbon-based nanomaterials, such as chirality, length, size, and the number of layers of depend on the alterations in the synthesis method, pressure, temperature, catalyst used, and electron field. The diameter of carbon nanostructures is an important factor for exploring their applications and properties. The size of SWNT is strongly associated to its formation method (i.e., compound of σ and α bonds and electron orbital re-hybridization). Alteration in bonding structure may encourage some properties of the SWCNTs, such as chemical, electronic, mechanical, optical, thermal, and elastic properties. All these properties depend on the size of nanotubes (diameter) and are complemented by their capillary behavior, in terms of ecological and agricultural applications. The thin diameter of the inner surface of nanotubes makes them unique to be explored in separation, novel molding, and size exclusion processes. The combination of long tubules and narrow diameters also indicate the extraordinary aspect ratios in nanotubular structures. The properties, such as large surface area and volume ratio differentiates these nanomaterials from their counter-microscale. The ratio of ΔG-surface/ΔG-volume accelerates, where ΔG signifies the change in free energy among the nanoscale structure and bulk materials (Jost et al., 2004; Fornasiero et al., 2008).

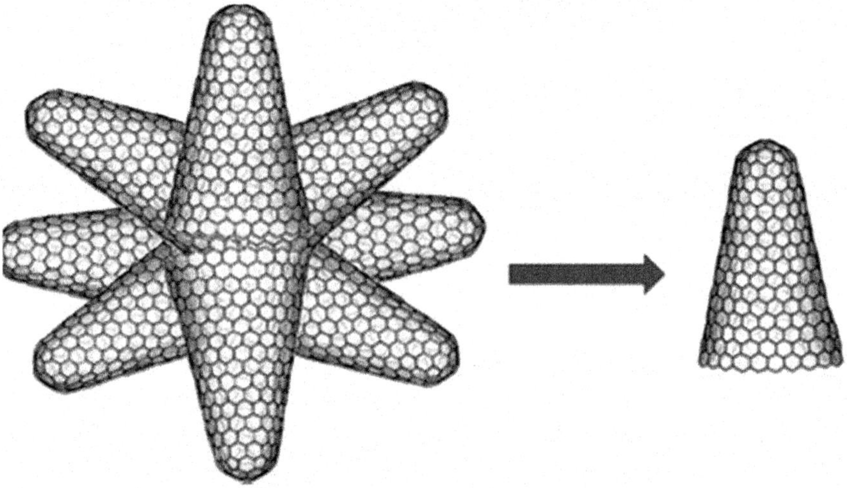

FIGURE 4.6 Structure of carbon-based nanohorns.

The surface area, size, and shape of carbon nanomaterials are extremely dependent on the solvent aggregation and chemical state. The electron characteristics, thermal characters, aggregation behavior, mechanical strength, and physical–chemical properties of the nanomaterials are altered by impurities adsorbed to the surface of nanomaterials (Jost et al., 2004; Fornasiero et al., 2008). The carbon-based nanomaterials' bonding configuration confers exceptional optical, conductive, and thermal properties which are very much useful in electronic industries. These new electronic properties also play an important role in ecological sensing devices and in novel ecological remediation practices for persistent organics. Highly tunable bonds gap, stability, low ionization capacity, high current-carrying potential, and efficient field emission properties are extremely the cited properties of SWCNTs and are also associated with diameter, chirality, length, and the number of concentric tubules. Many hypothetical along with practical works state that bond gaps are reliant on diameter and chirality of the nanotubes. Armchair structures signified by (n,n) tubes are metallic and self-governed by tube arc and diameter. In nanotubes, the carbon atoms are patterned in zigzag or helical forms, which act as semiconductors. The ionization potential of SWCNTs is obligatory to stimulate an electron from the ground state to the excited state (Patel et al., 2018; Nasir et al., 2018).

Molecular interaction and sorption: The carbon-based nanomaterials are usually steady with physicochemical properties and the sorption and molecular interaction are due to concepts, including hydrophobicity, adsorption, and electrostatics. The molecular theories can explain the chemical and physical processes at nanoscale. The potential energies produced by the interaction between carbon bonds (for both van der Waals attractive forces and Pauli repulsion) are initiated due to overlapping of electron orbitals at short separation distances. The hydrophobicity and capillarity contribute to the adsorption behavior and orientation of sorbates in microporous carbons. The adsorption studies of carbon nano also demonstrate the properties, such as high adsorption capacity, rapid equilibrium rates, minimal hysteresis in dispersion, and low sensitivity to pH. Hydrophobicity mainly states the strength between water–particle interaction and water–water interactions to particle–particle interaction. The hydrophobic properties of CNTs and C-60 encourage drying and removal of interstitial water over low scale (length) and stimulate their accumulation state.

The previous works states that the capillary forces present in nanotubes are extremely strong, and taking molecules from vapor and liquid phases into the molecular "tubes" by means of van-der-waals attractive forces and dipolar (persuaded dipole interactions with polar molecules) (Patel et al., 2018; Nasir et al., 2018).

In ecological studies, the adsorptive capacity of carbon nano has comprehensive inferences for the removal of contaminant or pollutants. Adsorption of environmental pollutants to sorbents, such as clay, NOM, and activated carbon plays a vital role in engineered and natural ecology. The sorption capacity of carbon nano as sorbents are restricted by activated energy of sorption bonds, thickness of surface-active sites, slow kinetics, and nonequilibrium of sorption in heterogeneous systems. The huge sizes of traditional sorbents have some limitations such as low porosity which complicates the subsurface remediation. On the other hand, the carbonaceous nanosorbents possess large surface area to volume ratio and precise and equal distributed porosity along with enlarge surface chemistry overcome most of these essential limitations.

4.5 ENVIRONMENTAL POLLUTANTS AND ITS DETECTIONS

Decontaminating soil, air, and water quality represents a crucial task of the century. Pollutants of environment (air, water, and soil) can be classified as inorganic (e.g., phosphates, nitrates, heavy metals, toxic gases, etc.) and synthetic organic compounds (e.g., VOC, endocrine disruptors, volatile gases, weedicides, pesticides, insecticides, herbicides and fertilizers, drugs, pharmaceutical, or forensic substances. Table 4.1 summarizes some typical environmental contaminants of air, water, and soil and all are influencing several injurious effects on human health, even if they are at very low level (Zhang et al, 2011).

With regard to air pollution, two forms of toxic gases are of main concern: VOC (volatile organic compounds) and inorganic gases. The inorganic toxic gases mainly include nitrogen oxide, sulfur oxide, hydrogen sulfide and other sulfur base gases, whereas, VOCs are main contaminants of air, released from the exhaust emission of vehicle, evaporation techniques of oil–gas industries, industries release, oil refining, storage and transportation, pharma waste management, waste material disposal, etc. which can evaporate effortlessly and disperse at room temperature and pressure due to their low boiling point and high vapor pressure in the environment. The SO_2 and NO_2 in the environment are released from sulfur and nitrogen-based impurities from combustion of fossil fuels, respectively have distinct foul odor. The 50% lethal dose (LC50) of NO_2 gas is stated to be 174 ppm for 60 min, whereas the maximum 50% lethal dose (LC50) of SO_2 is 110 ppm on exposure for 30 min. VOCs are mainly defined as any compound of carbon, excluding ammonium carbonate, carbon monoxide, carbonic acid, carbon dioxide, and metallic carbides and carbonates. The release of VOCs in air has led to hazardous ecological glitches, for example, mutagenic, pathogenic, and sometimes very fatal diseases (Zhang et al., 2011; Thamri et al., 2016; Patel et al., 2018; Nasir et al., 2018). In water pollution, numerous f heavy metal and inorganic ions enter the surface of water as well as ground/waste water along with several organic mixtures, such as, herbicides, pesticides, polycyclic aromatic hydrocarbons, endocrine-disrupting compounds (EDCs), pharmaceuticals, personal care products (PPCPs), and dyes which work as contaminants are released into water source.

TABLE 4.1 Examples of Some Pollutants of Environmental.

Pollutant	Due to	Affect
Products of benzene	Components of solvent, hand wash, industrial chemical, gasoline	Water and air (indoor)
Cyclohexane	Components industrial chemical	Air, soil
Toluene	Disinfectant solvent and pharma industrial chemical	Atmospheric air and water
Acetone	Personal care products, solvent, pharma industrial chemical	Air (outdoor/indoor)
Trichloroethylene	Industrial chemical, solvent, household flush	Air, water
NO_2	Fuel, fossil burning, metal casting, industrial exhaust	Air and atmosphere
Sulfur dioxide	Combustion procedures for sulfur-containing minerals and industry outlets	Air
Hydrogen sulfide	Petrochemical and pharma industry, hospital waste.	Atmosphere, water
CO	Automobile exhaust, insufficient combustion of carbonaceous materials, petroleum and pharma Industries	Atmosphere and air
Ammonia	Hospital/medical, gas and chemical effluents of industry, pharma industry outlets	Air, soil, atmosphere
Cobalt	Domestic waste, Industrial waste, pharma waste, nuclear power waste and mineral mining,	Soil, water, Atmosphere,
Cd (cadmium)	Domestic waste, industrial effluent, Pharma- industries, mining of minerals	Soil, water, atmosphere
Mercury	Medical/hospital waste, aquaculture, industrial effluent waste, pharma industries	Atmosphere, water, soil
Pb (Lead)	Petroleum industries waste, pharma industry effluent, vehicle exhaust, mineral mining,	Air, water, soil, atmosphere
Nickel	Pharma and other industrial effluent waste, mineral mining waster	Water, soil
Chromium	Pharma, industrial waste, fossil fuel burning	Air, atmosphere, water
Phenol	Domestic cleaner, solvent, industrial chemical, medicinal waste	Soil, water
Bisphenol	Pharma/hospital waste, industrial waste chemical	Water, soil
2,4-Dichlorophenol	Disinfectant, insecticides, pesticide and pharmaceutical effluent and intermediate waste, solvents	Soil, water

TABLE 4.1 *(Continued)*

Pollutant	Due to	Affect
Perfluorooctanoic acid	Pharma waste, industrial emulsifier, surfactant,	Water, atmosphere
Atrazine	Insecticide, pesticide	Water, atmosphere
Phthalate	Pharma waste, plasticizer waste	Water, atmosphere
Methylene blue dye	Hospital waste, dyeing, chemical analysis and detection, medical	Atmosphere, water
Methyl orange dye	Hospital waste, dyeing, chemical analysis and detection, medical	Water, atmosphere
Rhodamine B dye	Hospital waste, chemical analysis, dyeing, and detection, medical, cosmetics industry	Water, atmosphere
Rhodamine 6G dye	Hospital waste, dyeing, medical, chemical analysis and detection	Water, atmosphere
Parathion	Agricultural pesticides	Soil, water
Malathion	Agricultural pesticides	Soil, water
Diazinon	Agricultural pesticides	Soil, water
Azinphos-methyl	Agricultural pesticides	Soil, water
Drugs for TB and Cancer	Pharma industries, hospital waste	Soil, water
Crude oil	Petroleum industries and vehicle waste	Soil, water, atmosphere
Biomolecules	Human waste, animal waste, microbial waste, agricultural waste	Soil, water, air

Water pollution was further characterized into six types by The United States Environmental Protection Agency (EPA), that is, biodegradable waste, sediment, plant nutrients, heat, radioactive pollutants and hazardous and toxic chemicals. Thus, contaminated water has key components, such as organic toxins, industrial effluents or discharges, pathogens, pharma industrial waste containing different types anions, along with heavy metals, etc. cannot be naturally degraded in the environment and tend to alter the natural properties of water (Liu et al., 2018; Zhang et al., 2011). The heavy metal ions, such as Zn^{2+}, Cu^{2+}, Cd^{2+}, Co^{2+}, and $Pb^{2+}$are extremely toxic even at traces to the humans, plants, aquatic-life, animals, birds, and the earth. The EPA and WHO have fixed the minimal/maximal acceptable contact or exposure limits of these toxicants in water (Xuan and Park, 2018; Philips et al., 2012). Another pollutant named as EDCs is very harmful and can interrupt the hormonal activities via interaction to hormonal receptors which mainly affects animals and mammals' reproductive systems. Furthermore, consequently, even low-level exposure of EDCs may cause diseases like cancer and diabetes.

In addition, PPCPs are another group of artificial organic compounds that have potential to disrupt the function of the endocrine system at trace level. Other than that, artificial dyes utilized in the bakery, textiles, food, paper, cosmetic industries, etc., can also cause hazardous effects on environment along with health hazard to humans when released into water source (Zhang et al., 2011). Pharmaceutical drugs and products, such as anti-spasm, antibiotics, antiseptics, personal care products, pain killers, disinfectants, toxoids, hormonal drugs, and muscle relaxants, etc., are often present in domestic wastewater, ponds, lakes, rivers, and other water bodies. They really cause adverse effects on humans as well as on the aquatic environment at very low rate or concentration, but lasting ingestion of these compounds may be very dangerous (Chen et al., 2018).

The pollution of Soil is mainly triggered when harmful compounds are entering the natural soil environment from the waste. The main contaminants of soil are heavy metals which are existing naturally in soil but barely at toxic stages. The key sources of pollutants in soil are form landfilling, insecticides, mining, manufacturing industrial effluent sites, industrial wastes, such as batteries, paint residues, pesticides, electrical wastes, etc., agricultural and municipal waste. Heavy metals are one of the most stimulating pollutants of soil because of their nondegradability nature. They stay in this environment for long time except selenium and mercury (they can be transformed and volatilized by microorganisms). In the case of soil contamination, OPP (organophosphate pesticides) which includes parathion, methyl parathion, malathion, diazinon, phosmet, triazophos, azinphos-methyl, oxydemeton, tetrachlorvinphos, dichlorvos, chlorpyrifos have major contribution. The OPP acquire toxicity from their ability to inhibit cholinesterase which mainly causes neurotoxicity, and also has the capacity to cause interruption to the endocrine system of the organisms. This is the known fact that sometimes pesticides are transformed in harmful compounds in the environment through biological, physical, and chemical interaction procedures (Ibrahim et al., 2016).

All these contaminants of soil, air, and water, such as pesticides, volatile organic compounds, toxic and noxious inorganic gases, personal care products, heavy metals, dyes, pharmaceutical products, endocrine-disrupting chemicals, etc. are enhancing environmental toxicants and also disturbing the equilibrium of the earth at alarming rate. Thus, their adsorption, detection, and elimination are required at great level, and for the same purpose, several carbon-based-nanomaterials are been explored and verified to be ideal, which includes graphene, carbon nanotubes, carbon dots, mesoporous carbon, diamond, etc. (Zhang et al., 2011; Ibrahim et al., 2016).

The detection and monitoring of environmental pollutants, needs devices that must be rapid in working, error-free, sensitive, stable, effective, effortless, and selective. The sensitive and rapid detection along with effective elimination of an alarming amount of persistent and developing pollutants of environment is a major task. Keeping the current scenario in loop, these problems can be better resolved by employing nanotechnology in addition to traditional approaches. Due to exceptional features of carbon nanomaterials, such as size, high surface area, photoelectronic, adsorption, and photocatalytic properties, they have developed to be significant resources in the analytical detection and remediation of pollutants of the environment.

For the detection purpose of these contaminants, recently, numerous sensors and biosensors based on carbon nanoparticles have been explored and they are further categorized on the basis of their type of pollutant detection.

4.5.1 TYPES OF SENSORS BASED ON CARBON NANOMATERIALS

The elimination of environmental contaminants is the primary need, the presence of these contaminants is to be determined. The vigorous and rapid sensors have been explored by various researchers to recognize contaminants at the molecular level, and they can improve the ecological safety. These sensors are coming in the loop when continuous monitoring devices are required to detect these pollutants at the trace level. All sensors are based on deferent principle based on their operating system. Following are the sensors based on carbon nanomaterials (Bezzon et al., 2019; Zhang et al., 2011)

1. Biosensors
2. Chemical sensors
3. Pressor sensors
4. Strain sensors
5. Flow sensors
6. Mass sensors
7. Temperature sensors

1. Biosensors: These sensors, also called nanobiosensors, are mainly used to monitor biological contaminants, that is, to recognize biomolecules. These sensors are mainly composed of biological constituents, such as oligo- or polynucleotide, proteins (e.g., enzymes, antibodies, cell receptors, etc.), microorganisms, or even entire tissues (Bezzon et al., 2019; Zhang et al., 2011).The combination of CNTs with biomolecules has given rise to hybridized systems, in which CNTs are working as nanolevel electronic elements (e.g., CNT-Field-effect transistors), as electrode elements (e.g., enzyme electrodes), and as attachment site or space, upon which biomolecules can be attached. Following are subtype of biosensor based on carbon nanomaterial (Yi et al., 2005; Bezzon et al., 2019).
    a. Amperometric biosensor
    b. Potentiometric transducers
    c. Chronoamperometric biosensor
    d. Conductimetric biosensor
    e. Optical biosensor
    f. Piezoelectric biosensor
        a. Amperometric biosensor: The principle of these biosensors is working on the measurement of current produced when a constant potential is applied. The current measured is correlated with the reduction/oxidation of an electrochemical species in terms of the percentage at which it is produced or consumed by a biological component which was attached at the surface of electrode.
        b. Potentiometric transducers: These sensors are mainly measuring the changes in potential, which is directly proportional to the concentration of the analyte present in the sample.

c. Chronoamperometric biosensors: In this type of sensors, enzymes are required. These enzymes do short-term reaction which is required to proceed before applying the potential step. In these types of biosensors, biocomponents are kept between two closely kept electrodes. The main principle of these type of biosensor is that the overall change occurs in reaction in the conductivity of solution induced by the production or consumption of ionic species.
d. Optical biosensors: The principle of these biosensors is based on the change in the optical phenomena, for instance, fluorescence, absorption, scattering, luminescence, refractive index, etc., which are formed when light is reflected on the surface of the sensor.
e. Piezoelectric transducer: These sensors convert one form of energy to another form by taking the benefit of piezoelectric characteristics of certain crystals or materials, such as QCM (Quartz Crystal Microbalance).

2. Chemical sensors: Chemical sensors mainly detect and screens the VOCs. Currently, the determination of gas samples (chemical components) is normally done with gas chromatography-mass spectrometry (GC-MS) (Jia et al., 2005), this technique, though extremely sensitive and selective, but at the same time, it has limitations too, such as highly costly, need skilled users and of limited portability. On the other hand, the CNTs, are useful constituents in chemical sensing due to their sensitiveness of their electrical conductance to the chemical analytes present in the sample of gas (Frazier et al., 2014). This is a prolific direction to enhancing the sensitivity and selectivity of these materials to a specific analyte. These sensors can work with metals, polymers, or small molecules in covalent or noncovalent manner. The application of these materials along with different substrates into devices can produce simple, rapid, potable, and low-power sensors, which may be capable of detecting wide range of contaminants or vapors at parts-per-million (ppm) concentrations. To prepare these types of sensors, solid composites of SWCNTs are mainly employed in combination with small molecules (PENCILs (**P**rocess **E**nhanced **N**ano**C**arbon for **I**ntegrated **L**ogic) as the sensing material, and graphite as electrodes. The DRAFT (**D**eposition of **R**esistors with **A**brasion **F**abrication **T**echnique) technique is used to deposit these materials on different type of substrates. It is already reported that

the chemical reactivity of SWCNTs can be significantly improved by providing strain on SWCNTs (Lobez et al., 2014). Although, the deposition of solid-sensing materials on the electrode surface makes carbon nano-based sensors simple, solvent-free, and easily accessible, but along with application, some limitations also exist. The thickness, location, size, and distribution of these sensors may result in the formation of conductive carbon "film" which is difficult to control and is limited to the substrate (e.g., surface roughness and distribution of cellulose fibers on the surface of paper) (Frazier et al., 2014). The experiment-based studies mention that mechanical alteration can alter the electrical conductivity of semiconducting and metallic CNTs (Wang et al., 2008). This property shows that the use of CNTs can develop highly sensitive electromechanical sensors.

3. Pressure sensors: The CNTs–and graphene are the most appropriate nanomaterials to be used in pressure-sensing technology. These pressure sensors are based on capacitive, inductive, and piezoresistive phenomena, and can be employed for regulation and screening of the pressure of different types practical applications (Karimov et al., 2011). Among all the three sensors (i.e., Piezoresistive pressure sensors, Inductive pressure sensors, and Capacitive pressure sensors), the piezoresistive pressure sensors demonstrate outstanding potential for the detection of real-time pollutants and are cost-effective, simple device structure, and easy signal gathering. Highly sensitive and flexible piezoresistive pressure sensors are based on micropatterned films with the coating of carbon nanotubes (Ali et al., 2018).
4. Strain sensors: Generally, the applications of nanocarbons in mechanical sensors are associated with the measurement of strain. As CNTs gain special attention as they demonstrate both piezoelectric and piezoresistive behaviors under pressure. Because of the above properties, CNTs are explored to be used in sensors, and usually, they are determining the variation in electrical resistance or voltage when exposed to external action. The strain sensors are mainly focused on piezoresistive effects, and electromechanical effect which is shown by most of the composite matrices and CNT arrangements. Piezoresistive strain behavior of CNTs has been thoroughly exploited in the form of sensors. The strain sensors are mainly explored in environmental pollutant detections. Thin film piezoresistive strain sensors in combination with single-walled and multiwalled carbon nanotubes serve as good substitutes for developing novel sensors and show promising potential. Carbon nanotube-based strain sensors

surrounded by structural materials can work as both multidirectional and multifunctional sensors with high strain resolution on nanoscale. The electromechanical properties of carbon nanotube as strain sensors display an exceptional application compared with the existing strain sensors due to a blend of outstanding electrical property and increased elastic moduli (Novoselov, 2004).

5. Flow sensors: These sensors are used for the quantification of the flow rate of a mobile gas and liquid. A flow sensor based on CNT is based on the formation of voltage/current in a bundle of SWCNTs when it comes in contact with the moving gas or liquid.
6. Mass sensing: The mechanism of these sensors is depending on the resonant frequency of a CNT–resonator when it is exposed to alteration in external loading or attached mass. Micromechanical resonators (microcantilevers) are based on the mechanism that the resonant frequency of the cantilever depends on the inverse square root of the cantilever mass. Hence, an alteration in the mass of the resonator is measured as peak in the frequency of the resonant. The small size and unique mechanism of the CNTs make these nanostructures capable candidate for substituting cantilever structures in a mass sensor. The limitations of these sensors are difficulty in measuring added variation in the resonant frequency, because of the added mass.
7. Temperature sensors: These sensors are mainly sensitive to detection of type and temperature of the gas in the environment. These are triple electrode-based sensor, have one cathode along with two extracting and accumulating electrodes, where CNTs are vertically aligned and are disposed on top of the cathode, and then the power is applied in between the electrodes. In this system, CNTs play a vital role to introduce electrical breakdown, subsequently, this type of systems do not change adsorption or desorption of gas molecules. Temperature of the gas affects the grade of ionization and released current that stimulates the accumulating current. This sensor can sense temperature range from 20°C to 110°C with a sensitivity of 22.72 μA/°C (N_2) and 4.74 μA/°C (air) both at 110°C (Song et al., 2017).

4.5.2 EXAMPLES OF DIFFERENT CARBON-BASED SENSORS FOR THE DETECTION OF POLLUTANTS

The application of carbon-based nanostructures as a gas sensor increased because of their structure and morphology, which allow the detection and quantification of gas. The CNTs have hollow structure and enhanced surface/vol ratio. This makes them an ideal model for the adsorption of gas molecules. CNTs can be applied in many systems in the form of gas sensor, such as in capacitance, ionization, sorption, resonance frequency shift, etc. Generally, these systems sensors are based on change in applied voltage/current concerning the quantity of adsorbed gas molecules onto the surface provided, and the observations are formed by electrical signals produced through the interaction between the sensor and gas molecules. As a gas sensor, the carbonaceous material play a very important role in the determination and detection of polluting gases, especially carbon monoxide (CO), hydrogen sulfide (H_2S), nitrogen oxide (NO_x), methane (CH_4), and ammonia (NH_3). GPN, GO, CNTs, and fullerene are some key carbon allotropic forms used in this kind of sensor.

Ricciardella et al. (2017) demonstrated that the mechanical exfoliation (ME) method can produce less-defective material of GOP, and a nominal level of defectiveness brings rapid interface during the contact time toward the gas.

Among all gases, the NO_2 gas is very harmful for human's respiratory system, if the exposure concentration is above the threshold limit. This is also accountable for acid rain. According to UEPA (US Environmental Protection Agency), the recommended yearly exposure concentration of NO_2 is about 53 ppb. Therefore, the detection of such low concentration of NO_2 gas requires a very sensitive system (sensor). On comparison with different methods, Seekaew et al. (2017) proved that the performance of Graphene(bilayered) gas sensor by using CVD method has high selectivity toward the NO_2 in contrast to ethanol (C_2H_5OH), CO_2, NH_3, H_2, and CO.

Other group of researchers (Wu et al., 2018) used hydrophobic rGO (reduced graphene oxide) for the analysis and detection of NO_2. It is well known that graphene has tremendously hydrophilic functional groups, such as carboxylic, epoxide, and hydroxyl, which decrease their conductivity and make these GO-based gas sensors very sensitive to humidity. The rise in defect area and surface area may expand gas adsorption and sensibility. These NO_2 sensor demonstrates high sensitivity at about 25.5 ppm^{-1} and a prominent low limit of detection at about 9.1 ppb.

Furthermore, in gases, NH_3 is required to be monitored due to its property of causing water (surface) pollution, acidification, and eutrophication of soil. NH_3 can be detected by highly sensitive NH_3 sensor by using laser ablation technique which is prepared by the combination of GPN and vanadium pentoxide (V_2O_5) (Kodu et al., 2017). Other than NO_2, many more gases exist which may act as air pollutant and are injurious to health, such as H_2S, which can be fatal due to loss of breathing. Graphene-based H_2S gas sensor is showing highly sensitive response and fast (~1s) along with rapid recovery time (~20s), and it can detect CH_4, CO_2, N_2, O_2 below the recommended level (100 ppm).

A constituent of natural gas, named as CH_4 can also be very harmful, it may cause suffocation and explosions when present in the environment exceeding the permitted level. Chen et al. (2018), formed a sensor based on Li^+ incapacitated CNTs which produces a CNT-Li^+ thin film, while exposure of CH_4 gas to CNT-Li^+, the Li^+ strengthen up and induces dipole interaction. The alteration on hybridization state of CNT (from sp^2 to sp^3) helps to link CH_4 to its p-bonds, which affect the system's resistance. With reference to above, this sensor demonstrates a high sensitivity to CH_4 (14.5% at 500 ppm), which makes this device an excellent choice.

Even though H_2 is not a polluting agent, its properties, such as high flammability making worries about its storage, production, and transportation. A sensor formed by graphene-coated CNTs (GCNTs) and adorned with nanoparticles of Pt were used for the detection of H_2. It is well known that Graphene–CNTs are used to promote a decent electrical conductivity along with high surface area, they are also good to be used in sensors. Though, these carbon nanomaterials are virtually insensitive while exposure to H_2 So, Pt nanos have been employed due to their high catalytic activity toward molecule of H_2 gas. GCNTs-based sensors are prepared using single-furnace catalytic technique for the detection of H_2 gas. Gas-sensing sensor displayed a steady and reproducible response to hydrogen and as a conclusion it is noted that the rise in temperature leads to reduced sensitivity (Jaidev et al., 2018).

The carbonaceous-gas sensor was also used to detect C_2H_5OH, O_2, and infrequent gases. The C_{60} FULLERENE–zinc oxide tetrapod materials were explored by some researchers for the detection of ethanol gas (Smazna et al., 2018).

Algadri et al. (2018) made sensor for sensing H_2 gas by applying dielectrophoresis technique (sensor with glass substrate and Pd electrodes), they produced multiwalled CNTs (MWCNTs) by using a combination of graphite and ferrocene, which were microwaved (heated) and further treated at room temperature with nitric acid. As a result, MWCNTs were deposited onto a substrate (glass) between the electrodes of Pd. As a result, this device demonstrated good sensitivity (max sensitivity about 239%) to the hydrogen gas when come in contact with the mixture of H_2-N_2 (20–1000 ppm) at 0.05 V. This technique also suggests that MWCNTs enhance the sensitivity of gas sensors by forming a passage for the current to move faster.

C_{20} fullerene are also being used as a gas sensor, specifically to detect rare and diatomic gas. Zhao et al. (2017) explored nonequilibrium Green's function formalism combined with density functional theory (DFT) for the analysis of steadiest adsorption structure and energy along with transport characteristics on C_{20} molecular intersections with diatomic gas. Outcomes of this displayed that O_2 and NO could be sensed selectively, whereas C_{20} also has limitation for sensing CO. Their study proved that C_{20}-fullerene demonstrate an excellent capacity to be explored as nanosensors for the detection of gas.

In addition to the detection and quantification of the above-mentioned gases, nanocarbon sensors can be explored to evaluate the temperature of ionized gases. Song et al. (2018) reported that these sensors can sense or determine temperature ranges of 20°C–110°C with a sensitivity of 4.74 μA/°C (air) and 22.72 μA/°C (N_2) (both at 110°C).

Carbon-based sensors also show countless viability to be utilized as pollutant detector. The nitrogen cycle is an essential component of many living beings, and nitrogen is derived from nitrates as important nutrient source and enters in water cycle too. Besides screening of NO_2 and other air gases, detection of nitrogen in water is a recent apprehension for human health. Water eutrophication is an ecological problem which reduces the oxygen rate, leading to death of fishes and other aquatic faunas. Alahi et al. (2018) reported about low-cost graphene–nitrate-sensor. In comparison to other sensor, this sensor is low cost, provides real-time response, and needs nominal time for sample preparation along with error value below 5%. Results obtained was confirmed by using UV-spectrometry. Chen et al. (2018) also established a sensor for nitrate, using rGO in combination with benzyl-tri-ethyl-ammonium chloride. This sensor can sense a low limit of 1.1 μg/L in a short fraction of period (2–7 s) and also shows selectivity against interfering ions (Cl^-).

Phenolic compounds are other major environment contaminants that need to be monitored. The boron-doped nanodiamond electrochemical sensors are in use for the detection of monophenols and bi-phenols. In this sensor, nanodiamonds were released onto GCEs (glassy carbon electrodes) and used for the detection of phenolic compounds. Little alteration with this method also helped in the detection of aromatic phenolic compound called hydroquinone (HQ) (Chen et al., 2017).

For the determination of organic and inorganic contaminants of water, fullerene-based nanocomposites were explored. Wei et al. (2018) proposed photocatalytic sensor, for the detection of organic compounds, which was made up of Au-TiO_2 nanoparticles and C_{60} fullerene (Au-TiO_2-C_{60}). In this,

TiO_2 was working as a regular catalyst material, Au–np (nanoparticles) was working as co-catalyst and plasmonic sensor, whereas C_{60} was explored as an accelerator for redox reaction. The Au nanoplasmonic peak (in redox reaction) was measured using surface plasmon spectroscopy. The frequency of plasmonic waves indicates the rate of reaction and was used to estimate the concentration of organic compounds in water. Due to its great accuracy and susceptibility, plasmonic sensor are observed to be good solution for the determination of water quality.

When heavy metals accumulate in the environment by food chains may cause many health issues, such as kidney and liver malfunctioning or failure to humans. Liu et al. (2018) formed graphene-based sensor for sensing of ions, such as Pb^{2+}, Cu^{2+}, and Cd^{2+}, using anodic stripping voltammetry technique. In this process, highly conductive electrodes coated with graphene films were employed for the analysis of heavy metals. Observations exhibited by this sensor are ranging from 5 to 400 μg/L and detection limit of 1.0 μg/L for Pd ion, 5.0 μg/L for Cu ion, and 0.5 μg/L for Cd ion.

A sensor based on reduced graphene oxide–CNT was explored by Xuan and Park (2018) along with gold substrate for the detection of metallic ions. In this process, the electrode was made up of bismuth (Bi) (environmentally). The sensor was based on silicon-coated biscuits coated with a film of polyimide and a layer of reduced graphene–CNT. The sensor's efficiency was measured by cyclic voltammetry technique for the detection of Pb ions and Cd ions. As an observation, this sensor showed its better significance in terms of sensitivity, surface area of working electrode, stability, and response time in comparison to other traditional sensors.

The GPN–quantum dots (QDs) were also explored in the form of fluorescence sensor for Pb^{2+}. In this method, carbon nanoparticles were immobilized with the mixture of amine-modified and 1-ethyl-3-(3-dimethyl-aminopropyl) carbodiimide. The combination of Graphene–QD and nanoparticles led to fluorescence reducing and retrieval. As a result, this sensor presented a good selectivity for Pb^{2+} than other metal ions, specifically Hg^{2+}.

Menacer et al. (2018) formed and reported about CNT-based sensor for the detection of acetone. In this SWCNT was explored as a sensor due to its extraordinary physical properties, such as length, diameter, chirality, and molecular arrangement onto the sensor surface. As a result, it was measured that the natural frequency of CNT resonator decreases with the increase in the number of acetone molecules and frequency increases with decrease of CNT dimension along with high chirality. This method is highly effective, sensitive, and inexpensive compared with other methods.

Lezi (2017) explored Graphene-Nafion-based sensor for the detection of caffeine. They evaluated the results with a conventional HPLC (high-performance liquid chromatography) method. The limit of detection was 0.021 µmol, with standard deviation of 2%. Study was conducted for other samples also such as Citric acid, maltose, sucrose, Na^+, K^+, Ca^{2+} at 0.1 µmol/L and found that these samples unaffected with the determination of 1.0 µmol/L of caffeine.

The quantity of water vapor or moisture in a gas is known as humidity, which is a critical constituent of industrial and lab processes, pharma industries, agriculture, food storage along with production, chemistry, textile industry, semiconductors. Therefore, in these methods, humidity control system is essential to attain appropriate outcomes. Li et al. (2018), also projected an extremely sensitive sensor for humidity based on graphene oxide with consistently spread MWCNTs, where MWCNTs were working as detecting agents. As we know van-der-Waals interactions can hamper the competence of carbonaceous material, the grapheme oxide was used in the mixture to promote the dispersion of MWCNTs. The electrodes were made up of a Si-wafer coated with silicon dioxide (SiO_2) and Au/Ti coats, and then the detecting film of MWCNT-GO was put onto the surface of the substrate. The researcher also reported that the sensor with the MWCNT/GO film has 13% more sensibility compared with the sensor with graphene oxide alone.

Graphene-based sensors are explored for the detection of humidity of the gases. Leng et al. (2018) constructed humidity sensor by using modified diamine–graphene oxide compounded with Nafion-polymer to form hybrid film. The result is observed as 10% lasting stability and linearity than the other counter methods.

The other widely in use sensor for humidity is associated with QMC because of its high digital frequency yield, low cost, and high resolution. This sensing system comprises of QCM coated with nanodiamond–MWCNTs hybrid film which is highly sensitive to humidity and helps in the detection of humidity (Yu et al., 2018).

Ding et al. (2018) applied QCM-based humidity sensor in association with C_{60}/GO. Graphene oxide (GO) has hydrophilic functional groups which expand its sensibility. While coming in contact with molecules of water the viscosity of (GO films increases, which may damage the sensor. Therefore, in order to reduce GO's aggregation, C_{60} was added which forms hydrophobic separation coatings between GO film sheets. Deposition of GO onto QCM electrode was achieved using drop-casting technique. After the comparative assessment, it was exposed that QCM humidity sensor coated with C_{60}/GO- was better than QMC coated with GO due to its stability, dynamic response, and recovery properties.

For humidity in gaseous form, Saha and Das (2018) applied fullerene-C_{60} along with nanopores of γ-alumina (Al_2O_3) thick films as a humidity sensor. AFM (atomic force microscopy) and SEM (scanning electron microscope) were used to check the porosity and surface area. The response, sensitivity, recovery, and hysteresis were performed as electrical characterization. Results revealed that the modified version of C_{60}-γ-Al_2O_3 sensor was capable of determining the low range of humidity in a gaseous atmosphere with low pressure and was extremely sensitive too.

Carbon nanostructures exhibit both piezoelectric and piezoresistive behaviors under strain. Because of these properties, CNTs were explored for developing strain sensors. Generally, a change in the electrical voltage of CNT composites are measured, subjected to peripheral action. The utmost substantial development in strain sensors are based on piezoresistive effects, and that is due to stronger electromechanical effect of most CNT and composite matrices. Yu et al. (2018) established a highly sensitive strain sensor. This sensor was mainly based on conductive–SBS (styrene–butadiene–styrene)–CNT fiber prepared by wet-spinning method. Ma et al. (2017) explored a method for the detection of real-time strain/stress of composites by evaluating piezoresistivity and electrochemical impedance changes in a CNT yarn grid of whole composite, without disturbing its mechanical properties. Arif et al. (2018) confirmed that the gauge factor has strong sensitivity to the strain rate while using CNTs–polydimethylsiloxane (PDMS) composite sensors. It also confirms that the gauge factor changes (increases) under higher strain rate. CNTs also improve the poling proficiency of piezoelectric composite-based sensors when used as additives. Zhao et al. (2016), stated the efficiency of CNTs in cement–sand-based piezoelectric composites, and are ranging from 0 to 0.9 vol.% of nanotubes.

The pressure sensors are relying upon piezoelectric and piezoresistive behaviors of the CNT-based composites, defining resistance or voltage across the sensor and regulating the response for the purpose of external stimulus. The flexibility of sensors is generally accomplished by PDMS (elastomer) which is used as structural material. These sensors are mainly relying on results that results on the alteration in the conductivity of CNT when exposed to external pressure. Gao et al. (2018) prepared a cost-effective wearable CNT–PDMS sensors by utilizing a technique called sand paper molding. These sensors can detect pressure, ranging from 5 to 50 kPa with stimuli–response period of 0.2 s along with stability of over 5000 cycles. Similarly, Giffney et al. (2017) used sensor based on MWCNTs–silicone rubber composite as a strain sensor, they demonstrated that the sensor is capable of stretching of strain up to 300% through absorbent of 11% hysteresis for the above said 300% maximum level of strain. Costa and Choi (2017) fabricated another strain sensor with inkjet printing followed by the 2D force sensors (made up of patterned CNTs on PDMS nanocomposite) for measuring the strain.

Other than ecological monitoring, carbon nano-GPN are also explored for making piezoresistive equipment. For the same, Dong et al. (2018) projected an easy approach for constructing piezoresistive sensor, which was based on hierarchical structure of sea sponges and conductive property of compounds of polydopamine-reduced graphene oxide along with A-nanowires for the measurement of pressure. In this, Sea sponge was employed to maintain the sustainably of structure when the sensor was under strain–stress, whereas, silver was working for the improvement of conductivity and strain sensitivity. As a result, its was obtained that this sensor has good reproducibility (more than 7000 loading/unloading cycles), fast response time (less than 0.54 ms), and high sensibility ($0.016\,kPa^{-1}$ at 0–40 kPa).

Drugs are known as one of the important environment contaminants which need to be identified and quantified before elimination. Nanocomposites based on nanodiamond, CNTs, graphene, and fullerene can be employed to detect various drugs, such as drugs used as anticancer drugs (flutamide, chlorambucil, etc.), illicit or illegal drugs (amphetamine, phenylpropanolamine, etc.), drugs for the treatment of tuberculosis (R-cinex, combitol, tranquilizers, pyrazinamide, etc.). Prasad et al. (2017) used C_{60}-based fullerene for improving sensitivity and conductivity of embossed polymeric-micelles and explored for the detection of these drugs. These electrochemical sensors were made by incorporating electroactive C_{60}-monoadduct-micellar (molecularly engraved polymer) attached on ionic liquid-based carbon ceramic electrode (IL-CCE). This method revealed the capacity of fullerene–nanomediator in the signal transduction, and allows it to be used as a selective system for the detection of drugs in aqueous samples. Similarly, Simioni et al. (2017) prepared modified-nanodiamonds (ND) with glassy carbon electrode (ND-GCE) for the determination of pyrazinamide (PZA), one of the utmost important and acceptable drug (antibiotics) for the treatment of tuberculosis.

Similarly, Farias et al. (2017) prepared MWCNT-based sensor for the detection or monitoring of flutamide (anticancer drug), in this, a noninvasive monitoring system for flutamide was prepared and functionalized for the detection of this drug. In this, MWCNT was working with-COOH, =O and -OH, on the glassy carbon electrodes and the reading was measured. MWCNT showed active catalytic capacity to reduce or oxidize the evaluating drug.

Ahmadi et al.'s (2018) theory proved fullerene to be considered as promising carbon nanomaterial to be explored as electroactive sensor. Fullerene has solid interaction with drugs, such as methyldopa, tyramine, enalapril, dextroamphetamine, metoprolol, and tolazoline. Similarly, Parlak and Alver (2017) studied Ge, Al, B, Ga, and Si-doped C_{60}-fullerene as sensor for detecting amantadine. Bashiri et al. (2017) used fullerene-based sensor for the detection of amphetamine (AA). Moradi et al. (2017) also used Si and Ai-doped C_{60}-fullerenes for the detection of drug named phenylpropanolamine (PPA). Rahimi-Nasrabadi et al. (2017) developed a sensor based on CNT-C_{60} in combination with 1-butyl-3-methylimidazolium tetrafluoroborate ionic liquid (IL) for detecting the drug named diazepam.

Ways to control health problems need to monitor biomedical contamination of the environment. For the same, different nanocarbon-based sensors have been employed, such as fullerene, graphene, and CNT–GPN hybrid, etc. For an instance, Sutradhar et al. (2017) also established a novel biosensor based on C_{60}–Au nanocomposite using 3-amino-capto-1,2,4-triazole for nonenzymatic detection method for glucose. These hybrid sensors were also shown as promising substitute for achieving the desired sensing property.

In concern, the Purdey et al. (2017) also developed sensor with organic fluorescent probe attached with nanodiamond and was used for the detection of hydrogen peroxide, labeled peroxy-nanosensor (PNS). Liu et al. (2018) made Quantum Dots-C_{60} fullerene sensor and used it for detecting nucleic acid or RNA–DNA from biomolecules. Li et al. (2017) established a fullerene-based biosensor for sensing ultrasensitive ATP (adenosine tri-phosphate), an important biological molecule accountable in metabolic activities of cells. Many researches also designed carbon-based-sensors for the detection of prostate antigen, L-histidine, proteins, biomarkers, and numerous of other biomolecules. CNT-FET (carbon-based field-effect transistors) was effectively applied as DNA sensors.

NDs along with CNTs could be easily and equally altered to be utilized in the form of nanocomposites and have numerous of biological applications. Properties of NDs, such as stability in biological fluids, optically transparent, increased physical adsorption, etc. make it as an ideal sensor for the detection of biological materials. NDs are also in use as intracellular magnetosensitive nanosensors. Kumar et al. (2017) developed a hybrid graphitized-ND-(GND)-based sensor for detecting urea and it was easy to measure, reproducible, and constant. Similarly, Tran et al. (2017) also stated about the CNT-FET-based DNA sensor for detecting virus named as influenza A virus.

4.6 CONCLUSION

Worldwide health is a concern, particularly in terms of air–water-borne diseases and causative organism (pathogens), and other possible toxic compounds or contaminants existing in the environment. Detection procedures with high sensitivity, selectivity, and rapidity are essential for screening and monitoring these pollutants in contest to safety guidelines of earth. By using diverse signal intensification and background reduction systems combined with carbon nanoparticles in different forms with sensors for detection offer selective and sensitive tools for screening numerous forms of air–soil–water pollutants.

This chapter emphasized on the detection of environmental pollutants and steps of preprocessing along with CNTs. Combination of sensors and CNTs developed as miniaturizations, ecofriendly, multiple detection, and inexpensive technique. The significant benefit comprises rapid consequences in terms of detection values, because these methods monitored changes (increase) in signal rather than targeting the analytes, which has greatly altered the paradigm of detection. These techniques conjugated with nanocarbon will accelerate the potential of existing technology and also offer robustness, specificity, sensitivity, self-cleaning and speediness, to accompaniment of required analysis, substitute the typical counter standards present, and also endorse the approachability of safe environment.

Nanotechnology provided an ultrasensitive detection method for screening various environmental pollutants, in which carbon nanostructure-based sensors are one of the pollutants detecting systems with extraordinary properties. Number of allotropic forms of carbon have been explored as sensors for sensing and eliminating environmental contaminants. Researches well proved the applications of different forms of carbon nanos, that is, fullerenes, ndiamonds, graphene, nanotubes, and hybrid assemblies due to their thermal, electric, chemical, and ano mechanical properties. They also proved that carbon nanostructures also have a high grade of specificity and selectivity when functionalized as detection sensors. Among all, nanodiamond is an excellent example of nanocarbon that has been functionalized as biosensor. On other hand, due to its cage-like morphology, fullerene is highly explored as biocompatible sensor and is useful in sensors which work on interactions (chemical and biological) between analysts and the device.

Nanosensors also exhibit the capacity of sensing microbial pathogens or chemical contaminants at very minute or trace levels, in addition, are very convenient to handle as portable device for real-time monitoring along with on-site. It is inexpensive, requires less labor, rapid, and time saving along with multiple contaminant detecting efficiency. Besides, advancement in nanotechnology also boosts a noble way of designing new techniques, keeping environment and sustainability in mind. The benefits of nanotechnology thus have numerous important roles in maintaining the environmental health well.

KEYWORDS

- **nanocarbon material**
- **carbon-based sensors**
- **environmental pollutants**
- **detection sensors**
- **carbon allotropes**

REFERENCES

Ahmadi, R.; Jalali Sarvestani, M. R.; Sadeghi, B. Computational Study of the Fullerene Effects on the Properties of 16 Different Drugs: A Review. *Int. J. Nano Dimension* **2018,** *9* (4), 325–335.

Alahi, M. E. E.; Nag, A.; Mukhopadhyay, S. C.; Burkitt, L. A Temperature-Compensated Graphene Sensor for Nitrate Monitoring in Real-Time Application. *Sens. Actuat. A Phys.* **2018,** *269*, 79–90.

Algadri, N. A.; Hassan, Z.; Ibrahim, K.; AL-Diabat, A. M. A High-Sensitivity Hydrogen Gas Sensor Based on Carbon Nanotubes Fabricated on Glass Substrate. *J. Electr. Mater.* **2018,** *47* (11), 6671–6680.

Ali, A.; Khan, A.; Karimov, K. S.; Ali, A.; Khan, A. D. Pressure Sensitive Sensors Based on Carbon Nanotubes, Graphene, and Its Composites. *Hindawi J. Nanomater.* **2018,** *18*, 1–12.

Arif, M. F.; Kumar, S.; Gupta, T. K.; Varadarajan, K. M. Strong Linear-Piezoresistive-Response of Carbon Nanostructures Reinforced Hyperelastic Polymer Nanocomposites. *Compos. A Appl. Sci. Manuf.* **2018,** *113*, 141–149.

Balamurugan, R.; Sundarrajan, S.; Ramakrishna, S. Recent Trends in Nanofibrous Membranes and Their Suitability for Air and Water Filtrations. *Membranes* **2011,** *1*, 232–248.

Bashiri, S.; Vessally, E.; Bekhradnia, A.; Hosseinian, A.; Edjlali, L. Utility of Extrinsic [60] Fullerenes as Work Function Type Sensors for Amphetamine Drug Detection: DFT Studies. *Vacuum* **2017,** *136*, 156–162.

Brame, J.; Li, Q.; Alvarez, P. J. J. Nanotechnology-Enabled Water Treatment and Reuse: Emerging Opportunities and Challenges for Developing Countries. *Trends Food Sci. Technol.* **2011,** *22*, 618–624.

Chen, X.; Huang, Z.; Li, J.; Wu, C.; Wang, Z.; Cui, Y. Methane Gas Sensing Behavior of Lithium Ion Doped Carbon Nanotubes Sensor. *Vacuum*, **2018a**, *154*, 120–128.

Chen, X.; Pu, H.; Fu, Z. Real-Time and Selective Detection ofNitrates in Water Using Graphene-Based Field-Effect Transistor Sensors. *Environ. Sci. Nano* **2018b**, *5* (8), 1990–1999.

Costa, T. H.; Choi, J. W. A Flexible Two Dimensional Force Sensor Using PDMS Nanocomposite. *Microelectr. Eng.* **2017,** *174*, 64–69.

Ding, X.; Chen, X.; Chen, X.; Zhao, X.; Li, N. A QCM Humidity Sensor Based on Fullerene/Graphene Oxide Nanocomposites with High Quality Factor. *Sens. Actuat. B Chem.* **2018,** *266*, 534–542.

Dong, X.; Wei, Y.; Chen, S.; Lin, Y.; Liu, L.; Li, J. A Linear and Large-Range Pressure Sensor Based on a Graphene/Silver Nanowires Nanobiocomposites Network and a Hierarchical Structural Sponge. *Compos. Sci. Technol.* **2018,** *155*, 108–116.

Duncan, T. V. Applications of Nanotechnology in Food Packaging and Food Safety: Barrier Materials, Antimicrobials and Sensors. *J. Colloid Interf. Sci.* **2011,** 363,1–24.

Eatemadi, A.; Daraee, H.; Karimkhanloo, H.; Kouhi, M.; Zarghami, N.; Akbarzadeh, A.; Abasi, M.; Hanifehpour, Y.; Joo, S. W. Carbon Nanotubes: Properties, Synthesis, Purification, and Medical Applications. *Nanosc. Res. Lett.* **2014,** *9*, 393.

Falcao, E. H. L.; Wudl, F. Carbon Allotropes: Beyond Graphite and Diamond. *J. Chem. Technol. Biotechnol.* **2007,** *82*, 524–531.

Farias, J. S.; Zanin, H.; Caldas, A. S.; Santos, C. C.; Damos, F. S.; de Cássia Silva Luz, R. Functionalized Multiwalled Carbon Nanotube Electrochemical Sensor for Determination of Anticancer Drug Flutamide. *J. Electr. Mater.* **2017,** *46* (10), 5619–5628.

Fereidoun, H.; Nourddin, M. S.; Rreza, N. A.; Mohsen, A.; Ahmad, R.; Pouria, H. The Effect of Long-Term Exposure to Particulate Pollution on the Lung Function of Teheranian and Zanjanian Students. *Pak. J. Physiol.* **2007,** *3* (2), 1–5.

Fornasiero, F.; Park, H. G.; Holt, J. K. Mechanism of Ion Exclusion by Sub-2nm Carbon Nanotube Membranes. *MRS Online Proc. Lib.* **2008,** *1106*, 303.

Frazier, K. M.; Mirica, K. A.; Walish, J. J.; Swager, T. M. Fully-Drawn Carbon-Based Chemical Sensors on Organic and Inorganic Surfaces. *Lab Chip* **2014,** *14* (20), 4059–4066.

Gao, Y.; Yu, G.; Tan, J.; Xuan, F. Sandpaper-Molded Wearable Pressure Sensor for Electronic Skins. *Sens. Actuat. A Phys.* **2018,** *280*, 205–209.

Georgakilas, V.; Perman, J. A.; Tucek, J.; Zboril, R. Broad Family of Carbon Nanoallotropes: Classification, Chemistry, and Applications of Fullerenes, Carbon Dots, Nanotubes, Graphene, Nanodiamonds, and Combined Superstructures. *Chem. Rev.* **2015,** *115* (11), 4744–4822.

Grobert, N.; Carbon Nanotubes—Becoming Clean. *Mater. Today*, **2007,** *10* (1), 28–35.

Hu, L.; Hecht, D. S.; Grüner, G. Carbon Nanotube Thin Films: Fabrication, Properties, And Applications. *Chem. Rev.* **2010,** *110*, 5790–5844.

Hu, Y.; Shenderova, O.; Brenner, D. Carbon Nanostructures: Morphologies and Properties. *J. Comput. Theor. Nanosci.* **2007,** *4*, 199–221

Ibrahim, R. K.; Hayyan, M.; AlSaadi, M. A.; Hayyan, A.; Ibrahim, S. Environmental Application of Nanotechnology: Air, Soil, and Water. *Environ. Sci. Pollut. Res.* **2016,** *23*, 13754–13788.

Jaidev, B. M.; Baro, M.; Ramaprabhu, S. Room Temperature Hydrogen Gas Sensing Properties of Mono Dispersed Platinum Nanoparticles on Graphene-Like Carbon-Wrapped Carbon Nanotubes. *Int. J. Hydrogen Energy* **2018,** *43* (33), 16421–16429.

Jeon, I. Y.; Chang, D. W.; Nanjundan, A. K.; Baek, J. B. *Functionalization of Carbon Nanotubes, Carbon Nanotubes–Polymer Nanocomposites*; Siva Yellampalli: IntechOpen, 2011.

Jia, L. W.; Shen, M. Q.; Wang, J.; Lin, M. Q. Influence of Ethanol-Gasoline Blended Fuel on Emission Characteristics from a Four-Stroke Motorcycle Engine. *J. Hazard. Mater.* **2005,** *123*, 29–34.

Jost, O.; Gorbunov, A.; Liu, X. J.; Pompe, W., Fink, J. Single-Walled Carbon Nanotube Diameter. *J. Nanosci. Nanotechnol.* **2004,** *4*, 433–440.

Karimov, K. S.; Abid, M.; Mahroof, M.; Tahir V_2O_4-PEPC Composite Based Pressure Sensor. *Microelectr. Eng.* **2011,** *88* (6), 1037–1041.

Karn, B.; Kuiken, T.; Otto, M. Nanotechnology and in Situ Remediation: A Review of the Benefits and Potential Risks. *Environ. Health Perspect* **2009,** *117*, 1823–1831.

Karousis, N.; Irene, S. M.; Christopher, P. E.; Nikos, T.; Structure, Properties, Functionalization, and Applications of Carbon Nanohorns. *Chem. Rev.* **2016,** *116* (8), 4850–4883.

Kim, J. Y.; Lee, J.; Hong, S.; Chung, T. D. Formaldehyde Gas Sensing Chip Based on Single-Walled Carbon Nanotubes and Thin Water Layer. *Chem. Commun.* **2011,** *47*, 2892–2894.

Kodu, M.; Berholts, A.; Kahro, T. Graphene Functionalised by Laser-Ablated V_2O_5 for a Highly Sensitive NH_3 Sensor. *Beilstein J. Nanotechnol.* **2017,** *8*, 571–578.

Kreyling, W. G.; Behnke, M. S.; Chaudhry, Q. A Complementary Definition of Nanomaterial. *Nano Today* **2010,** *5* (3), 165–168.

Kumar, V.; Mahajan, R.; Kaur, I; Kim, K.-H.; Simple and Mediator-Free Urea Sensing Based on Engineered Nanodiamonds with Polyaniline Nanofibers Synthesized in Situ. *ACS Appl. Mater. Interf.* **2017,** *9* (20), 16813–16823.

Leng; Luo, D.; Xu, Z.; Wang, F. Modified Graphene Oxide/Nafion Composite Humidity Sensor and Its Linear Response to the Relative Humidity. *Sens. Actuat. B Chem.* **2018,** *257*, 372–381.

Lezi, N. Fabrication of a "Green" and Low-Cost Screen-Printed Graphene Sensor and Its Application to the Determination of Caffeine by Adsorptive Stripping Voltammetry. *Int. J. Electrochem. Sci.* **2017,** *12*, 6054–6067.

Li, M.-J.; Zheng, Y.-N.; Liang, W.-B.; Yuan, R.; Chai, Y.-Q. Using p-type PbS Quantum Dots to Quench Photocurrent of Fullerene-Au NP@MoS_2 Composite Structure for Ultrasensitive Photoelectrochemical Detection of ATP. *ACS Appl. Mater. Interf.* **2017,** *9* (48), 42111–42120.

Li, X.; Chen, X.; Chen, X.; Ding, X.; Zhao, X. High-Sensitive Humidity Sensor Based on Graphene Oxide with Evenly Dispersed Multiwalled Carbon Nanotubes. *Mater. Chem. Phys.* **2018,** *207*, 135–140.

Liu, S.; Wu, T.; Li, F.; Zhang, Q.; Dong, X.; Niu, L. Disposable Graphene Sensor with an Internal Reference Electrode for Stripping Analysis of Heavy Metals. *Analy. Methods* **2018a,** *10* (17), 1986–1992.

Liu, Y.; Kannegulla, A.; Wu, B.; Cheng, L.-J. Quantum Dot Fullerene-Based Molecular Beacon Nanosensors for Rapid, Highly Sensitive Nucleic Acid Detection. *ACS Appl. Mater. Interf.* **2018b,** *10* (22), 18524–18531.

Lobez, J. M.; Han, S. J.; Afzali, A.; Hannon, J. B. Surface Selective One-Step Fabrication of Carbon Nanotube Thin Films with High Density. *ACS Nano* **2014,** *8* (5), 4954–4960.

Loos, M. Nanoscience and Nanotechnology. In *Carbon Nanotube Reinforced Composites (Chapter)*; Loos, M., Ed.; William Andrew Publishing: Oxford, **2015**; pp 1–36.

Ma, X.; Dong, Y.; Li, R. Monitoring Technology in Composites Using Carbon Nanotube Yarns Based on Piezoresistivity. *Mater. Lett.* **2017,** *188*, 45–47.

Mauter, M. S.; Elimelech, M. Environmental Applications of Carbon-Based Nanomaterials. *Environ. Sci. Technol.* **2008,** *42* (16), 5843–5859.

MeiJiao, L.; Jing, L.; Yang, X. Y.; Zhang, C. A.; Yang, J.; Hu, H.; WANG, X. B. Applications of Graphene-Based Materials in Environmental Protection and Detection. *Chinese Sci. Bull.* **2013,** *58* (22).

Menacer, F.; Dibi, Z.; Kadri, A. Modeling a New Acetone Sensor Based on Carbon Nanotubes Using Finite Elements and Neural Network. *Eur. Phys. J. Plus* **2018,** *133* (6), 238.

Nasir, S.; Hussein, M. J.; Zainal, Z.; Yusof, N. A. Carbon-Based Nanomaterials/Allotropes: A Glimpse of Their Synthesis, Properties and Some Applications. *Material* **2018.**

Niu, X.; Zhong, Y.; Chen, R.; Wang, F.; Liu, Y.; Luo, D. A "Turn-on" Fluorescence Sensor for Pb^{2+} Detection Based on Graphene Quantum Dots and Gold Nanoparticles. *Sens. Actuat. B Chem.* **2018,** *255*, 1577–1581.

Novoselov, K. S. Electric Field Effect in Atomically Thin Carbon Films. *Science* **2004,** *306*, 666–669.

Parlak, C.; Alver, Ö. A Density Functional Theory Investigation on Amantadine Drug Interaction with Pristine and B, Al, Si, Ga, Ge Doped C60 Fullerenes. *Chem. Phys. Lett.* **2017,** *678*, 85–90.

Patel, K.; Singh, R.; Kim, H. W. Carbon Based-Nanomaterials as an Emerging Platform for Theragnostic. *Mater. Horiz.* **2018.**

Philips, M. F.; Gopalan, A. I.; Lee, K.-P. Development of a Novel Cyano Group Containing Electrochemically Deposited Polymer Film for Ultrasensitive Simultaneous Detection of Trace Level Cadmium and Lead. *J. Hazard. Mater.* **2012,** *237–238*, 46–54.

Prasad, B. B.; Singh, R.; Kumar, A. Synthesis of Fullerene (C60-Monoadduct)-Based Water-Compatible Imprinted Micelles for Electrochemical Determination of Chlorambucil. *Biosens. Bioelectr.* **2017,** *94*, 115–123.

Purdey, M. S.; Capon, P. K.; Pullen, B. J. An Organic Fluorophore-Nanodiamond Hybrid Sensor for Photostable Imaging and Orthogonal, on-Demand Biosensing. *Sci. Rep.* **2017,** *7* (1), 15967.

Rahimi-Nasrabadi, M.; Khoshroo, A.; Mazloum-Ardakani, M. Electrochemical Determination of Diazepam in Real Samples Based on Fullerene-Functionalized Carbon Nanotubes/Ionic Liquid Nanocomposite. *Sens. Actuat. B Chem.* **2017,** *240*, 125–131.

Rao, C. N. R.; Seshadri, R.; Govindaraj, A.; Sen, R. Fullerenes, Nanotubes, Onions and Related Carbon Structures. *Mater. Sci. Eng. R Rep.* **1995,** *15* (6), 209–262.

Ricciardella, F.; Vollebregt, S.; Polichetti, T. Effects of Graphene Defects on Gas Sensing Properties towards NO_2 Detection. *Nanoscale* **2017,** *9* (18), 6085–6093.

Saha, D.; Das, S. Development of Fullerene Modified Metal Oxide Thick Films for Moisture Sensing Application. *Mater. Today Proc.* **2018,** *5* (3), 9817–9825.

Seekaew, Y.; Phokharatkul, D.; Wisitsoraat, A.; Wongchoosuk, C. Highly Sensitive and Selective Room-Temperature NO_2 Gas Sensor Based on Bilayer Transferred Chemical Vapor Deposited Graphene. *Appl. Surf. Sci.* **2017,** *404*, 357–363.

Shanthilal, J.; Bhattacharya, S. Nanoparticles and Nanotechnology in Food. In *Conventional and Advanced Food Processing Technologies*; Wiley, 2014; pp 567–594.

Simioni, N. B.; Silva, T. A.; Oliveira, G. G.; Fatibello-Filho, O. A Nanodiamond-Based Electrochemical Sensor for the Determination of Pyrazinamide Antibiotic. *Sens. Actuat. B Chem.* **2017,** *250*, 315–323.

Smazna, D.; Rodrigues, J.; Shree, S.; Buckminsterfullerene Hybridized Zinc Oxide Tetrapods: Defects and Charge Transfer Induced Optical and Electrical Response. *Nanoscale* **2018,** *10* (21), 10050–10062.

Song, H.; Zhang, Y.; Cao, J. Sensing Mechanism of an Ionization Gas Temperature Sensor Based on a Carbon Nanotube Film. *RSC Adv.* **2017,** *7* (84), 53265–53269.

Sun, Y. P.; Zhou, B.; Lin, Y. Quantum-sized Carbon Dots for Bright and Colorful Photoluminescence. *J. Am. Chem. Soc.* **2006,** *128* (24), 7756–7757.

Sutradhar, S.; Jacob, G. V.; Patnaik, A. Structure and Dynamics of a dl-Homocysteine Functionalized Fullerene-C60-Gold Nanocomposite: A Femtomolar l-Histidine Sensor. *J. Mater. Chem. B* **2017,** *5* (29), 5835–5844.

Thamri, A.; Baccar, H.; Struzzi, C.; Bittencourt, C.; Abdelghani, A.; Llobet, E.; MHDA-Functionalized Multiwall Carbon Nanotubes for Detecting Non-Aromatic VOCs. *Sci. Rep.* **2016,** *10* (6), 35130.

Tran, T. L.; Nguyen, T. T.; Huyen Tran, T. T.; Chu, V. T.; Thinh Tran, Q.; Tuan Mai, A. Detection of Influenza A Virus Using Carbon Nanotubes Field Effect Transistor Based DNA Sensor. *Phys. E Low-Dimension. Syst. Nanostruct.* **2017,** *93*, 83–86.

Troshin, P. A.; Lyubovskaya, R. N. Organic Chemistry of Fullerenes: The Major Reactions, Types of Fullerene Derivatives and Prospects for Practical Use. *Russian Chem. Rev.*, **2008,** *77* (4), 323–369.

Vinícius, D. N. B.; Montanheiro, T. L. A.; de Menezes, B. R. C.; Ribas, R. G.; Righetti, V. A. N.; Rodrigues, K. F.; Thim, G. P. Carbon Nanostructure-based Sensors: A Brief Review on Recent Advances. *Adv. Mater. Sci. Eng.* **2019,** *21*.

Wang, F.; Gu, H.; Swager, T. M.; Carbon Nanotube/Polythiophene Chemiresistive Sensors for Chemical Warfare Agents. *J. Am. Chem. Soc.* **2008,** *130*, 5392–5393.

Wei, Z. N.; Yang, X.-X.; Mo, Z.-H.; Leng, F. Photocatalytic Sensor of Organics in Water with Signal of Plasmonic Swing. *Sens. Actuat. B Chem.* **2018,** *255*, 3458–3463.

Wu, J.; Tao, K.; Miao, J.; Norford, L. K. Three-Dimensional Hierarchical and Superhydrophobic Graphene Gas Sensor with Good Immunity to Humidity. *Proc. 2018 IEEE Micro Electro Mech. Syst. (MEMS)* **2018,** 901–904, Belfast, UK.

Xu, X.; Ray, R.; Gu, Y. Electrophoretic Analysis and Purification of Fluorescent Single-Walled Carbon Nanotube Fragments. *J. Am. Chem. Soc.* **2004,** *126* (40), 12736–12737.

Xuan, X.; Park, J. Y. A Miniaturized and Flexible Cadmium and Lead Ion Detection Sensor Based on Micro-Patterned Reduced Graphene Oxide/Carbon Nanotube/Bismuth Composite Electrodes. *Sens. Actuat. B Chem.* **2018,** *255*, 1220–1227.

Yi, H.; Wu, L. Q.;Bentley, W. E.; Ghodssi, R.; Rubloff, G. W.; Culver, J. N.; Payne, G. F. Biofabrication with Chitosan. *Biomacromolecules* **2005,** *6 (*6), 2881–2894.

Yu, S.; Wang, Y.; Xiang, H.; Zhu, L.; Tebyetekerwa, M.; Zhu, M. Superior Piezoresistive Strain Sensing Behaviors of Carbon Nanotubes in One-Dimensional Polymer Fiber Structure. *Carbon* **2018a**, *140*, 1–9.

Yu, X.; Chen, X.; Li, H.; Ding, X. A High-Stability QCM Humidity Sensor Coated with Nanodiamond/Multiwalled Carbon Nanotubes Nanocomposite. *IEEE Trans. Nanotechnol.* **2018b**, *17* (3), 506–512.

Zhang, Y.; Niu, Q.; Gu, X.; Yang, N.; Zhao, G. Recent Progress of Carbon Nanomaterials for Electrochemical Detection and Removal of Environmental Pollutants. *Nanoscale* **2019,** *11* (25), 11979–12398.

Zhao, W.; Yang, C.; Zou, D.; Sun, Z.; Ji, G. Possibility of Gas Sensor Based on C20 Molecular Devices. *Phys. Lett. A* **2017,** *381* (21), 1825–1830.

Zuo, J.; Jiang, T.; Zhao, X.; Xiong, X.; Xiao, S.; Zhu, Z. Preparation and Application of Fluorescent Carbon Dots. *J. Nanomater.* **2015,** 1–13.

CHAPTER 5

Applying Noncarbon-Based Nanomaterials to Detect Environmental Contaminants

KANNAN DEEPA, ASHISH KAPOOR, and PRABHAKAR SIVARAMAN*

Department of Chemical Engineering, SRM Institute of Science and Technology, Kattankulathur 603203, Tamil Nadu, India

Corresponding author. E-mail: prabhaks@srmist.edu.in

ABSTRACT

Nanomaterials possess remarkable optical, electrical, thermal, mechanical, and catalytic properties which can be efficiently used for sensing a wide variety of substances. It is necessary to periodically assess the contaminant levels in our environment both qualitatively and quantitatively using inexpensive, rapid, and sensitive techniques/tools. The high reactivity of the nanomaterials make them potential analytical tools for monitoring several types of environmental contaminants ranging from heavy metal ions, organic pollutants, toxic gases, to pathogenic microorganisms. In this regard, different varieties of carbon-based and noncarbon based nanomaterials have been investigated for their detection capabilities. Herein, the usage of noncarbon based metals, metal oxides, semiconductor nanoparticles/quantum dots, composite nanomaterials, etc., as sensing platforms are discussed with emphasis on the predominant mechanisms behind contaminant-nanomaterial interactions.

Nanotechnology for Environmental Pollution Decontamination: Tools, Methods, and Approaches for Detection and Remediation. Fernanda Maria Policarpo Tonelli, Rouf Ahmad Bhat, & Gowhar Hamid Dar (Eds.)

5.1 INTRODUCTION

The ecosystem is greatly impaired due to the effects of environmental pollution from industrial and agricultural activities. A wide range of contaminants ranging from heavy metal ions, pesticides pathogenic microorganisms, and gaseous pollutants cause severe damage to the environment and have adverse effects on human health. Most of these contaminants resist degradation and hence accumulate in the environment. This global crisis of environmental pollution necessitates the development of detection tools and techniques of utmost precision and efficacy. The tools and techniques used for detection have to be inexpensive, simple to operate, selective, and sensitive as well (Su et al., 2012). The techniques based on colorimetry, chromatography, fluorimetry, and voltammetry have been extensively used for monitoring these contaminants (Aragay et al., 2011; Su et al., 2012). However, most traditional analytical tools such as high performance liquid chromatography and inductively coupled plasma-based spectroscopy are expensive, highly sophisticated, and do not support in situ analysis.

Different types of sensors have been developed to detect specific chemical or biological entities. Of these, nanomaterial-based sensors have garnered attention in the past decade owing to their high surface-area to volume ratio, reactivity, and enhanced catalytic and adsorptive abilities. It is also possible to fine tune the sensing configurations based on recognition phenomena at the nanoscale (Arduini et al., 2020). For brevity, this chapter is limited to noncarbon-based nanomaterials to exclude the large class of nanomaterials based on carbon and its allotropes such as carbon nanotubes, graphene, and graphene oxide, nanodiamonds, fullerenes, and carbon dots (Fig. 5.1). A wide variety of noncarbon-based nanomaterials including metal and metal oxide nanoparticles, bimetallic nanoparticles, quantum dots, nanocomposites, and organic nanoparticles, which are used for the detection of contaminants are discussed in this chapter.

5.2 DIFFERENT TYPES OF NONCARBON-BASED NANOMATERIALS

The major classes of noncarbon-based nanomaterials dealt in this chapter (Fig. 5.1) are metals and metal oxide nanoparticles, bimetallic nanoparticles, quantum dots, nanocomposites, and organic nanoparticles (Sudha et al., 2018; Saleh, 2020). Besides, the applications of ceramic, zeolite, and silica-based nanomaterials for the detection of environmental contaminants are discussed briefly (Fig. 5.2).

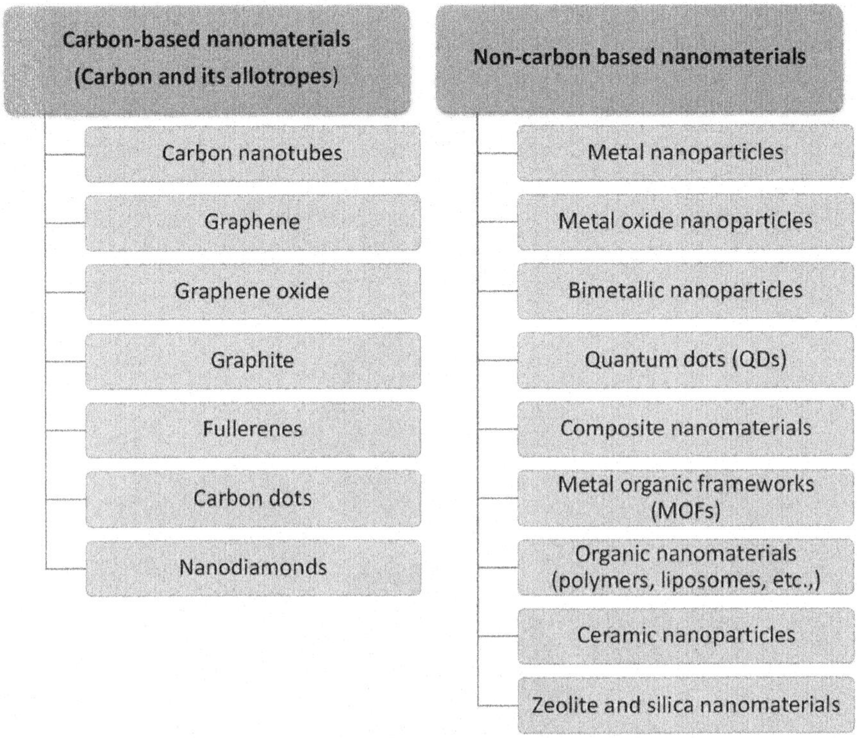

FIGURE 5.1 Classification of nanomaterials.

Noble metal nanoparticles such as Ag, Au, and Pt have received maximum consideration due to their high conductivity, biocompatibility, provisions for surface functionalization, and their size- and shape-related optical, electronic, and catalytic properties (Buzea and Pacheco, 2017). Metal oxide nanoparticles such as TiO_2, Fe_2O_3, Fe_3O_4, ZnO, etc. are excellent chemical sensors with interesting catalytic properties due to their surface properties influencing the bandgap energy of materials (Saleh and Fadillah, 2019). Bimetallic nanoparticles comprising of two different metals exhibit extraordinary chemical stability and reactivity, which are further dictated by their composition, morphology, and size distribution (Saleh, 2020). The synthesis procedures for bimetallic nanoparticles are relatively complex depending on the type of nanoparticles, that is, core-shell, alloy, and contact aggregates (Sharma et al., 2019).

Quantum dots are semiconductor nanomaterials with distinct photophysical and chemical characteristics (Alivisatos, 1996; Ahmad et al., 2012). Quantum dots (QDs)-based immunoassays have been used for the detection of pathogenic microorganisms by tagging them with biomolecules such as nucleic acid sequences, proteins, antibodies, etc. (Alivisatos, 2000; Dubertret, 2005). When such QD-immobilized biomolecules interact with complementary nucleic acid sequences or antigens specific to the target pathogen, the bioconjugation or hybridization event is recorded by a transducer and converted to measurable signals (Jin et al., 2011; Ahmad et al., 2012). The QD probes display high sensitivity and are excellent fluorophores compared to organic dyes (Ferancová and Labuda, 2008; Sharma and Mehata, 2020).

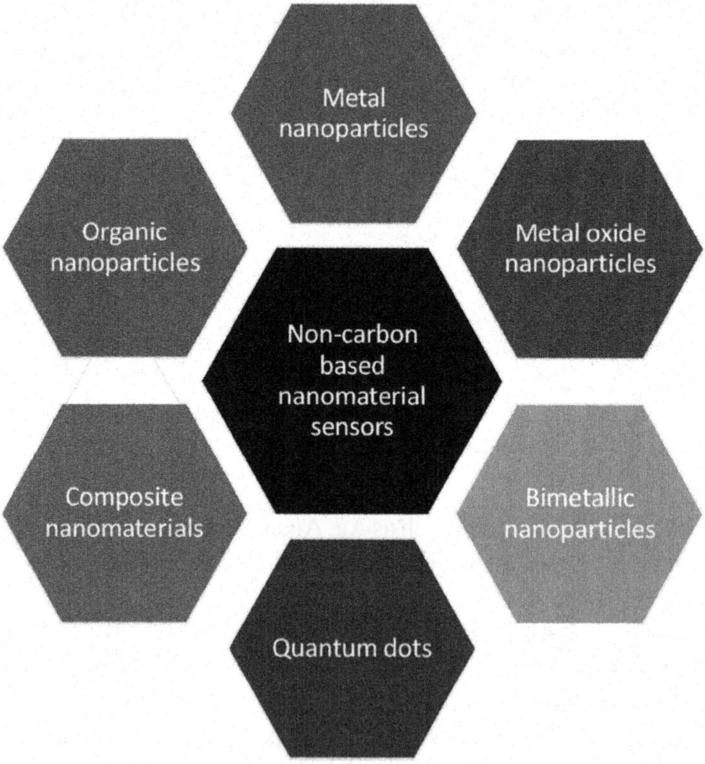

FIGURE 5.2 Noncarbon-based nanomaterial sensors.

Nanocomposites are multiphase nanomaterials comprising of nanostructures/nanoparticles as building blocks embedded in a supporting matrix. The common nanocomposite materials include polymer-based nanocomposites, nonpolymer-based nanocomposites, metal/metal nanocomposites, metal/ceramic nanocomposites, ceramic/ceramic nanocomposites, polymer/ceramic nanocomposites, carbon/metal nanocomposites, carbon/polymer nanocomposites, and carbon/ceramic nanocomposites (Sudha et al., 2018). Nanoscale metal organic frameworks (MOFs) are porous crystalline coordination networks assembled from inorganic nodes (metal ions or clusters) connected by organic linkers. Due to their high porosity, they are suitable candidates for sensing and capture of a number of pollutants (Wang et al., 2018a; Cai et al., 2020; Zhao, 2020).

It is necessary to differentiate the carbon- and its allotrope-based nanomaterials such as single-and multiwalled carbon nanotubes, graphene, graphene oxides, etc., from the category of organic nanomaterials composed of lipids, carbohydrates and other polymers. Our discussion here will be restricted to the latter category of organic nanomaterials. Polymeric nanoparticles, liposomes, polymeric micelles, and dendrimers are the widely studied organic nanomaterials for the detection of notorious compounds. Polymeric nanomaterials are solid particles composed of either natural or synthetic polymers. They are generally biocompatible, biodegradable, and exhibit greater structural integrity, stability, and controlled release characteristics (Hadinoto et al., 2013; Sudha et al; 2018; Saleh, 2020). The other types of organic nanomaterials used specifically for detecting contaminants are discussed in upcoming sections.

5.3 MECHANISMS OF DETECTION OF CONTAMINANTS

Analytical nanotechnology refers to the application of nanomaterials as sensors for the detection of chemical and biological entities. The mechanisms of detection of the contaminants are discussed later with a mention of microfluidic platforms incorporating one of these mechanisms but on a miniature scale. The interactions between the sensor (nanomaterial) and the analyte (contaminants) could be based on colorimetric absorption, enhanced Raman scattering, electrochemical, or fluorescent signalling.

5.3.1 COLORIMETRIC DETECTION

The use of colorimetric sensors for monitoring environmental contaminants is simple, economic, fast, and allow on-site detection without the need for large sophisticated equipment (Prosposito et al., 2020). Noble metal colloidal solutions, especially those of gold and silver have very high extinction coefficients and exhibit vibrant hues of colors in the visible region. The interaction between the target analyte and the nanoparticles leads to visible changes in color of the colloids, depending on the state of aggregation (Vilela et al., 2012). The extent of aggregation of the nanoparticles is found to be proportional to the amount of analyte, thus enabling quantitative detection of the contaminants (Fig. 5.3). The detection limit of such nanomaterial-based colorimetric assays is in the nanomolar range which is much better compared to conventional spectrophotometric assays without the nanoparticles (Tiwari and Prakash, 2018). The sensitivity of the nanoparticles to changes in the local environment surrounding the nanoparticles are responsible for the success of the nanomaterial-based colorimetric estimation (Alzahrani, 2020).

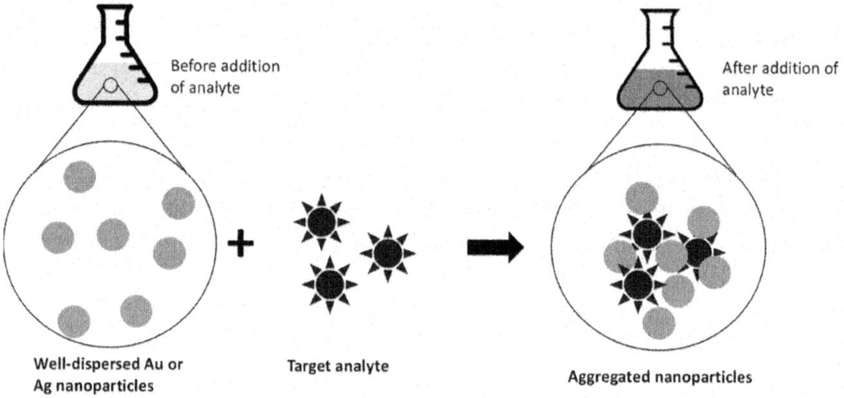

FIGURE 5.3 Colorimetric sensor based on Au and Ag nanoparticles.

5.3.2 ENHANCEMENT OF SERS SIGNAL

Surface enhanced Raman spectroscopy (SERS) is a spectroscopic technique that allows for highly sensitive structural detection of analytes at very low concentration by amplifying the electromagnetic fields generated by the excitation of localized surface plasmons (Sharma et al., 2012). The weaker Raman signals from molecules are amplified by orders of 10^{10}–10^{15} facilitating detection of even single molecules (Kneipp et al., 1999; Pieczonka and Aroca, 2008). Molecules that are adsorbed onto rough nanostructured surfaces exhibit enhanced Raman scattering (Fig. 5.4). One criteria for SERS is that the choice of laser wavelength has to correspond with that of the SERS metal used. Nanoscale gold, silver, and copper are the best enhancers of electromagnetic fields and hence the choice of material for SERS substrates. Also, metal nanoparticles allow tuning of sensitivity of the analysis through proper control of shape and size of nanoparticles (Tian et al., 2014). It is already established that noble metal nanoparticles of different sizes and shapes can be obtained by controlling the reaction parameters such as pH, temperature, precursor ion concentration, etc., (Sau and Rogach, 2010; Jamkhande et al., 2019).

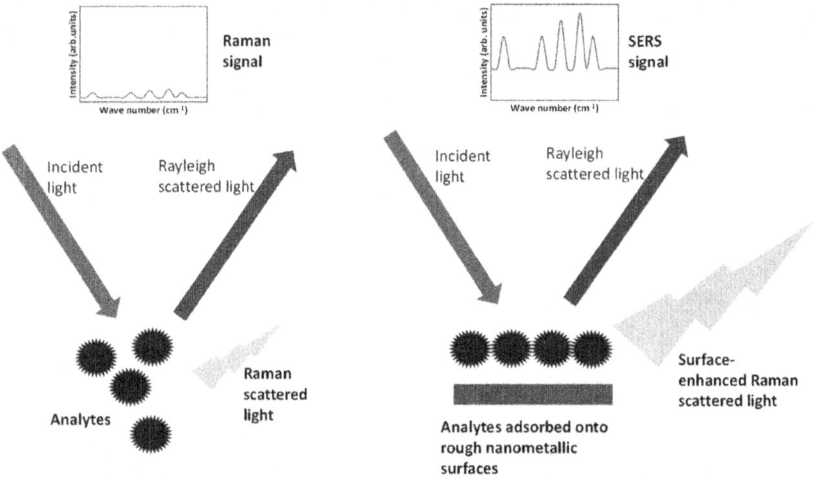

FIGURE 5.4 Raman scattering vs. surface-enhanced Raman scattering.

When SERS was used for detecting heavy metals, organic and inorganic pollutants, some of the lowest limits of detection were reported. Hence, SERS is a promising tool for monitoring pollutants at environmentally valid concentrations. Though a number of benchtop Raman spectrometers have been successfully employed for detecting heavy metals such as mercury, pharmaceuticals, perfluorinated compounds, and pesticides, similar sensitivity is yet to be achieved in portable Raman spectrometers for in field applications (Ong et al., 2020).

5.3.3 ELECTROCHEMICAL DETECTION

Electrochemical techniques of detection allow on site and real time monitoring of the contaminants. The nanomaterials used to modify the electrodes determine the sensitivity and selectivity of the sensors depending on the size of the nanostructures and the electron transfer areas (Fig. 5.5). Variations in material of the electrodes/electrode surfaces have resulted in different rates of electron transfer and hence different strengths of electrochemical signals (Khan et al., 2015; Huang et al., 2017). Metal nanoparticles modified electrodes act as arrays of microelectrodes, which offer better mass transport and signal-to-noise ratio, low limits of detection, and avoid solution resistance as much as possible in comparison with conventional macroelectrodes (Aragay et al., 2011). Nanomaterial-modified electrodes can bring together the features of catalysis, conductivity, and biocompatibility to enhance signal transduction. Also, they can amplify chemical or biological recognition events through distinct signal tags for selective and sensitive detection (Kumar et al., 2017).

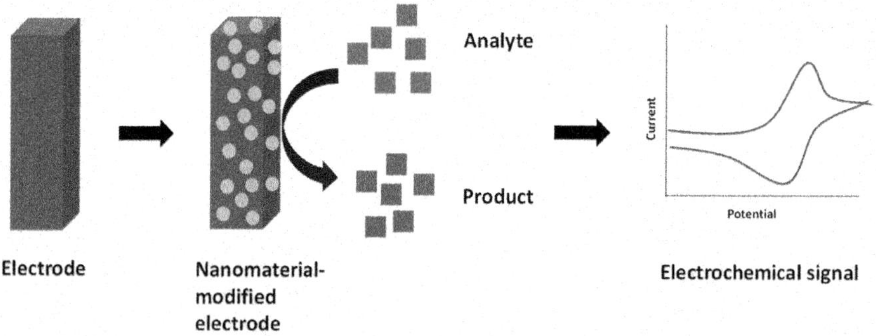

FIGURE 5.5 Electrochemical detection using nanomaterial-modified electrodes.

5.3.4 FLUORESCENCE EMISSION

The fluorescence of nanomaterials is a size-dependent photophysical property manifested as a result of many inter- and intra-molecular charge transfer processes occurring between the fluorophore and the analyte. The interactions between the fluorophore and the analyte include fluorescence resonance energy transfer (FRET), intramolecular charge transfer (ICT), photoinduced electron transfer (PET), metal to ligand charge transfer (MLCT), and ligand to metal charge transfer (LMCT). The fluorescence signal after the interaction process could either be enhanced or quenched. A Stern-Volmer plot represents these changes in fluorescence after interaction with the analyte (Walekar et al., 2017). In general, the sensitivity of a fluorescent probe increases with the specific surface area or surface to mass (S/M) ratio of the sensing materials (Fig. 5.6). Owing to their high S/M ratio and adsorbability, nanomaterials have gained immense attention for use as fluorescent sensors (Borgå et al., 2005; Wang and Meng, 2017).

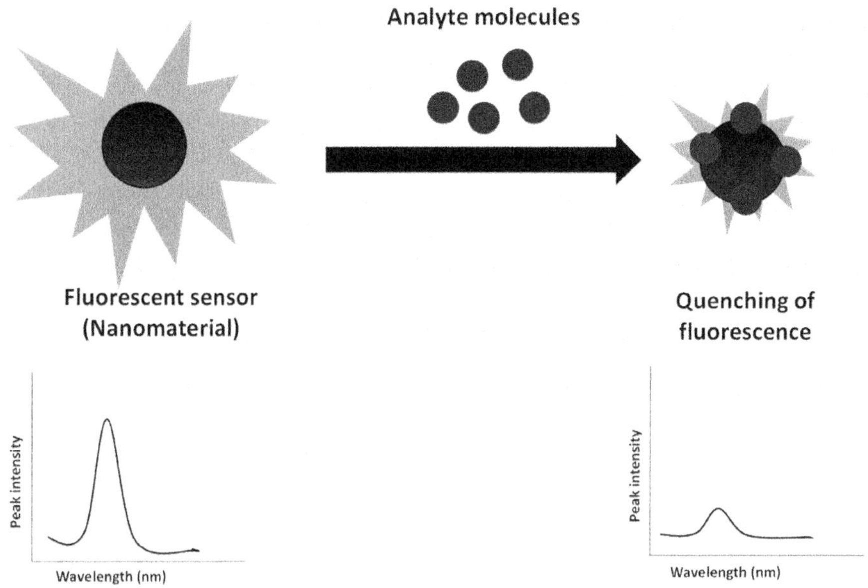

FIGURE 5.6 Process of fluorescence quenching.

5.3.5 *MICROFLUIDIC LAB-ON-A-CHIP PLATFORMS*

Lab-on-a-chip (LOC) devices are miniaturized sensing devices, which facilitate significant reduction in reagent volumes, easy operation, reduced cost, portability, and in situ analysis (Govindarajalu et al., 2019). Both electrochemical and optical sensing devices have been reported in this regard. The electrochemical LOC platforms work on the principle of generating an electrical signal due to the interaction between the analyte and the electrode surface (Pol et al., 2017). Electrochemical sensing (voltammetric and potentiometric) methods integrated with micro-electromechanical systems allow simultaneous detection of the contaminants along with other environmental quality parameters (Jang et al., 2011). Optical LOC platforms are based on colorimetric changes due to chemical reactions or interactions between the analyte and a reagent (Pol et al., 2017).

5.4 DETECTION OF VARIOUS ENVIRONMENTAL CONTAMINANTS

Different types of nanomaterial-based sensors have been developed for monitoring a wide range of contaminant substances present in different sources. Most of these sensors have a common design: Nanomaterial + Receptor + Transducer. The receptors bound to the nanomaterials (antibodies, aptamers) are highly selective and sensitive to the target contaminant while the transducer signals the presence of the analyte in the form of an optical, colorimetric, or electrochemical response (Su et al., 2012; Willner and Vikesland, 2018; Rastogi et al., 2020). These sensors have been used to target pathogenic microorganisms, heavy metals, and a number of organic contaminants such as chlorinated compounds, organic solvents, pesticides and food contaminants. Table 1.1 presents some representative data for the different nanomaterials used to detect a plethora of contaminants along with the mechanisms and sensitivity of each case study.

5.4.1 PATHOGENIC MICROORGANISMS

It is necessary to detect and identify the pathogenic microorganisms present in air, water, and food to avoid outbreaks of infectious diseases. In the past, such infectious diseases have resulted in high morbidity and mortality rates and require periodical monitoring and surveillance of air, water, and food samples (Baldursson and Karanis, 2011; Bhardwaj et al., 2019). Pathogenic microorganisms of every type ranging from bacteria, viruses, protozoa, and fungi have been reported to cause infectious diseases. The conventional methods that have been used to detect microorganisms are culture-based analysis using physical identification and chemical assays, molecular techniques such as PCR and DNA microarrays, and lastly, immunological methods such as ELISA. The culture-based methods have low sensitivity and are time-consuming while the molecular and immunological methods are expensive and do not differentiate viable from nonviable microorganisms (Bhardwaj et al., 2019).

Recently, nanobiosensors have been developed and successfully implemented for detecting microorganisms with very high sensitivity due to efficient biomolecular loading on the surfaces of the nanoparticles. The various optical properties of nanomaterials such as absorbance, reflectance, resonance, and luminescence have been utilized to develop optical biosensors to detect pathogens. Bui et al. (2015) have used gold nanoparticles conjugated with cysteine-loaded liposomes to detect even single cells of *Salmonella, Listeria,* and *E. coli* O157 cells in water and food samples. The nanoliposomes would burst in the presence of these microorganisms leading to aggregation of the gold nanoparticles manifested as visual change in color of the reaction mixture from red to dark blue (Bui et al., 2015). A receptor-free generic sensor was developed using polyaniline nanoparticles where the color changed from blue

TABLE 1.1 Nanomaterials for the Detection of Various Environmental Contaminants.

S. No.	Type of nanomaterial	Target analyte (chemical/biological)	Mechanism of detection	Limit of detection	References
1.	Gold nanoparticles conjugated with cysteine-loaded liposomes	*Salmonella, Listeria,* and *E. coli* O157	Colorimetry	6.7 aM (attomolar)	Bui et al. (2015)
2.	Polyaniline nanoparticles	*E. coli*	Colorimetry	10^6 cells/ml for measurement time of 2 h	Thakur et al. (2015)
3.	Silver nanorod arrays	*Escherichia coli, E. coli* O157:H7, *E. coli* DH5α, *Staphylococcus aureus, S. epidermidis, Salmonella typhimurium*, and bacteria mixtures	SERS	10^8 CFU/mL	Chu et al. (2008)
4.	Gold nanoparticles conjugated with aptamers	*Escherichia coli* ATCC 8739	FRET	3 CFU/mL	Jin et al. (2017)
5.	Quantum dot –streptavidin conjugates	Human immunodeficiency virus	Lab-on-a-chip	–	Kim et al. (2009)
6.	Silica nanoparticles-Rubpy-IgG	*Xanthomonas axonopodis* pv. *vesicatoria*	Fluorescence quenching	10^3 CFU/mL	Yao et al. (2009)
7.	Silica nanoparticles-dye- IgG	S. pullorum and S. Gallinarum	Agglutination test	4 × 103 to 4 × 109 CFU/mL	Yu et al. (2015)
8.	Bismuth nanoparticles	Cd^{2+}, Pb^{2+}	Stripping voltammetry using modified electrodes	10 µg/L	Yang et al. (2013)
9.	Bismuth nanopowder electrode	Cd, Zn	Anodic stripping voltammetry	Cd: 0.15 µg/L Zn: 0.07 µg/L	Lee et al. (2007)
10.	Ag@CuS, Au@PtS2, Au@HgS, Ag@Ag2S NPs, and Ag@CuS core@shell nanorods	Hg^{2+}	SERS	0.1 ppb	Bao et al. (2019)

TABLE 1.1 *(Continued)*

S. No.	Type of nanomaterial	Target analyte (chemical/biological)	Mechanism of detection	Limit of detection	References
11.	Gold nanoparticles and alkanethiols	Hg^{2+}, Ag^{+}, Pb^{2+}	Colorimetry (aggregation)	Nanomolar range	Hung et al. (2010)
12.	Gold nanoparticles on carbon ceramic electrodes	CN^{-}	Square wave voltammetry	0.09 µM	Shamsipur et al. (2017)
13.	Fe_2O_3 nanoparticles	NO_2	Electrical resistance	10–200 ppm	Navale et al. (2013)
14.	Glassy carbon electrodes modified with zinc oxide nanoparticles and silver doped zinc oxide nanoparticles	$Cd^{2}+$, Pb^{2+}	Cyclic voltammetry	3.5–3.8 nM	Nagaraju et al. (2017)
15.	NiO/ZnO PN heterojunction	Triethylamine gas	Chemiresistive gas sensor	100 ppm	Ju et al. (2014)

to green when protons were generated in its microenvironment as a result of metabolic activity of *E. coli* cells (Thakur et al., 2015).

Nanomaterials conjugated with Raman reporter molecules such as organic dyes and bioreceptor molecules such as antibodies and nucleic acid are used as SERS tags to detect analytes of interest. Such label-based SERS tags are not suited for in situ, high throughput recognition of pathogens and the organic dye requires increased reactant volumes and preparation times. Another approach is a label-free one which involves direct electrostatic adhesion of the pathogen to the nanomaterials. The label-free method allows for differentiation of viable from nonviable bacteria as well (Chu et al., 2008; Zhou et al., 2015; Bhardwaj et al., 2019).

Quantum dots, metal organic frameworks, and polymeric nanoparticles have been used as fluorescent tags to detect microorganisms using fluorescence measurements. These fluorescent nanomaterials functionalized with bioreceptor molecules specific to the pathogen can be analyzed by flow cytometry and/or a fluorescence spectrophotometer. A FRET-based sensor was developed using gold nanoparticles conjugated with aptamers as the "acceptor" and upconversion nanoparticles functionalized with complementary DNA as the "donor." Due to spectral overlap between the donor and acceptor molecules, fluorescence quenching results. However, when the target pathogen was present, the aptamers preferentially bind to the microorganism, thus reclaiming the upconversion fluorescence (Jin et al., 2017).

Silica nanoparticles have large surface areas, high stability, biocompatibility, and can be functionalized with a number of receptor molecules (Tang et al., 2007; Mokhtarzadeh et al., 2017). An organic dye, tris-2, 2' -bipyridyl dichlororuthenium (II) hexahydrate (Rubpy), incorporated into the core of silica nanoparticles were conjugated with the secondary antibody of goat anti-rabbit immunoglobulin G (IgG) to detect the plant pathogen, *Xanthomonas axonopodis* pv. *vesicatoria* that causes bacterial spot disease in *Solanaceae* (Yao et al., 2009).

5.4.2 HEAVY METALS

Anthropogenic activities from chemical industry, mining and combustion of fossil fuels are some of the major contributing sources for heavy metal contamination. The harmful effects of heavy metals to the environment are mainly due to their high reactivity and high atomic densities. Contamination from heavy metal ions such As, Cu, Pb, Hg, Ni, Cd, and Zn have long-term detrimental effects on human health. The techniques currently employed for the detection of trace levels of heavy metals are atomic absorption spectroscopy (AAS), atomic fluorescence spectroscopy (AFS), inductively coupled plasma—optical emission spectroscopy, inductively coupled plasma—mass spectroscopy, and inductively coupled plasma—atomic emission spectroscopy (ICP-OES, ICP-MS, and ICP-AES respectively) and X-ray fluorescence spectroscopy (XPS). However, these instruments require skilled technical personnel, sample preparation steps and are expensive and nonportable (Kumar et al., 2017). Nanomaterials-based sensing tools for toxic heavy metal ions work on optical, biological, and electrochemical sensing strategies. The chemical recognition of heavy metal ions through receptor molecules in the detection system is the basis for sensing. Besides the receptor (biological or heavy metal ionophores), there is an immobilization/transducing platform that imparts stability and sensitivity to the device (Aragay et al., 2011). The additional feature of surface modification of the nanomaterials allows for better selectivity of the sensor.

Bismuth nanoparticles were used to modify conventional glassy carbon electrodes to detect Pb^{2+} and Cd^{2+} by stripping analysis with detection limits as low as 10 μg/L (Yang et al., 2019). Bao et al. (2019) have created core-shell nanoparticles of Ag@CuS, Au@PtS2, Au@HgS, Ag@Ag2S, and Ag@CuS and achieved SERS signalling of Hg^{2+} ions (Bao et al., 2019). A selective colorimetric assay was developed by Hung et al. (2010) to detect aqueous Hg^{2+}, Ag^{+}, and Pb^{2+} ions using label- free gold nanoparticles and alkanethiols. The degree of alkanethiol induced aggregation was monitored using absorbance data in a simple UV−visible spectrophotometer. Carbon ceramic electrodes modified with gold nanoparticles were used to detect cyanide concentrations as low as 0.09 μM by square wave voltammetry (Shamsipur et al., 2017). Glassy carbon electrodes modified with zinc oxide nanoparticles and silver doped zinc oxide nanoparticles were used to detect lead and cadmium ions (Nagaraju et al., 2017). Lu et al. (2018) have presented a comprehensive review of the detection of heavy metal ions in the environment by voltammetry. The effects of deposition potential, time, types of buffer solution, and pH on the stability, sensitivity, reproducibility, and anti-interference ability of the sensor during the simultaneous detection of several heavy metal ions are discussed.

5.4.3 ORGANIC CONTAMINANTS AND GASES

Persistent organic pollutants (POPs) infamously called as "forever chemicals" resist chemical and biological degradation and hence tend to bioaccumulate in humans and persist in the environment as well. POPs such as aldrin, endrin, toxaphene, heptachlor (DDT), and poly-chlorinated biphenyls (PCBs) are released through the bulk use of pesticides, solvents, and other industry chemicals (Zhang and Fang, 2010). Fe_2O_3 nanoparticles were successfully utilized as NO_2 sensor by electrical resistance measurements with a limit of detection of 10–200 ppm (Navale et al., 2013). MOFS-derived sensors have been used to detect a number of organic and inorganic gaseous contaminants such as acetone, ethanol, benzene, toluene, xylene, formaldehyde, and hydrogen sulphide (Wang et al., 2018b).

5.5 FUTURE PROSPECTS

Nanomaterials have been widely used as sensors for a number of biological and chemical compounds with very low limits of detection than those possible with conventional spectrophotometric and chromatographic methods. Many reports have provided comparative analysis of nanomaterial-based sensors and found differences in orders of magnitude with respect to the sensitivity of the analysis. The biosensing approaches have incorporated receptors such as DNA, proteins, antibodies, etc. to enhance the selectivity of the process but these methods are tedious, require adequate sample preparation steps and are time-consuming. Existing challenges in nanomaterial-based sensing are interferences in real time applications, cost-effectiveness, pH- and temperature-effects, improvement in response time, and widening the detection scope of the sensors. Also, the elimination of false positive results in an analysis is possible by arranging dual modes of signal transduction, say by generating both an optical and electrochemical response from the same sensor.

KEYWORDS

- **environmental contaminants**
- **nanomaterials**
- **metal oxides**
- **bimetallic nanoparticles**
- **electrochemical detection**

REFERENCES

Ahmad, F.; Pandey, A. K.; Herzog, A. B.; Rose, J. B.; Gerba, C. P.; Hashsham, S. A. Environmental Applications and Potential Health Implications of Quantum Dots. *J. Nanopart. Res.* **2012,** *14* (8), 1038.

Alivisatos, A. P. Semiconductor Clusters, Nanocrystals, and Quantum Dots. *Science* **1996,** *271* (5251), 933–937.

Alivisatos, P. Colloidal Quantum Dots. From Scaling Laws to Biological Applications. *Pure Appl. Chem.* **2000,** *72* (1–2), 3–9.

Alzahrani, E. Colorimetric Detection Based on Localized Surface Plasmon Resonance Optical Characteristics for Sensing Of Mercury Using Green-Synthesized Silver Nanoparticles. *J. Analy. Methods Chem.* **2020,** *2020*.

Aragay, G.; Pons, J.; Merkoçi, A. Recent Trends in Macro-, Micro-, and Nanomaterial-Based Tools and Strategies for Heavy-Metal Detection. *Chem. Rev.* **2011,** *111* (5), 3433–3458.

Arduini, F.; Cinti, S.; Scognamiglio, V.; Moscone, D. Nanomaterial-Based Sensors. In *Handbook of Nanomaterials in Analytical Chemistry*; Hussain, C. M., Ed.; **2020**; pp 329–359.

Baldursson, S.; Karanis, P. Waterborne Transmission of Protozoan Parasites: Review of Worldwide Outbreaks–an Update 2004–2010. *Water Res.* **2011,** *45* (20), 6603–6614.

Bao, H.; Zhang, H.; Zhou, L.; Fu, H.; Liu, G.; Li, Y.; Cai, W. Ultrathin and Isotropic Metal Sulfide Wrapping on Plasmonic Metal Nanoparticles for Surface Enhanced Ram Scattering-Based Detection of Trace Heavy-Metal Ions. *ACS Appl. Mater. Interf.* **2019,** *11* (31), 28145–28153.

Bhardwaj, N.; Bhardwaj, S. K.; Bhatt, D.; Lim, D. K.; Kim, K. H.; Deep, A. Optical Detection of Waterborne Pathogens Using Nanomaterials. *TrAC Trends Analy. Chem.* **2019,** *113*, 280–300.

Borgå, K.; Fisk, A. T.; Hargrave, B.; Hoekstra, P. F.; Swackhamer, D.; Muir, D. C. Bioaccumulation Factors for PCBs Revisited. *Environ. Sci. Technol.* **2005,** *39* (12), 4523–4532.

Bui, M. P. N.; Ahmed, S.; Abbas, A. (2015). Single-Digit Pathogen and Attomolar Detection with the Naked Eye Using Liposome-Amplified Plasmonic Immunoassay. *Nano Lett.* **2015,** *15* (9), 6239–6246.

Buzea, C.; Pacheco, I. Nanomaterials and Their Classification. In *EMR/ESR/EPR Spectroscopy for Characterization of Nanomaterials*; Springer: New Delhi, 2017; pp 3–45.

Cai, X.; Xie, Z.; Li, D.; Kassymova, M.; Zang, S. Q.; Jiang, H. L. Nano-Sized Metal-Organic Frameworks: Synthesis and Applications. *Coord. Chem. Rev.* **2020,** *417*, 213366.

Chu, H.; Huang, Y.; Zhao, Y. Silver Nanorod Arrays as a Surface-Enhanced Raman Scattering Substrate for Foodborne Pathogenic Bacteria Detection. *Appl. Spectrosc.* **2008,** *62* (8), 922–931.

Dubertret, B. DNA Detectives. *Nat. Mater.* **2005,** *4* (11), 797–798.

Farbod, F.; Mazloum-Ardakani, M. Typically Used Nanomaterials-Based Noncarbon Materials in the Fabrication of Biosensors. In *Electrochemical Biosensors*; Elsevier, 2019; pp 99–133.

Govindarajalu, A. K.; Ponnuchamy, M.; Sivasamy, B.; Prabhu, M. V.; Kapoor, A. A Cellulosic Paper-Based Sensor for Detection of Starch Contamination in Milk. *Bull. Mater. Sci.* **2019,** *42* (6), 255.

Hadinoto, K.; Sundaresan, A.; Cheow, W. S. Lipid–Polymer Hybrid Nanoparticles as a New Generation Therapeutic Delivery Platform: A Review. *Eur. J. Pharm. Biopharm.* **2013,** *85* (3), 427–443.

Huang, X.; Liu, Q.; Yao, S.; Jiang, G. Recent Progress in the Application of Nanomaterials in the Analysis of Emerging Chemical Contaminants. *Analy. Methods* **2017,** *9* (19), 2768–2783.

Hung, Y. L.; Hsiung, T. M.; Chen, Y. Y.; Huang, Y. F.; Huang, C. C. Colorimetric Detection of Heavy Metal Ions Using Label-Free Gold Nanoparticles and Alkanethiols. *J. Phys. Chem. C* **2010,** *114* (39), 16329–16334.

Jamkhande, P. G.; Ghule, N. W.; Bamer, A. H.; Kalaskar, M. G. Metal Nanoparticles Synthesis: An Overview on Methods of Preparation, Advantages and Disadvantages, and Applications. *J. Drug Deliv. Sci. Technol.* **2019,** *53*, 101174.

Jang, A.; Zou, Z.; Lee, K. K.; Ahn, C. H.; Bishop, P. L. State-of-the-Art Lab Chip Sensors for Environmental Water Monitoring. *Measure. Sci. Technol.* **2011,** *22* (3), 032001.

Jin, B.; Wang, S.; Lin, M.; Jin, Y.; Zhang, S.; Cui, X.; Lu, T. J. Upconversion Nanoparticles Based FRET Aptasensor for Rapid and Ultrasenstive Bacteria Detection. *Biosens. Bioelectr.* **2017,** *90*, 525–533.

Ju, D.; Xu, H.; Qiu, Z.; Guo, J.; Zhang, J.; Cao, B. Highly Sensitive and Selective Triethylamine-Sensing Properties of Nanosheets Directly Grown on Ceramic Tube by Forming NiO/ZnO PN Heterojunction. *Sens. Actuat. B Chem.* **2014,** *200*, 288–296.

Khan, S. B.; Asiri, A. M.; Akhtar, K.; Rub, M. A. Development of Electrochemical Sensor Based on Layered Double Hydroxide as a Marker of Environmental Toxin. *J. Indust. Eng. Chem.* **2015,** *30*, 234–238.

Kim, Y. G.; Moon, S.; Kuritzkes, D. R.; Demirci, U. Quantum Dot-Based HIV Capture and Imaging in a Microfluidic Channel. *Biosens. Bioelectr.* **2009,** *25* (1), 253–258.

Kneipp, K.; Kneipp, H.; Itzkan, I.; Dasari, R. R.; Feld, M. S. Ultrasensitive Chemical Analysis by Raman Spectroscopy. *Chem. Rev.* **1999,** *99* (10), 2957–2976.

Kumar, P.; Kim, K. H.; Bansal, V.; Lazarides, T.; Kumar, N. Progress in the Sensing Techniques for Heavy Metal Ions Using Nanomaterials. *J. Indust. Eng. Chem.* **2017,** *54*, 30–43.

Lee, G. J.; Lee, H. M.; Rhee, C. K. Bismuth Nano-Powder Electrode for Trace Analysis of Heavy Metals Using Anodic Stripping Voltammetry. *Electrochem. Commun.* **2007,** *9* (10), 2514–2518.

Lu, Y.; Liang, X.; Niyungeko, C.; Zhou, J.; Xu, J.; Tian, G. A Review of the Identification and Detection of Heavy Metal Ions in the Environment by Voltammetry. *Talanta* **2018,** *178*, 324–338.

Mokhtarzadeh, A.; Eivazzadeh-Keihan, R.; Pashazadeh, P.; Hejazi, M.; Gharaatifar, N.; Hasanzadeh, M.; de la Guardia, M. Nanomaterial-Based Biosensors for Detection of Pathogenic Virus. *TrAC Trends Analy. Chem.* **2017,** *97*, 445–457.

Nagaraju, G.; Prashanth, S. A.; Shastri, M.; Yathish, K. V.; Anupama, C.; Rangappa, D. Electrochemical Heavy Metal Detection, Photocatalytic, Photoluminescence, Biodiesel Production and Antibacterial Activities of Ag–ZnO Nanomaterial. *Mater. Res. Bull.* **2017,** *94*, 54–63.

Navale, S. T.; Bandgar, D. K.; Nalage, S. R.; Khuspe, G. D.; Chougule, M. A.; Kolekar, Y. D.; Patil, V. B. Synthesis of Fe_2O_3 Nanoparticles for Nitrogen Dioxide Gas Sensing Applications. *Ceram. Int.* **2013,** *39* (6), 6453–6460.

Ong, T. T.; Blanch, E. W.; Jones, O. A. Surface Enhanced Raman Spectroscopy in Environmental Analysis, Monitoring and Assessment. *Sci. Total Environ.* **2020,** 137601.

Pieczonka, N. P.; Aroca, R. F. Single Molecule Analysis by Surfaced-Enhanced Raman Scattering. *Chem. Soc. Rev.* **2008,** *37* (5), 946–954.

Pol, R.; Céspedes, F.; Gabriel, D.; Baeza, M. Microfluidic Lab-on-a-Chip Platforms for Environmental Monitoring. *TrAC Trends Analy. Chem.* **2017,** *95*, 62–68.

Prosposito, P.; Burratti, L.; Venditti, I. Silver Nanoparticles as Colorimetric Sensors for Water Pollutants. *Chemosensors* **2020,** *8* (2), 26.

Saleh, T. A. Nanomaterials: Classification, Properties, and Environmental Toxicities. *Environ. Technol. Innov.* **2020,** 101067.

Saleh, T. A.; Fadillah, G. Recent Trends in the Design of Chemical Sensors Based on Graphene–Metal Oxide Nanocomposites for the Analysis of Toxic Species and Biomolecules. *TrAC Trends Analy. Chem.* **2019,** *120*, 115660.

Sau, T. K.; Rogach, A. L. Nonspherical Noble Metal Nanoparticles: Colloid-Chemical Synthesis and Morphology Control. *Adv. Mater.* **2010,** *22* (16), 1781–1804.

Shamsipur, M.; Karimi, Z.; Tabrizi, M. A. A Novel Electrochemical Cyanide Sensor Using Gold Nanoparticles Decorated Carbon Ceramic Electrode. *Microchem. J.* **2017,** *133*, 485–489.

Sharma, B.; Frontiera, R. R.; Henry, A. I.; Ringe, E.; Van Duyne, R. P. SERS: Materials, Applications, and the Future. *Mater. Today* **2012,** *15* (1–2), 16–25.

Sharma, G.; Kumar, A.; Sharma, S.; Naushad, M.; Dwivedi, R. P.; ALOthman, Z. A.; Mola, G. T. Novel Development of Nanoparticles to Bimetallic Nanoparticles and Their Composites: A Review. *J. King Saud Univ.-Sci.* **2019,** *31* (2), 257–269.

Su, S.; Wu, W.; Gao, J.; Lu, J.; Fan, C. Nanomaterials-Based Sensors for Applications in Environmental Monitoring. *J. Mater. Chem.* **2012,** *22* (35), 18101–18110.

Sudha, P. N.; Sangeetha, K.; Vijayalakshmi, K.; Barhoum, A. Nanomaterials History, Classification, Unique Properties, Production and Market. In *Emerging Applications of Nanoparticles and Architecture Nanostructures*; Elsevier, 2018; pp 341–384.

Tang, D.; Yuan, R.; Chai, Y. Magnetic Control of an Electrochemical Microfluidic Device with an Arrayed Immunosensor for Simultaneous Multiple Immunoassays. *Clin. Chem.* **2007,** *53* (7), 1323–1329.

Thakur, B.; Amarnath, C. A.; Mangoli, S. H.; Sawant, S. N. Polyaniline Nanoparticle Based Colorimetric Sensor for Monitoring Bacterial Growth. *Sens. Actuat. B Chem.* **2015,** *207*, 262–268.

Tian, F.; Bonnier, F.; Casey, A.; Shanahan, A. E.; Byrne, H. J. Surface Enhanced Raman Scattering with Gold Nanoparticles: Effect of Particle Shape. *Analy. Methods* **2014,** *6* (22), 9116–9123.

Tiwari, M.; Prakash, R. Colorimetric Detection of Picric Acid Using Silver Nanoparticles Modified with 4-amino-3-hydrazino-5-mercapto-1, 2, 4-triazole. *Appl. Surf. Sci.* **2018,** *449*, 174–180.

Vilela, D.; González, M. C.; Escarpa, A. Sensing Colorimetric Approaches Based on Gold and Silver Nanoparticles Aggregation: Chemical Creativity Behind the Assay. A Review. *Analy. Chim. Acta* **2012,** *751*, 24–43.

Walekar, L.; Dutta, T.; Kumar, P.; Ok, Y. S.; Pawar, S.; Deep, A.; Kim, K. H. Functionalized Fluorescent Nanomaterials for Sensing Pollutants in the Environment: A Critical Review. *TrAC Trends Analy. Chem.* **2017,** *97*, 458–467.

Wang, H.; Lustig, W. P.; Li, J. Sensing and Capture of Toxic and Hazardous Gases and Vapors by Metal–Organic Frameworks. *Chem. Soc. Rev.* **2018a,** *47* (13), 4729–4756.

Wang, X. F.; Song, X. Z.; Sun, K. M.; Cheng, L.; Ma, W. MOFs-Derived Porous Nanomaterials for Gas Sensing. *Polyhedron* **2018b,** *152*, 155–163.

Willner, M. R.; Vikesland, P. J. Nanomaterial Enabled Sensors for Environmental Contaminants. *J. Nanobiotechnol.* **2018,** *16* (1), 1–16.

Yang, H.; Li, J.; Lu, X.; Xi, G.; Yan, Y. Reliable Synthesis of Bismuth Nanoparticles for Heavy Metal Detection. *Mater. Res. Bull.* **2013,** *48* (11), 4718–4722.

Yao, K. S.; Li, S. J.; Tzeng, K. C.; Cheng, T. C.; Chang, C. Y.; Chiu, C. Y.; Lin, Z. P. Fluorescence Silica Nanoprobe as a Biomarker for Rapid Detection of Plant Pathogens. In *Advanced Materials Research*, Vol. 79; Trans Tech Publications Ltd., 2009; pp 513–516.

Yu, H.; Zhao, G.; Dou, W. Simultaneous detection of Pathogenic Bacteria Using Agglutination Test Based on Colored Silica Nanoparticles. *Cur. Pharm. Biotechnol.* **2015,** *16* (8), 716–723.

Zhang, L.; Fang, M. Nanomaterials in Pollution Trace Detection and Environmental Improvement. *Nano Today* **2010,** *5* (2), 128–142.

Zhao, Y. A New Era of Metal–Organic Framework Nanomaterials and Applications, 2020.

Zhou, H.; Yang, D.; Ivleva, N. P.; Mircescu, N. E.; Schubert, S.; Niessner, R.; Haisch, C. Label-Free in Situ Discrimination of Live and Dead Bacteria by Surface-Enhanced Raman Scattering. *Analy. Chem.* **2015,** *87* (13), 6553–6561.

PART III
Nanotechnology and Nanobiosensors

CHAPTER 6

Introduction: Nanozymes—Nouvelle Vague of Artificial Enzymes

MANMEET KAUR[1*], INDERPAL KAUR[2], and GAUTAM CHHABRA[3]

[1]*Department of Microbiology, Punjab Agricultural University, Ludhiana 141001, India*

[2]*Department of Biochemistry, Punjab Agricultural University, Ludhiana 141001, India*

[3]*School of Agricultural Biotechnology, Punjab Agricultural University, Ludhiana 141001, India*

[*]*Corresponding author. E-mail: manmeet1-mb@ pau.edu*

ABSTRACT

In evolutionary processes, natural enzymes perform essential functions. Nanozymes which are "catalyzed nanoparticles containing enzyme-like features" have gained significant attention as a budding scientific area of "artificial enzymes." They also incur through feasible drawbacks, such as "substantial rate"; "limited effectiveness," and "reusing concerns" while they are spectacular. Scientists have long been accused to exploring synthetic zyme that emulates to address these limitations. In the next century, following the disclosure of "ferromagnetic nanomaterials by integrated peroxidase-like reaction," a considerable number of nanozyme research have steadily grown. Nanozymes are such form of nanostructures with "enzymatic catalytic properties" and have the benefits of "minimal price"; "better resistance," and "dimensional stability" over the "commercial" natural enzymes. A detailed evaluation of potential catalytic mechanisms will contribute to the production

Nanotechnology for Environmental Pollution Decontamination: Tools, Methods, and Approaches for Detection and Remediation. Fernanda Maria Policarpo Tonelli, Rouf Ahmad Bhat, & Gowhar Hamid Dar (Eds.)

of innovative and relatively inexpensive nanozymes, and the strategic issue is the intellectual effect of nanozyme activities. The "categorization"; "catalyzed pathway"; "activity monitoring," and scientific reports for nanozyme processing in the fields such as environmental protection have also been prioritized over the past couple of decades.

6.1 INTRODUCTION

Enzymes are primarily made up of proteins as potent biocatalysts, although some are catalytic RNA molecules (Bornscheuer et al., 2012). Although conventional chemical catalysts or industrial catalysts are frequently used to catalyze the modification of biomolecules under brutal conditions, such as "elevated temperatures"; "high pressure"; "organic solvents," and "intense pH enzymes" under light irradiation, these reactions are specifically excluded (Behrens et al., 2012). Natural enzymes are widely used because of their high catalytic efficiency and substrate specificity in "manufacturing industries," "pharmaceutical," and "epigenetic fields." Although desirable, intrinsic shortcomings such as "exorbitant prices of scheduling and purification," "lower efficiency of processing"; "acuity of catalytic process to global threats," and uncertainties of recovery are often apparent (Yan, 2018). These drawbacks all restrict their further applications in "food manufacturing," "microbiology," "bioimaging," "sustainable development." Researchers have long been committed to the conceptualization of artificial enzyme mimics in order to address these disadvantages. Research has consistently shown that a number of materials can serve as "artificial enzymes" with function and structure close to those of "fullerenes," "cyclodextrins," "polymers." "dendrimers," "porphyrins," "metal complexes," and certain "biomolecules" (Gong et al., 2015).

A lot of researchers on nanostructured materials "artificial enzymes (nanozymes)" have indeed been gaining traction since the inventors of iron oxide nanoparticles as peroxidase exemplifies in 2007 (Wang et al., 2018). One type of nanocrystals with "nano-scale sizes upto 100 nm" and enzymatic catalyzed attributes is "nanozymes." The attributes of chemically synthesized catalysts and biocatalysts are effectively incorporated with nanozymes. It is possible to classify them into two classes (Zhou et al., 2017):

- ➢ Nanomaterials have intrinsic enzymatic catalytic properties that demonstrate a mechanism identical to enzymes that catalyze the identical biocatalytic reactions.

- Nanomaterials, called "nanomaterial hybrid enzymes," "reorient enzymes," or "enzymatic catalytic groups." With the help of nanoparticles, modified enzymes or enzymatic catalytic groups can offer outstanding fortitude as well as strength.

In 2004, Pasquato, Scrimin, and their colleagues coined the term "nanozymes" to characterize the imitates of "Au nanoparticle-based transphosphorylation" arising out of the actualization of "triazacyclonane-functionalized thiols" on the exterior of "au nps." Than, Wang, and Wei described "nanozymes" as "nanomaterials with enzyme-like features" in their comprehensive analysis published in 2013. The resurgence of interest in nanozymes might be attributed to "natural enzymes" because of their unique traits. Nanozymes are unique in many ways, such as their "size-tunable catalytic activities," "large alteration," and "bioconjugation surface area," "multiple functions in addition to catalysis," "smart reaction to external stimuli," etc. (Table 6.1). Nanozymes are discussed because numerous "enzyme-mimicking behaviors" have been demonstrated by several nanomaterials (Wei and Wang, 2013).

6.2 CLASSIFICATION OF NANOENZYMES

In metabolic pathways, a number of enzymes were found. They engage individually or together in complicated biosynthetic mechanisms and play a role in the emergence of life (Peters et al., 2014). It is of enormous effect to analyze their alternative methods, stressing the importance of metabolic pathways in the life system and their profoundly embedded disadvantages. Many nanomaterials have acted as possible enzymatic recruits for implementations due to the continued endeavors of researchers (Liu et al., 2017). Nanozymes are graded according to two categories:

- family of oxidoreductase, such as "oxidase," "peroxidase," "catalase," "superoxide," "dismutase," and "nitrate reductase"
- family of hydrolase, including "nuclease," "esterase," "phosphatase," "protease," and "silicatein" (Table 6.2). Graphene and carbon nanotubes, for instance, have been shown to have exemplary peroxidase-like properties in the presence of H_2O_2 to catalyze the oxidation of several substrates, such as "3,3,"5,5"-tetramethylbenzidine (TMB)" and "2,2"-azino-bis(3-ethylbenzothiazoline-6-sulfonic acid) (ABTS)."

TABLE 6.1 Nanozymes Properties with Artificial and Natural Enzymes (Wei and Wang, 2013).

S. No.	Nanozymes	Natural enzymes	Conventional artificial enzymes
1	Low cost	High catalytic output	Low cost
2	Resilience to extreme conditions	Elevated selectivity	Resilience against harsh environments
3	Strong degree of stability	Large specificity of substrate	High stability
4	Easy for mass manufacturing	Three structures with proportions	Simple for mass production
5	Long-term storage	Excellent biocompatibility	Long-term storage
6	Action tunable	Action tunable	Tunable intervention
7	Multifunctioning	Reasonable design through the engineering of proteins	Uniform dimensions
8	Convenient for alteration (such as bioconjugation)	–	Established molecular mimics structure
9	Self-assembling	–	–
10	Intelligent reaction to external stimuli	–	–

TABLE 6.2 Nanomaterials as Enzyme Mimics with Applications.

S. No.	Enzymes	Nanomaterial	Applications	References
1	Oxidase	N-CNMs	Cancer therapy	Fan et al. (2018)
		Ag	Detection	Wang et al. (2015)
		ZnO	Detection	Biparva et al. (2014)
		MnO_2	Immuno-detection	Liu et al. (2017)
2	Glucose oxidase	Au	Detection	Zhou et al. (2016)
		Au-MOF	Detection	Huang et al. (2017)
3	Nitric oxide synthase	Grapheme-hemein	Antithrombosis	Xue et al. (2014)
4	Horseradish peroxidase	Cu	Detection	Hu et al. (2013)
		Pt-Cu	Phenol degradation	Li et al. (2018)
		MnO_2	Immuno-assays	Liu et al. (2012)
		MoS_2	Antibacteria	Yin et al. (2016)
5	Superoxide dismutase	Pd	Antioxidant	Ge et al. (2016)
		Mn_3O_4	Antioxidant	Singh et al. (2017)

6.3 NANOENZYMES CATALYTIC PATHWAYS

While nanozymes have vital consideration attention in recent decades, there seems to be no qualitative interpretation of their "catalytic and kinetic mechanisms." The "catalytic mechanisms" based on the enzyme groups can be evaluated.

6.3.1 OXIDASE FAMILY

a) **Glucose Oxidase:** In 2004, Rossi and colleagues suggested that in the presence of O_2 "naked Au nanoparticles" could catalyze $C_6H_6O_2$ and produce "gluconic acid with H_2O_2." Insulation panel have said that there is no superior catalytic capability of other metal nanoparticles to react with "glucose oxidation." The mechanism of molecular formation for gold catalysis is still seen as a "bonding of moisturized carbohydrate moiety with Au particle surface" might form "electron-rich gold species" based on the recruitment "essence of alkali as well as the iteration of H_2O_2" and "effectively triggering oxygen through nucleophilic assault" (Fig. 6.1) (Comotti et al., 2004).

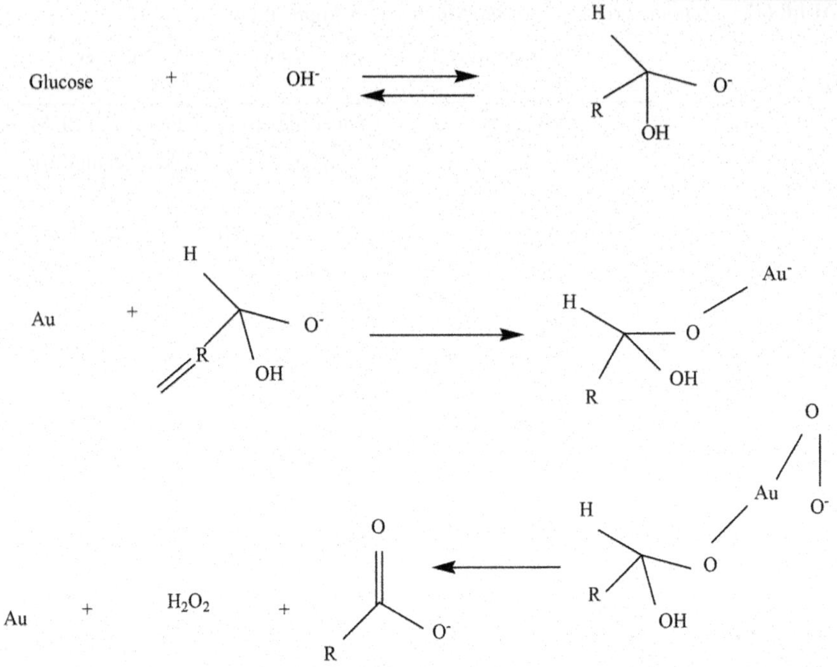

FIGURE 6.1 Catalytic mechanism of gold nanoparticles (glucose oxidase mimics).

b) **Sulfite oxidase:** A Mo-containing enzyme with "Cytochrome c" since an "acceptor of electrons" will catalyze toxic "sulfite sulphate oxidation." In intracellular detoxification procedures, sulfite oxidase is essential, although the depletion can lead to certain ailments (Hundallah and Jabari, 2016). Tremel et al. (2014) showed that "MoO_3 nanoparticles" had an affirmative "sulfite oxidase-like behavior" that could function under defined circumstances as a potential therapist for above oxidase deficiency. The potential catalytic activity of "MoO_3 nanoparticles" was described in such a way that {SO_3 $^{2-}$} has been coupled to MoO_3 nanoparticles and then, with a decrement from "Mo^{VI} to Mo^{IV}," oxidized to [SO_4 $^{2-}$]. [Mo^{IV}] was reoxidized by the electron - accepting K_3 by one-electron steps to the initial valence state "[$Fe(CN)_6$]."

6.3.2 PEROXIDASE FAMILY

a) **Peroxidase:** Many nanomaterials based on carbon have so far been discovered to disport acceptable "peroxidase-like properties."

However, their catalytic mechanism is seldom debated. Qu et al. finally took the exemplar of "graphene quantum dots (GQDs)" to unearth the rate equation (Sun et al., 2015). Individuals reported that perhaps the "-C=O and-O=CO" groups will sometimes accomplish as that of the "catalytic site" and "surface site," consequently, through most of juxtaposition with evaluated facts and theoretical calculation. The reinterpretation of these categories would greatly enhance the thermal stability of GQDs. That exclusion, mostly on the opposite hand, for the "-C-OH groups" might restrict the enzyme - mediated properties. Unusually, Yang, Perrett, and colleagues found that "iron oxide nanoparticles" might comprise the oxidation of a sequence of substances as imitates of peroxidase to catalyze (Gao et al., 2007). In keeping with the "kinetic observations of an equilibrium point," the activation energy based on "Lineweaver-Burk plots" was identical. Such model implies that somehow a "ponging overall reaction" may follow the kinetic model of nanomaterials. To create intermediate "•OHO4," iron oxide can interact with the very first hydrogen peroxide substratum. "Trimethyl benzoic," "H+" from either the recipient of hydrogen will then absorb the "•OH" emitted.

b) **Glutathione peroxidase:** Glutathione peroxidase is a type of enzymatic antioxidants whose cysteine-selenium is catalytic center (Wirth, 2015). Qu, Ren, and co-workers rationally developed an innovative nanozyme with a spectacular "glutathione peroxidase-like antioxidant" potential for cytoprotection by assembling glucose oxidase and Se. A "ping-pong reaction" mechanism to catalyze H_2O_2 decomposition, with respect to "glutathione peroxidase," adopted by "glucose oxidase-selenium nanozyme," a molecule of H_2O_2 interacted to create intermediary "Se-oxide with nanoselenium." The obtained intermediate then "oxidized GSH" further to form " GSSG," while the "Se" intermedi would retrograde to its initial state and another molecule of H_2O_2 then react with the portion of "nanoselenium" (Huang et al., 2017).

c) **Haloperoxidase:** In 2012, the Tremel finding shows that "Vanadium oxide" nanowires would act as "vanadium haloperoxidase" in the underwater biome to resist biofilms (Natalio et al., 2012). "Vanadium oxide" nanowires can catalyze "halide ions" in the presence of hydrogen peroxidase, and generate "hypohalous acids." The "hypohalous acids" produced, thus shielding ships against underwater microbial colonization, will induce oxidative stress to marine organisms. They found that the accessible simultaneous oscillator planes of vanadium oxide nanowires seemed to have a localized vanadium coordination configuration similar with that of the active site of normal "vanadium haloperoxidase" by investigating the reaction.

6.3.3 CATALASE

CeO_2 nanostructures could act in a "redox-state-dependent manner" as catalase photosensitizer; and elevated Ce^{4+} levels should contribute to catalyzed ability (Pirmohamed et al., 2010). Initiated by particles acceleration and oxygen, "Ce^{4+}" had first been diminished by hydrogen peroxidase molecule to create "Ce^{3+}" and then further combined with the oxygen vacancy site and oxidized "Ce^{3+} to Ce^{4+}" with the release of water (Celardo et al., 2011).

6.3.4 OTHERS

Gao and colleagues have carried out a number of experiments on the catalytic function of metallic nanoparticles (Shen et al., 2015). It seemed that the "oxidase-like activity" of metals, such as "gold," "silver,"; "Platinum," and "Palladium" may depend on the dissociation of the high oxygen reactions assisted by the numerical simulation and descriptive statistical based on the surfaces of such nanoparticles. The catalyzed pathways were largely due to "high oxygen protonation" and "biosorption" and "OH radicals reconfiguration" for "super oxide dismutase-like abilities" on the interfaces. Furthermore, metal nanomaterials" "peroxidase methods" and "catalase-like activities" too were studied (Li et al., 2015). They indicated that the inherent properties of materials were enzymatic processes. This was primarily due to the normal "hydrogen peroxide decomposition" on the materials due to their "peroxidase-like action" demonstrated throughout the oxidizing atmosphere. Although on their interfaces, the "acid-like decomposition of hydrogen peroxidase" was attributable to the "catalase-like behavior" demonstrated under normal conditions. Under simple conditions can the preadsorbed "OH groups" be positively established, triggering the transformation of catalase and peroxidase interaction.

6.4 TUNING OF THE "NANOENZYMES" CATALYTIC ACTIVITIES

The behaviors of nanozymes, analogous to natural enzymes, can be altered by many factors, such as "temperature," "ambient atmosphere," "pH," and "metal ions." The "steric effect" caused by light will disrupt the association between nanozymes and their corresponding substrates, leading to a decrease in reactants. Furthermore, the catalytic efficiency of nanozymes could also be affected by thermal influenced temperature along with pH varies. The consequences mostly on activities of nanozymes are intuited in the catalysis on "pH," "size," "structure," "ions or molecules," "surface modification classes," "morphology," "substrate selectivity," and "temperature" in order to assess the effects of light irradiation (Puvvada et al., 2012).

6.4.1 SIZE

It is excellently documented that the size of materials determines the "catalytic efficiency of nanomaterials." Yang, Perrett, and colleagues researched the catalytic activity of "Fe_3O_4 nanoparticles" of various sizes showed that smaller size nanoparticles had the highest peroxidase-like property, while the minimum catalytic activity was observed in the largest size nanoparticles. Perhaps it is because it is possible to combine precarious nanoparticles with a higher surface-to-volume ratio with specific substrates (Gao et al., 2007). The catalytic properties of nanozymes could be controlled mostly on basis of size. Fan et al. proposed that "size-dependent glucose oxidase" was equivalent to "catalytic properties of gold nanoparticles" (Luo et al., 2010).

6.4.2 MORPHOLOGY

Nanozyme biocatalytic production can often be tuned by changing the structure and also the substrate. Two "Palladium nanomaterials" were developed by the Yin and Chen groups: nanocubes and octahedrons" (Ge et al., 2016). They showed that "Palladium octahedrons" with lower energy density had increased "catalase & superoxide dismutase-like properties" than that of "Palladium nanocubes" using the "Electron spin resonance spectroscopy." Quite few sample experiments showed that under the same condition," Palladium octahedrons" had a greater ability to extract "reactive oxygen species" compared favorably to "Palladium nanocubes." The conceptual guesstimate was contrary to the results of analytics. In an attempt to reduce oxidative stress with excellent catalytic properties, research may increase the efficiency of nanozymes (Puvvada et al., 2012).

6.4.3 SURFACE MODIFICATION

Surface modification, ranging "coated layer," "surface ion," and "functional group" in order to comply with extant literature, indeed directly impact the catalyzed abilities of nanozymes (Wang et al., 2012). Different "functional groups" of modified "AuNPs may display various catalytic enzymatic activities." "Citrate-modified AuNPs" have "glucose oxidase-like characteristics," while "cysteine-modified AuNPs" can act as "peroxidase imitators" (Lin et al., 2013). In 2012, the Lin and Chen groups concentrated on the variations in the "peroxidase-like reliability of gold particles" with various coating changes. "Simple amino-modified" as well as "citrate-modified gold nanoparticles" have been identified for analysis in research work. They found that "uncoated gold nanoparticles" exhibited "promising properties" relative to the other gold particles through a separate analysis.

The catalyzed cores of nanozymes and the corresponding substrates can be separated by implementing surface coatings, leading to decreased catalytic efficiency. The Perez group systematically analyzed the influence of "polymer thickness" and the "modification group" on the "oxidase-like property" on the existence of nanoceria. Control tests showed that the thin "poly (acrylic acid) coated" nanoceria had a comparatively high oxidase-like property compared to the considerably "thicker dextran wrapped" nanoceria. It might be because, especially in comparison to a "heavier dextran surface," the "thinner poly (acrylic acid) surface" was permeable; that would allow the adsorbent surface to be transferred from the nanoplate's core surface. The incredibly catalytic activity of natural enzymes, such as the "catalysts activity center" and the "substrate binding site" is due to the inherent catalytic structure (Asati et al., 2009).

6.4.4 COMPOSITION

The catalytic activity of nanozymes could also be monitored by altering the nanoparticle portion of the materials (Xu et al., 2015). Furthermore, the injecting of many aspects into nanozymes was shown to be an effective way of controlling the actions of nanozymes. A biocompatible method to optimizing the catalytic reactions of nanozymes was speculated by Qu, Ren, and co-workers, driven by the "configuration and mediated process of natural enzymes." "The "iron porphyrin" is identified as the "catalytic nucleus," while the "eroded heme tip" represents the "substrate binding site.

"Fe^{3+}doped polymeric nanospheres (Fe3+-MCNs)" have been engineered to imitate the structure.

The "peroxidase-like behavior" might be controlled appropriately by altering the concentrations of nanostructures. Afterwards, "Palladium-Platinum" alloy nanodots were synthesized and evaluated mainly mostly on substrate of "gold nanoparticles." The "gold-Palladium Platinum" alloy nanowires formed in the presence of oxygen could act as an oxidase mimic for catalyzing the oxidation of tetramethyl benzidine. The presence of the amino-acid oxidation portion of Palladium in the "Platinum anoparticle" could promote the effectiveness of the process effectively. The catalytic potential of the gold–palladium–platinum alloy nanorode has greatly affected the elevation of the Palladium nanocomponent. As the "Palladium to Platinum" proportion for tetramethyl benzoic oxidation increased to five, the" Au-PdPt alloy "nanorod" oxidase-like activity increased significantly. The required alloying of palladium and platinum together sponsored a crucial factor to alter the oxidase-like behavior of "gold-Palladium-Platinum" nanoparticles (Sang et al., 2018).

6.4.5 CONSTRUCTING HYBRID NANOPARTICLES

To present, the prominence of prototype nanomaterials has gained due to certain "well-defined processes," "improved efficiency" (Lee et al., 2013). Catapulted by all this, when producing hybrid nanoparticles, nanozymes will obtain supervisory catalyzed abilities. Although "gold nanoparticles" can work as" peroxidase imitates" effective; their catalytic performance is often restricted by ambient pH (Li et al., 2015).

Hybrid nanozymes from "GO-AuNCs" were produced by the Qu and Ren groups to achieve optimum catalytic efficiency over a broad pH range (Tao et al., 2013). In their operation, glucose oxidase was used as the regulator "peroxidase-like action" to amplify the "lysozyme-coated gold nanoclusters." Under a weak base environment, "tetramethyl benzoic" had a remarkable similarity to "diamine" that resulted in low aqueous viscosity. Glucose oxidase may also be used for the effectively sorption surface tetramethyl benzoic, mainly with specific properties such as high "surface area to volume ratio" and "good affinity" for electrostatic repulsion. This could facilitate the tendency to function with relevant substrates with the active "gold nanocomposites" sites, causing an increase in "AuNCs peroxidase-like ability." The accumulating hybrid nanozymes can thus serve as strong peroxide over a very wide pH range (Wang et al., 2009).

6.4.6 pH AND TEMPERATURE

The accompanying pH also may affect the nanozymes" catalytic activity; for example, under acidic conditions, gold nanoparticles may be considered candidates for peroxidase, while gold nanoparticles may demonstrate catalytic characteristics comparable to "catalase or superoxide dismutase" under neutral and alkaline conditions (Li et al., 2015). In addition, the groups of Gu and Zhang reported that iron oxide nanoparticles can act as peroxidase accentuates to catalyze iron oxide nanozymes, and will exhibit a catalase-like capacity to generate detrimental hydrogen peroxidase in a favorable physiological conditions. Likewise, temperature may also have a significant impact on nanozymes" catalytic performance (Chen et al., 2012).

6.4.7 MOLECULES OR IONS

Earlier studies have consistently shown that ions may represent as effectors, as well as some molecules, to regulate the catalytic performance of nanozymes. In order to activate or hinder the processes of nanozymes, certain ions or molecules may be used. For example, the catalytic ability of citrate-capped gold nanoparticles could be increased by "mercury," "lead," and "bismuth." The citrate resistance on the layer of "gold nanoparticles" could limit these to the related material if these organic compounds endured. Due to the strong affinity between both the substrate of gold nanoparticles and metal0, the decreased metal could reabsorb on the particle surfaces, allowing the surface properties and the "peroxidase-like characteristics" of "gold nanoparticles" to alter.

While nanozymes provide unprecedented thermostability, it is nevertheless hard to fathom biocatalytic processes under elevated temperature. Perhaps this is attributable to the fact that, under extreme temps, the heat dissipation resilience of oxidative products including "ABTS•+" can sometimes provoke the biocatalytic reaction to operate inadequately. In order to tackle this issue, Qu, Ren, and coworkers utilized organic solvent as the transmitter to instigate "EMSN-AuNPs" reactions over ambient humidity. Ionic liquid is an irresistible fluid, which can adequately sustain the material "ABTS•+." In conjunction, with the assistance of adsorbent, the nanozyme may display enhanced electrochemical capacity. This phenomenon can depend on the coulomb interaction among "organic solvent and material," "anions and cations" as well as the "higher solvation strength." In addition, as an appropriate improving agent, "adenosine triphosphate (ATP)" may be used to perform biocatalysts processes throughout a broad temperature range. Especially in comparison with organic solvent, "ATP" has the leverage of being affordable and conveniently usable, highly transferable to other nanozymes, etc. These studies would encourage the creation of suitable operators to improve the catalytic performance of nanozymes (Lin et al., 2014).

6.4.8 LIGHT

As lighting demonstrates multispectral precision, the design of light-modulated devices could be used as a meaningful method to track the catalyzed energy consumption of nanozymes. For example, by manipulating the isomers "metallo-supramolecular molecules" ($[Fe_2L_3]^{4+}$) mostly on surface of carboxyl-functioned nano graphene sheets (GO-COOH). Again, this dynamic system will carry out the intracellular perception of hydrogen peroxidase in PC-12 cells with the aid of an isomers component and the "peroxidase-like action" of the "graphene oxide" component (Li et al., 2012).

6.5 NANOMATERIALS DEPENDENT ON CARBON FOR NANOENZYMES

In several areas, such as "carbon nanotubes (CNTs)," "fullerene," "grapheme," and their respective derivatives, large applications for carbon-based nanoparticles have already been identified. They have been extensively studied in order to mimic different natural enzymes due to their fascinating catalytic activities.

6.5.1 FULLERENE AND DERIVATIVES AS IMITATORS OF NUCLEASES

Nuclease induces the phosphodiester bond cleavage between two nucleotides in a nucleic acid. Immaculate fullerenes, like "C_{60}," really aren't hydrophilic, rendering it difficult for them all to imitate the enzymes of the aqueous. By some of the implementation of hydrophilic moieties, fullerenes have indeed been assuaged. The Nakamura group developed and analyzed its photo-induced biochemical activity with water-soluble C_{60-1} (Tokuyama et al., 1993). Oddly, they found that "fullerene carboxylic acid (i.e., C_{60-1})" oxidatively divides DNA under light irradiation. As "C_{60-1}" could never entangle to the genetic material to still be sliced, the cleavage was trivial (Boutorine et al., 1994). The targeted cleavage was accomplished by building a triplex at guanine-rich regions. There are other ways of making preferential cleavage feasible. For example, by centralizing fullerenes with DNA intercalators, the modified fullerene derivatives can improve DNA cleavage activity, such as "acridine" compared to the regular fullerene. Previous research has already shown that water-soluble derivatives of carbon (fullerene) can mimic the nuclease. (Yamakoshi et al., 1996).

6.5.1.1 FULLERENE AND DERIVATIVES AS IMITATED BY SUPEROXIDE DISMUTASE

"Reactive oxygen species (ROS)" perform both positive as well as negative roles in life processes. Even if it wasn"t properly managed, "tissue damage" and "associated inflammation" may be caused by superoxide radicals. In existence, superoxide dismutase was created to catalyze the hydroxyl radical dispute into "hydrogen peroxidase" and "molecular oxygen" and thus protecting living systems from harm sustained by anion-induced superoxide. In order to overcome the limitations of natural superoxide dismutase, major attempts have been made to create superoxide dismutase imitates such as "reduced stability and high cost." Clearly influenced by the initial detection that C_{60} could behave as a radical sponge, "two polyhydroxylated C_{60} (i.e., $C_{60}(OH)_{12}$ and $C_{60}(OH)_nO_m$, n = 18–20, m = 3–7 hemicetal groups)" evaluated the neuroprotective activities, the groundbreaking work of Choi and colleagues established the suoeroxide dismutase-mimicking activities of fullerenes and has been successfully implemented (Krusic et al., 1991).

6.5.2 GRAPHENE AND DERIVATIVES

Oxidizing the substrates with hydrogen peroxidase into oxidized materials is catalyzed by peroxidase. The products usually have to be either fluorescent or illuminated, making it simpler to produce peroxidase for a wide range of molecular diagnostic and biological research. Yan and colleagues initially proposed the "peroxidase-mimicking behaviors" of nanomaterials with nanoparticles of iron oxide (Gao et al., 2007). A variety of intriguing nanomaterials to mimic peroxidase have been studied. Among them, when mimicking peroxidase, "grapheme" and its various derivatives displayed moments of magic. Graphene and its derivatives can be loosely divided into two classes of peroxidase-imitating activities. The activities of the first form are primarily derived from "graphene or derivatives" of that form. The activities for the second type are either focused on "graphene-assembled catalysts" or on "the decorated catalysts" synergistic action as well as the graphene-assembled catalysts.

6.5.2.1 GRAPHENE AND ITS DERIVATIVES AS PEROXIDASE MIMICS

While pure graphene with no alterations is not really highly soluble in water, the "peroxidase-mimicking" behavior has been thoroughly researched with graphene derivatives. The inherently peroxidase-mimicking behavior of carbonyl group modification graphene oxide was reported by Qu and coworkers (Song et al., 2010). The process was first illustrated by catalytic oxidation of tetramethyl benzoic with hydrogen peroxidase in the presence of "GO-COOH." Kinetic research showed that "GO-COOH" has a "higher binding affinity" compared to normal peroxidase for tetramethyl benzoic. Interestingly, the reactions of "GO-catalyzed COOH" went through a "ponging phase," it is around the same as natural peroxidase.

No mechanism responsible for enzymatic performance was suggested, even though it was recommended that the transfer of electrons from carboxylated glucose oxidase to hydrogen peroxidase may be possible. These functional moieties such as "Hydroxyl; Ketone; Carboxyl; Epoxide may play a pivotal role in their enzyme-mimicking activities. To specifically disconnect certain functional moieties, multiple analytes were used to explore the fairly huge moieties liable for a graphene derivative"s "peroxidase mimicking" activity called "GQDs." "Phenylhydrazine (PH)," "benzoic anhydride (BA)" and "2-bromo-1-phenylethanone (BrPE)" will react with above three groups selectively, respectively (Fig. 6.2) (Sun et al., 2015).

6.5.2.2 CARBON NANOTUBES AS PEROXIDASE MIMICS

The peroxidase-mimicking behavior of "single-walled carbon nanotubes (SWNTs)" was demonstrated by Qu and colleagues. Via "catalytic oxidation of tetramethyl benzoic" with hydrogen peroxidase, the behavior of "SWNTs" was prosecuted. Instead of just "SWNTs" themselves, the imitating sequence may be responsible for a minor mixture of solid catalyst particulates, since metal catalysts often appear to develop SWNTs. "Sonic-assisted processing with modified acids" (i.e., a mixture of concentrated sulfuric and nitric acids) was carried out to completely remove metal residues in order to solve these problems (i.e., Cobalt). Incredible "SWNTs" and treated "SWNTs" have just never demonstrated any major differences in respective catalyzed processes. This acknowledged that the peroxidase-mimicking activity of "SWNTs" was from the "SWNTs" themselves, but unlike metal residues. They developed a colorimetric method for DNA detection by exploring the various affinities of "ssDNA" and "dsDNA" with "SWNTs."

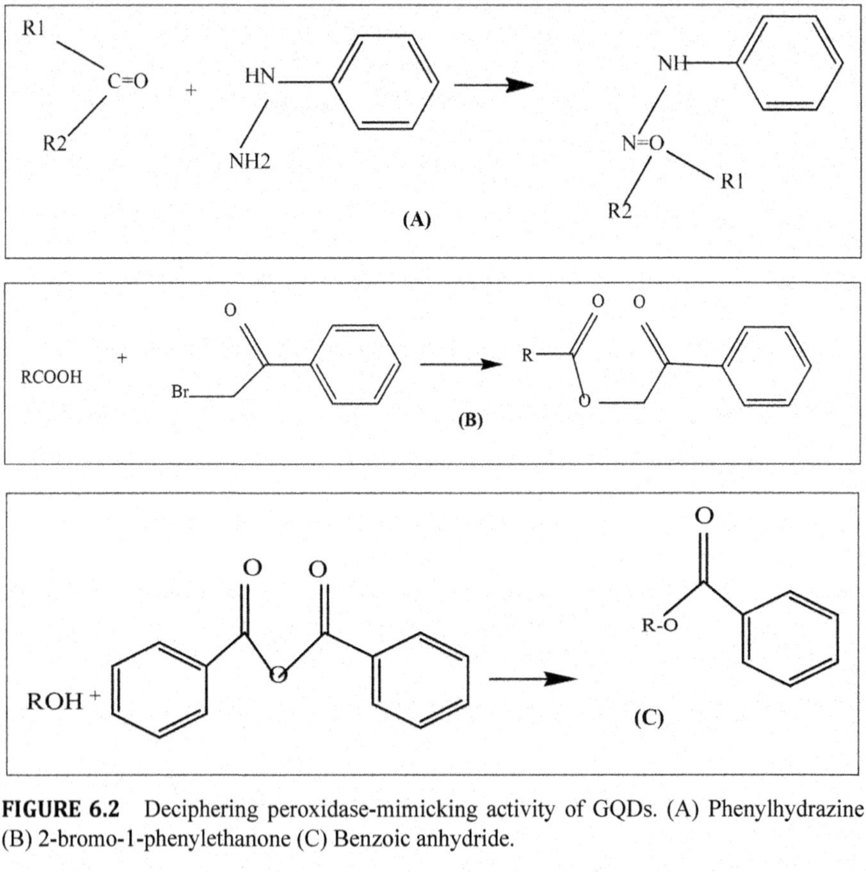

FIGURE 6.2 Deciphering peroxidase-mimicking activity of GQDs. (A) Phenylhydrazine (B) 2-bromo-1-phenylethanone (C) Benzoic anhydride.

Zhu and colleagues claim that the efficacy of iron in helical carbon nanotubes has played a key role in experimental catalysts processes. The significantly larger the "helical carbon nanotube" iron content, the stronger the helical carbon nanotube "peroxidase-mimicking" activity. Its activity was already optimum for the "helical carbon nanotube" with the lowest possible iron number than that of "MWNTs (multiwalled CNTs)" (Cui et al., 2011).

6.5.2.3 CERTAIN CARBON NANOMATERIALS AS IMITATORS OF PEROXIDASE

The "peroxidase-mimicking activities" of "carbon nanohorns" and "carbon nanodots" were explored in several classes (Xu et al., 2015). Xu and colleagues have shown that "carboxyl-functionalized single-walled carbon nanohorns" have an operation similar to "peroxidase" (Zhu et al., 2015). A particular colorimetric method for glucose was developed when the nanozyme was then combined with glucose oxidase. "Carbon nanodots" with a mean range of 2.0 nm also exhibited "peroxidase-mimicking behavior" and were used for "glucose sensing." The peroxidase-like activity of "selenium-doped graphitic carbon nitride nanosheets" was illustrated and further investigated when triggered with "xanthine oxidase" (Q) for "xanthine" prediction.

6.6 NANOMATERIALS OF METAL WITH CATALYTIC MONOLAYERS (TYPE 1)

Metal nanomaterials (such as gold, silver, etc.) with self-assembled nanoparticles (particularly conjugated monolayers) are being thoroughly researched due to various potential vital importance to nanoscience (Love et al., 2005). One might assume that if catalyzed ligands get introduced into another nanosheets, the metallic nanoparticles shielded by the monolayers would be thermodynamically stable. These functionalized metal oxides, of course, can require enzyme-like catalysis and are therefore called nanozymes.

Such nanozymes may be further split into three subgroups, as per the molecules used during the monolayers (Prins, 2015):

- "Alkanethiol," with catalytic moieties, ended
- In addition, the catalytic moieties would be further integrated on the nanoparticles except "catalytic terminal alkanethiol."
- Catalyzed "thiolated" targeting ligands

6.6.1 ALKANETHIOL-PROTECTED AUNPS WITH CATALYTIC TERMINAL MOIETIES

To reorient the gold nanoparticle structure, Scrimin and co-workers used "alkanethiol" re-instated with catalyzed ligands. The accessed gold nanoparticles configurations are therefore taken as a serious "metallonuclease (i.e., RNase)" (Fig. 6.3) imitate to crystallize the "transphosphorylation of 2-hydroxypropyl p-nitrophenyl phosphate (HPNPP)" (Fig. 6.4). The noncatalytic "transphosphorylation of HPNPP," once "gold naparticle-1 nanozyme" was used, that process was aggravated by several over than four times higher. Special, that much objected "zinc-1 (i.e., the 1, 4,7-triazacyclonane (TACN) and zinc ion)" combined with both the decommissioned catalyzed molecule; "gold nanoparticle-1 nanozyme" evidenced more than 600 times the rate of momentum. The "gold nanoparticles" were altered whenever "alkanethiol reinstated by NH^{4+}," the constructed AuNP-based frameworks were practically uncooperative. This regulatory discovery shows that "gold nanoparticle-1 nanozyme" catalytic activity would be about the integrated nanoparticles instead of just the "gold nanoparticle core." The cleavage of ribonucleic acid dinucleotides (such as "polyadenylation," "guanine & uracil nucleotides") may also be catalyzed by the "gold nanoparticle-1" nanozyme (Manea et al., 2004).

Subsequently, the shaped catalyzed substitutions "TACN-Zn^{2+} complex" is semiotic. The cooperativity was suggested by the sigmoid function gradient for catalytic properties depending on zinc charge density. Researches and empirical evidence really have perpetuated the co-operative effect. Two adjacent catalyzed ligands were theorized to be selected to develop a catalyzed pocket for collaborative catalytic reactions. The ones in natural enzymes were emulated by such a catalytic pocket. Additionally, due to the extreme proximity effect, a "0.4 unit dissociation constant" decrease throughout the catalogued "TACN-Zn^{2+}" complexes also could contribute to higher absorbance. Lastly, the ferocious "Au-S interaction" chose to make it quite convenient to carry out the self-assembly. In addition, it endowed the structured gold nanoparticle-based catalyzed substances with much more better durability particularly in relation with micelle-based systems.

Apart from that, its activity could be regulated analytically. Although the catalytic activity of "gold nanoparticle-1" zyme has been in the complicated "Zn^{2+}," when using a "Zn^{2+} chelating reagent" the activity can be peripherally transitioned apart and can therefore be recovered by re-adding "Zn^{2+}." Their functional groups have quite a limited "dielectric constant (ε)" for natural metallonucleases. As illustrated in eq 6.1 underneath, at membrane proteins of low conductivity, the electrostatic interaction between both the enzyme and its substrate might be preferred. Modest conductivity should therefore be integrated into functional groups to imitate a very unique micro-environment (Pieters et al., 2012).

$$\text{Electrostatic} \propto \frac{(Q1 \times Q2)}{(\varepsilon \times r_{1,2})} \tag{6.1}$$

To verify the research results, gold nanoparticle-based nanozymes with varying concentrations have been configured and their metallonuclease emulation tendencies were investigated. At reduced polarities, "gold nanoparticle-3" and "gold nanoparticle-5" increases more rapidly, whereas "gold nanoparticle-2" and "gold nanoparticle-4" contains fewer activity at higher polarities. Consequently, the catalytic activities were well connected with the polarities of the installed monolayers. The decreased polarities would significantly raise the electrostatic interaction between both the dianionic transfer state as well as the active sites of the nanozyme (i.e., Zn^{2+} complex), thus improving the catalytic properties. This research also showed that the catalytic properties of the nanozymes can be modified by attenuating the integrated monolayers (Diez et al., 2014).

6.6.1.1 ALKANETHIOL-PROTECTED AUNPS AS OTHER ENZYME MIMICS

Besides altering the terminal "TACN" ligand of the alkanethiol sequence to other functional ligands, establishing other enzyme imitates is comparatively easy. A "DNA topoisomerase" imitate has also been formed, by assembling "BAPA"-terminated alkanethiol on gold nanoparticles. The cleavage of "BNP (bis-p-nitrophenyl phosphate)," a deoxyribonucleic acid model substratum into "MNP (p-nitrophenyl phosphate)" and "p-nitrophenolate" with "gold nanoparticle-6 zyme" was exacerbated by thousand folds in the acoustic cleavage mixture. Furthermore, DNA molecules ("pBR 322 plasmid DNA") were appropriately ligated with gold nanoparticle-66 nanozyme (Bonomi et al., 2008). DNase might also be mimicked through the use of gold nanoparticles modified with "cerium (IV)" complex revoked alkanethiol monolayers. Through using "gold nanoparticles-Cerium (IV)-based nanozymes," the frequency of acceleration for both the "BNP" cleavage was observed upto million times. The outstanding electrochemical activity too was due to the novelty of the nanozyme, like cooperative catalytic reactions (Bonomi et al., 2010).

6.6.1.2 RNASE (DNASE) MIMICS USING OTHER SUPPORTING CORES

As gold nanoparticles served mainly as that of the representing center, they could be replaced by certain materials. Numerous researches have already shown that by implementing the catalyzed ligands (such as "TACN-Zn^{2+} complex" on polymer, "silica," it is able to capture "RNase" and "DNase" mimics. The formation and purifying of such enzyme imitates is far more costly and time consuming relative to basic gold sulfide chemistry. Those certain noble metal nanoparticles may also be used as the enhancing base in hypothesis. Even so, no research was published, that may be due to the efficient oxidation of AgNPs (Savelli and Salvio 2015).

6.6.2 THIOLATED BIOMOLECULES-PROTECTED AUNPS

Multiple gold nanoparticles-based nanozymes are being formed through incorporating "thiolated"-biological molecules on the "gold nanoparticle cores." A zyme has also been formulated to imitate the "RNA-induced silencing complex (RISC)" process connectivity. A "RISC" may abide to the equivalent intention "ssRNA" and then shotgun it"s own activity by triggering target RNA cleavage with a nuclease integrated within the "RISC" in combination with the assistance of a regulatory "ssRNA," Both "nucleases" and "regulatory ssRNA" were co-assembled with a gold nanoparticle core to emulate the attributes of a "RISC." In the accessibility of a "targeted ssRNA" including "objective ssRNA" the "regulatory ssRNA" installed could generate a duplex and thus interact it to the configured nuclease.

The nanozyme"s "anti-HCV" ("hepatitis C virus") effectiveness has been evaluated to exemplify the proposition using the "HCV replicon tissue culture" method ("FL-Neo cell line"). The replication of the HCV was distorted with nanozyme treatment. The in vivo assessment of the nanozyme was then screened using a "pluripotent stem cells." The therapy of nanozymes led, extraordinarily, to a decrease in "HCV- RNA" of even more than 99%. Considering the non-detectable cellular cell adhesion response, the engineered nanozyme might be used as an efficient biomedicine for infectious disorders and diseases. Those certain appropriate methodologies have been developed for "RNA interference-independent identification of transcription" (Rouge et al., 2015).

6.6.3 AUNPS WITH NONCOVALENTLY ASSEMBLED CATALYTIC MOIETIES COVERED BY ALKANETHIOL

Just about few other experiments have already shown "catalytic ligands" may be generated noncovalently on (or in) the gold nanoparticles coated by alkanethiol. The peptide incision was enabled by electrochemically binding the two peptide molecules to the "trimethylammonium-functionalized gold nanoparticle." So, both the substrate and the catalyzed peptide are non-covalently designed to functionalize "trimethylammonium" and the non-covalent configuration was mainly driven by electro - hydraulic correlations (Zaramella et al., 2012).

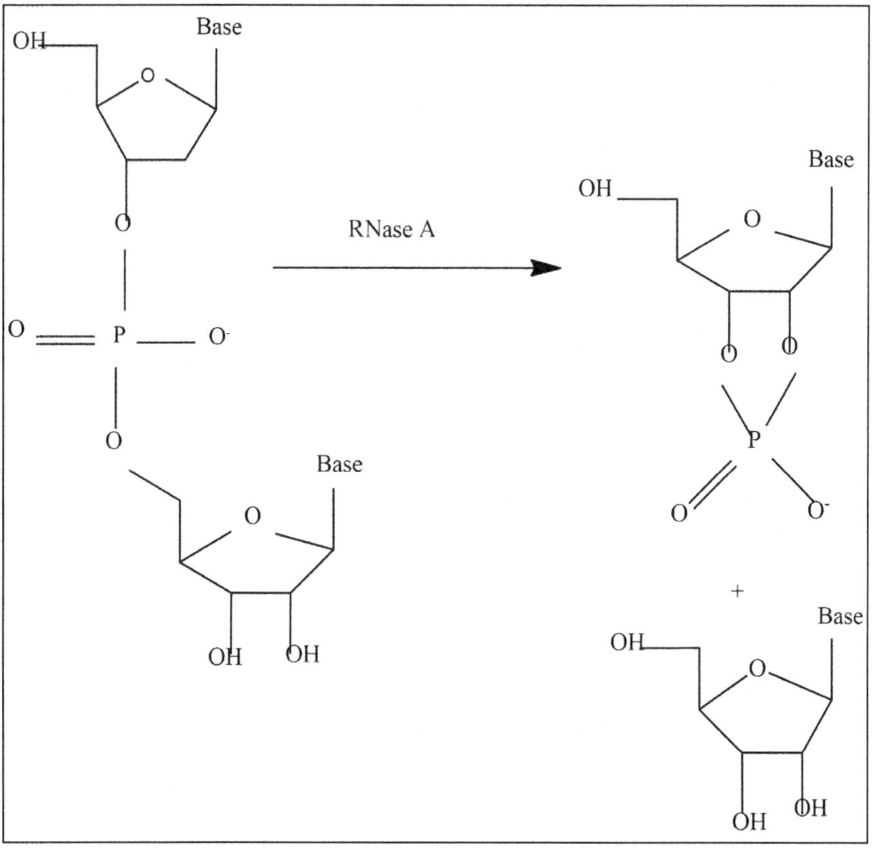

FIGURE 6.3 Transphosphorylation catalyzed by RNase.

6.7 NANOENZYMES APPLICATIONS IN THE ENVIRONMENT

With development of industry, the environmental quality concerns have arisen. Human factors, such as unwarranted stormwater drains waste water and waste gas, are mainly responsible for environmental contamination. It can modify the essence of the entire environment, which can harm and destroy the environmentally friendly and life form holistic growth (For example, "Japan's minamata disease" was caused by "Showa Denko Corporation's" dumping of unregulated Hg-containing liquid waste into the shore (Takeuchi et al., 1962). Air degradation has been a systemic concern before now, but there has been a belief that attempts to protect and maintain the ecosystem ought to be strengthened. Since before the modern era, biotechnology has become more and more popular with people as a crossover between chemistry and biology. Thus far, biotechnology has also been widely used in agricultural processing, pharmacology, forest products, etc. Applications of IoT reveal potential the pollution crisis in the atmosphere. In general, enzymes and protein engineering seems to have a vital impact on environmental pollution remediation (Karam and Nicell, 1997).

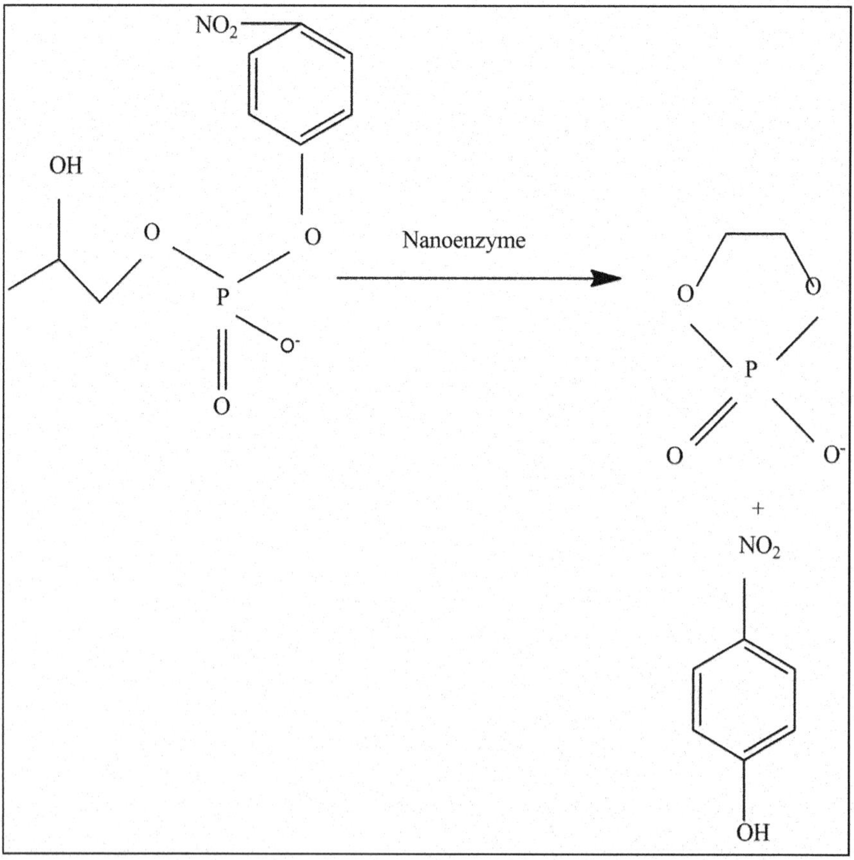

FIGURE 6.4 Transphosphorylation of HPNPP catalyzed by AuNP-1 nanozyme.

a) **Wastewater depletion of chemical contaminants:** Organic pollution has become one of the significant forms of emissions among diverse contaminants. Conventional physiochemical techniques also require complex procedures and costly equipment. Klibanov and co-workers used peroxidase for the first time since the early 1980s to catalyze the disintegration of phenols and aromatic compounds in waste water. Afterwards the, growing attention was paid with the use of natural enzymes for treating wastewater. "Horse radish peroxidase" is the most commonly applied enzyme of these enzymes. To produce highly reactive •OH, HRP will catalyze H_2O_2. The OH obtained will then oxidize several organic contaminants into aqueous coagulated products, such as esters and polyamines. While natural enzymes have many avails, their limitations all restrict their potential ecological treatment applications. Nanozymes have shown promise in the treatment of environmental pollutants, as mimics of natural enzymes. Fe_3O_4 zymes have virtues, such as "cheaper price," "better governance," and "cycling speed" compared to natural HRP. Fe_3O_4 nanoparticles could, therefore, be more appropriate for the removal of pollutants (Shahwan et al., 2011). Needless to mention, iron oxide nanoparticles are versatile for inevitably accumulating as a consequence of elevated thermal conductivity. Carbon nanotubes, just behind a comparatively small and stringent pH range, could have efficient electrochemical ability. All these curb their future uses in the removal of pollutants. Huang et al. used multiwalled carbon nanotubes carbon nanotubes to build "iron oxide" hybrid nanoparticles as the frameworks to achieve these objectives (Wang et al., 2014).

Furthermore, in order to spur pollutants destruction, plenty other peroxidase-like nanomaterials have also been recorded (Huang et al., 2015). Promptly, the Qu and Ren groups revealed that "Fe^{3+}-MCNs" nanozymes can theoretically forecast biological extraction efficiency for HRP itutes. Lacquers would also be used, analogous to HRP, in treating wastewater. The developed "Cu/GMP metal organic framework" will continue to accelerate polyphenol substrates" as a laccase mimic. These phenolic compounds, as substantial toxic substances, may trigger global pollution. The "Cu/GMP MOF" structure including its inherent laccase-like operation will thus prove responsible for serious sustainability issues (Liang et al., 2017).

b) **Chemical warfare agents degrading:** "Chemical Warfare Agents (CWA)" are chemical phenomena that are aimed at preventing highly toxic and widespread devastation. Between them, biological devices, a type of phosphate bond like "CWA," have some of the most toxic metals known to exist. They can reach the body via indirect techniques, namely mucous epithelial cells. These other nerves gases, when mixed with acetylcholine, can trigger tyrosine kinase function, resulting in autonomic nervous dysfunction. Recent western military developments, like the "Syrian military war" also led to targeted measures to seek efficient ways to tear down such dangerous materials rapidly. Alternatives to date have concentrated largely on personal security, massive destruction of chem weapons processing and prevention of "CWA" leakage (Mondloch et al., 2015). The potential to destroy with above "CWA" is provided by certain strong stratified compounds, such as modified carbon black and metal atoms. Even so, such compounds often suffer with adverse effects, such as poor absorption potential and versatility of elimination. Thus it is critically necessary to investigate new technologies for the effective decomposition processes of nerve agents and preparations (Cannard, 2006).

c) **Inhibiting the formation of biofilm:** On the surface of the "vessel," "propeller," "hold," and other ship regions in the underwater atmosphere, microorganisms can reabsorb, stimulating the production of "marine biofouling." Flow of water resilience could be impaired by the prevalence of microbes on ships, thus increasing fuel usage and greenhouse gases. "Marine biofouling" has earned considerable attention (Braithwaite and McEvoy, 2004). A workable alternative is the manufacture of anti-fouling coatings to avoid cell adherence. "Tributyltin-free antifouling coatings" have been used to virtually eliminate aquatic biofouling by "metal composite or biocide dependent paints" (Callow and Callow, 2011). Even so, such substances can also have complications such as "metal ions leakage" and "bacterial resistance" that can trigger certain ecological harm. Throughout the ocean setting, "vanadium haloperoxidase " may be produced by other seaweeds to facilitate the oxidation of "halogens" to regenerate analogous "hypohalous acids" with the help of hydrogen peroxidase (Carter et al., 2003).

Tremel and co-workers found, inspired by these unique properties, that "CeO_{2-x} nanorods" had inherent haloperoxidase-like activity.

In the presence of H_2O_2, the "bromination reaction" of signalling compounds could be catalyzed by "CeO_{2-x} nanorods," leading to bacterial quorum sensing inhibition. This control capacity of the quorum sensing bacteria was comparable to "haloperoxidases" and "halogenases of natural or artificial vanadium." "CeO_{2-x} nanorods" have the advantages such as "low toxicity" and "excellent catalytic efficiency" compared to the widely used "cuprous oxide" antifouling agent. They also maintained a high antifouling ability to effectively inhibit the adhesion of microorganisms while applying CeO_{2-x} nanorods to paints. Taken together, while nanozymes have great advantages in catalyzing the "degradation of organic contaminants," "chemical warfare agents," etc., great efforts are still needed to achieve practical applications in environmental treatment, such as the enhancement of catalytic activities (Herget et al., 2017).

6.8 CONCLUSION AND FUTURE ASPECTS

Over the past decade, nanozyme investigations have been performed worldwide and nanozyme application has also been applied to "biotechnology," "medications," "agricultural production," "pollution prevention," and several other sectors. Nanozymes have multifunctional cloud components in the field of environmental nanotechnology, from toxin recognition to detection. In addition to driving the precision and strength of natural enzymes, nanozymes may independently operate on unfavorable climatic variables that can influence biotic processes, that is, "interpretations in pH and temperature." There are also a few restrictions on nanozymes, and since nanozymes surpass innumerable static constraints that are efficient for natural enzymes, several nanozyme catalytic activities are indeed significantly cheaper than traditional natural enzymes. Currently, nanozyme research focuses mainly on redox enzyme mimics, such as "catalase," "oxidase," "superoxide dismutase," "peroxidase" with much less emphasis on other metabolites that may be important for the degradation of many composite materials. In this respect, although the productivity of the extraction of particulates is immense, the expense of industrialized technology may be beneficial in increasing carbon-emission treatment methods.

It is common for nanozymes to demonstrate satisfactory accuracy in limited experiments, but their integration into environmental science is still restricted, mainly because catalyzed nanozyme technologies have a strong development of flexibility and an extended lifetime if these limitations can be brought to light, nanozyme technologies can gain tremendous benefits for the contaminant treatment industry. The progress of nanozyme technologies almost seems appealing, but subsequent research on nanozymes appears rare. Examination of nanoparticles is growing in the environmental biotechnology viewpoint, and recent nanozymes start popping up relevant theories encompassing how and when to effectively extract contaminants in unfavorable conditions, and then how to delegitimize rather obdurate polymeric materials. In multilevel incentives for industrialized technology and community health, execution of nanozyme technology in the disciplines of "atmosphere," "genetics," "farming," and "pharmaceutics" may emerge.

KEYWORDS

- **nanozymes**
- **enzyme mimics**
- **carbon and metal oxidized based nanoparticles**
- **catalytic nanomaterials**
- **artificial enzymes**
- **catalytic mechanisms**
- **environmental protection**

REFERENCES

Asati, A.; Santra, S.; Kaittanis, C.; Nath, S.; Perez, J. M. Oxidase-like activity of Polymer-Coated Cerium Oxide Nanoparticles. *Angew. Chem.; Int. Ed.* **2009,** *48*, 2308–2312.

Behrens, M.; Studt, F.; Kasatkin, I.; Kühl, S.; Havecker, M.; Äbild-Pedersen, F.; Zander, S.; Girgsdies, F.; Kurr, P.; Kniep, B. L. et al. The Active Site of Methanol Synthesis over Cu/ZnO/Al_2O_3 Industrial Catalysts. *Sci.* **2012,** *336*, 893–897.

Biparva, P.; Abedirad, S. M.; Kazemi, S. Y. ZnO Nanoparticles as an Oxidase Mimic-Mediated Flow-Injection Chemiluminescence System for Sensitive Determination of Carvedilol. *Talanta* **2014,** *130*, 116–121.

Bonomi, R.; Scrimin, P.; Mancin, F. Phosphate Diesters Cleavage Mediated by Ce (IV) Complexes Self-Assembled on Gold Nanoparticles. *Org. Biomol. Chem* **2010,** *8*, 2622–2626.

Bonomi, R.; Selvestrel, F.; Lombardo, V.; Sissi, C.; Polizzi, S.; Mancin, F.; et al. Phosphate Diester and DNA Hydrolysis by a Multivalent, Nanoparticle-Based Catalyst. *J. Am. Chem. Soc.* **2008,** *130*, 15744–15745.

Bornscheuer, U. T.; Huisman, G. W.; Kazlauskas, R. J.; Lutz, S.; Moore, J. C.; Robins, K. Engineering the Third Wave of Biocatalysis. *Nature* **2012,** *485*, 185—194.

Boutorine, A. S.; Tokuyama, H.; Takasugi, M.; Isobe, H.; Nakamura, E.; Helene, C. Fullerene-Oligonucleotide Conjugates: Photoinduced Sequence-Specific DNA Cleavage. *Angew. Chem.Int. Ed.* **1994,** *33*, 2462–2465.

Braithwaite, R. A.; McEvoy, L. A. Marine Biofouling on Fish Farms and Its Remediation. *Adv. Mar. Biol.* **2004,** *47*, 215–252.

Callow, J. A.; Callow, M. E. Trends in the Development of Environmentally Friendly Fouling-Resistant Marine Coatings. *Nat. Commun.* **2011,** *2*, 244.

Cannard, K. The Acute Treatment of Nerve Agent Exposure. *J. Neurol. Sci.* **2006,** *249*, 86–94.

Carter-Franklin, J. N.; Parrish, J. D.; Tschirret-Guth, R. A.; Little, R. D.; Butler, A. Vanadium Haloperoxidase-Catalyzed Bromination and Cyclization of Terpenes. *J. Am. Chem. Soc.* **2003,** *125*, 3688– 3689.

Celardo, I.; Pedersen, J. Z.; Traversa, E.; Ghibelli, L. Pharmacological Potential of Cerium Oxide Nanoparticles. *Nanoscale* **2011,** *3*, 1411–1420.

Chen, Z. W.; Yin, J. J.; Zhou, Y. T.; Zhang, Y.; Song, L. N.;Song, M. J.; Hu, S. L.; Gu, N. Dual Enzyme-Like Activities of Iron Oxide Nanoparticles and Their Implication for Diminishing Cytotoxicity. *ACS Nano* **2012,** *6*, 4001–4012.

Cheng, H. J.; Zhang, L.; He, J.; Guo, W. J.; Zhou, Z. Y.; Zhang, X. J.; Nie, S. M.; Wei, H. Integrated Nanozymes with Nanoscale Proximity for in Vivo Neurochemical Monitoring in Living Brains. *Anal. Chem.* **2016,** *88*, 5489–5497.

Comotti, M.; Della Pina, C.; Matarrese, R.; Rossi, M. The Catalytic Activity of "Naked" Gold Particles. *Angew. Chem.; Int. Ed.* **2004,** *43*, 5812–5815.

Cui, R.; Han, Z.;Zhu, J.-J. Helical Carbon Nanotubes: Intrinsic Peroxidase Catalytic Activity and Its Application for Biocatalysis and Biosensing. *Chem. Eur. J.* **2011,** *17*, 9377–9384.

Diez-Castellnou, M.; Mancin, F.; Scrimin, P. Efficient Phosphodiester Cleaving Nanozymes Resulting from Multivalency and Local Medium Polarity Control. *J. Am. Chem. Soc.* **2014,** *136*, 1158–1161.

Fan, K. L.; Xi, J. Q.; Fan, L.; Wang, P. X.; Zhu, C. H.; Tang, Y.; Xu, X. D.; Liang, M. M.; Jiang, B.; Yan, X. Y. et al. In Vivo Guiding Nitrogen-Doped Carbon Nanozyme for Tumor Catalytic Therapy. *Nat.Commun.* **2018,** *9*, 1440.

Gao, L. Z.; Zhuang, J.; Nie, L.; Zhang, J. B.; Zhang, Y.; Gu, N., et al. 2007). Intrinsic Peroxidase-Like Activity of Ferromagnetic Nanoparticles. *Nat. Nanotechnol.* **2007,** *2*, 577–583.

Ge, C. C.; Fang, G.; Shen, X. M.; Chong, Y.; Wamer, W. G.; Gao, X. F.; Chai, Z. F.; Chen, C. Y.; Yin, J. J. Facet Energy Versus Enzyme-Like Activities: The Unexpected Protection of Palladium Nanocrystals against Oxidative Damage. *ACS Nano* **2016,** *10*, 10436—10445.

Gong, L.; Zhao, Z. L.; Lv, Y. F.; Huan, S. Y.; Fu, T.; Zhang, X. B.; Shen, G. L.; Yu, R. Q. DNAzyme-Based Biosensors and Nanodevices. *Chem. Commun.* **2015,** *51*, 979–995.

Herget, K.; Hubach, P.; Pusch, S.; Deglmann, P.; Götz, H.; Gorelik, T. E.; Gural"skiy, I. A.; Pfitzner, F.; Link, T.; Schenk, S. et al. Haloperoxidase Mimicry by CeO2-x Nanorods Combats Biofouling. *Adv. Mater.* **2017,** *29*, 1603823.

Hu, L. Z.; Yuan, Y. L.; Zhang, L.; Zhao, J. M.; Majeed, S.; Xu, G. B. Copper Nanoclusters as Peroxidase Mimetics and Their Applications to H_2O_2 and Glucose Detection. *Anal. Chim. Acta* **2013,** *762*, 83–86.

Huang, Y. Y.; Liu, C. Q.; Pu, F.; Liu, Z.; Ren, J. S.; Qu, X. G. A GO-Se Nanocomposite as an Antioxidant Nanozyme for Cytoprotection. *Chem. Commun.* **2017,** *53*, 3082–3085.

Huang, Y. Y.; Ran, X.; Lin, Y. H.; Ren, J. S.; Qu, X. G. Self Assembly of an Organic-Inorganic Hybrid Nanoflower as an Efficient Biomimetic Catalyst for Self-Activated Tandem Reactions. *Chem. Commun.* **2015,** *51*, 4386–4389.

Huang, Y.; Zhao, M. T.; Han, S. K.; Lai, Z. C.; Yang, J.; Tan, C. L.; Ma, Q. L.; Lu, Q. P.; Chen, J. Z.; Zhang, X. et al. Growth of Au Nanoparticles on 2D Metalloporphyrinic Metal-Organic Framework Nanosheets Used as Biomimetic Catalysts for Cascade Reactions. *Adv. Mater.* **2017,** *29*, 1700102.

Hundallah, K.; Jabari, M. Sulfite Oxidase Deficiency. *Neurosciences* **2016,** *24*, 376–378.

Karam, J.; Nicell, J. A. Potential Applications of Enzymes in Waste Treatment. *J. Chem. Technol. Biotechnol.* **1997,** *69*, 141–153.

Krusic, P. J.; Wasserman, E.; Keizer, P. N.; Morton, J. R.; Preston, K. F. Radical Reactions of C60. *Sciences* **1991,** *254*, 1183–1185.

Lee, M. S.; Lee, K.; Kin, S. Y.; Lee, H.; Park, J.; Choi, K. H.;Kim, H. K.; Kim, D.-G.; Lee, D.-Y.; Nam, S.-W. et al. High Performance, Transparent, and Stretchable Electrodes Using Graphememetal Nanowire Hybrid Structures. *Nano Lett.* **2013,** *13*, 2814–2821.

Li, J. H.; Li, X. N.; Feng, W. P.; Huang, L.; Zhao, Y.; Hu, Y.; Cai, K. Y. Octopus-like Pt Cu Nanoframe As Peroxidase Mimic for Phenol Removal. *Mater. Lett.* **2018,** *229*, 193–197.

Li, J. N.; Liu, W. Q.; Wu, X. C.; Gao, X. F. Mechanism of pH-Switchable Peroxidase and Catalase-Like Activities of Gold, Silver, Platinum and Palladium. *Biomaterials* **2015,** *48*, 37–44.

Li, M.; Yang, X. J.; Ren, J. S.; Qu, K. G.; Qu, X. G. Using Graphene Oxide High Near-Infrared Absorbance for Photothermal Treatment of Alzheimer"s Disease. *Adv. Mater.* **2012,** *24*, 1722–1728.

Liang, H.; Lin, F. F.; Zhang, Z. J.; Liu, B. W.; Jiang, S. H.; Yuan, Q. P.; Liu, J. W. Multicopper Laccase Mimicking Nanozymes with Nucleotides as Ligands. *ACS Appl. Mater. Interf.* **2017,** *9*, 1352–1360.

Lin, Y. H.; Huang, Y. Y.; Ren, J. S.; Qu, X. G. Incorporating ATP into Biomimetic Catalysts for Realizing Exceptional Enzymatic Performance over a Broad Temperature Range. *NPG Asia Mater.* **2014,** *6* (e114).

Lin, Y. H.; Li, Z. H.; Chen, Z. W.; Ren, J. S.; Qu, X. G. Mesoporous Silica-Encapsulated Gold Nanoparticles as Artificial Enzymes for Self-Activated Cascade Catalysis. *Biomaterials* **2013,** *34*, 2600–2610.

Liu, B. W.; Liu, J. W. Surface Modification of Nanozymes. *Nano Res.* **2017,** *10*, 1125–1148.

Liu, J.; Meng, L. J.; Fei, Z. F.; Dyson, P. J.; Jing, X. N.; Liu, X. MnO_2 Nanosheets as an Artificial Enzyme to Mimic Oxidase for Rapid and Sensitive Detection of Glutathione. *Biosens. Bioelectron.* **2017,** *90*, 69–74.

Liu, X.; Wang, Q.; Zhao, H. H.; Zhang, L. C.; Su, Y. Y.; Lv, Y. BSA-Templated MnO_2 Nanoparticles As Both Peroxidase and Oxidase Mimics. *Analyst* **2012,** *137*, 4552–4558. Pesticides. *Anal. Chem.* **2012,** *84*, 9492–9497.

Long, Y. J.; Li, Y. F.; Liu, Y.; Zheng, J. J.; Tang, J.; Huang, C.Z. Visual Observation of the Mercury-Stimulated Peroxidase Mimetic Activity of Gold Nanoparticles. *Chem. Commun.* **2011,** *47*, 11939–11941.

Love, J. C.; Estroff, L. A.; Kriebel, J. K.; Nuzzo, R. G.; Whitesides, G. M. Self-Assembled Monolayers of Thiolates on Metals as a form of nanotechnology. *Chem. Rev.* **2005,** *105*, 1103–1169.

Luo, W. J.; Zhu, C. F.; Su, S.; Li, D.; He, Y.; Huang, Q.; Fan, C.H. Self-Catalyzed, Self-Limiting Growth of Glucose Oxidase-Mimicking Gold Nanoparticles. *ACS Nano* **2010,** *4*, 7451–7458.

Manea, F.; Houillon, F. B.; Pasquato, L.; Scrimin, P. Nanozymes: Gold-Nanoparticle-Based Transphosphorylation Catalysts. *Angew. Chem.-Int. Ed.* **2004,** *43,* 6165–6169.

Mondloch, J. E.; Katz, M. J.; III Isley, W. C.; Ghosh, P.; Liao, P. L.; Bury, W.; Wagner, G. W.; Hall, M. G.; DeCoste, J. B.; Peterson, G. W. et al. Destruction of Chemical Warfare Agents Using Metalorganic Frameworks. *Nat. Mater.* **2015,** 14, 512–516.

Natalio, F.; Andre, R.; Hartog, A. F.; Stoll, B.; Jochum, K. P.; ́ Wever, R.; Tremel, W. Vanadium Pentoxide Nanoparticles Mimic Vanadium Haloperoxidases and Thwart Biofilm Formation. *Nat. Nanotechnol.* **2012,** *7,* 530–535.

Peters, R. J. R. W.; Marguet, M.; Marais, S.; Fraaije, M. W.; van Hest, J. C. M.; Lecommandoux, S. Cascade Reactions in Multicompartmentalized Polymersomes. *Angew. Chem. Int. Ed.* **2014,** *53,* 146–150.

Pieters, G.; Pezzato, C.; Prins, L. J. Reversible Control over the Valency of a Nanoparticle-Based Supramolecular System. *J. Am. Chem. Soc.* **2012,** *134,* 15289–15292.

Pirmohamed, T.; Dowding, J. M.; Singh, S.; Wasserman, B.; Heckert, E.; Karakoti, A. S.; King, J. E. S.; Seal, S.; Self, W. T. Nanoceria Exhibit Redox State-Dependent Catalase Mimetic Activity. *Chem. Commun.* **2010,** *46,* 2736–2738.

Prins, L. J. Emergence of Complex Chemistry on an Organic Monolayer. *Acc. Chem. Res.* **2015,** *48,* 1920–1928.

Puvvada, N.; Panigrahi, P. K.; Mandal, D.; Pathak, A. Shape Dependent Peroxidase Mimetic Activity towards Oxidation of Pyrogallol by H_2O_2. *RSC Adv.* **2012,** *2,* 3270–3273.

Qian, F. M.;Wang, J. M.; Ai, S. Y.; Li, L. F. As a New Peroxidase Mimetics: The Synthesis of Selenium Doped Graphitic Carbon Nitride Nanosheets and Applications on Colorimetric Detection of H_2O_2 and Xanthine. *Sens. Actuat. B Chem.* **2015,** *216,* 418–427.

Ragg, R.; Tahir, M. N.; Tremel, W. Solids Go Bio: Inorganic Nanoparticles as Enzyme Mimics. *Eur. J. Inorg. Chem.* **2016,** 1906–1915.

Rouge, J. L.; Sita, T. L.; Hao, L.; Kouri, F. M.; Briley, W. E.; Stegh, A. H. et al. Ribozyme-Spherical Nucleic Acids. *J. Am. Chem. Soc.* **2015,** *137,* 10528–10531.

Sang, Y. J.; Huang, Y. Y.; Li, W.; Ren, J. S.; Qu, X. G. Bioinspired Design of Fe3+-Doped Mesoporous Carbon Nanospheres for Enhanced Nanozyme Activity. *Chem.—Eur. J.* **2018,** *24,* 7259–7263.

Savelli, C.; Salvio, R. Guanidine-Based Polymer Brushes Grafted Onto Silica Nanoparticles as Efficient Artificial Phosphodiesterases. *Chem. Eur. J.* **2015,** *21,* 5856–5863.

Shahwan, T.; Abu Sirriah, S.; Nairat, M.; Boyaci, E.; Eroğlu, A. E.; Scott, T. B.; Hallam, K. R. Green Synthesis of Iron Nanoparticles and Their Application as a Fenton-Like Catalyst for the Degradation of Aqueous Cationic and Anionic Dyes. *Chem. Eng. J.* **2011,** *172,* 258–266.

Shen, X. M.; Liu, W. Q.; Gao, X. J.; Lu, Z. H.; Wu, X. C.; Gao, X. F. Mechanisms of Oxidase and Superoxide Dismutation-Like Activities of Gold, Silver, Platinum, and Palladium, and Their Alloys: A General Way to the Activation of Molecular Oxygen. *J. Am. Chem. Soc.* **2015,** *137,* 15882–15891.

Singh, N.; Savanur, M. A.; Srivastava, S.; D"Silva, P.; Mugesh, G. A Redox Modulatory Mn_3O_4 Nanozyme with Multi-Enzyme Activity Provides Efficient Cytoprotection to Human Cells in a Parkinson"s Disease Model. *Angew. Chem. Int. Ed.* **2017,** *56,* 14267–14271.

Song, Y.; Qu, K.; Zhao, C.; Ren, J.; Qu, X. Graphene Oxide: Intrinsic Peroxidase Catalytic Activity and Its Application to Glucose Detection. *Adv. Mater.* **2010,** *22,* 2206–2210.

Sun, H. J.; Zhao, A. D.; Gao, N.; Li, K.; Ren, J. S.; Qu, X. G. Deciphering a Nanocarbon-Based Artificial Peroxidase: Chemical Identification of the Catalytically Active and Substrate-Binding Sites on Graphene Quantum Dots. *Angew. Chem. Int. Ed.* **2015,** *4,* 7176–7180.

Takeuchi, T.; Morikawa, N.; Matsumoto, H.; Shiraishi, Y. A Pathological Study of Minamata Disease in Japan. *Acta Neuropathol.* **1962,** *2*, 40–57.

Tao, Y.; Lin, Y. H.; Huang, Z. Z.; Ren, J. S.; Qu, X. G. Incorporating Graphene Oxide and Gold Nanoclusters: A Synergistic Catalyst with Surprisingly High Peroxidase-Like Activity Over a Broad pH Range and Its Application for Cancer Cell Detection. *Adv. Mater.* **2013,** *25*, 2594–2599.

Tokuyama, H.; Yamago, S.; Nakamura, E.; Shiraki, T.; Sugiura, Y. Photoinduced Biochemical-Activity of Fullerene Carboxylic-Acid. *J. Am. Chem. Soc.* **1993,** *115*, 7918–7919.

Wang, C.; Xu, C. J.; Zeng, H.; Sun, S. H. Recent Progress in Syntheses and Applications of Dumbbell-Like Nanoparticles. *Adv. Mater.* **2009,** *21*, 3045–3052.

Wang, G. L.; Jin, L. Y.; Wu, X. M.; Dong, Y. M.; Li, Z. J. Labelfree Colorimetric Sensor for Mercury(II) and DNA on the Basis of Mercury(II) Switched-On The Oxidase-Mimicking Activity of Silver Nanoclusters. *Anal. Chim. Acta.* **2015,** *871*, 1–8.

Wang, H.; Jiang, H.; Wang, S.; Shi, W. B.; He, J. C.; Liu, H.; Huang, Y. M. Fe_3O_4-MWCNT Magnetic Nanocomposites as Efficient Peroxidase Mimic Catalysts in a Fenton-Like Reaction for Water Purification without pH Limitation. *RSC Adv.* **2014,** *4*, 45809–45815.

Wang, H.; Wan, K. W.; Shi, X. H. Recent Advances in Nanozyme Research. *Adv. Mater.* **2018,** 1805368.

Wang, N.; Zhu, L. H.; Wang, D. L.; Wang, M. Q.; Lin, Z. F.; Tang, H. Q. Sono-Assisted Preparation of Highly-Efficient Peroxidase Like Fe_3O_4 Magnetic Nanoparticles for Catalytic Removal of Organic Pollutants with H_2O_2. *Ultrason. Sonochem.* **2010,** *17*, 526–533.Wang, S.; Chen, W.; Liu, A.-L.; Hong, L.; Deng, H. H.; Lin, X. H. Comparison of the Peroxidase-Like Activity of Unmodified, Aminomodified, and Citrate-Capped Gold Nanoparticles. *Chem. Phys. Chem.* **2012,** *13*, 1199–1204.

Wei, H.; Wang, E. K. Nanomaterials with Enzyme-Like Characteristics (Nanozymes): Next-Generation Artificial Enzymes. *Chem. Soc. Rev.* **2013,** *42*, 6060–6093.

Wirth, T. Small organoselenium compounds: more than just glutathione peroxidase mimics. *Angew. Chem.; Int. Ed.* **2015,** 54, 10074–10076.

Xu, X. J.; Hu, L. F.; Gao, N.; Liu, S. X.; Wageh, S.; Al-Ghamdi, A. A.; Alshahrie, A.; Fang, X. S. Controlled Growth from ZnS Nanoparticles to ZnS-CdS Nanoparticle Hybrids with Enhanced Photoactivity. *Adv. Funct. Mater.* **2015a,** *25*, 445–454.

Xu, Z.-Q.; Lan, J.-Y.; Jin, J.-C.; Dong, P.; Jiang, F.-L.; Liu, Y. (2015). Highly Photoluminescent Nitrogen-Doped Carbon Nanodots and Their Protective Effects Against Oxidative Stress on Cells. *ACS Appl. Mater. Interf.* **2015a,** *7*, 28346–28352.

Xue, T.; Peng, B.; Xue, M.; Zhong, X.; Chiu, C. Y.; Yang, S.; Qu, Y. Q.; Ruan, L. Y.; Jiang, S.; Dubin, S. et al. Integration of Molecular and Enzymatic Catalysts on Graphene for Biomimetic Generation of Antithrombotic Species. *Nat. Commun.* **2014**, *5*, 3200.

Yamakoshi, Y. N.; Yagami, T.; Sueyoshi, S.; Miyata, N. Acridine Adduct of 60 Fullerene with Enhanced DNA-Cleaving Activity. *J. Org. Chem.* **1996,** *61*, 7236–7237.

Yan, X. Y. Nanozyme: a new type of artificial enzyme. *Prog. Biochem. Biophys.* **2018,** *45*, 101–104.

Yin, W. Y.; Yu, J.; Lv, F. T.; Yan, L.; Zheng, L. R.; Gu, Z. J.; Zhao, Y. L. Functionalized Nano-MoS_2 with Peroxidase Catalytic and Near-Infrared Photothermal Activities for Safe and Synergetic Wound Antibacterial Applications. *ACS Nano.* **2016,** *10*, 11000–11011.

Zaramella, D.; Scrimin, P.; Prins, L. J. Self-Assembly of a Catalytic Multivalent Peptide-Nanoparticle Complex. *J. Am. Chem. Soc.* **2012,** *134*, 8396–8399.

Zhou, H.; Han, T. Q.; Wei, Q.; Zhang, S. S. Efficient Enhancement of Electrochemiluminescence from Cadmium Sulfide Quantum Dots by Glucose Oxidase Mimicking Gold Nanoparticles for Highly Sensitive Assay of Methyltransferase Activity. *Anal. Chem.* **2016,** *88*, 2976–2983.

Zhou, Y. B.; Liu, B. W.; Yang, R. H.; Liu, J. W. Filling in the Gaps between Nanozymes and Enzymes: Challenges and Opportunities. *Bioconjugate Chem.* **2017,** *28*, 2903–2909.

Zhu, S. Y.; Zhao, X. E.; You, J. M.; Xu, G. B.; Wang, H. Carboxylic-Group-Functionalized Single-Walled Carbon Nanohorns as Peroxidase Mimetics and Their Application to Glucose Detection. *Analyst*, **2015,** *140*, 6398–6403.

Zhou, L., Han, L. X., Wu, G., Zhang, N. S., [illegible] Enhancement of [illegible] from [illegible] Calcium [illegible] [illegible] Nanoparticles [illegible] Journal [illegible] [illegible] Materials [illegible] [illegible]

Zhou, Y. [illegible] Wu, [illegible] Yang, K. [illegible] Liu, [illegible] Y. [illegible] Between [illegible] Nanoparticles and [illegible] Characteristics and Challenges. *Nanomaterials* [illegible] 2017, [illegible] 001

Zhu, S. [illegible] Zhao, X. [illegible] You, [illegible] Wang, H. [illegible] [illegible] [illegible] [illegible] and Their Application [illegible] [illegible] 2015, [illegible]

CHAPTER 7

Potential of Nanobiosensors for Environmental Pollution Detection: Nanotechnology Combined with Enzymes, Antibodies, and Microorganisms

TAMOGHNI MITRA[1], SAURAV KUMAR SAHOO[2], ARPITA BANERJEE[1], ANJANI KUMAR UPADHYAY[1*], and KAZI NAHID HASAN[1]

[1]*School of Biotechnology, Kalinga Institute of Industrial Technology, Bhubaneswar, Odisha, India*

[2]*MITS School of Biotechnology, Bhubaneswar, Odisha, India*

[*]*Corresponding author. E-mail: upadhyayanjanikumar6@gmail.com*

ABSTRACT

Environmental pollutants are hazardous to humans, and the challenges for environmental and analytical chemistry are detecting directly the pollutants, such as heavy metals, pesticides, and toxins from waste streams and the monitoring of soil and aquatic conditions. Due to the lack of tools with broad detection constraints and the lack of expensive facilities, the current approaches being implemented to perform real-time assay and monitor contaminated specimens were restricted. Strategies have been exerted in the development of awareness technology and therefore fast, low-cost, responsive sensors are required. The increasing focus on nanomaterials with advanced optoelectrical properties has contributed to the further introduction

Nanotechnology for Environmental Pollution Decontamination: Tools, Methods, and Approaches for Detection and Remediation. Fernanda Maria Policarpo Tonelli, Rouf Ahmad Bhat, & Gowhar Hamid Dar (Eds.)

of biosensors with new applications. Nanobiosensors are used for monitoring pollution levels by analyzing the ultra-sensitivity of contaminants swiftly. For example, nano-biosensors have certain distinctive features of a small scale, compact, powerful, tangible, vulnerable, and relatively low cost. Therefore, nano-biosensors may seem a powerful response to traditional methods of analysis since these allow extremely responsive, active, and high-frequency detection of the presence of contaminants without detailed analysis.

7.1 INTRODUCTION

Globally, there is a significant concern about the overregulated or unregulated release of environmental pollutants, for example, harmful heavy metals, antibiotics, and pesticides. New prototypes are therefore urgently needed to detect their presence in terrestrial and aquatic environments. Conventional methods like chromatography have some bottlenecks, that is, lack of selectivity and sensitivity, long and specialized pretreatment of samples that can theoretically lead to time-consuming procedures and detection at the low sample level. In this case, using the bioelements in conjunction with green nanomaterial functionality allows for the development of nano-biomaterials to track the environment. In contrast with conventional methodologies, these extraordinary characteristics of nano-biosensors make them extremely responsive and economic instruments for the environmental monitoring and the pollutant detection. These are generally categorized in two classes, that is, bioreceptors and transductors. The biosensors based on bioreceptors are often classified as enzymes, proteins, antibodies, bacteria, and DNA based. Biosensors based on the methods of transduction are grouped into three kinds, that is, optical, calorimetric, electrochemical, and mass-based biosensors. Electrochemical biosensors have proved useful for the identification of small sample volumes, low concentrations of biological components, and even miniature analytical instruments.

A biosensor identifies any bioproduct by combinations of biological identifying features and a physical or chemical transducer. These biosensors majorly consist of three components: biological marker, transducer, and electronics for signal processing. For different applications, different biosensors are being developed, including environmental protection, food quality control, bioprocess control, agriculture, the medical and military, and pharmaceutical sectors specifically. These generally function at five different levels: (1) bioreceptors, binding the specific form to the sample, (2) an electrochemical interface where explicit biological processes happen inducing a sign, (3 a transducer that changes over the particular biochemical response in an electrical sign, (4) a sign processor for changing over the electronic sign into a significant physical parameter, lastly, and (5) a legitimate interface to show the outcomes to the operator (Luz et al., 2013). Biosensors can be classified by combining the fundamental concepts of signal transduction and biorecognition components. In the overall structure of a biosensor, the biorecognizing part responds to the objective compound, and the transducer changes the organic response to the discernible sign that can be quantifiable electrochemically, optically, acoustically, mechanically, calorimetrically, or electronically and afterward compared to the analyte concentration. Enzymes, antibodies, microorganisms, biological tissue, and organelles are classified in the biological components (Touhami, 2014).

The production of nanoscale biosensors that are highly sensitive and flexible has been motivated by developments in nanotechnology. The ultimate aim of nano-biosensors, at the level of the molecule or cell, is used for the identification of a biochemical and biophysical signal correlated with a particular disorder. As such, a nano-biosensor is regularly an insightful analytical system that incorporates a biological recognition particle immobilized on the surface of a transducer (Jain et al., 2010; Verma et al., 2010). The nano-biosensors have diversified applications including molecular diagnostics that can be combined with developments like lab-on-chip to make the process simpler. They include microorganism identification in multiple samples, metabolite tracking in the body fluids, and cancer tissue detection. Their portability gives them an advantage for the pathogenesis of cancer applications.

There are various methodologies for making the next generations of nano-biosensor systems: (1) the utilization of a totally new class of nanomaterial for detecting purposes, (2) new immobilization techniques, and (3) the new nanotechnological approaches. The major benefits of nano-biosensors for environmental purposes are portability, miniaturization, and function compared with traditional analytical techniques. In biological/ecological quality assessment or chemical testing for inorganic and organic priority toxins, nano-biosensors may be utilized as environmental quality monitoring instruments (Salgado et al., 2011). Continued tracking, onsite activity, and capacity to measure contaminants in complex matrices with limited samples are major characteristics provided by nano-biosensors. A huge number of fertilizers, herbicides, pesticides, insecticides, pathogens, temperature, soil, and pH can be sensed effectively and monitored to promote sustainable agriculture to improve crop production that can be done with the help of nano-biosensors (Sekhon, 2014). Also, production is threatened regularly by rodents, weeds, and diseases impacting the relative agricultural economy; plants, therefore, must be covered by the proper intervention. By monitoring the conditions of land and plant growth in large areas, nanostructured biologic sensors may lead to smart farming as well as by detecting contagious diseases in crops before visible symptoms (Antonacci et al., 2018).

7.2 ATTRIBUTES AND KEY ELEMENTS OF NANO-BIOSENSORS

With the emerging new scientific findings and discoveries in the 21st century, biosensors and nano-biosensors with the assistance of nanotechnology have gotten perhaps the best innovation that is effective in environmental contamination remediation. A crucial section of biosensing is the transduction instruments that are liable for changing over the reactions of bioanalyte communications in a recognizable and reproducible manner using the change of unequivocal biochemical reaction energy into an electrical structure. Nanomaterials are splendid occupants in this measurement as they have a high surface region to volume proportions that permit the surface to be used in a prevalent and indisputably more practical way. Also, their electromechanical properties are splendid assets for biosensor development (Malik et al., 2013). As of late, analysts have utilized an incorporated methodology by joining electronics, computers, nanoscience, and biological science to make biosensors with remarkable detecting capacities that show uncommon spatial and fleeting goals and unwavering quality. Nanosensors having inactive or immobilized bioreceptor probes those target the analyte molecules very selectively are named as the nano-biosensors. Molecular analysis is now getting easier for the nano-biosensors that are being used along with some other technologies like laboratory functions in a single integrated circuit (i.e., lab-on-a-chip). Detection of various microorganisms, pathogens, and chemicals such as urea, pesticides, glucose, and monitoring the metabolites, are the applications of nano-biosensors. For their ease in performance, few points should be kept in mind while designing nano-biosensors that ultimately makes them characteristically efficient (Rai et al., 2012; Bhattarai and Hameed, 2014);

- Analytes are the chemicals of interest in an analytical procedure, and nano-biosensors should be profoundly explicit with the end goal of the analyses; for example, a sensor should have the option to recognize the analyte and any "other" material. The best example of this characteristic is the connection between an immobilized antibody and an antigen that is profoundly explicit.

- Under normal storage conditions, the nano-biosensors ought to be stable for their best performance.
- Reproducibility, then again, is the capacity of a biosensor to yield indistinguishable final results, no matter how many times the experiment is repeated. This is dictated by the exactness and precision of the transducer or electronic parts in a biosensor.
- The reaction time ought to be negligible.
- The physical parameters such as blending (stirring), pH, and temperature shouldn't affect specific interactions between analytes.
- In a clinical setting, sensitivity and linearity are the essential attributes of biosensors that can't be neglected and ought to be managed with the most extreme consideration. Sensitivity alludes to the most minimal identification breaking point of an analyte by a biosensor that may go from nanogram per milliliter to even femtogram per milliliter. Then again, linearity speaks to the precision of the obtained output inside a working range where the convergence of the analyte in the sample is directly proportional to the measured signal.
- The reactions acquired should be exact, pinpoint, error-free, and linear over the valuable analytical range and be liberated from the electrical commotion.
- The nano-biosensors ought to be modest, convenient, and fit for being utilized by semiskilled administrators.
- "Being minuscule, biocompatible, non-poisonous, and non-antigenic" should be the characteristic of the nano-biosensors.

A biosensor can be depicted as a detecting system or sensing device or an estimation framework planned explicitly for the assessment of a material that utilizes the interactions between biological agents and afterward surveying these interactions into a meaningful structure with the help of transduction and electromechanical interpretation. The major components that help in the smooth functioning of a nano-biosensor are bioreceptor, transducers, detectors, and most importantly, nanomaterials (Fig. 7.1).

"Detecting a biologically specific material" (e.g., antibodies, proteins, enzymes, immunological molecules, etc.) is the primary purpose behind a biosensor. To recognize the target material, a template is required that is satisfied by a "biologically sensitive material" that helps in the making of a bioreceptor. Using the example of an antibody screened by antigens and

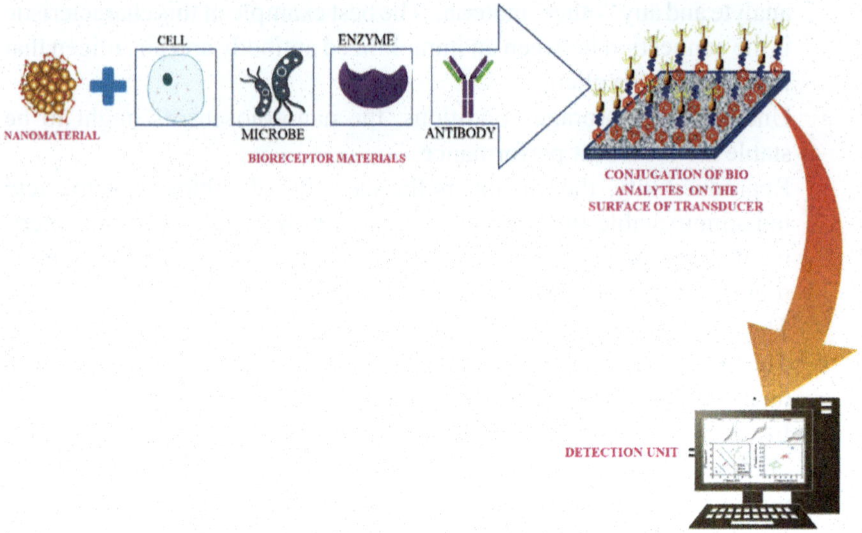

FIGURE 7.1 Schematic of nano-biosensor development for pollution detection application.

vice versa, and proteins being screened using its correlated substrates, the phenomenon of bioreceptors can be well understood (Malik et al., 2013). The probes or the biologically sensitized elements including enzymes, nucleic acids, receptors, lectins, antibodies, molecular imprints, tissue, microbes, organelles, and so forth are either naturally inferred materials (biologically derived materials) or biomimic components that send signals from analytes of interest to transducers (Rai et al., 2012). Changing over the connection of bioanalyte and its comparing bioreceptor into an electrical structure is the fundamental capacity of the transducer. The name itself portrays the word "trans" that signifies change and "ducer" that signifies energy, and thus, a transducer changes over one sort of energy into another type. The principal structure is biochemical in nature as it is produced by the particular connection between the bioanalyte and bioreceptor while the subsequent structure is typically electrical. This change of biochemical reaction into an electrical sign is brought through the transducer. Then, the third part is the detector system that gets the electrical sign from the transducer part and amplifies it reasonably so the corresponding reaction can be read and studied appropriately. Nano-biosensors have nanomaterials as their major constituent that is defined as material at any rate one dimension more modest than 100 nm. Nanomaterials are classified into (1) nanofilms and coatings (<100 nm in one-dimension), (2) nanotubes and wire (<100 nm in two-dimension), and (3) nanoparticles (<100 nm in three-dimension).

Because of their unfathomably little size, nanomaterials show remarkable highlights, chemically and physically. The nanosensors are tiny devices, with measurements in the order of one billionth of a meter, equipped for identifying and reacting to chemical, physical, and biological stimuli. This capacity of nanosensors can be utilized usefully for ecological investigation by using them for the recognition of poisons, microorganisms, modern and natural toxins, pesticides, heavy metals, allergens, and so forth utilizing various components. Different nanomaterials have been explored to dissect their properties and ongoing applications as nano-biosensors. These nanomaterials are nanoparticles, quantum dots, nanotubes, or other organic nanomaterials. These nanomaterials can add either to the bioreceptor component or to the transducer or to both nanosensors; nanoprobes and other nanosystems have upset the fields of biological and chemical analysis in so many example frameworks (Kuswandi, 2019).

7.3 DIFFERENT APPROACHES TO NANO-BIOSENSORS AND THEIR ROLES

7.3.1 NANO-BIOSENSORS INTEGRATED WITH ENZYMES

Progression in the enzyme-based nano-biosensing framework is associated with the level of progress in affectability toward target atom recognition (Ghormade et al., 2011). Due to their unparalleled optical, electrochemical properties, nanomaterials are highly appreciated as inexorable candidates for nano-biosensor development (Sassolas et al., 2012). Also, their certain features like biocompatibility, larger surface area to volume ratio, ease of separation, etc., hold huge importance to get incorporated into a genetically modified enzyme (Verma et al., 2008). Thus, nanotechnology has given a critical effect on conventional enzyme immobilized innovation (da Silva et al., 2014). Enzymatic nano-biosensor was created by utilizing atomic modeling techniques. This nano-biosensor was created utilizing functionalization of atomic force microscopy tips with acetyl co-catalyst A carboxylase. The location was checked by estimating the forces between the immobilized enzyme and the herbicide analyte. The subatomic displaying procedure was utilized to advance the enzyme adsorption concentrates on the tips of atomic force microscopy and the degree of recognition (Franca et al., 2011). Apart from this, biorecognition is the salient feature of every biosensor. Enzymatic nano-biosensor has an enzyme bioreceptor that has a high level of substrate particularity. Choice of enzyme relies upon the basic boundaries, for example, thermal stability and high turnover number (Verma et al., 2016). The minimal detection limit of contaminants in the food sample was properly evaluated by the enzymatic nano-biosensor. Along these lines, the enzymatic nano-biosensor opens up new trends in the estimation of the food quality in an ultrasensitive manner (Antiochia et al., 2004). This cutting-edge technology is also effective in the agricultural industry where the most common disadvantage is the excessive usage of pesticides (Ghormade et al., 2011). Nano-biosensors can detect small traces of pesticide residues that ultimately help maintaining the quality of the soil so that crop yield would not get hindered. This organophosphorus pesticide has a huge affinity for acetylcholinesterase enzymes that catalyze a hydrolytic reaction regulating the function of the neurotransmitter acetylcholine (Periasamy et al., 2009). Subsequently, nano-biosensors based on the recognition part of acetylcholinesterase have been exploited to locate the pesticide residues (Table 7.1). Another important sector is the food industry where detection of microorganisms in the food samples that are harmful to gut flora has become easier with the intervention of nanotechnology (Dasgupta et al., 2015). The incorporation of nanomaterials in the biosensor assembly for getting higher specificity for the analyte and also for faster transfer of electrons into the enzyme has shown a huge impact in the food industry to combat many difficult situations at ease (Perez-Lopez and Merkoci, 2011). Industrially, useful enzymes such as β-galactosidase, glucose oxidase, fructose dehydrogenase, and pyranose oxidase are used as molecules of biorecognition to check the carbohydrate molecules such as lactose/lactulose, glucose, and fructose in food samples (Ozdemir et al., 2010). After analyzing many factors, it has been concluded that enzymatic nano-biosensor can be utilized as a cost-effective platform for monitoring environmental pollution at the early stage. Performance of the enzymatic nano-biosensor can overcome any other expensive bioanalytical system. To sum up, enzymatic nano-biosensors hold great potential for those industries that are accountable in daily life.

7.3.2 NANOMATERIAL-DEPENDENT FLUORESCENT BIOSENSORS FOR MONITORING ENVIRONMENTAL POLLUTANTS

Fluorescence is a phenomenon in which light is emitted by a substrate that has absorbed any electromagnetic radiation or light. This property of fluorescence is exploited to develop nano-biosensors that can sense pollution in the environment. Fluorescence-based sensors use changes in anisotropic, intensity, emission lifetime alterations, absorption, or emission wavelength. Change of fluorescence intensity is mostly used, as they are generally quite

TABLE 7.1 Enzymatic Nano-biosensors for Agricultural Industry.

Nano-bioconjugation system	Role	References
Magnetic NPs, carbon nanotube, and zirconium nanoparticle are immobilized covalently on transducer with the enzyme, acetylcholinesterase	Helps in the detection of pesticide, dimethoate at ultrasensitive level	Gan et al. (2010)
Carbon nanotube, electrostatically assembled with multienzyme consortium on the superficial layer of transducer	Detects organophosphorus pesticide paraoxon and non-organophosphorus pesticide at the limit of 0.5–1 μM	Zhang et al. (2015)
Silica nanoparticle integrated with organophosphate hydrolase in silica	Contributes in the detection for organophosphate paraoxon	Ramanathan et al. (2009)
Acetylcholinesterase is entrapped in nanoliposome	Detects pesticides, paraoxon, and dichlorvos at ultrasensitive level	Vamvakaki and Chaniotakis (2007)
Covalent immobilization of carbon nanotube-aflatoxin oxidase on the surface of transducer	Detection of aflatoxin	Li et al. (2011)
Nanocomposite of ZnO nanoparticle, chitosan, carbon nanotube, and polyaniline integrated with xanthine oxidase	Detection of xanthine	Devi et al. (2012)
Gold nanoparticle along with β-galactosidase, physically adsorbed on the surface of transducer	Helps in the detection of E. coli	Miranda et al. (2011)

sensitive, easy to use, and adaptable to various analytes and systems. The ecosystem is heavily damaged by the use of pesticides. Recently, there is a trend of detecting organophosphorus pesticides using the fluorescence DNA biosensors that are mainly based upon immunochemical traits such as recognition of DNA, or RNA aptamers, antigen–antibody reactions, and enzyme–substrate reactions. For the enzymatic detection of these pesticides, acetylcholinesterase (AChE) is generally used as a catalytic hydrolyzer of acetylcholine releases, thiocholine, which contains a thiol group and is reactive toward nanomaterials and fluorophores. Hence in turn fluorescent or colorimetric signals can be received that can detect and quantify the presence of pesticide. Antibiotics are life-saving drugs that are used to treat different diseases but when these antibiotics are present in the environment it poses a threat as it can lead to mutation in bacteria and develop antibiotic resisting properties in them, so it is important to detect them even if present in trace amounts. These can effectively be detected with the aid of fluorescence-based nanosensors. Enrofloxacin (ENR), an antibiotic of the fluoroquinolones group, is simply detected by the use of common fluorescence of ENR. Graphite oxide acts as a quencher dye for its detection. In the presence of aptamer, ENR goes under a conformational change that hinders the absorption of the aptamer–analyte pair. To detect the presence of chloramphenicol, hairpin aptamers are being used that comprises a specific chloramphenicol recognized sequence. Tetrakis (4-carboxyphenyl) porphyrin (TCPP) combined with copper nanosheets help in the synthesis of MOF (metal organic frames). MOF acts as a quenching agent and has properties that are the same as organic 2D nanostructures (graphene).

Phenolic compounds such as pentachlorophenol, 2,4 di-chlorophenol, bisphenol, and 2,4,6-trichlorophenol are most common in the environment and pose threat to human health. Detecting them in the ecosystem and removing them are the need of the hour, and the dual fluorescent-colorimetric system can be used to detect BPA (Bisphenol A), a type of harmful phenolic compound. The process of detection involves the use of a very high salt concentration that enhances the agitation of gold nanoparticles and produces a colorimetric signal that further gives rise to fluorescence (Gaviria-Arroyave et al., 2020).

A life-threatening heavy metal that is very common in sewage is mercury. Detection of this metal and removal of it from the environment are necessary. So, fluorescence can be detected by using mercury-specific oligonucleotides (MSOs). Following the nature of mercury fluorescent sensors, it can be detected in six ways as follows: binary probes, "label-free" probes, molecular beacon, chain reaction probes, G-quadruplex probes, and nanoparticle-using probes. These types of sensors can detect even the slightest amount of Hg, if present (Hu et al., 2020). Arsenic, a highly toxic heavy metal generally found in water, is a hazardous compound that adversely affects human health. Sensitivity toward arsenic and its detection are improved by the use of nanomaterials. Arsenic that is present in As(III) cannot be identified directly as it is transformed to As (III), which is further identified by anodic stripping voltammetry (ASV) method. Arsenic can bind to some enzymes like other heavy metals which in turn influence the catalytic activity. This principle of inhibition of enzymes is being exploited to develop biosensors for arsenic detection. Zymolytic-modified nanomaterials as signal indicators are the most common arsenic sensing method. With the use of natural enzymes, the substrate gets decomposed which in turn changes the signal of the nanomaterial indicator. As the level of arsenic increases again, it suppresses the catalytic activity of decomposition that leads to the recovery of the signal indicator. A fluorescent nanoprobe of arsenate was developed by the Qiu's group by gathering guanosine monophosphate, Tb(III), and carboxylated CdSe/ ZnS QDs (Xu et al., 2020).

7.3.3 FUNCTION OF NANO-BIOMATERIALS AS A DETECTION SIGNAL FOR ENVIRONMENTAL POLLUTION

Synthesis of nanomaterials can be executed by a variety of techniques whether it be physical, chemical, or biological. Specifically, when the biological method is used, it makes use of varied microorganisms, plant extract, agricultural waste, and biomolecules. Biomolecules when combined with the features of nanomaterials give rise to the component known as nanobiomaterials. And hence these biomaterials combined with nanosensors are widely being used nowadays in the detection of plant infection and other environmental contaminants. Moreover, nowadays the trend is to use green methods for the diagnosis of heavy metals as it is more environment-friendly and nontoxic in nature as well as they have a comparatively easy recovery rate and easy synthesis process (Rasheed et al., 2019; Iqbal and Bilal, 2020). The commonly used secondary metabolites for environmental monitoring purposes are proteins, alkaloids, carbohydrates, and phenolic compounds. Nano-biomaterials that are synthesized based on these secondary metabolites contain a metallic precursor (i.e., silver, Ag or gold, Au), which is indicated by the change in the color of the solution. The color change occurs, as the metallic atoms get nucleated (Shah et al., 2015). Gathering the targeted ions from the environment with the help of microorganisms and transforming it into metal ions with the aid of different biomolecules like sugars, enzymes, and protein that are secreted from the microorganism is the basic pathway for the synthesis of nanoparticles through a biological process. The synthesis can be intracellular or extracellular depending upon where the nanoparticle is synthesized. The nature of the nanoparticles depends upon factors, such as pH, type of microorganism, and temperature (Arun et al., 2013).

Being water is the most essential resource on the earth for any organism, it is quite necessary to inspect the quality of water regularly and distinguish the presence of any sort of harmful toxins, pharmaceutical waste, heavy metals, or any other kind of natural or anthropogenic waste. As a possible solution to this problem, newly designed nano-biosensors are formed that can efficiently control and detect the presence of toxic metals (e.g., Hg, Cd, Pb, Se). Au and Ag nanoparticles in combination with biomolecules have given rise to a new type of colorimetric sensor that is much more cost-effective, simple, and facilitates real-time detection of the target (Kulkarni and Muddapur, 2014; Parra-Arroyo et al., 2020).

Heavy metals being another major concern nowadays; their detection and remediation are prioritized. Mercury (Hg), a heavy metal found in water bodies, causes extensive damage to the neurological system. Previously, it has been observed that microorganisms can convert ionic mercury to methylmercury when released into the sea. It was also concluded that the migration of Hg from contaminated soil landfills and sediments to water bodies, streams, lakes, and underground water occurs due to the strong binding action between Hg and the organic matter which in turn enters through the food chain. In a study by Hongwei Luo et al., photochemical methylation, photo-oxidation, photodegradation, and photo-reduction are some photochemical behavior exhibited by Hg. A colorimetric assay was used for the detection of Hg^{2+} ions that comprise an amalgam of nano-biomaterials based on Ag nanoparticles. In this process, the nanoparticles are inculcated slowly in the solution containing the ionic mercury at room temperature with pH ranging between (3.73 and 11.18). A change in the coloration is observed that reveals the presence of ionic mercury (Luo et al., 2020).

Peptide-based gold nanoparticle probes are used to detect other metal ions such as Co^{2+}, Pt^{2+}, Pb^{2+}, and Pd^{2+}. It is achieved with the help of Flg-A3 fusion peptide (-Asp-Tyr-Lys-Asp-Asp-Asp-Asp-Lys-Pro-Ala-Tyr-Ser-Ser-Gly-ProAla-Pro-Pro-Met-Pro-Pro-Phe-), which acts as a multifunctional group. To synthesize and stabilize the Au nanoparticles in an aqueous solution of tetrachloroauric ($HAuCl_4$) generally, use A3 (AYSSGPAPPMPPF). The formation of the metal ion complex through non-covalent interaction takes place in the region of the N-terminal of the peptide (-Asp-Tyr-Lys-AspAsp -Asp-Lys-) (Slocik et al.,2005; Alam et al., 2015).

Genetic mutation and cancer are some common diseases related to the consumption of heavy metals in one or other ways. Chromium is one such metal, released in the environment through effluents and enters the water bodies and, in turn, enters the food chain. *Xanthocerassorbifolia* extract (XT-AuNP), a green agent, is used to stabilize gold nanoparticles that helps in the recognition of ionic chromium level. The only drawback of this method is that it can detect chromium ions at a concentration of not more than 3 μM. Although these nano-biosensors are highly effective in detecting Cr^{3+} in the environment (Fan et al., 2009). Anacardium occidentale leaf extract helped to develop another method to detect chromium. It was found that the nano-biosensor hence developed was highly selective and extremely sensitive toward Cr^{6+} ions with a limit of 1 μM. The test was carried out as the secondary metabolites of the extract helped in the conversion of Cr^{6+} ions to Cr^{3+}. The change in the color of the sample confirms the presence of Cr^{3+} (Balavigneswaran et al., 2014). Alternatively, Ag-nanospheres (C-SNSs) that are generally synthesized at varied pH values (4.5 and 11.5) are used for colorimetric detection of Cr^{3+} and Mn^{2+}. In this process, the C-SNSs solution is added to a solution that has a high concentration of Cr^{3+} ions at a pH of 4.5; later, a change in the coloration from yellow to white indicates the presence of Cr^{3+} ions. Although when the pH is maintained at 11.5, it was observed a change in color from yellow to brown that causes an alteration in the morphology from nanospheres to square pyrimity when Mn^{2+} ions are present (Joshi et al., 2016).

Nano-biosensors are comparatively more sensitive, selective, rapid, and convenient to use. Hence, it can also be used for the detection of pesticides combined with high-performance liquid chromatography (HPLC), liquid gas chromatography (GC), and mass spectroscopy. Several types of nano-biosensors are presently employed for detecting pesticides, whole-cell (bacterial, fungal, and algal cell-based) nano-biosensors, immuno (electrochemical and fluorescence detection technique), enzyme (electrochemical and fluorescence detection technique), and nucleic (electrochemical detection technique) nano-biosensors (Kumar and Arora, 2020).

A study conducted by Babolghani and Manesh based on the use of nucleic techniques brought forward the simulation and experimental analysis of a DNA/Cu2O-GS nanostructure. Particularly, this helps to detect polycyclic aromatic hydrocarbons (PAHs) and are categorized as DNA-based nano-biosensors. PHAs (polyhydroxyalkanoates) are known to be carcinogenic, toxic, and mutagenic agents and are generally represented as soil pollutants. DNA/Cu_2O-GS nanostructure can be used as a field-effect transistor (FET) for the detection of PHAs (Babolghani and Mohammadi-Manesh, 2019).

7.3.4 DEVELOPMENT OF NANO-BIOSENSORS FOR DETECTING PHENOLIC COMPOUNDS

In the past decades, there have been several kinds of research that were based upon the construction of nano-biosensors to detect phenolic compounds. The main point of benefit for constructing these biosensing platforms is that it can work in a microenvironment for stabilizing biorecognition elements at the electrode surface. With the use of several nanomaterials, nanoparticles assorted with polymers laccase sensors can be constructed. These nano-biosensors were found to be effective in the detection of hydroquinone and other phenolic compounds (Upan et al., 2016). The appearance of metal oxides in polymeric composite based on Fe_3O_4 magnetic nanoparticles increases the steadiness of the biosensor. An amalgam of Nafion and gold nanoparticles showed an increase in the stability of laccase biosensors for detecting hydroquinone. The stability of the laccase biosensor can also be enhanced by using several immobilization techniques, such as integration of laccase in ZnO sol-gel using chitosan is one of the techniques. Stabilizing the laccase enzyme with the use of laccase laser printing technology provides us with much more valid scientific data in case of catechol determination than using any other techniques (Touloupakis et al., 2014; Li et al., 2016; Qu et al., 2015).

Tyrosinase enzyme is often used as a biorecognition element for the detection of p-cresol or phenol. They have a high sensitivity and elevated range of affinity for these analytes; it is stabilized with the help of Nafion, a kind of polymer that shows an enhancement in the stability of the biosensors, glutaraldehyde, or by the use of the sol-gel method. Gold nanoparticles (AuNPs) are commonly used for the development of these tyrosine biosensors and they have a sensitivity of 15.7 μA/ppm as the transfer of electrons between the electrode surface and enzyme is very rapid. It was found that all p-cresol detecting tyrosine nano-biosensors can monitor the level of phenols. Quantum dots when mixed with chitosan or with the addition of copper oxide and mesoporous silica material give the lowest range for detection of p-cresol (Nurul Karim and Lee, 2013; Li et al., 2017; Han et al., 2015).

The widespread use and continuous release of bisphenol-A (BPA) in the environment have become the new environmental pollutant that directly or indirectly affects the aquatic ecosystem. Hence, they are specially monitored by using aptamer-based electrochemical biosensors. Generally, glassy carbon is used to construct the electrochemical transducers for these aptamers. The electrode surface is generally altered by gold nanoparticles embellished with carbon nanotubes or nanocomposite film of NH_2 functionalized by Fe_3O_4 or by gold nanoparticles and nanocomposites and immobilized graphene by the formation of thiol-gold (s-Au) bonds. BPA at a very low concentration (0.056 nM) can be detected by using nanoporous gold film (NPGF) that is attached to a glassy carbon electrode (Beiranvand and Azadbakht, 2017; Zhu et al., 2015). A recently introduced strategy to selectively detect BPA based on the enzyme deoxynucleotidyl transferase that creates a bridge on the surface of the electrode when there is no bisphenol A present, ultra-low levels of bisphenol A can be detected by electrochemical aptasensors that are developed by molecularly imprinted pyrrole, and electrodeposition of gold nanoparticles (AuNPs) with the thiolated DNA sequence (p-63) and free bisphenol A complex (Ensafi et al., 2018).

The development of nano-biosensors for the detection of phenolic compounds with high sensitivity, rapid response, and long-term stability is presently an area of significant research activity. But at the same time, the process of detection still suffers from some disadvantages such as diffusion limitations, lack of affinity, and weak stability (Table 7.2).

TABLE 7.2 Advantages and Disadvantages of Using Nano-Biosensores for the Detection of Pollutants.

Pros	Cons
Easy transportation	Inefficient in case of numerous pollutants
Thorough diagnosis of environment	Reusability issue
Small size	Can be hindered by the chemical factors
Ecofriendly	Unclear strategies for its scalability
Sensitivity is high toward the contaminants	Detection limit is limited for transformed products
Economical	
Detection of more than one pollutant at a time	

7.3.5 NANOWIRE SENSORS: A UNIT OF ELECTROCHEMICAL BIOSENSOR

Nanowire biosensors are a class of nano-biosensors of which the significant detecting segments are made of nanowires covered by natural particles or biological molecules, for example, polypeptides, DNA molecules, filamentous bacteriophages, and fibrin proteins. A bionanowire is a 1D (1-dimensional) nanostructure like a fibril, which measures some nanometers or less and do not have a restricted length (Touhami, 2014). Since the surface properties of bio-nanowires are effortlessly altered with the help of chemical or biological molecular ligands, nanowires can be enriched with practically any potential compound or biological subatomic acknowledgment unit, making the wires independent of analyte. This transduces the occasion of chemical binding on their surface into an adjustment in conductance of the nanowire with extraordinary sensitivity, real-time, and quantitative fashion. Boron-doped silicon nanowires (SiNWs) are being utilized to make profoundly sensitive, real-time electrically based sensors for biological and synthetic species (Rai et al., 2012). One-dimensional nanowire, nanobelts, nanosprings, and nanotubes have become the focal point of an escalated research in biosensing because of their extraordinary properties and potential of creating high-density nanoscale systems. The nanowires can be utilized for both effective vehicles of electrons and optical excitation, and these two components making them basic to the capacity and mix of nanoscale systems. Indeed, they are the littlest measurement structures that can be utilized for the effective vehicle of electrons, and hence, they are basic to the capacity and joining of these nanoscale devices. Their electrical properties are unequivocally affected by minor perturbations in light of their high surface-to-volume proportion and tunable electron transport properties because of the quantum confinement effect. One of the great possibilities for the improvement of enzyme/protein-based biosensors is the CNT because of its unique kind of mechanical, electrocatalytic, and electric properties. Researchers have utilized CNT/Nafion-based electrodes for immobilization glucoseoxidase (GOx) chemical for a sensitive recognition of glucose. This CNT/Nafion complex was set up by scattering solubilized CNTs in Nafion arrangement onto an electrode surface. CNT-based biosensor offers considerably more noteworthy signals, particularly at low potential, mirroring the electrocatalytic action of CNTs. Such a low potential activity of CNT-based biosensor brings about a wider linear range and a quick reaction time (Touhami, 2014; Hernandez-Vargas et al., 2018).

7.3.6 MICROORGANISM-BASED NANO-BIOSENSORS

Agriculture networks have been broken to revolutionize them by pathogen identification, high productivity checkup and high agriculture production monitoring services. The nano-biosensor is a particular sensor that manufactures a separate transducer package for different transmitters (physical, chemical, biological, electrochemical, etc.). They are classified, whether optically, piezoelectric, or mechanical because of their transduction theory. Biosensors were also characterized on the basis of their most sensitive carrier/recognition substances and may include the immunosensors, the biosensors, and the enzyme biosensors. Biologically originated (especially bacterial) biosensors are now being produced by integrating different NPs (Ag, Au, Cu, Zn, etc.) into microbes (bacteria, virus, and fungus). They are specialists in characterization of diseases in plants and cactus environments; these NP bases (nanowires, nanoformulated compounds, and nanoencapsulated beds) have a crucial role to play in clean strategies connected with pesticides and insecticides build-up in agriculture. Researchers identified in the diagnosis for several pathogenic bacteria the function of NPs. Further, it is important to explore various types of packages that allow us to design their introduction as point-of-care systems and complex devices. Recently (Bucur et al., 2018), the function of biosensors was reviewed on the basis of enzymatic inhibition (microbial origin). These biosensors could help to detect the quantitative toxicity of a variety of insecticides currently in use, such as organophosphorus compounds, carbamate compounds, etc.; these insecticides are almost forbidden or stated to have the highest risk as reported by the European Food Protection Authority (EFSA) for food (Justino et al., 2017; Mocan et al., 2017; Oliveira et al., 2018; Peiyan et al., 2018). Gui et al., 2017, revealed that Pseudomonas putida (BMM-PL) could be used as a whole cell-based biosensor. They concluded the role of these bacteria in the identification of phenanthrene in polluted soil. It is also necessary to choose the type of host cell. A biosensor's accuracy and time-response can be highly influenced by the part of its host cells that serve as the expression module for detection. As metabolism, genome, and cell composition are of elevated resemblance between eukaryotic-based sensors and the host body, around 85% of the currently used metals identification of whole cell biosensors are eukaryotes (Magrisso et al., 2008). In multiple environmental study sample, Hernández-Sánchez and their associates developed different full cell biosensors utilizing multiple host cells but a similar recombinant control method for the discovery of monocyclic aromatic compounds (Hernández-Sánchez et al., 2016). The biosensor from Alcanivorax borkumensis SK2 was reported to have fewer particulate tolerances while greater to show the better efficiency for detecting the low concentrations of polluted substances of the seawater sample. As a result of their high resistance at high temperatures until it was saturated, the Pseudomonas putida DOT-T1E biosensor was considered to be the best choice for environmentally highly polluted conditions (Espinosa-Urgel et al., 2015). A biosensor based on *Alcanivorax borkumensis* specialized in assimilating linear alkanes, and exhibited a four-fold lower detection sensitivity toward the fuel octane (0.5 μM) when compared to biosensors that used *Escherichia coli* as a vehicle. When measuring low concentrations of pure alkanes or petrol in samples, this performance improvement was clearly evident (Sevilla et al., 2015). Brutesco et al., 2016, prepared a functional biosensor based on Deinococcus deseri (an environmental bacterium that is tolerant to desiccation and radiation) and reported that after 7 days of storage, these sensors were able to detect arsenite. A series of whole cell biosensors have also been prepared from several *E. coli* strains to detect nickel in drinking water.

7.4 DNA NANOSENSORS—AN IDEAL APPROACH FOR THE DETECTION OF POLLUTANTS

Pathogen contamination in the environment has led to some serious contagious disease outbreaks that can cause epidemics or even pandemics. Through some studies to detect pathogens in the environment, some techniques have been introduced like traditional biosensor and nanosensor, out of which DNA-based nanosensor is being used particularly to identify pathogens like *E. coli, Vibrio cholera*, methicillin-resistant *Staphylococcus aureus* (MRSA), *Aspergillus Bacillus subtilis*, and *Candid. V. cholera*, a waterborne microbe can be detected by employing selective binding of O1 OmpW (outer membrane protein W) gene with two DNA probes- Magnetic NP-probe1-O1 OmpW-fluorescein amidite (FAM) probe2-AuNP complex. The association of DNA-pollutants is unique since specific pollutants contribute to a specific toxic reaction. The capacity of the DNA of changing specificity with the change in its structure and sequence is beneficial for the sensing of diverse analytes. The most widely used nanomaterials for the production of DNA nanosensor are AuNPs and AgNPs. As optical nanosensors, magnesium oxide, iron oxide, QDs, and manganese oxide NPs have been used and can also be investigated for vast numbers of unexplored analytes to design DNA nanosensors as an alternative to routine complex analytical assays (Alahi and Mukhopadhyay, 2017). The descent biocompatibility, extensive availability, water-soluble property, and fluorescence emission based on DNA sequence are all the major advantages of using DNA nanosensors (Song et al., 2019).

Pesticides are virulent to living beings. The categorization of pesticides is difficult as it involves about 100 classes with a cumulative number of over 800. Organophosphorus, carbamates, neonicotinoids, and triazine are the major groups that are detected using DNA-based nanosensors. There is a certain number of approaches available for the detection of pesticides that includes both conventional and sensor-based methods. For example, HPLC, GC-MS, and other analytical methods are commonly used for a minute number of pesticides whereas DNA nanosensors have been used with huge amounts of pesticides to detect (Willner and Vikesland, 2018). The colorimetric DNA nanosensor is designed for the identification of acetamiprid in celery and green tea leaves. Salt-induced aggregation happens with the aptamer binding to ssDNA-AuNPs and the solution looks purple. Acetamiprid availability contributes to the solubilization of ssDNA-AuNPs due to the development of the red-colored aptamer-acetamiprid complex (Fei et al., 2015). Aggregation of ssDNA aptamer-AgNPs may be selectively caused by phorate, an organophosphate pesticide. Aggregation leads to a decrease in the rate of UV-visible absorption and a change of the color of NP solution from brown to colorless (Li et al., 2018). For the detection of triazophos in apple, water, rice, turnip, and cabbage samples with a sensitivity almost similar to the ELISA method, a bio-barcode amplification-based competitive immunoassay was developed. Thiolated ss-oligonucleotide as barcodes and mAb as recognition elements in AuNPs were subjected to a laboratory sample containing an ovalbumin-pesticide-hapten (hapten-OVA) functionalized pesticide and magnetic microparticle (MMP) probe. Detection is concluded by AuNPs catalyzed Ag staining. The graycolor intensity of Ag relies on the concentration of triazophos (Du et al., 2018).

GMO has transgene-containing modified DNA and several other sequences such as selection markers and promoter. Similarly, mutant and resistant species contain sequences of DNA that are distinct from regular organisms. The sequences that are complementary to these particular sequences are being used as bioreceptors to have DNA nanosensor specificity. Fluorescence intensity recovery is used for quantifying *Mycobacterium tuberculosis* resistance to rifampicin. Specific FAM-labeled probe for a mutant DNA sequence with rifampicin resistance to *M. tuberculosis* is adsorbed on graphdiyne's nanosheet surface. The nanosheet quenches the FAM probe's fluorescence. The FAM probe is released from the nanosheet in the presence of target ssDNA to form a structure with target DNA (Chang et al., 2019). For the resistance of microbes such as MRSA (methicillin-resistant *Staphylococcus aureus*) to antibiotics, the penicillin-binding protein-encoding gene is responsible. In the presence of intercalating fluorescent dye SYBR green I, MRSA containing mecA target ssDNA interacts with carboxyfluorescein (FAM)-ssDNA-GO and eliminates FAM ssDNA to form double-stranded mecA-ssDNA FAM-SYBR green

I-ssDNA complex. SYBR green is later integrated as the polymerase lies in the solution expands the target DNA utilizing FAM ssDNA as a template leading to a rise in the fluorescence intensity (Ning et al., 2016).

The heavy metal pollutants most widely detected include mercury, arsenic, lead, chromium, and cadmium. Toxic metals like the above can cause damage multiple organs and disrupt the metabolomics in humans. Heavy metal ions' precise binding to the molecules of bioreceptor DNA allows the properties of nanomaterials or DNA to change. Samples of hot spring and seawater have the fluorescence quenching technique, which was used for the identification of sulfide ions (S^{2-}) (Chen et al., 2011). Cytosine-rich DNA has an Ag+ affinity that has been used for the visual identification of Ag+ in tap and river water samples to design the DNA-based lateral flow test. To prevent non-specific encounters with intervening iodide ions (I^+), the sensor includes the addition of sodium peroxydisulfate. In the test sample, the presence of arsenic [As (III)] eliminates aptamer shielding the pores of MSN filled with rhodamine B, contributing to a rise in fluorescence strength. As (III) is a carcinogenic pollutant. Aptamer-controlled crystal violet synthesis is used for designing luminescence resonance energy transfer (LRET)-based nanosensor in the research to detect As (III) (Chen et al., 2020). The rate of resonance Rayleigh scattering (RRS) increases immediately with the increase of nanomaterial size that depends on the concentration of As (III) (Wu et al., 2012). DNA-AuNP nanosensors have also been used to simultaneously detect multiple metal ions. In wastewater samples, the sensor will directly detect Ag^+, Cd^{2+}, Cu^{2+}, Hg^{2+}, Pb^{2+}, Zn^{2+}, Mn^{2+}, Cr^{3+}, and Sn^{4+} (Tan et al., 2016). There have also been studies of a DNA nanosensor-based identification of contaminant dyes, explosives, and toxins. The broad ssDNA consists of two thin, complementary DNA strands, one side is made of a scaffolder that carries luminescence acceptors, and the other side is supported by a π-donor 6-hydroxy-l-DOPA base that contains the pair. Dopamine, with CuNPs, can enable the fluorescence exemption of dsDNA-CuNPs via a photo-induced digital electron transfer mechanism. Optical and electrical-to chemical sensors with NP-base to detect mycotoxin have been used for mutagenic nitrosamines, namely *N*-nitrosodiethanolamine (Majumdar et al., 2020).

7.5 IMMUNO-BASED NANO-BIOSENSOR

Conventional ELISA (enzyme-linked immunosorbent assay) assays utilize immunological reagents to identify bacteria (Swanink et al., 1997). Following similar principles, an assortment of detecting stages have been planned that fuse antibody (Ab) to improve compactness, decrease examination time, and disentangle recognition (Guner et al., 2017). Most nanosensing stages depend on AuNPs adjusted with Ab (antibody) and detection depends on estimations of the surface plasmon reverberation (SPR) or the color change related to aggregation/deaggregation upon target binding (Choi et al., 2020). Antibodies explicit for these microbes were immobilized over a gold layer or AuNPs stored on the gold layer utilizing 16-mercaptoundecanoic acid and carbodiimide coupling between the acid group on the Au surface and the amine buildup of the antibody. Estimations of the adjustment in resonance angle and refractive index with various bacteria concentrations gave a detection cutoff of 103 CFU/mL when AuNPs were utilized when contrasted with 104 CFU/mL without NPs. Singh et al. (2009) have announced an immunosensor for recognition of *E. Coli* utilizing Au nanorods functionalized with *E. coli* Ab and two-photon Rayleigh scattering (TPRS) spectroscopy as an identification method. In presence of *E. coli* O157:H7 bacterium, the adjusted nanorods tie to *E. coli* causing conglomeration or aggregation which brought about an expansion in the TPRS signal. The investigation took 15 min and the LOD was 50 CFU/mL.

An electrochemical immunosensor with magnetic separation for identification of *Bacillus* and *E. coli* O157:H7 utilizing trifunctional NPs of immuno-attractive/polyaniline core/shell (c/sNP) was planned. The NP framework contains antibodies as a particular bioreceptor for microbes, a magnetic moiety to upgrade detachment and fixation, and polyaniline as an electrically conductive material to improve the conductivity for electrochemical estimations. For indicating a current reduction with expanding bacteria concentration, cyclic voltammetry and amperometry were utilized as recognition methods (Vu et al., 2020).

Different works revealed a low-cost effort paper-based innovation in which nitrocellulose paper was changed with immunological reagents against microbes and AuNPs for discovery. It has been revealed that a multiplex paper-based immunosensor for the location of *Pseudomonas aeruginosa* and *Staphylococcus aureus* in which microbial antibodies were joined to AuNPs on nitrocellulose paper. The measure was created as a portable strip reader and had the option to distinguish 500–5000 CFU/mL. In different models, electrochemical immunosensors for Salmonella were planned to utilize graphene quantum spots (GQDs) (Ye et al., 2017). In different works, Deisingh and Thompson misused the utilization of Raman spectroscopy on nano-designed surfaces for the detection of bacteria in food and ecological investigation (Deisingh and Thompson, 2004). Other platforms use silica NPs as immobilization platforms for the detection of *Escherichia coli*

(Mathelié-Guinlet et al., 2016). Immobilization of the biosensing component on nanomaterials has appeared to improve sub-atomic acknowledgement and increase selectivity. 3D demonstration was utilized for nano-manipulations and anticipating the selectivity toward various analyses. The current advancement status exhibits that immunosensors and aptasensors can possibly improve the presentation of gadgets for pathogenic microorganism identification, and this methodology can resolve a conceivably huge number of difficulties in bioassays (Cho et al., 2020; Jeddi and Saiz, 2017). Nonetheless, during the immobilization cycle, the absence of direction of the antibodies or aptamers, which may bring about irregular conjugation with the objective of interest, is a basic issue that actually should be tended to. Extra difficulties are issues of explicitness, some because of the presence of vague adsorption that requires the advancement of appropriate materials and techniques to improve selectivity, empower site-explicit direction of bioreceptors on surfaces, and forestall vague adsorption (Mustafa et al., 2017).

7.6 A GUIDE TO POLLUTION DETECTION: CURRENT TRENDS AND NOVEL APPLICATIONS

Nanotechnology is a recent field of science that deals with nanomaterials. These nanoparticles, when integrated with sensors, can detect many types of contaminations whether it be in the soil, food, or the environment. Nano-biosensors also play an important role in the field of medicine and therapeutics.

In recent advancements, nano-biosensors have proved to be an effective tool to detect pesticides because of their nature, that is, its high sensitivity, selectivity, and fast response by reacting with the target to develop a signal. It is expected that very soon in the future, detection of pesticides will be carried out with the help of commercialized nanoparticles that are integrated with bioelements. The development of pesticide detecting nano-biosensors has shown a great advancement and is quite reliable (Christopher et al., 2020).

Nowadays, agricultural products contain a much higher amount of pesticide than approved. These develop serious health disorders hence detecting them is important. Recently array type and unimolecular nanomaterial-based nano-biosensors are designed for the detection of pesticides. Carbon nanotubes, iMono-based nano-biosensors, and nanoparticles are generally being used for the detection of smaller levels of organophosphorus pesticide. Quantum dot (QD) based fluorescent nano-biosensors also help to detect pesticides in soil. Acetamiprid binding DNA aptamers are currently being used to detect insecticides with the aid of QD.

Very rapid and sensitive detecting systems with high molecular precision are used to detect contamination and pathogens present in the soil. Nanoparticle-based nano-biosensors are a boon for the detection of these contaminants at a very low scale because of their enormous surface area. Ralstonia solanacearum is a bacterium that harms potato cultivation and is a very serious issue from the agricultural point of view. Aurum-based nanoparticles are used to develop nano-biosensors that can detect even a very small amount (15 ng) of the genomic DNA of Ralstonia solanacearum if present in the soil. Au nanorods based nanosensors have been developed to detect the virus in orchid plantations. Heavy metals that are present in soil and water cause havoc damage to the human system and environment. Hence, it is important to identify them and eliminate them from nature. Integration of nanomaterials with electrochemical biosensors helps in the detection of these heavy metals in the soil like the presence of mercury can be detected with the aid of AuNP integrated amperometric biosensors. Recently, a carbon nanotube-based single-walled electrochemical biosensor has been developed that can effectively detect the presence of Hg in water (Salouti and Derakhshan, 2020).

There is a rapid growth in the development of biosensors to detect environmental pollution, and it is aided by the new advancement in the field of nanotechnology. Cost-effective, highly sensitive, and portable DNA nano-biosensors are a good option for monitoring the quality of water. This serves as a solution to many new challenges arising from the detection of water pollution (Soukarié et al., 2020). The high concentration of chromium consumption leads to mutation and develops cancer. It is generally present in the effluents that go and mix with the water bodies. In a recent study, it was found that ionic chromium can be recognized with the aid of Au nanoparticles that have been secured by Xanthoceras sorbifolia extract (XT-AuNP). Ag nanospheres (C-SNSs) that are manufactured at distinct pH values are now helping to detect the presence of chromium in water (Aguilar-Pérez et al., 2020). Lead is another heavy metal that is generally found in the industrial sewage and water bodies where this sewage is drained out. It is extremely important to detect them and eliminate them hence a recently developed cantilever nano-biosensor is used (Rigo et al., 2020). A newly designed alternative for the detection of heavy metals is the quantum dots and carbon nanodots based system that is based on Förster resonance energy transfer (FRET). It mainly consists of carbon nanodots as acceptor components and GQD as donors. The recent trend applies SPR nanosensor to detect the presence of zinc. The use of zinc ion imprinted poly (2-hydroxyethyl methacrylate-N-methacryloyl- (L)-histidine methyl ester) SPR nanosensor to detect the presence of zinc in the water-based fluid is common nowadays. It is fast, cost-efficient, and much more effective. Even to detect temperature the newly developed temperature nano-biosensors made of carbon dots are being used.

Currently, there is a trend to use nano-biosensors to detect environmental viruses. Detection of these viruses using a single-walled carbon nanotube field-effect-transistor (swCNT-FET) is common. This recent technique helps to deactivate the airborne viruses with the help of engineered water nano-structures (EWNS). The presence of bacteriophage viruses in water bodies is detected by the use of a molecularly imprinted polymer (MIP) nanoparticle-based affinity system (Debnath and Das, 2020).

7.7 CONCLUSION

Significant developments have been made in the biotechnology and nano-technology sectors to get rid of harmful pollutants. Globally, air contamination in multiple media is a serious health issue. The need for rapid biosensor identification will grow due to ever growing public health issues about the damage that environmental emissions will cause. The pivotal part of the new biosensing method is nanomaterials. The future of nano-biosensors will depend on the performance of advanced technologies that are evolving. The nano-biosensors have a lot of characteristic features like, high surface area to volume proportions, good electrochemical properties, needs normal storage conditions, accuracy of the transducer, pH, temperature, sensitivity, and linearity. There are several approaches to nano-biosensors that include enzymatic nano-biosensors, nanomaterial-based fluorescent biosensors, nano-biomaterials as a detection signal for environmental pollution, nano-biosensors specifically designed for detecting phenolic compounds, nanowire sensors, etc. The process of detection still suffers from some disadvantages such as diffusion limitations, lack of affinity, and weak stability, and it can be overcome with the significant research activity in the area of high sensitivity, rapid response, and long-term stability.

ACKNOWLEDGMENT

The authors gratefully acknowledge the support of Ms. Mousumi Meghamala Nayak for her valuable suggestions and helping us in formatting the manuscript.

KEYWORDS

- **nano-biosensors**
- **nanomaterials**
- **environmental monitoring**
- **pollutant detection**
- **pesticide**
- **integrated sensing platforms**

REFERENCES

Aguilar-Pérez, K. M.; Heya, M. S.; Parra-Saldívar, R.; Iqbal, H. M. Nano-Biomaterials in-Focus as Sensing/Detection Cues for Environmental Pollutants. *Case Studies Chem. Environ. Eng.* **2020,** *2*, 100055.

Alahi, M. E.; Mukhopadhyay, S. C. Detection Methodologies for Pathogen and Toxins: A Review. *Sensors* **2017,** *17* (8), 1885.

Alam, M. N.; Chatterjee, A.; Das, S.; Batuta, S.; Mandal, D.; Begum, N. A. Burmese Grapefruit Juice Can Trigger the 'Logic Gate'-Like Colorimetric Sensing Behavior of Ag Nanoparticles towards Toxic Metal Ions. *RSC Adv.* **2015,** *5* (30), 23419–23430.

Antiochia, R.; Lavagnini, I.; Magno, F. Amperometric Mediated Carbon Nanotube Paste Biosensor for Fructose Determination. *Anal. Lett.* **2004,** *37* (8), 1657–1669.

Antonacci, A.; Arduini, F.; Moscone, D.; Palleschi, G.; Scognamiglio, V. Nanostructured (Bio) Sensors for Smart Agriculture. *TrAC Trends Anal. Chem.* **2018,** *98*, 95–103.

Arun, P.; Shanmugaraju, V.; Ramanujam, J. R.; Prabhu, S. S.; Kumaran, E. Biosynthesis of Silver Nanoparticles from Corynebacterium sp. and Its Antimicrobial Activity. *Int. J. Curr. Microbiol. App. Sci.* **2013,** *2* (3), 57–64.

Babolghani, F. M.; Mohammadi-Manesh, E. Simulation and Experimental Study of FET Biosensor to Detect Polycyclic Aromatic Hydrocarbons. *Appl. Surf. Sci.* **2019,** *488*, 662–670.

Balavigneswaran, C. K.; Kumar, T. S. J.; Packiaraj, R. M.; Prakash, S. Rapid Detection of Cr(VI) by AgNPs Probe Produced by Anacardium Occidentale Fresh Leaf Extracts. *Appl. Nanosci.* **2014,** *4* (3), 367–378.

Beiranvand, S.; Azadbakht, A. Electrochemical Switching with a DNA Aptamer-Based Electrochemical Sensor. *Mater. Sci. Eng. C.* **2017,** *76*, 925–933.

Bhattarai, P.; Hameed, S. Basics of Biosensors and Nanobiosensors. Nanobiosensors: From Design to Applications. **2014,** 374–403.

Brutesco, C.; Prévéral, S.; Escoffier, C.; Descamps, E. C. T.; Prudent, E.; Cayron, J.; Dumas, L.; Ricquebourg, M.; Adryanczyk-Perrier, G.; Groot, A. D. Bacterial Host and Reporter Gene Optimization for Genetically Encoded Whole Cell Biosensors. *Environ. Sci. Pollut. Res. Int.* **2016,** *24*, 1–14.

Bucur, B.; Munteanu, F. D.; Marty, J. L.; Vasilescu, A. Advances in Enzyme-Based Biosensors for Pesticide Detection. *Biosensors* **2018,** *8*, 128.

Chang, F.; Huang, L.; Guo, C.; Xie, G.; Li, J.; Diao, Q. Graphdiyne-Based One-Step DNA Fluorescent Sensing Platform for the Detection of Mycobacterium Tuberculosis and Its Drug-Resistant Genes. *ACS Appl. Mater. Interf.* **2019,** *11* (39), 35622–35629.

Chen T, Wang, H.; Wang, Z.; Tan, M. Construction of Time-Resolved Luminescence Nanoprobe and Its Application in As (III) Detection. *Nanomaterials* **2020,** *10* (3), 551.

Chen, W. Y.; Lan, G. Y. Chang, H. T. Use of Fluorescent DNA-Templated Gold/Silver Nanoclusters for the Detection of Sulfide Ions. *Anal Chem.* **2011,** *83* (24), 9450–9455.

Cho, I. H.; Kim, D. H.; Park, S. Electrochemical Biosensors: Perspective on Functional Nanomaterials for on-Site Analysis. *Biomater. Res.* **2020,** *24* (1), 1–12.

Choi, J. H.; Lee, J. H.; Son, J.; Choi, J. W. Noble Metal-Assisted Surface Plasmon Resonance Immunosensors. *Sensors* **2020,** *20* (4), 1003.

Christopher, F. C.; Kumar, P. S.; Christopher, F. J.; Joshiba, G. J.; Madhesh, P. Recent Advancements in Rapid Analysis of Pesticides Using Nano Biosensors: A Present and Future Perspective. *J. Clean. Prod.* **2020,** 122356.

da Silva, A. C.; Deda, D. K.; Bueno, C. C.; Moraes, A. S.; Da Roz, A. L.; Yamaji, F. M.; Prado, R. A.; Viviani, V.; Oliveira, O. N.; Leite, F. L. Nanobiosensors Exploiting Specific Interactions between an Enzyme and Herbicides in Atomic Force Spectroscopy. *J. Nanosci. Nanotechnol.* **2014,** *14* (9), 6678–6684.

Dasgupta, N.; Ranjan, S.; Mundekkad, D.; Ramalingam, C.; Shanker, R.; Kumar, A. Nanotechnology in Agro-Food: From Field to Plate. *Food Res. Int.* **2015,** *69*, 381–400.

Debnath, N.; Das, S. Nanobiosensor: Current Trends and Applications. In *NanoBioMedicine*; Springer: Singapore, 2020; pp 389–409.

Deisingh, A. K.; Thompson, M. Biosensors for the Detection of Bacteria. *Can. J. Microbiol.* **2004,** *50*, 69–77.

Devi, R.; Yadav, S.; Pundir, C. S. Amperometric Determination of Xanthine in Fish Meat by Zinc Oxide Nanoparticle/Chitosan/Multiwalled Carbon Nanotube/Polyaniline Composite Film Bound Xanthine Oxidase. *Analyst* **2012,** *137*, 754–759.

Du, P.; Jin, M.; Zhang, C.; Chen, G.; Cui, X.; Zhang, Y. et al. Highly Sensitive Detection of Triazophos Pesticide Using a Novel Bio-Barcode Amplification Competitive Immunoassay in a Microwell Plate-Based Platform. *Sens. Actuat. B Chem.* **2018,** *256*, 457–464.

Ensafi, A. A.; Amini, M.; Rezaei, B. Molecularly Imprinted Electrochemical Aptasensor Forthe Attomolar Detection of Bisphenol A. *Microchim. Acta.* **2018,** *185* (5), 265.

Espinosa-Urgel, M.; Serrano, L.; Ramos, J. L.; Fernández-Escamilla, A. M. Engineering Biological Approaches for Detection of Toxic Compounds: A New Microbial Biosensor Based on the Pseudomonas Putida TtgR Repressor. *Mol. Biotechnol.* **2015,** *57*, 1–7.

Fan, Y.; Liu, Z.; Zhan, J. Synthesis of starch-stabilized Ag nanoparticles and Hg 2þ Recognition in Aqueous Media. *Nanosc. Res. Lett.* **2009,** *4* (10), 1230–1235.

Fei, A.; Liu, Q.; Huan, J.; Qian, J.; Dong, X.; Qiu, B.; Mao, H.; Wang, K. Label-Free Impedimetric Aptasensor for Detection of Femtomole Level Acetamiprid Using Gold Nanoparticles Decorated Multiwalled Carbon Nanotube-Reduced Graphene Oxide Nanoribbon Composites. *Biosens. Bioelectron.* **2015,** *70*, 122–129.

Franca, E. F.; Leite, F. L.; Cunha, R. A.; Oliveira Jr, O. N.; Freitas, L. C. Designing an Enzyme-Based Nanobiosensor Using Molecular Modeling Techniques. *Phys. Chem. Chem. Phys.* **2011,** *13* (19), 8894–8899.

Gan, N.; Yang, X.; Xie, D.; Wu, Y.; Wen, W. A. Disposable Organophosphorus Pesticides Enzyme Biosensor Based on Magnetic Composite Nanoparticles Modified Screen Printed Carbon Electrode. *Sensors* **2010,** *10*, 625–638.

Gaviria-Arroyave, M. I.; Cano, J. B.; Peñuela, G. A. Nanomaterial-Based Fluorescent Biosensors for Monitoring Environmental Pollutants: A Critical Review. *Talanta Open* **2020,** 100006.

Ghormade, V.; Deshpande, M. V.; Paknikar, K. M. Perspectives for Nano-Biotechnology Enabled Protection and Nutrition of Plants. *Biotechnol Adv.* **2011,** *29*, 792–803.

Gui, Q.; Lawson, T.; Shan, S.; Yan, L.; Liu, Y. The Application of Whole Cell-Based Biosensors for Use in Environmental Analysis and in Medical Diagnostics. *Sensors* **2017,** *17*, 1623.

Guner, A.; Çevik, E.; ¸Senel, M.; Alpsoy, L. An Electrochemical Immunosensor for Sensitive Detection of *Escherichia coli* O157:H7 by Using Chitosan. MWCNT, Polypyrrole with Gold Nanoparticles Hybrid Sensing Platform. *Food Chem.* **2017,** *229*, 358–365.

Han, E.; Yang, Y.; He, Z.; Cai, J.; Zhang, X.; Dong, X. Development of Tyrosinase Biosensor Based on Quantum Dots/Chitosan Nanocomposite for Detection of Phenolic Compounds. *Anal. Biochem.* **2015,** *486*, 102–106.

Hernández-Sánchez, V.; Molina, L.; Ramos, J. L.; Segura, A. New Family of Biosensors for Monitoring BTX in Aquatic and Edaphic Environments. *Microb. Biotechnol.* **2016,** *9*, 858–867.

Hernandez-Vargas, G.; Sosa-Hernández, J. E.; Saldarriaga-Hernandez, S.; Villalba-Rodríguez, A. M.; Parra-Saldivar, R.; Iqbal, H. Electrochemical Biosensors: A Solution to Pollution Detection with Reference to Environmental Contaminants. *Biosensors* **2018,** *8* (2), 29.

Hu, J.; Wang, D.; Dai, L.; Shen, G.; Qiu, J. Application of Fluorescent Biosensors in the Detection of Hg (II) Based on T-Hg (II)-T Base Pairs. *Microchem. J.* **2020,** 105562.

Iqbal, H. M.; Bilal, M. Time to Automate the Microbial Detection and Identification: The Status Quo. *J. Pure Appl. Microbiol.* **2020,** *14* (1), 01–03.

Jain, Y.; Rana, C.; Goyal, A.; Sharma, N.; Verma, M. L.; Jana, A. K. Biosensors, Types and Applications. In *Proceedings of International Conference on Biomedical Engineering and Assistive Technology*; Jalandhar, India, 2010; pp 1–6.

Jeddi, I.; Saiz, L. Three-Dimensional Modeling of Single Stranded DNA Hairpins for Aptamer-Based Biosensors. *Sci. Rep.* **2017,** *7*, 1178.

Joshi, P.; Nair, M.; Kumar, D. PH-Controlled Sensitive and Selective Detection of Cr(III) and Mn(II) by Using Clove (*S. aromaticum*) Reduced and Stabilized Silver Nanospheres. *Anal. Methods* **2016,** *8* (6), 1359–1366.

Justino, C.; Duarte, A.; Rocha-Santos, T. Recent Progress in Biosensors for Environmental Monitoring: A Review. *Sensors* **2017,** *17*, 2918.

Kulkarni, N.; Muddapur, U. Biosynthesis of Metal Nanoparticles: A Review. *J. Nanotechnol.* **2014**.

Kumar, V.; Arora, K. Trends in Nano-Inspired Biosensors for Plants. *Mater. Sci. Energy Technol.* **2020,** *3*, 255–273.

Kuswandi, B. Nanobiosensor Approaches for Pollutant Monitoring. *Environ. Chem. Lett.* **2019,** *17*, 975–990.

Li, G.; Sun, K.; Li, D.; Lv, P.; Wang, Q.; Huang, F.; Wei, Q. Biosensor Based on Bacterial Cellulose-Au Nanoparticles Electrode Modified with Laccase for Hydroquinone Detection. *Colloids Surf. A Physicochem. Eng. Asp.* **2016,** *509*, 408–414.

Li, H.; Hu, X.; Zhu, H.; Zang, Y.; Xue, H. Amperometric Phenol Biosensor Based on a New Immobilization Matrix: Polypyrrole Nanotubes Derived from Methyl Orange as Dopant. *Int. J. Electrochem. Sci.* **2017,** *12*, 6714–6728.

Li, S. C.; Chen, J. H.; Cao, H.; Yao, D. S.; Liu, D. L. Amperometric Biosensor for Aflatoxin B1 Based on Aflatoxin Oxidase Immobilized on Multiwalled Carbon Nanotubes. *Food Control* **2011,** *22* (1), 43–49.

Li, X.; Shi, J.; Chen, C.; Li, W.; Han, L.; Lan, L. et al. One-Step, Visual and Sensitive Detection of Phorate in Blood-Based on a DNA– AgNC Aptasensor. *New J Chem.* **2018,** *42* (8), 6293–6298.

Luo, H.; Cheng, Q.; Pan, X. Photochemical Behaviors of Mercury (Hg) Species in Aquatic Systems: A Systematic Review on Reaction Process, Mechanism, and Influencing Factor. *Sci. Total Environ.* **2020,** *720*, 137540.

Luz, R. A.; Iost, R. M.; Crespilho, F. N. Nanomaterials for Biosensors and Implantable Biodevices. In *Nano Bioelectrochemistry*; Springer: Berlin, Heidelberg, 2013; pp 27–48.

Magrisso, S.; Erel, Y.; Belkin, S. Microbial Reporters of Metal Bioavailability. *Microb. Biotechnol.* **2008,** *1*, 320–330.

Majumdar, S.; Thakur, D.; Chowdhury, D. DNA Carbon-Nanodots Based Electrochemical Biosensor for Detection of Mutagenic Nitrosamines. *ACS Appl. BioMater.* **2020,** *3* (3), 1796–803.

Malik, P.; Katyal, V.; Malik, V.; Asatkar, A.; Inwati, G.; Mukherjee, T. K. Nanobiosensors: Concepts and Variations. *Int. Sch. Res. Notices*. **2013**.

Mathelié-Guinlet, M.; Gammoudi, I.; Beven, L.; Moroté, F.; Delville, M. H.; Grauby-Heywang, C.; Cohen-Bouhacina, T. Silica Nanoparticles Assisted Electrochemical Biosensor for the Detection and Degradation of *Escherichia coli* Bacteria. *Procedia Eng.* **2016,** *168*, 1048–1051.

Miranda, O. R.; Li, X.; Garcia-Gonzalez, L.; Zhu, Z. J.; Yan, B.; Bunz, U. H. F.; Rotello, V. M. Colorimetric Bacteria Sensing Using a Supramolecular Enzyme-Nanoparticle Biosensor. *J. Am. Chem. Soc.* **2011,** *133*, 9650–9653.

Mocan, T.; Matea, C. T.; Pop, T.; Mosteanu, O.; Buzoianu, A. D.; Puia, C.; et al. Development of Nanoparticle-Based Optical Sensors for Pathogenic Bacterial Detection. *J. Nanobiotech.* **2017,** *15*, 114.

Mustafa, F.; Hassan, R. Y. A.; Andreescu, S. Multifunctional Nanotechnology-Enabled Sensors for Rapid Capture and Detection of Pathogens. *Sensors* **2017,** *17*, 2121.

Ning, Y.; Gao, Q.; Zhang, X.; Wei, K.; Chen, L. A Graphene Oxide-Based Sensing Platform for the Determination of Methicillin-Resistant *Staphylococcus aureus* Based on Strand-Displacement Polymerization Recycling and Synchronous Fluorescent Signal Amplification. *J. Biomol. Screen.* **2016,** *21* (8), 851–857.

Nurul Karim, M.; Lee, H. J. Amperometric Phenol Biosensor Based on Covalent Immobilization of Tyrosinase on Au Nanoparticle Modified Screen-Printed Carbon Electrodes. *Talanta* **2013,** *116*, 991–996.

Oliveira, J. L. D.; Campos, E. V. R.; Pereira, A. E. S.; Pasquoto, T.; Lima, R.; Grillo, R. et al. Zein Nanoparticles as Eco-Friendly Carrier Systems for Botanical Repellents Aiming Sustainable Agriculture. *J. Agric. Food Chem.* **2018,** *66*, 13301340.

Ozdemir, C.; Yeni, F.; Odaci, D.; Timur, S. Electrochemical Glucose Biosensing by Pyranose Oxidase Immobilized in Gold Nanoparticle-Polyaniline/AgCl/Gelatin Nanocomposite Matrix. *Food Chem.* **2010,** *119*, 380–385.

Parra-Arroyo, L.; Parra-Saldivar, R.; Ramirez-Mendoza, R. A.; Keshavarz, T.; Iqbal, H. M.; Laccase-Assisted Cues: State-of-the-Art Analytical Modalities for Detection, Quantification, and Redefining "Removal" of Environmentally Related Contaminants of High Concern. In *Laccases in Bioremediation and Waste Valorisation*; Springer: Cham, 2020; pp 173–190.

Peiyan, Y.; Xin, D.; Yan, Y. Y.; Qing-Hua, X. Metal Nanoparticles for Diagnosis and Therapy of Bacterial Infection. *Adv. Healthc. Mater.* **2018,** *7*, 1701392.

Perez-Lopez, B.; Merkoci, A. Nanomaterials Based Biosensors for Food Analysis Applications. *Trends Food. Sci. Technol.* **2011,** *22*, 625–639.

Periasamy, A. P.; Umasankar, Y.; Chen, S. M. Nanomaterials-Acetylcholinesterase Enzyme Matrices for Organophosphorus Pesticides Electrochemical Sensors: A Review. *Sensors* **2009,** *9* (6), 4034–4055.

Qu, J.; Lou, T.; Wang, Y.; Dong, Y.; Xing, H. Determination of Catechol by a Novel Laccase Bio-Sensor Based on Zinc-Oxide Sol-Gel. *Anal. Lett.* **2015,** *48*, 1842–1853.

Ramanathan, M.; Luckarift, H. R.; Sarsenova, A.; Wild, J. R.; Ramanculov, E. R.; Olsen, E. V.; Simonian, A. L. Lysozyme-Mediated Formation of Protein-Silica Nano-Composites for Biosensing Applications. *Colloids Surf B Biointerf.* **2009,** *73*, 58–64.

Rasheed, T.; Nabeel, F.; Adeel, M.; Rizwan, K.; Bilal, M.; Iqbal, H. M. Carbon Nanotubes-Based Cues: A Pathway to Future Sensing and Detection of Hazardous Pollutants. *J. Mol. Liq.* **2019,** *292*, 111425.

Rigo, A. A.; de Cezaro, A. M.; Martinazzo, J.; Ballen, S.; Hoehne, L.; Steffens, J.; Steffens, C. Detection of Lead in River Water Samples Applying Cantilever Nanobiosensor. *Water Air Soil Pollut.* **2020,** *231*, 1–11.

Salgado, A. M.; Silva, L. M.; Melo, A. F. Biosensor for Environmental Applications. In *Environmental Biosensors*; Somerset, V., Ed.; IntechOpen: Rijeka, 2011; pp 3–16.

Salouti, M.; Derakhshan, F. K. Biosensors and Nanobiosensors in Environmental Applications. In *Biogenic Nano-Particles and Their Use in Agro-Ecosystems*; Springer: Singapore, 2020; pp 515–591.

Sassolas, A.; Blum, L. J.; Leca-Bouvier, B. D. Immobilization Strategies to Develop Enzymatic Biosensors. *Biotechnol. Adv.* **2012,** *30*, 489–511.

Sekhon, B. S. Nanotechnology in Agri-Food Production: An Overview. *Nanotechnol. Sci. Appl.* **2014,** *7*, 31–53.

Sevilla, E.; Yuste, L.; Rojo, F. Marine Hydrocarbonoclastic Bacteria as Whole-Cell Biosensors for n-Alkanes. *Microb. Biotechnol.* **2015,** *8*, 693–706.

Shah, M.; Fawcett, D.; Sharma, S.; Tripathy, S. K.; Poinern, G. E. J. Green Synthesis of Metallic Nanoparticles via Biological Entities. *Materials* **2015,** *8*, 11.

Singh, A. K.; Senapati, D.; Wang, S.; Griffin, J.; Neely, A.; Naylor, K. M.; Varisli, B.; Kalluri, J. R.; Ray, P. C. Gold Nanorod Based Selective Identification of *Escherichia coli* Bacteria Using Two-Photon Rayleigh Scattering Spectroscopy. *ACS Nano* **2009,** *3*, 1906–1912.

Slocik, J. M.; Stone, M. O.; Naik, R. R. Synthesis of Gold Nanoparticles Using Multifunctional Peptides. *Small* **2005,** *1* (11), 1048–1052.

Song, C.; Xu, J.; Chen, Y.; Zhang, L.; Lu, Y.; Qing, Z. DNA-Templated Fluorescent Nanoclusters for Metal Ions Detection. *Molecules* **2019,** *24* (22), 4189.

Soukarié, D.; Ecochard, V.; Salomé, L. DNA-Based Nanobiosensors for Monitoring of Water Quality. *Int. J. Hyg. Environ. Health.* **2020,** *226*, 113485.

Swanink, C. M.; Meis, J. F.; Rijs, A. J.; Donnelly, J. P.; Verweij, P. E. Specificity of a Sandwich Enzyme-Linked Immunosorbent Assay for Detecting *Aspergillus galactomannan. J. Clin. Microbiol.* **1997,** *35*, 257–260.

Tan, L.; Chen, Z.; Zhao, Y.; Wei, X.; Li, Y.; Zhang, C.; et al. Dual Channel Sensor for Detection and Discrimination of Heavy Metal Ions Based on Colorimetric and Fluorescence Response of the AuNPs-DNA Conjugates. *Biosens. Bioelectron.* **2016,** *85*, 414–421.

Touhami, A. Biosensors and Nanobiosensors: Design and Applications. *Nanomedicine* **2014,** *15*, 374–403.

Touloupakis, E.; Chatzipetrou, M.; Boutopoulos, C.; Gkouzou, A.; Zergioti, I. A Polyphenol Bio-Sensor Realized by Laser Printing Technology. *Sens. Actuat. B Chem.* **2014,** *193,* 301–305.

Upan, J.; Reanpang, P.; Chailapakul, O.; Jakmunee, J. Flow Injection Amperometric Sensor with a Carbon Nanotube Modified Screen-Printed Electrode for Determination of Hydroquinone. *Talanta* **2016,** *146,* 766–771.

V. Rai.; S. Acharya.; N, Dey. Implications of Nanobiosensors in Agriculture. *J. Biomater. Nanobiotechnol.* **2012,** *3* (2A), 315–324.

Vamvakaki, V.; Chaniotakis, N. A. Pesticide Detection with a Liposome-Based Nano-Biosensor. *Biosens. Bioelectron.* **2007,** *22* (12), 2848–2853.

Verma, M. L.; Kanwar, S. S.; Jana, A. K. Bacterial Biosensors for Measuring Availability of Environmental Pollutants. In *Proceedings of International Conference on Biomedical Engineering and Assistive Technology*; Jalandhar, India, 2010; pp 1–7.

Verma, M. L.; Chauhan, G. S.; Kanwar, S. S. Enzymatic Synthesis of Isopropyl Myristate Using Immobilized Lipase from *Bacillus cereus* MTCC-8372. *Acta Microbiol. Immunol. Hung.* **2008,** *55,* 327–342.

Verma, M. L.; Puri, M.; Barrow, C. J. Recent Trends in Nanomaterials Immobilised Enzymes for Biofuel Production. *Crit. Rev. Biotechnol.* **2016,** *36,* 108–119.

Vu, Q. K.; Tran, Q. H.; Vu, N. P.; Anh, T. L.; Le Dang, T. T.; Matteo, T.; Nguyen, T. H. H. A Label-Free Electrochemical Biosensor Based on Screen-Printed Electrodes Modified with Gold Nanoparticles for Quick Detection of Bacterial Pathogens. *Mater. Today Commun.* **2020,** 101726.

Willner, M. R.; Vikesland, P. J. Nanomaterial Enabled Sensors for Environmental Contaminants. *J. Nanobiotechnol.* **2018,** *16* (1), 1–6.

Wu, Y.; Zhan, S.; Xing, H.; He, L.; Xu, L.; Zhou, P. Nanoparticles Assembled by Aptamers and Crystal Violet for Arsenic (III) Detection in Aqueous Solution Based on a Resonance Rayleigh Scattering Spectral Assay. *Nanoscale* **2012,** *4* (21), 6841–6849.

Xu, X.; Niu, X.; Li, X.; Li, Z.; Du, D.; Lin, Y. Nanomaterial-Based Sensors and Biosensors for Enhanced Inorganic Arsenic Detection: A Functional Perspective. *Sens. Actuat. B Chem.* **2020,** 128100.

Ye, W.; Guo, J.; Bao, X.; Chen, T.; Weng, W.; Chen, S.; Yang, M. Rapid and Sensitive Detection of Bacteria Response to Antibiotics Using Nanoporous Membrane and Graphene Quantum Dot (GQDs)-Based Electrochemical Biosensors. *Materials* **2017,** *10,* 603.

Zhang, Y.; Arugula, M. A.; Wales, M.; Wild, J.; Simonian, A. L. A Novel Layer-by-Layer Assembled Multi-Enzyme/CNT Biosensor for Discriminative Detection between Organo-phosphorus and Non-Organophosphorus Pesticides. *Biosens. Bioelectron.* **2015,** *67,* 287–295.

Zhu, Y.; Zhou, C.; Yan, X.; Yan Y.; Wang, Q. Aptamer-Functionalized Nanoporous Gold Film for High-Performance Direct Electrochemical Detection of Bisphenol A in Human Serum. *Anal. Chim. Acta.* **2015,** *883,* 81–89.

CHAPTER 8

Nanobiosensors Containing Noncarbon-Based Nanomaterials to Assess Environmental Pollution Levels

VINARS DAWANE[1], SATISH PIPLODE[2], VISHNU K. MANAM[3], PRABODH RANJAN[4], and ABHISHEK CHANDRA[5]

[1]School of Environment and Sustainable Development, Central University of Gujarat, Gandhinagar 382030, Gujarat, India

[2]Department of Chemistry, Sahid Bhagat Singh Govt. P. G. College, Pipariya, Hoshangabad 461775, Madhya Pradesh, India

[3]Department of Plant Biology and Biotechnology, Unit of Algal Biotechnology and Nanobiotechnology, Pachaiyappa's College, University of Madras, Chennai 600030, Tamil Nadu, India

[4]Department of Chemical Engineering, Indian Institute of Technology Madras, Chennai 600036, Tamil Nadu, India

[5]Department of Chemistry, HVHP Institute of Post Graduate Studies and Research, Kadi Sarva Vishwavidhyalaya, Kadi 382715, Gujarat, India

ABSTRACT

The world of sensing technology is expanding into new horizons owing to their exceptional selectivity and sensitivity based application. These remarkable sensing progresses have been positively influenced by the conjugation of biology, material sciences, and nanotechnology into an advanced sensing technology. The combination of biology with nanotechnology has allowed

Nanotechnology for Environmental Pollution Decontamination: Tools, Methods, and Approaches for Detection and Remediation. Fernanda Maria Policarpo Tonelli, Rouf Ahmad Bhat, & Gowhar Hamid Dar (Eds.)

the sensing technology to enable faster, cheaper, as well as more reliable evaluations and applications in various scientific domains. Thus, a new era in sensing has been developed into nanobiosensing technology. Environmental monitoring, various processes assessment, detections of contaminates, and complex environmental forensics are among the most promising applications of nanobiosensors, and thus, a an important area of research. In this regard, the nanobiosensors containing noncarbon nanomaterials such as metals, metal oxides, their composites, nonmetals other than carbon, and their conjugations have been given significant attention due to their remarkable potential in the area of environmental science and research.

This chapter therefore highlights the role and importance of noncarbon-based nanobiosensors, their recent advancements, and major factors which have made the noncarbon-based nanobiosensors a preferential and successful choice in nanobiosensing technology. This chapter intensively highlights the environmental applications of noncarbon-based nanobiosensors and key future challenges associated with it.

8.1 INTRODUCTION

Increasing industrialization, urbanization, and usage of hazardous substances in chemical processing as well as their by-products have put our environment and life at high risk (Alharbi et al., 2018; Hill, 2020). Increasing pollution level in the soil, water, and air bodies pose peril to life as well as the environment (Shaheen et al., 2020). The increasing pollution level has triggered the need to develop cutting-edge technologies to mitigate pollution from the environment and ecosystem (Liu et al., 2020). In this regard, sensing technology has proved to be advantageous than others because it can check and control environmental pollution (Sugunan and Dutta, 2010; Li et al., 2015). Particularly, nanotechnology has become an exceptional choice in various filed of science and technologies (Baruah and Dutta, 2009; Li et al., 2015; Manam, 2015; Chandra, 2017; Chandra and Singh, 2018) including environmental science and pollution cleanup technology (Piplode, 2017; Piplode and Dawane, 2020; Chandra et al., 2020) (refer to Figure 8.1).

In the last few decades, biosensors have taken a center stage in sensing technologies and the use of nanotechnology along with various nanomaterials have made this field a more fascinating and hot topic (Abdel-Karim et al., 2020). These nanomaterials can be classified into two broad categories such as carbon containing and noncarbon containing (Li et al., 2015;

Nandita et al., 2020). The noncarbon containing materials may be various metals, their oxides, composites, as well as elements that are nonmetals other than carbon. Their application in controlling environmental pollution is a very comprehensive subject as the changes and dynamism of environmental conditions are extremely rapid (Yaraki and Tan, 2020). In biosensing application of nanotechnology, selected nanomaterials incorporated with a biosensing mechanism is used to detect targeted compounds. This has created a remarkable enhancement in various scientifically burning domains such as analysis of pollutants and pesticides in the soil, presence of toxic materials and their intermediates, detection of parasitic pathogens and their toxins in the drinking water, contamination of dyes and heavy metals in water bodies, and the assessment of weather conditions or humidity/temperature monitoring (Li et al., 2015; Mehrotra, 2016) (refer Figure 8.2).

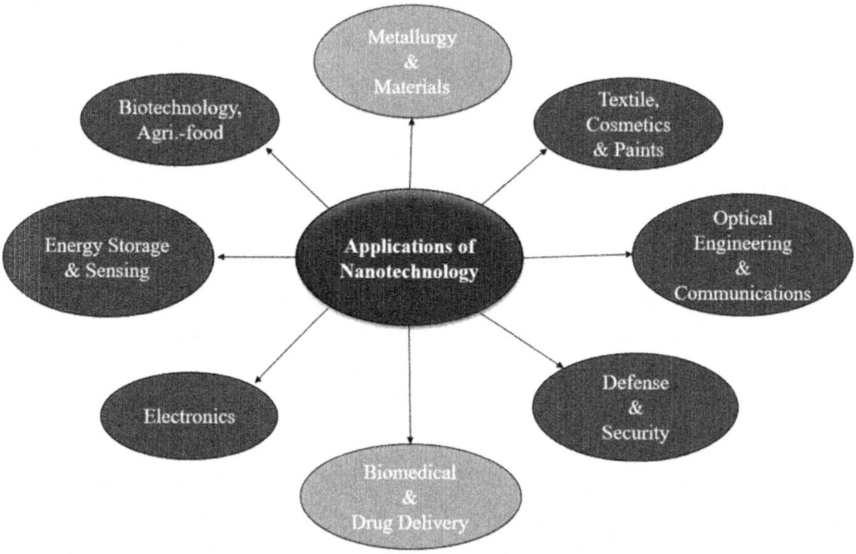

FIGURE 8.1 The various applications of nanotechnology in field of science and technologies.

A general sensor mechanism comprises receptor functions and transducer functions (refer Figure 8.3). In principle, the bioanalyte of interest binds on to bioreceptors which modulates the physiochemical activity of the binding process of a typical biosensor (Viswanathan, 2009; Malik et al., 2013) and nanomaterial it is involved in, thereby enhancing the whole process for better detection or signal enrichment. Nanobiosensors have been developed using specific recognition molecules such as DNA, antibody, aptamer, enzymes, protein, etc. (Bisen et al., 2010), integrated on a surface to make it sensitive toward selective targets such as environmental pollutants and to generate a unique as well as enhanced signals. These enhanced signals may have various forms such as electrochemical, optical responses (Malik et al., 2013), colorimetric signals or fluorescent responses (Hoon et al., 2020). Some of the applications of nanobiosensors include biotoxin detection, carcinogen detection, GMO detection, gas detection, agrochemical residue detection, pathogen detection, sugar detection, and allergen detection (D'Souza et al., 2017).

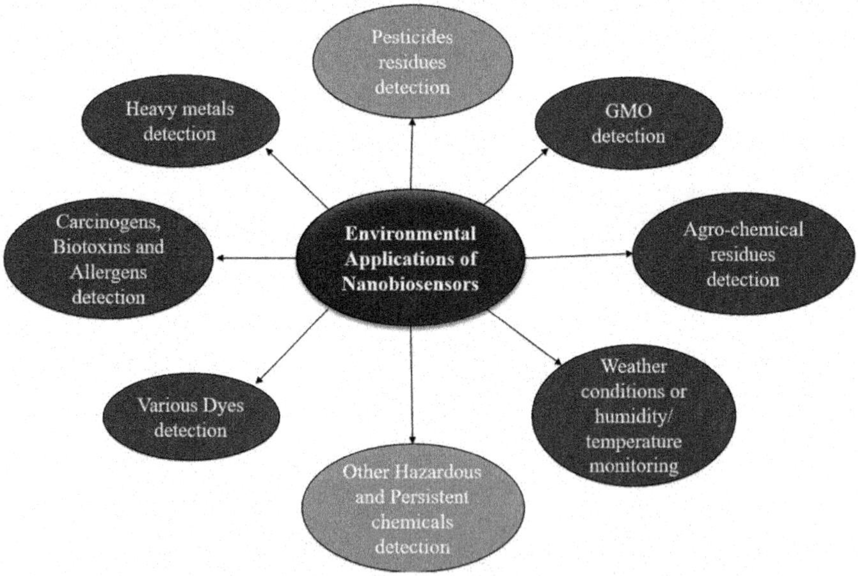

FIGURE 8.2 Various environmental applications of nanobiosensors.

The environmental application of nanobiosensors includes detection of environmental pollution and toxicity, agricultural monitoring, groundwater screening, and ocean monitoring (Kumar and Guleria, 2020). Thus, this chapter highlights the nanobiosensing materials, especially the noncarbon-based nanomaterials in biosensing technology and their various applications in environmental pollution assessment specifically since the last two decades. The chapter also includes the challenges in nanobiosensor technology as well as emphasizes the prospects of nanobiosensing technology under the lens of noncarbon-based nanomaterials.

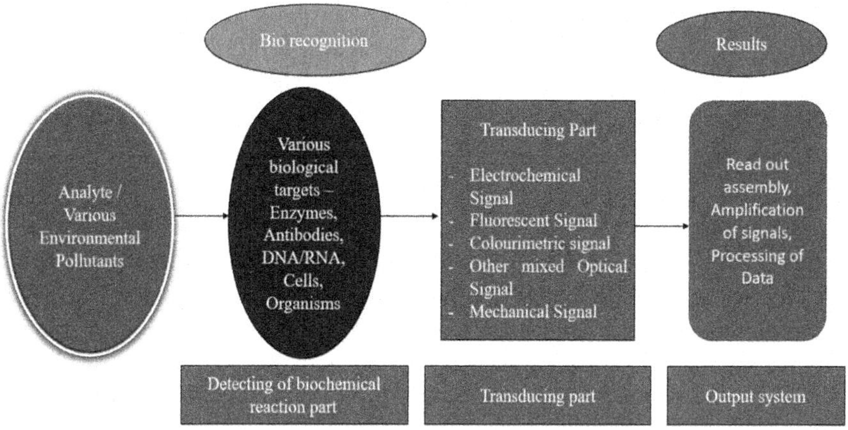

FIGURE 8.3 The general mechanistic approaches involved in a sensor/nanobiosensor.

8.2 NONCARBON-BASED NANOBIOSENSORS AS ADVANTAGIOUS AND PREFERENTIAL CHOICE

Factors such as sensitivity, wide-scale application flexibility, toxicities, biocompatibility, and cost effectiveness determine the quality of sensor and must be considered while designing a sensor. In this regard, use of noncarbon-based sensor could prove to be a better alternative as it can overcome these issues. Such nanomaterials are acquiring immense consideration in the field of sensing due to their application in mitigating environmental pollution (Malik et al., 2013).

Noncarbonbased nanobiosensors, such as metallic nanobiosensors, have been a very preferential choice for diverse types of principles-based sensor making such as colorimetric, fluorescence-based, and dynamic light scattering (DLS) based, as well as their various combinations (Kanan et al., 2009). Their remarkable surface functionality is useful in the colorimetric evaluation and are promising as signal enhancer in DLS processes to build various nano-level DLS sensing applications. Even metal-based fluorescent nanobiosensors composed of <2 nm nanoclusters have been developed for making ultra-small biotemplate versions (Yaraki and Tan, 2020). The fluorescence emission of the nanobiosensors such as metal and semiconductor nanomaterials, quantum dots, silica, as well as polymer nanomaterials can give narrow peak spectra which in turn can be adjusted by varying their particle size distribution during production. This gives flexibility because multiple emission spectra can now be achieved for noncarbon nanomaterials by varying their particle size distribution; which in turn makes it possible to build multi-colorimetric or multiplex assay using a single light source only (Goldman et al., 2004; Koedrith et al., 2015). It can be done along with additional benefits such as very high intensity, photostability, resistance toward photo bleaching as well as good biocompatibility (Koedrith et al., 2015).

As a noncarbon material which is mostly metallic nanomaterials have good magnetic properties (Koedrith et al., 2015); thus recently the use of magnetic beads along with fluorescent nanomaterials have helped to manage the critical limitations of low fluorescence intensity problems by focusing the targeted samples during the process and is advantageous for fluorescent-based materials in biosensing technology. They have been applicable as open-valent scaffolds for supramolecular gatherings as well as multipurpose artificial platforms for superficial coatings such as various chemical integrations toward the aptamers, antibodies, and other sensitive bioreceptor targets to prepare biosensor (Koedrith et al., 2015).

Various types of carbon-based nanomaterials in biosensing technology have used single/ multiwalled carbon nanotubes and graphene. They have various advantages which include huge surface–to-volume proportion, low charge-carried thickness, electrical conductivity enhancements (Shirsat et al., 2007; Fortunati, 2019; Norizan et al., 2020) excellent conducting and electrocatalytic properties, high thermal conductivity, better mechanical flexibility (Viswanathan, 2009; Li et al., 2017; Aziz, 2007) fast electron transfer, and good biocompatibility (Liang, 2017; Han, 2017; Trindade, 2019). Despite these, the major limitations of carbon-based nanobiosensors include narrow surface to interface with biological constituents, nonspecific adsorption of other biomolecules, problematic management during sensor construction progression, challenging chemical functionalization, requirement to functionalize the surface for increasing biocompatibility, and hard to dissolve in water especially graphene (Scida et al., 2011).

No doubt, the carbon-based nanomaterials have a varied array of monitoring applications such as gas sensing, heavy metals analysis, biomolecules, and toxic chemicals evaluations (Doria et al., 2012) but on the other hand, the noncarbon-based nanomaterials also have larger physicochemical, spectral, and optical features that help in enhancing the performance of a new class of biosensors (Rackauskas et al., 2017). For example, metal oxide nanowires are developed as a promising class of sensing nanomaterials due to their flexible manufacturing methods and chemical stability (Munawar et al., 2019; Abdel-Karim et al., 2020). Thus limitations of carbon-based nanobiosensors have led the pathway for the noncarbon-based nanobiosensors in biosensing technology (Abdel-Karim et al., 2020). But the optimum use of biosensing technology has been obtained by the combination of both carbon- and noncarbon-based nanomaterials in the biosensing technology (Wang et al., 2019; Abdel-Karim et al., 2020; Noah, 2020). The noncarbon-based nanomaterials have their advantages as well as limitations compared with carbon-based nanomaterials. Some of the advantages of noncarbon-based nanomaterials include effective electron transfer and increase in surface-to-volume ratio, superior conductivity (Ma et al., 2016; Zhang, 2016), good biocompatibility, easy functionalization, high pore volume, and surface area, as in mesoporous silica nanoparticle, which remains a salient feature and a lacking parameter in carbon-based nanomaterials (Fan, 2016; Afsharan, 2016; Yin, 2017; Munawar et al., 2019). Other compensations include decent electron transference and great loading capability (Cao, 2013), distinct surface properties, flexibility of size and shape (Wang et al., 2017; Wang et al., 2017), speedy response and extraordinary electrocatalytic ability (Bahadir et al., 2016; Fang, 2016), improved charge transfer and stability (El-Said and Choi, 2014), as well as other decencies (Rahman et al., 2005; Tertiş, 2017; Taylor, 2017). These above qualities have made noncarbon based nanomaterials a versatile choice in the applications of biosensing technology. Although the noncarbon-based nanomaterials have huge and enormous advantages and have limited drawbacks such as instability of their electrical properties under high salt concentration, stability shifts under signal amplification, tough to make disciplined scattered size dispersal, negative correlation of electrostatic potential with distance, time consuming when used with coated surfaces, thin-film morphological dissimilarity, and worries regarding material constancy and stable concentration (Díez, 2018; Lu et al., 2020). Therefore depending on the application, a combination of both carbon and noncarbon-based nanomaterials in biosensing technology can be developed into a robust and a versatile biosensor for the welfare of the environment and humans (Wang et al., 2019).

In the case of environmental pollutants such as various gas detection and monitoring, both carbon-based materials (Sayago et al., 2008; Lee et al., 2018; Han et al., 2019) and noncarbon-based materials (Afzal et al., 2012; Kwon et al., 2017) have been giving fascinating opportunities for their use in sensing technology (Sharma et al., 2018). The noncarbon-based nanomaterials have gained significant attention (Wang et al., 2017) and most of the times, is a preferential choice of material for specific environmental applications comparatively (Kanan et al., 2009). It has been receiving attention in the construction of gas sensor strategies (Ponzoni et al., 2005; Wang et al.,

2014; Jaroenapibal et al., 2018) as well as for future potential in sensors/biosensors/nanobiosensor technology. Pollutants that were previously promising targets for carbon-based sensor applications are providing better detections when the sensing carbon material has been doped with noncarbon materials (Penza et al., 2008; Ueda et al., 2008; Kanan et al., 2009).

The use of various metals (Ag, Au, Pt, Pd, Cu, Co, etc.) in nanotechnological applications (Maduraiveeran and Jin, 2017) have been preferred in biosensors because of their easy synthesis, attractive biological-based/green alternative fabrication (Manam and Murugesan, 2020A), and economically cheap and scalability (Manam and Murugesan, 2020B). The large surface area has been the most important property of metallic nanoparticles preference, which helps in immobilization of the biomolecules efficacy followed by unique electron transfer quality, high catalytic activity, and enhanced biocompatibilities (Cai, 2016; Chang et al., 2016). The exceptional antimicrobial activities of silver nanoparticles (Manam and Murugesan, 2014, 2020A,B) paved a way for its applications in biosensing technology and portable water technology as well as the ease of fabrication by using different biological medium also gave an additional advantage over other type materials (Li et al., 2011; Manam and Murugesan, 2014; Aritonang et al., 2019; Manam and Murugesan, 2020A). In this regard, silver nanomaterials have been considered important among different noble metal nanoparticles because they have been subjected to new manufacturing practices along with unusual subsequent morphologies and behaviors. These silver nanoparticles have various advantages such as to accommodate more active sites on their surface due to their large surface area which allows an easier passage for electrons to flow more effectively (Yu, 2011; Shahdost-fard and Roushani, 2017). The addition of silver nanoparticles to reduce graphene oxide dramatically enhances the electrical conductivity and the restoration of structural imperfections of the reduced graphene oxide (Cho, 2018). Thus, the addition of noncarbon-based nanomaterials, especially silver nanoparticles, to carbon-based nanomaterials enhances the specificity and electrical conductivity during the biosensing process.

The use of various metal oxides such as WO_3, Bi_2O_3, ZnO, SnO_2, and TeO_2 are remarkable choices for nitrogen oxide sensing applications. Likewise, WO_3, TiO_2, SnO_2, as well as chromium, copper, and iron-based oxides have exceptional capabilities in sensing applications of various pollutants such as sulphur dioxide, H_2S, ozone sensing, NH_3 sensing, and other volatile organic compounds evaluation (Kannan et al., 2009). Various metal oxyhalides have been also found to be impressive candidate for assessment of hazardous and persistent pollutants (Piplode et al., 2020; Piplode and Dawane, 2020) and candidate for potential future environmental sensor applications (He et al., 2020). Thus noncarbon-based materials are indispensable tools right now in the environmental sensing technology.

8.3 SELECTED CASE STUDIES ON NANOBIOSENSORS AND THEIR VARIOUS ENVIRONMENTAL APPLICATIONS

Environmental monitoring and forensics have always been a critical task (Luo and Yang, 2019) due to their extremely complex and dynamic nature of environmental processes, interconnected cycles, and presence of different types of chemicals along with diverse toxicities (Lega and Teta, 2016; Erickson, 2020; Buckley et al., 2020). The following selected examples have been highlighting the use of noncarbon-based materials in the making of nanobiosensors and their potential environmental applications. The case studies have also been covering the use of specific nanomaterials along with the biosensor toward a targeted specific environmental pollutant assessment and a detailed evaluation of various performance attributes of prepared nanobiosensor such as linearity range of detection, the lower limit of detection as well as the sensitivity values.

8.3.1 METAL-BASED NANOBIOSENSORS

Metal-based nanobiosensors are significant candidates for various environmental applications and their recent advancements have made them potent for advanced sensing. Their great biocompatibility with other biological tissues (Wu et al., 2019), unique optical properties because of their localised surface plasmon responses (Ryu and Ha, 2020; Yaraki et al., 2020), and their remarkable photoluminescence properties due to the quantum confinement effect (Goswami et al., 2016; Badshah et al., 2020) make them useful for various applications (Rizwan et al., 2018). These significant qualities of metal nanomaterials help make colorimetric or dynamic light scattering as well as fluorescence sensing principle type of sensors fabrication (Yaraki and Tan, 2020).

Metal nanomaterials, specifically the Silver nanomaterials, have been associated with various biomedical applications (Manam and Murugesan, 2013, 2014, 2017), and recent bioinformatics (Manam and Murugesan, 2017) made them an exceptional nanomaterial tool in biosensing technology. Similarly, gold nanoparticles are also considered as a remarkable electrochemical biosensing platform due to their rare biocompatibility and biomolecular functionality (Lin, 2012).

In one study, Lin et al. (2006) has developed a biosensor by using Au nanoparticles-acetylcholinesterase (AChE) for the paraoxon detection and evaluated the localized surface plasma response for investigation. They reported that 1–100 ppb paraoxon can be easily detected by using this method along with 0.234 ppb as the detection limit under optimum conditions. Valera et al. (2008) developed conductometric-based biosensor by using gold nanoparticles for the detection of atrazine. They reported linear ranges from 0.1 to 1 μ/gL for atrazine in microenvironment along with all the necessary technical aspects detailing. In a very specific study on pesticide detection, Lisa et al. (2009) developed gold nanoparticles based biosensor for DDT detection. They reported a very simple method for the detection of DDT in nanogram level. The lowest detection limit has been found around 27 ng/mL by using this method. In a fiber optical sensor study, Lin and Chung (2009) used gold nanoparticles modified optical fiber based biosensor for Cd^{2+} detection. They reported detection linear range from 1 to 8 ppb for Cd^{2+} along with the 0.16 ppb lowest detection limit by this method. In a real water pollution evaluation study, Li et al. (2011) developed Ag-based biosensor for As^{3+} detection. They used surface-enhanced Raman scattering for detection purpose. They reported 4–300 ppb and 0.76 ppb linear detection range and lowest detection limit for As^{3+} ions, respectively. In another evaluation, Chen et al. (2011) used Au nanoparticles based biosensor for methyl parathion (MP) detection. They used a square-wave voltammetry technique for the detection of MP and reported 0.001–5.0 μg/mL and 0.3 ng/mL linearity range and detection limit respectively, under optimum experimental conditions. Wu et al. (2012) described a biosensor for the detection of As^{3+} by using aggregated gold nanoparticles along with the combination of colorimetric and scattering techniques for analysis. They reported 1–1500 ppb and 40 ppb linear detection range and lowest detection limits, respectively, by naked eye visualization. They also detected 0.6 ppb and 0.77 ppb values for colorimetric assay and response scattering assay, respectively, for a real water sample. In a typical soil sample evaluation for acetamiprid pesticide detection, Shi et al. (2013) used Au-based biosensor and reported the linear range between 75 nM and 7.5 μM along with 5 nM lowest detection limit for acetamiprid pesticide. They highlighted the sensitivity of the method by explaining that pesticides, having a similar structure including imidacloprid and chlorpyrifos also coexist with acetamiprid, could not interfere with the detection process using this method. Liu et al. (2014) developed a gold-based biosensor for atrazine detection and reported 0.05–0.5 ng/mL linear detection range along with the low detection limit of 0.016 ng/mL, under optimum conditions. Guan et al. (2014) reported poly (gamma-glutamic acid) functionalized gold nanoparticles enhanced biosensor for Hg^{2+} detection in mineral water sample along with PGA and CTAB to build biosensor. They reported 0.01–10 μM and 50–300 μM two linear detection ranges along with 1.9 nM lowest detection limit. Kumar et al. (2014) developed biofunctionalized Au nanoparticles for Hg^{2+}, Pb^{2+}, and Cd^{2+} detection. They used *Hibiscus Sabdariffa* (Gongura) leaves and stem extract for the synthesis of Au nanoparticles. This was reported for the first time as biofunctionalized Au NPs for toxic metal detection. Au nanoparticles were prepared by using the leave extract exhibited high sensitivity without selectivity, whereas Au nanoparticles that have been prepared by using stem extract exhibit selectively sensing for only Hg^{2+} ion. Wei and Wang (2015) used biosensor based on gold, fourth-generation poly (amidoamine) [PMAM-G4] and AChE for carbaryl detection by using Cyclic and Linear Voltammetry methods for pollutants detection. Effect of various operation parameters such as enzyme amount, the concentration of glutaraldehyde, and inhibition time has also been discussed in this study. They highlighted the biosensor's linear range from 1.0 to 0.9 μmol/L along with 0.032 μmol/L lowest detection limit. In a study to detect organothiophosphate insecticide, Dou et al. (2015) developed a unique nanobeacon construction (gold nanoparticles functionalized with DNA aptamer and FAM as fluorophore) based biosensor for isocarbophos detection. With the use of this "turn off" strategy based biosensor they reported 10 μ/L lowest detection limit for isocarbophos. Mura et al. (2015) developed Au nanoparticles based biosensor for nitrate detection in water. They also used cysteamine and citrate for biosensor development to evaluate the nitrate profile in the water. They detected precisely 35 ppm of nitrate in the water by using this method along with very easy detection with the naked eyes. Deng et al. (2016) used colorimetric pH indicator gold biosensor for evaluation of urea, urease, and urease inhibitor. They highlighted the use of 3, 3', 5, 5' tertamethylbenzinide-H_2O_2 to make the biosensor and gold nanoparticles as catalyst for sensing method. They reported 0.5 μM and 1.8 U/L detection limit for urea and urease. They used urine and soil sample for urea and urease detection, respectively. In one attempt to make gold-based nanobiosensor for highly persistent and hazardous pesticide detection, Lang et al. (2016) developed Au nanorod based biosensor for paraoxon and dimethoate pesticide detection. Both pesticides belong to the organophosphate class of pesticides. The results indicated the 1–5 nM linear range and 0.7 and the lowest detection limit for paraoxon as well as 5 nM–1 μM detection linear range and 3.9 nM lowest detection limit for dimethoate. They further enhanced this method by comparing the effect of Au–Ag hetero-structures nanorod based biosensor under the same experimental conditions and concluded that the Au nanorod based nanobiosensor exhibited higher electrocatalytic properties as compared with Au–Ag based nanobiosensor. In a typical application of biosensor to detect Cu^{2+} in water, Peng et al. (2016) developed Au-based biosensor and studied the rhodamine b hydrazine (RBH) ions used for biosensing application by utilizing the EIS method for detection of pollutant. The prepared AuNPs-RBH biosensor exhibited 0.1 pM to 1 nM linear range along with 12.5 fM lowest detection limit for Cu++ in water. In a promising work to develop a versatile biosensor for the detection of surface water pollutants, Tan et al. (2016) developed a biosensor by using Au nanoparticles for detecting all the nine heavy metals. It was a first report where all the nine heavy metals (Ag^{+}, Hg^{2+}, Cr^{3+}, Sn^{4+}, Cd^{2+}, Cu^{2+}, Pb^{2+}, Zn^{2+}, and Mn^{2+}) were detected with a very low detection limit in nano range with 100% identification accuracy. In this study 30 nM detection limit was observed for Hg^{2+} ions. This sensor exhibited superior performance as compared with other previous sensors with the use of real water samples along with dynamic and static quenching processes. Li et al. (2017) reported using an Ag nanocluster based biosensor for Hg^{2+} detection along with the use of single-stranded DNA for the construction of the biosensor. They reported 0–150 nM linear detection range and 4.5 nM as the lowest detection limit. In an attempt to develop a fast, simple, selective, cost-effective, and reusable biosensor, Skotadis et al. (2017) reported a novel platinum-based biosensor for Pb^{2+} ion detection by using Pt- DNAzyme biosensor along with the 10 nM value of lowest detection limit of the sensor. In a colorimetric sensing based study for pesticides evaluation by using nanoparticles, aptamers, and cationic peptide, Bala et al. (2017) developed an Au nanoparticle based biosensor for malathion detection. They reported a very simple method to detect malathion. In the absence of malathion, Au nanoparticles remain red and in the presence of malathion, the Au nanoparticles turned into blue color. The linear range of 0.01–0.75 nM has been observed for malathion. A lowest detection limit of 1.94 pM has been reported in this work. In a recent work, Madianos et al. (2018) developed a sensor/biosensor for detecting acetamiprid and atrazine by using platinum nanoparticles. It was a first report on biosensor-based aptamers evaluation for atrazine detection. They reported linear detection ranges from 10 to 100 and 1 pM lowest detection limit for acetamiprid as well as 100 pM–1 μM linear range and 10 pM lowest detection limits for atrazine in this study. In a recent attempt for making "turn on" strategy based biosensor by using fluorescent response energy transfer (FRET) phenomena for pesticides detection, Cheng et al. (2018) fabricated an Au-based biosensor for detection of chlorpyrifos with diazinon and malathion simultaneously. They also used QDs and aptamer to construct the biosensor. They reported a lowest detection of 0.73 mg/L for chlorpyrifos.

8.3.2 METAL OXIDE BASED BIOSENSOR AND ENVIRONMENTAL APPLICATIONS

Metal oxide sensors have been widely studied and are among most sensitive sensor materials for specific environmental applications (Kayhomayun et al., 2020) such as specific reduction/oxidation gas sensors available for the commercial, industrial, as well as home sensing systems applications (Korotcenkov, 2007; Kanan et al., 2009; Miller et al., 2014; Dey, 2018). Eectrochemical performance, catalytic ignition, or resistance variation. based systems have been associated with making metal oxide-based biosensors (Kanan et al., 2009; Yang et al., 2017). Metal oxide-based biosensors exhibit unique benefits such as cost-effectiveness, portable size, measurement easiness, robustness, ease of creation, and very sensitive low detection limits up to ppm levels along with the exceptional durability and resistance toward poisoning (Kanan et al., 2009; Korotcenkov and Cho, 2017). Thus they have gained significant popularity in sensing technology and research domains (Kayhomayun et al., 2020). Various noncarbon containing or metal oxides such as TiO_2, SnO_2, ZnO, WO_3 have been useful in the environmental sensing applications (Rim et al., 2017, Wu et al., 2019) as well as others such as In_2O_3, CuO, ZrO_2, iron oxide, CeO_2, MnO_2, antimony oxide, IrO, and CoOx. The competitiveness of metal oxides is based on various characteristics such as chemical stability, light excitation/light conversion, and surface-to-volume proportion (Zhang and Gao, 2019; Erban and Enesca, 2020). Some selected examples are discussed here.

8.3.2.1 TIO_2-BASED BIOSENSOR AND ENVIRONMENTAL APPLICATIONS

TiO_2 is a promising n-type semiconductor a candidate for numerous applications like photo catalysis, photovoltaic, or energy storing as well as biosensing due to their high chemical stability, biocompatibility, morphological resourcefulness, etc. With such understanding, in a very genuine study, Yu et al. (2003) invented a novel amperometric biosensor based on TiO_2 for phenol detection. In this study, tyrosinase and TiO_2

sol–gel matrix has been used for the sensing environmental application. They reported that a short time of less than 5 s is required to detect the phenol in the environment. The method can easily detect phenol with linear ranges from 1.2×10^{-7} to 2.6×10^{-4} M along with a low detection limit of 1.2×10^{-7} M. The nanobiosensor exhibited 103 μA/mM sensitivity in a sol–gel matrix. In an industrially applicable study, Cheng et al. (2008) reported a biosensor based on TiO_2 for the detection of lactic acid. In this method, nano-TiO_2 lactate dehydrogenase electrode is used for the detection of lactic acid by using the Cyclic Voltammetry (CV) technique. They reported that the lactic acid range from 1 to 20 μmol/L can be easily detected by using this method along with a 0.4 μmol/L lowest detection limit. In another study, Chen et al. (2009) developed a novel biosensor using TiO_2 for sensing of urea by using potentiometric techniques for the investigation. They evaluated the effect of various processing/concentration parameters such as pH, the concentration of urea, and ionic strength and reported a highly sensitive and very quick detection of urea (25 s) in the study. This method showed a wide linear range from 8 to 3 and 5.0 μM linear ranges and a lowest detection limit, respectively. Wu et al. (2009) used TiO_2 in an amperometric method for the detection of nicotine. They reported that this method has a sensitivity of 31.35 μA/mM/cm^2 within the range of 0–5 μM along with a value of 4.9 for the lowest detection limit for nicotine. In another study, Kafi et al. (2011) designed a biosensor on the glassy carbon electrode by using Hb, TiO_2, and methylene blue dye to detect H_2O_2. They used amperometric method for detection purpose. In this biosensor study, the role of immobilized Hb has been recognized as an important role in H_2O_2 detection. Prepared biosensor showed linear ranges from 2×10^{-7} to 1.6×10^{-4} M and a very low detection limit of 8×10^{-8} M. Recent studies explored the use of TiO_2 in improving the selection in electrochemical biosensing (Dai et al., 2020), better photocatalytic systems (Wang et al., 2020), and other applications (Nadzirah et al., 2020; Jafari et al., 2020).

8.3.2.2 SNO$_2$-BASED BIOSENSOR AND ENVIRONMENTAL APPLICATIONS

SnO_2 is also a great candidate for diverse applications and plays a key role in biosensing technology due to its salient features such as a great surface area, worthy biocompatibility, less poisonousness, admirable chemical stability, and catalytic action along with easy and flexible synthesis approaches (Dong and Zheng, 2014; Zhou et al., 2013; Lavanya et al., 2012). In a very sensitive sensor preparation work, Li et al. (2010) designed a mediator-free HRP/

Sb-doped SnO_2 nanowire-based biosensor to detect H_2O_2 using thermal evaporation method for sample preparation along with electrochemical impedance spectroscopy (EIS) method for detection purposes. They reported that the HRP/Sb-doped SnO_2 biosensor exhibits higher efficiency in comparison with the undoped HRP/ SnO_2 biosensor. Low detection limit for H_2O_2 has been observed as 0.8 μM by this method. In another study for sample pollutant detection, Liu et al. (2010) used SnO_2-based biosensor. In this study, SnO_2 nanorods array was prepared by using hydrothermal synthesis method and grown directly on an alloy substrate. They reported that the prepared nanorod array based biosensor can detect H_2O_2 with a wide linear range from 0.8 to 35 μM along with a low detection limit of 0.2 μM. The method showed very good sensitivity (379 $\mu A/mM/cm^2$). Recently, Song et al. (2020) developed a very sensitive biosensor by using 13 nm mesoporous SnO_2 hierarchical architectures to accurately assess H_2S under a high-humid atmosphere at ppb level.

8.3.2.3 ZnO-BASED BIOSENSOR AND ENVIRONMENTAL APPLICATIONS

ZnO is a wider platform for biosensing application. Its various forms such as ZnO nanorods (Zong and Zhu, 2018), nanoparticles (Lei et al., 2011), and ZnO nanocone arrays (Yuea et al., 2020) are remarkable candidates for various types of sensors in the biomedical and environmental domains. Zhao et al. (2009) developed a biosensor based on ZnO for the detection of p-cresol, 4-chlorophenol, and phenol. In that study, tyrosinase, boron-doped nanocrystalline diamond, 3 aminopropyltriethoxysilan (APTES), and tetraethoxysilane (TEOS) was used to develop the biosensor. They reported a linear range from 1 to 175 μM with a detection limit of 0.1 μM for p-cresol, linear range from 1 to 150 μM with a detection limit of 0.25 μM for 4-chlorophenol, and linear range from 1 to 150 μM with a detection limit of 0.20 μM for phenol under optimum conditions. Again in a remarkable study, Zhao et al. (2014) designed an electrochemical ZnO biosensor for to detect L-lactic acid. In this study, ZnO nanowires were prepared by using chemical vapor deposition method and prepared ZnO nanomaterial of 500 nm size. The results highlighted that the prepared ZnO nanowires successfully detected the L-lactic acid with a wide linear range of 12 μM to 1.2 mM along with the sensitivity of 25.6 $\mu A/mM/cm^2$ sensitivity. In a typical biosensor experiment for the detection of urea, Rahmanian and Mozaffari (2015) designed a novel biosensor by using ZnO along with polyvinyl alcohol, F-doped SnO_2, urease enzyme, and cyanuric chloride for designing and developing the biosensor. The CV and EIS methods were used for evaluation purpose. They reported that using ZnO-based biosensor urea can be easily detected with linear ranges of 5.0 to 125 mg/dL along with a detection limit of 3.0 mg/dL under experimental conditions. Sundarmurugasan et al. (2016) used ZnO nanosphere based biosensor for immediate recognition of monocrotophos (MCP) and dichlorvos (DDVP). They used orange samples for the study and reported 0.036–0.012 nM detection limit for MCP and DDVP, respectively. Ahmad et al. (2017A) reported a very sensitive ZnO based biosensor for phosphate detection. They prepared a Field Effect Transistor (FET) type biosensor and used pyruvate oxidase and ZnO nanorods on SiO_2/Si substrate for the fabrication. They synthesized the ZnO nanorods by using a hydrothermal method and used the CV method to detect pollutants. They concluded the detection of phosphate with wide linear ranges from 0.1 to 7 μM along with 0.05 μM and a lowest detection limit for phosphate under optimum conditions. The biosensor displayed high sensitivity (80.57 μA/mM/cm^2) toward phosphate detection. Ahmad et al. (2017B) prepared the simple one-step low-temperature route-assisted ZnO biosensor for the detection of uric acid. They reported a sensitivity of 239.62 μA/mM/cm^2 for detecting uric acid and the detection time was only 2 s along with the linear range of 0.01 to 4.56 μM and 5 nM as the low detection limit. Pabbi and Mittal (2017) developed an electrochemical biosensor based on ZnO quantum dots (QDs) for the detection of acephate pesticide. In that study, *chlorella sp.* algae, surface-active alkaline phosphate, and SiO2 were used to prepare a sensitive biosensor with the help of amperometric method to detect pesticides in the concentration ranges from 10^{-11} to 10^{-3} M of acephate. Zhang et al. (2014) studied the detection of glucose and hydrazine using Zn−ZnO integrated electrode. In that study, ultrathin ZnO nanofilms were used as the biosensor. They reported a wide linear range of 0.5 μM to 14.2 mM along with 0.5 μM as the lowest detection limit for hydrazine. Recently, Theerthagiri et al. (2019) reviewed various biological, environmental, and energy-related applications of ZnO.

8.3.2.4 WO_3-BASED BIOSENSOR AND ENVIRONMENTAL APPLICATIONS

WO_3 is an n-type semiconductor based metal oxide with significant applications such as various bioelectronics, neural assemblies, and interfaces as well as clever biosensing devices exfoliation (Liu et al., 2015; Santos et al., 2016; Li et al., 2019; Zhang et al., 2020). Deng et al. (2009) developed a third-generation biosensor based on WO_3/cytochrome c for H_2O_2 detection.

This unique system has been reported as free from anodic as well as cathodic interferences. Cyclic Voltammetry results revealed that the prepared biosensor exhibit activity with a wide linear detection range from 3×10^{-7} to 3×10^{-4} M for H_2O_2 within 5 s along with a low detection limit of 2.4×10^{-7} M. Harihran et al. (2011) developed a biosensor using nanocrystalline WO_3 for L-DOPA or levodopa detection. In this method, WO_3 was prepared by using a simple microwave-assisted synthesis method and CV method for sensing purpose. They reported that at pH 7.2, L-DOPA can be easily detected with linear ranges from 1×10^{-7} to 1×10^{-6} M along with 12×10^{-8} or 120 nM as the lowest detection limit for L-DOPA. In a very remarkable study for making a mediator-free biosensor, Liu et al. (2015) reported WO_3 nanowires (NWS) based on direct electron transfer hemoglobin with high length diameter ratio for nitrate detection by the use of hydrothermal synthesis method along with amperiometric method for the detection of pollutant. The biosensor showed a wide linear range from 1 to 400 μM of nitrate in the sample along with 0.28 μM as the lowest detection for nitrate in the study. In a doping study, Zhou et al. (2017) developed Na-doped WO_3 biosensor for bisphenol A detection by the use of a hydrothermal method for synthesis. The prepared Na-doped WO_3 had a rod-like morphology. They used the Differential Pulse Voltammetry (DPV) method for detection of pollutant. The biosensor showed a wide linear range from 0.08 to 22.5 μmol/L along with 0.028 μmol/L detection limit under optimum experimental conditions. In very recent work, Feng et al. (2018) utilized a flower-like WO_3 hierarchical structure for the detection of aflatoxin B1 (AFB1). They prepared a WO_3 material with large surface area and porous hierarchical structure which was able to detect AFB1 in the environment. This WO_3 material was synthesized using hydrothermal method. They used the photoelectrochemical method for detection of the pollutant. They reported linear ranger from 1 to 120 pg/mL for AFB1 along with a lowest detection limit of 0.28 pg/mL under experimental conditions.

8.3.2.5 OTHER METAL OXIDES BASED BIOSENSORS AND THEIR ENVIRONMENTAL APPLICATIONS

With the beneficial properties such as decent toxicity, compatibility with biological entities, unique super magnetism, catalytic advantages, synthetic mimetic applications, and wild electron transfer ability, the other metal oxides are also receiving significant attention in the field of biosensing technology (Wang and Tan, 2007; Yang, 2014). In a report on zirconium oxide-based nanomaterial for environmental applications, Liu et al. (2003)

used ZrO_2-based biosensor for H_2O_2 detection. They used sol–gel method for ZrO_2 synthesis. They concluded a highly sensitive nanobiosensor (sensitivity of 111 μA/mM/cm^2) with the detection linear ranges from 2.5 $\times 10^{-7}$ to 1.5 $\times 10^{-4}$ mol/L in a short time (10 s). In a similar kind of work, Liu et al. (2004) made ZrO_2-based material and used hemoglobin, ZrO_2,and DMSO for making a sensor for H_2O_2 detection. They reported a fine linear range from 1.5 to 30.2 μM along with 0.14 μM as the lowest detection by the electrocatalytic response. In an attempt for making a versatile sensor for pesticide/insecticide detection in vegetables, Yang et al. (2005) fabricated ZrO_2/chitosan-based biosensor and evaluated the sensor's activity on various pesticides such as acetyl thiocholine, phoxim, malathion, and dimethoate. They reported linear ranges of 9.9 $\times 10^{-6}$ to 2.03 $\times 10^{-3}$ M, 6.6 $\times 10^{-6}$ to 4.4 $\times$ 10^{-4} M, 1 $\times 10^{-8}$ to 5.9 $\times 10^{-7}$ M, and 8.6 $\times 10^{-6}$ to 5.2 $\times 10^{-4}$ M, respectively. In a study to make manganese oxide based biosensor, Yao et al. (2006) reported nano-MnO_2 based biosensor for H_2O_2 detection. The amperometric technique has been applied to detect samples and reported a linear range from 1.2 $\times 10^{-7}$ to 2.0 $\times 10^{-3}$ M along with 8 $\times 10^{-8}$ M low detection limit for H_2O_2. The method showed a sensitivity of 2.66 $\times 10^{-5}$ μA/mM/cm^2. This MnO_2-based method has been used in toothpaste and hair dyes for the detection of H_2O_2. Lu et al. (2006) developed antimony oxide bromide (AOB) biosensor for H_2O_2 detection. They used hydrothermal method for the preparation of rod-like structures of AOB nanomaterial in the range of 50 nm. They used chitason and horseradish peroxidase (HRP) for biosensor making. In a very comprehensive work with iron oxide nanomaterials, Hrbac et al. (2007) used five types of ferric oxides (hematite, magnetite, amorphous Fe_2O_3, beta Fe_2O_3, and ferrihydrite) for H_2O_2 detection. Persian blue is used to increase the applicability of these ferric oxide materials. They reported lowest detection limit of 2 $\times 10^{-5}$ M for H_2O_2. Iron oxide showed detection up to 8.5 μM of H_2O_2 in a very short time (less than 3 s) under experimental conditions. Salimi et al. (2008) developed CoOx-based biosensor for nitrate detection. They used the amperometric method for analysis. They reported a linear range of 1–30 μM and 0.2 μM as the lowest detection limit along with 10.5 nA/μM sensitivity for nitrate. Kaushik (2009) used chitosan and Fe_3O_4-based biosensor for pyrethroid pesticides/insecticides, such as, cypermethrin and permethrin detection. In this biosensor, calf thymus deoxyribose nucleic acid (ssCT-DNA) and indium-tin-oxide (ITO) electrodes are used for biosensor deposition. DPV technique has been applied for investigating the biosensor. The results showed a linear range from 0.0025 to 2 ppm and 1 to 300 ppm for CM and PM, respectively. The biosensor response times was found to be only 25 and 40 s for CM and PM, respectively. Ansari et al.

(2009A) reported about the detection of H_2O_2 based on nanocrystalline CeO_2 biosensor. In that study, amperometric technique was used. They reported a linear range for H_2O_2 from 1.0 to 170 μM using HRP/nano CeO_2/ITO electrode. Jindal et al. (2012) reported a sensing method for detecting uric acid using CuO thin film. In this method, p-type CuO was used without any reagent or external method to detect uric acid. They used Cu and photometric assay for the detection study. They reported the linear range around 0.05–1.0 mM and response around 2.7 mA/mM for the method under experimental conditions. Dong et al. (2012) worked on LaF_3-doped CeO_2 biosensor fabrication for nitrite (NO_2^-) detection. They used myoglobin for detection using the amperometric technique. Their results show the linear range from 5 to 4650 μM for nitrite detection in a short time (5 s) along with 0.2 μM as the lowest detection limit. Yagati et al. (2013) developed CeO_2-based biosensor for amperometric detection of H_2O_2. They used CV and DPV for electrochemical behavior of porous CeO_2 nanostructured film. They reported 0.6 μM lower detection limit and up to 3 μM linear range for H_2O_2 in that study along with a short time (8 s) for detection of H_2O_2. In a typical study for phenols and pesticides detection, Mayorga-Martinez et al. (2014) described an IrO-based biosensor for catechol and chlorpyrifos detection. In this method, 0.08 and 0.003 μM detection limit has been observed for catechol and chlorpyrifos, respectively. They used electrochemical and amperometric methods for detection of pollutants. In another pesticide detection work, Jeyapragasam and Saraswathi (2014) developed Fe_3O_4–chitosan biosensor for carbofuran pesticide detection by using the CV technique to detect pesticide in a vegetable sample (cabbage). They observed 3.6×10^{-9} M low detection limit for carbofuran. The authors cross-checked whether the results are reproducible or not by using a comparative HPTLC method and found similar results. In another phenol detection work, Sarika et al. (2017) designed alfa Fe_2O_3 based biosensor for amperometric detection of catechol in the microenvironment. They used laccase enzyme to detect nanobiocomposite biosensor. They reported a biosensor that has linear detection range from 8 to 800 μM along with the low detection limit of 4.28 μM for catechol.

8.3.3 COMPOSITE NANOBIOSENSORS

8.3.3.1 METAL-METAL COMPOSITE BASED BIOSENSOR

In an attempt to make a composite nanomaterial-based biosensor for organophosphate pesticide and carbamate pesticide as well as specific nerve agent, Upadhyay et al. (2009) developed a gold–platinum bimetallic nanoparticles based biosensor for paraoxon ethyl, Aldicard, and Sarin (nerve agent) detection. They used CV and EIS techniques for the analysis. The results obtained indicated the linearity ranges of 150–200, 40–60, 40–50 nM for paraoxon, aldicard, and sarin, respectively. Cu/Ag nanocluster-based biosensor was developed by Su et al. (2010) for the detection of Cu^{2+} ions in real environmental samples of pond water and soil. They reported 5–200 nM detection linear range along with 2.7 nM as the lowest detection limit by using this method. In a comparative study between laboratory as well as real lake water samples, Song et al. (2016) developed Au–Ag core-shell nanoparticles based biosensor for As^{3+} ions detection. They reported 0.5–10 ppb linear range and 0.1 ppb as the lowest detection limit for As^{3+} ions in water sample and concluded a lab-to-land transfer technology with satisfactory results.

8.3.3.2 METAL–METAL OXIDE BASED BIOSENSOR

In a genuine attempt to detect organophosphate pesticide sensing, Zhao et al. (2013) developed Au and Fe_3O_4 composite biosensor for methyl parathion detection. They highlighted the additional qualities of a fabricated biosensor Au nanoparticles have qualities such as high surface-area-to-volume ratio, greater loading capacity, spontaneous electron transfer, and enzyme stabilizing ability. They reported linear range of 0.5 to 1000 ng/mL along with 0.1 ng/mL detection limit. Li et al. (2015) developed Ag/ZnO based biosensor for Pb^{2+} ion detection. They used lake water and human serum as samples. They reported 5×10^{-12} to 4×10^{-6} M linear range for Pb^{2+} along with 9.6×10^{-13} M as the lowest detection limit for Pb^{2+} ion. Ibrahim et al. (2018), in their recent work, reported the first-ever application of Au-In_2O_3-chitosan nanocomposites as biosensor for sensitively detecting (64.57 μAμmol/L/cm^2) antimycotin cycloperoxolamine (CPX) using CV and Square Wave Voltammetry (SWV). The CPX detection followed a linear trend in the range of 0.199–16.22 μmol/L, with 6.64×10^{-9} mol/L as the limit of detection (LOD). For detecting H_2O_2, Wu et al. (2018) used mesoporous TiO_2 as the biosensor. They used rose petal and P123 dual template for the synthesis of TiO_2. In this method, Pt had introduced electrodeposition method on TiO_2 to develop Pt/TiO_2 modified material for H_2O_2 detection. They used the CV method for detecting and reported the ranges from 5 μM to 8 mM along with very low detection limit (1.65 μM).

8.3.3.3 METAL OXIDE-METAL OXIDE COMPOSITE BIOSENSOR

Zhang et al. (2007) developed a hydroquinone (HQ) biosensor by using Fe_3O_4–SiO_2 nanoparticles to make an oxide–oxide nanocomposite to detect HQ. They reported a linear trend in the range of 1×10^{-7} to 1.375×10^{-4} M, with 1.5×18^{-8} M as the LOD. They compared the obtained results of the biosensor with HPTLC results and concluded similar reproducibility. Njagi et al. (2008) reported biosensor based on the amperiometric method by using CeO_2 and TiO_2/CeO_2 for sensitive phenol detection under the presence and absence of oxygen. They reported detection limit of 9×10^{-9} M and 5.6×10^{-9} M in presence (LOD = 86 mA/M) and absence (LOD = 65 mA/M) of oxygen, respectively. They also tested this biosensor for dopamine detection. Ansari et al. (2009B) used TiO_2–CeO_2 based biosensor for urea detection. They constructed a nanocomposite film which was 23 nm in size. Urease and glutamate dehydrogenase (GLDH) were used in that study. They reported that the prepared TiO_2–CeO_2 nanocomposite film exhibit linear response range of 10 to 700 mg/dL. They highlighted the detection of urea with very low detection limit (0.166 μm) within 10 s along with the sensitivity of 0.9165 $\mu A/M/cm^2/mM$. Li et al. (2012) studied the multifunctional biosensor by making a WO_3–TiO_2 modified electrode for photoelectro catalytic evaluation of norepinephrine (NEP) and riboflavin (VB2) in the solutions. To selectively detect NEP and VB2 using a hybrid thin film of WO_3–TiO_2, prepared in ITO electrode; techniques like EIS, DPV, and CV were used. An NEP detection using the CV technique followed a linear trend in the range of 3.23×10^{-6} to 3.88×10^{-4} M, with 1.07×10^{-6} M as the LOD. A VB2 detection using CV technique also followed a linear trend in the range of 3.23×10^{-7} M to 4.0×10^{-5} M, with 1.87×10^{-7} M as the LOD. All the experiments were carried out at neutral pH by using Xe lamp. Srivastava et al. (2013) used a mediator-free microfluid biosensor for detection of urea. In this biosensor, urease, glutamate dehydrogenase, TiO_2, and ZrO_2 were used. They reported that this biosensor has linear ranges of 5 to 100 mg/dL for urea detection, with 0.07 mg/dL and 2.74 μA/[logmM]/cm as the LOD and sensitivity, respectively. Zhang et al. (2013) developed an immunobiosensor by using Fe_3O_4 and TiO_2 for the detection of organophosphorus (OPs) agents by Surface Plasmon Response (SPR) and SWV methods for investigation. They reported a range of 0.02–10 nM along with 0.01 nM lowest detection limit for OPs. In another "turn on" strategy based biosensor study, Wu et al. (2016) developed AIE fluorogen-SiO_2-MnO_2 sandwich nanocomposite-based biosensor for organophosphorus pesticide (paraoxon) detection.

They used fluorescence strip for a visual detection of paraoxon along with thiocholine as the turn on switch. They reported 1–100 mg/L as the linear range and 1 mg/L as the lowest detection limit.

8.3.3.4 OTHER NONCARBON-BASED BIOSENSORS

Ji et al. (2005) reported CdSe–ZnS-based biosensor for paraoxon detection for the first time by Circular Dichroism (CD) and Photo Luminance (PL) techniques. They reported 10^{-8} M as the detection limit under experimental conditions. Vinayak et al. (2009) worked on CdTe-based biosensor for the detection of herbicide (2,4 dichlorophenoxyacetic acid) using the fluoroimmunoassay technique. They used *N*-(3-dimethylaminopropyl)-*N*-ethylcarbodiimide hydrochloride (EDC), *N*-hydroxysuccinimide (NHS), anti-2,4 D, IgG-antibody, and mercaptopropionic acid also for making the sensor and reported a promising detection limit of 250 pg/ML for 2,4 D in the solution. Wu et al. (2010) worked with a biosensor based on CdSe–ZnS quantum dots for Cu^{2+} and Pb^{2+} detection. They reported 0.2 and 0.05 nM as the lowest detection limit for Pb^{2+} and Cu^{2+}, respectively, with a 25-min detection time. Fan et al. (2014) developed CdSe–TiO_2 based biosensor for beta-Estradiol (E-2) in water sample. They reported a 0.05–12-pM wide linear range for E-2 along with 33 fM detection limit at optimum condition. These photoelectrochemical(PEC) method based biosensor's results also show good reproducibility with HPLC results. Zhao et al. (2015) reported TiO_2/CdS-based biosensor for Hg^{2+} detection by using PEC method. They used rhodamine 123, probe DNA, and target DNA for constructing biosensors and reported 10 fM–200 nM as the detection range, with 3.3 fM as the LOD. Shtenberg et al. (2015) used porous Si (PSi) nanostructure based biosensor for tracing heavy metals (Ag^{2+}, Pb^{2+}, Cu^{2+}) in solution by enzymatic activity inhibition. They reported 60–120 ppb detection linear range for three metals along with the sensitivity order of $Ag^{2+} > Pb^{2+} > Cu^{2+}$. They compared this biosensor with another gold nanoparticles based biosensor by using ICP-AES techniques and concluded good reproducibility of the results along with decent agreement. For arsenic evaluation in river and tap water samples, Ravikumar et al. (2018) developed MoS_2 nanosheets based biosensor by using coprecipitation method for As^{3+} detection along with FRET phenomena. They reported 18 nM as the lowest detection limit for As^{3+} with excellent recovery in the environmental system. Hu et al. (2019A) developed Cd-Te QDs based biosensor for four different organophosphorus pesticides (paraoxon, parathion, dichlorvos, and deltamethrin) detection. They also used AchE for the construction of nanobiosensor. They reported a linear range from 10^{-5} to 10^{-12} M for all the pesticides (paraoxon, parathion, dichlorvos, and deltamethrin) and with LOD as 1.2, 0.94, 11.7 and 0.38 pM, respectively. Very recently, He et al. (2020) reported CdSe–CdS quantum dot based biosensor for Hg^{2+} ions detection. They reported a linear range from 1×10^{-2} to 1×10^{-6} M and 1×10^{-13} lowest detection limit for their nanobiosensor.

8.3.3.5 NONCARBON/CARBON COMPOSITES

In a typical pesticide detection study in vegetable samples, Jiao et al. (2017) reported GO/Fe_3O_4-based biosensor for chlorpyrifos detection. They also used chitosan and carbon black for constructing the nanobiosensor and reported a linear detection range of 0.1–105 ng/mL with 0.033 ng/mL as the LOD for chlorpyrifos. Again in the pesticides detection work, Yang et al. (2013) developed NiO/caboxylic-reduced grapheme-based biosensor for carbofuran, methyl parathion, and chlorpyrifos detection. AchE and Nafion carbon electrodes were used in this analysis. They reported a linear range of 1×10^{-13} to 10^{-10} and 10^{-10} to 10^{-8} for methyl parathion and chlorpyrifos, respectively, along with 5×10^{-14} as the lowest detection limit for both methyl parathion and chlorpyrifos pesticides. In the case of carbofuran, two linear ranges 10^{-12} to 10^{-10} M and 10^{-10} to 10^{-8} M along with 10^{-13} m as the lowest detection limit were observed. Hu et al. (2013) developed porphyrin decorated Au/graphene nanocomposite based biosensor for HQ detection. They used white light mediated PEC method for the study and reported 4.6 nM as the lowest detection limit along with 20–240 nM linear detection range for HQ. Liu et al. (2013) reported Au-PTA-TiO_2 nanotube based biosensor for H_2O_2 detection by using CV, SWV, and chronoamperometry methods for evaluation. They reported linear range of 65–2600 μM along with 5 μM and 18.1×10^{-3} μA/μM as the lowest detection limit and sensitivity, respectively. Jiang et al. (2015) developed Ag/N-doped graphene (NG) based biosensor for acetamiprid detection by using one-step thermal treatment method for synthesis. They compared the activity of Ag/NG, Ag, and NG under the same experimental conditions and reported superior activity of Ag/NG over Ag and Ng. They reported a linear detection range of 1×10^{-13} to 5×10^{-9} M, with 3.3×10^{-14} M as the LOD. In a versatile biosensor making study, Liu et al. (2015) developed Pd–Co alloy nanoparticles embedded carbon fiber (PdCo-CNF) based biosensor for H_2O_2 and nitrate detection by using EIS, CV, and DPV methods. They reported 0.2–23.5 μM and 0.1 μM linear detection range and lowest detection limit for H_2O_2 respectively. In the case of nitrate, they reported two linear ranges of 0.4–30 μM and 30–400 μM along with 0.2 μM as the lowest detection limit. In a study on detecting a broad spectrum neonicotinoid insecticide by making a "turn on" strategy, Lin et al. (2016) developed multiwalled carbon nanotubes (MWCNT), ZnS−Mn QDs and aptamers based biosensor for acetamiprid pesticide detection. They reported visual detection of the pollutant along with 0–150 nM linear range and 0.7 nM lowest detection limit. In another pesticide (organophosphorus pesticide) detection study, Chun et al. (2016) developed a nitrogen-doped carbon dot and gold nanoparticle based biosensor for paraoxon detection along with the use of AchE for the construction of nanobiosensor and evaluated the activity by using the FRET phenomena. They reported a wider linear detection range of 10^{-4} to 10^{-9} g/L, with 10^{-9} g/L (3.6×10^{-12} mol/L) as the LOD for paraoxon. Again with the pesticide detection work, Mogha et al. (2016) developed ZrO_2/RGO-based biosensor for chlorpyrifos detection. They used hydrothermal method for ZrO_2/RGO nanocomposite synthesis. Amperiometric method was used for electrochemical study, and the detection range was found to be linear in the concentration range: 10^{-13} to 10^{-9} M and 10^{-9} to 10^{-4} M, with 10^{-13} M as the LOD. For H_2O_2 sensing, hollow TiO_2- reduced graphene oxide microsphere of special morphology were synthesized by Liu et al. (2017) wherein they used amperiometric method for detection purpose. They reported a wider linear detection range of 0.1–360 μM, with 10 nM as the LOD. In an organic thiophosphate fungicide detection study, Arvand and Mirroshandel (2017) developed a "turn on" strategy oriented L-cysteine capped ZnS QDs and GOs based biosensor for edifenphos detection by using the FRET phenomena. They also used environmental and agricultural samples and reported 5×10^{-4} to 6×10^{-3} mg/L linear detection range along with 1.3×10^{-4} mg/L lowest detection limit for edifenphos. Wang et al. (2018) developed CDs and AuNPs based biosensor for acetamiprid detection. They used FRET and inner filter effect (IFE) phenomena for detection and reported 1.08 μg/L lowest detection limit for acetamiprid. They reported checking thesensitivity of this biosensor toward dimethoate, chlorpyrifos, and dichlorvos pesticides. They used real vegetable (tomato, cucumber and cabbage) samples for this study. Chen et al. (2018) developed a CQDs-MoS_2-based biosensor for bisphenol S detection. They developed a type of fluorescent aptamers by using the FRET phenomena, in which CQDs and MoS_2 worked as emitter and absorber, respectively. They reported 0.05–2 μM linear detection range for Bisphenol S along with 2 nM lowest detection limit under optimum conditions. In an experiment for antibiotic detection, Yang et al. (2018) developed a metal−organic framework (Cu-TCPP) based biosensor for chloramphenicol (CAP) analysis. They used a double-stranded DNA and hairpin probe aptamer for the construction of biosensor as a type for fluorescent aptamer by using the FRET phenomena. They reported 0.001–10 ngm/L linear detection range along with 0.3 pgm/L as the lowest detection limit under optimum experimental conditions. In a recent work on drug residue detection, Okoth et al. (2018) developed Au–graphene-doped CdS-based biosensor for diclofenac detection. This study was based on PEC method using a visible light source and they reported 1–150 nM linear detection range and 0.78 nM lowest detection limit. Hu et al. (2019B) first time reported the simultaneous detection of acetaminophen (APAP) and p-hydroxyamenophen (P-HAP) by using SnO_2 carbon nanofiber composite. They synthesized doped SnO_2 carbon nanofiber composite by using electrospinning technology and used EIS and DPV techniques for detection purpose. They reported 0.50–700 μM and 0.20–50 μM linear detection range for APAP and P-HAP, respectively, along with 0.086 and 0.033 μM detection limit for APAP and P-HAP, respectively. Both actual sample and serum environment samples have been used and significant results were obtained. Very recently, Li et al. (2020) developed a gold-doped carbon-dot based biosensor for Pb^{2+} ions detection by using microwave conditions for the synthesis. They reported 0.0005–0.46 μmol/L linear detection range and 0.25 nmol/L lowest detection limit. They also compared these results with the undoped carbon-dot material and concluded superior activities of Au-doped carbon-dot material.

8.4 KEY CHALLENGES INVOLVED WITH NANOBIOSENSING TECHNOLOGY

Recent advancements in nanobiosensing technology have resulted in a broad array of nanobiosensors, exhibiting diverse applications. Therefore in recent times, the sensing domain has widened and more nanobiosensors are being explored to handle some of the significant constraints in fabricating a trustworthy and cost-effective biosensor. Thus, the key challenges for advanced nanobiosensing technology is directly associated with its cost, linearity, sensitivity, multifunctionality, response to selectivity, response time, detection mechanism, long lastingness, and toxicity issues (refer Figure 3). Another, daunting task is to deal with their ability to detect trace amount of analyte diverse samples of soil, water bodies, biofluids, and air (Hahn et al., 2012). Also, the detection limit and response or recovery time are highly fluctuating and can affect the sensor efficiency. Moreover for a specific analyte, the data achieved from a biosensor must have an acceptable precision so that it can be reproduced within a certain range. A biosensor must avoid the systematic error to make sure that the accuracy of the results and its expected value are static. Therefore, adequate standard samples must be used for calibration. Since most of the biosensing applications are either clinical or nonclinical,it would be a good idea to evaluate their size dependent toxicity (Hahn et al., 2012; Munawar et al., 2019; Martin-Gracia et al., 2020; Bhalla et al., 2020).

On the other hand, as it is well known that the toxicity of the nanobiosensors/ or nanosensors highly depends on their size, their applications have been relatively critical. Therefore, various questions such as why these noncarbon materials are used, what fabrication mediums will be needed to obtain them, what characteristics do they have that can allow them to embody the intent, what level of engineering will be required to control and assess all the influencing parameters involved in the sensing technology, how we will be able to evaluate, design, and manipulate the processes involved, how long-lasting materials or products will be, will they maintain the same quality or desired catalytic activities when shifted to another medium or isolated from their original mediums as well as an assessment of their further applicability and fate with the respect of toxicities have always been asked (Li et al., 2015).

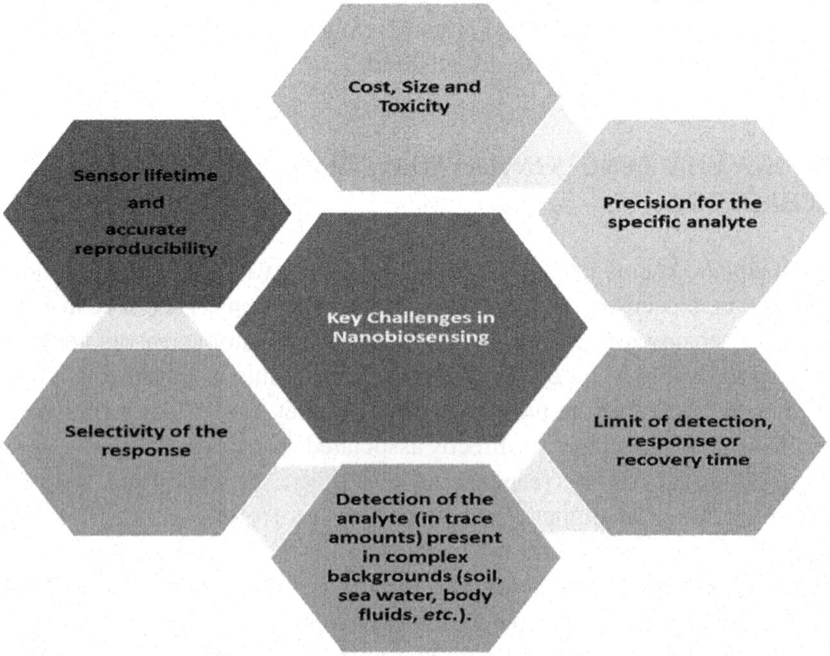

FIGURE 8.4 A brief illustrations of the vital challenges associated with nanobiosensor.

Other challenging factors associated with the nanobiosensors are the components and the structure of the sensor mechanism because the key benefits of any sensor is also directly associated with them. The materials, structure of receptor, or bioreceptor depend on the kind of samples that are being analyzed, the mechanism of detection involved, the type of transducer for signal conversion, and the detector for the response capturing are points that influence any design (Li et al., 2015). Additionally, flexibility and sensitivity are complementary qualities for the nanobiosensors and carbon as well as noncarbon nanomaterials, and both have shown great potential in resolving these issues. The excellent conductive properties of metallic nanomaterials have further allowed in improving the flexibility. Searching and making of multifunctional nanomaterials that can incorporate with biological sensing will surely enhance the results (Wen et al., 2020). Despite the simplicity and wider sensing application, majority of the sensing mechanisms have not yet been fully understood because of the complexity involved in the various parameters which can affect biosensors sensitivity (Kannan et al., 2009).

Receptor and transducer are the two significant functions which are generally involved in the sensing mechanism. They along with the other factors such as adsorption–desorption kinetics, physicochemical properties, surface property, Gibbs free energy, thermodynamic, and kinetic stability (Korotcenkov, 2005; Kanan et al., 2009) are the most important domains involved in sensing mechanism. Since nanobiosensing is a very new technology and a high level of engineering has been involved in the formation of the compact as well as the sensitive sensors, the cost is still a very critical challenging step. But due to the advancements of material sciences and engineering as well as the availability of new cheap materials are creating a potential hope for the future with cost-saving options (Li et al., 2015).

8.5 FUTURE ASPECTS FOR NONCARBON-BASED NANOBIOSENSING TECHNOLOGY

For a highly selective and sensitive onsite analysis of an analyte, the emergence of nanotechnology and its applications in different areas (refer Figure 8.1) have provided several functional nanomaterials as a promising solid substrate platform (Holzinger et al., 2014). Applicability of nanomaterials to be used as biosensor provides a remarkable prospect of developing new-generation nanobiosensing techniques (Li et al., 2015). Applications of biosensor or noncarbon-based biosensor technologies for environmental applicability are still in their budding stage and are facing huge challenges due to the inherent features of environmental analysis. Nanomaterials or noncarbon-based nanomaterials are evaluated in the development of biosensors for environmental usages (pollution mitigation). The engineered nanocomposite along with versatile nanostructures (particles, rods, wires, and tubes) makes the modified biosensor more sensitive and flexible for the analysis of analyte/pollutants. A paper-based device provides platform for a potential portable, miniaturized, economical, and user-friendly biosensor/or nanocomposite/or nanobiosensor. These types of biosensors are meeting the need of onsite detection of environmental samples or pollutants (i.e., soil, river water, seawater, air, etc.). Recently, incorporation of nanomaterial, nanocomposite inside these biosensors brings new tactics for improving their analytical performances. While many biosensors have been developed and tested till date against a wide range of environmental contaminants in laboratory, only a few biosensors are currently being used. Thus, it is clear that more advances and efforts are needed to bridge innovation that will play very crucial role in the development of efficient, automated, real-time biosensors with high throughput analysis (HTA) of environmental pollutants. Indeed, it is the need of the hour to make such an effort to adopt crucial technologies to mitigate environmental pollutants. There are several tools and techniques that are available or being developed to improve the nanomaterial-based nanobiosensors by improving the selected dimensions such as their electrochemical dimension, optical dimension, and magnetic dimensions. These modulations allow developing a single-molecule biosensors along with high throughput biosensor arrays (Pandey et al., 2008). One more important future prospect has been associated with the modulation of the electrodes surface which is directly linked with their interfaces, their preparative materials, characterization, and constraints, control on mechanisms specifically for the presently noncarbon-based sensing technology (Malhotra and Ali, 2018). The improvement in the signal-to-noise proportion, augmentation in transduction processes, and strengthening of the signals have been directly linked with major future requirements. Biological molecules and nanomaterials exhibit special structures as well as unique functions, therefore their awareness is still a future task in the field of noncarbon-based sensing technology (Zhang et al., 2009). A better scientific control on the mechanisms as well as the interaction sciences between the biomolecules present in the surface of the prepared biosensor and the nanomaterials are essential to explore the current utilities as well as to fabricate the new generation of biosensors/nanobiosensors that in turn advance the creation of future robust and variable noncarbon-based sensing technology (Malhotra and Ali, 2018). The combination of both carbon-based and noncarbon-based nanobiosensors have found their applications in various domains such as biomedical, agricultural, and environmental sectors. Nanomaterial-based biosensors (both carbon and noncarbon-based) showed eye-catching prospects, which will be largely applied in various medical diagnosis (Srivastava et al., 2020), food analysis, and process control (Zhang et al., 2020) as well as environmental evaluations in the near future (Verma and Rani, 2020; Şensoy and Muti, 2020).

8.6 CONCLUSION

Noncarbon nanobiosensor based nanobiosensing technology holds immense future potential in environmental-related technologies because they open a window of promising research domain, much needed for the betterment and welfare of humans and the environment. The noncarbon-based nanomaterials will be an ideal choice, with exceptional advantage in the enhancement of sensing technology along with precise sensitivity, broad flexibility, reduced toxicities, and costs as well as broader linear ranges and very low detection limits of pollutants in various environmental samples. The future of nanobiosensing will be in the hands of advanced materials such as composites and hybrids as well as the combination of carbon with noncarbon materials to make this technology a futile, robust, and versatile tool in the biosensor application at various fields. More research should be done and emphasis given to noncarbon-based nanobiosensors as there are limited studies done although they possess remarkable advantages in creating versatile nanobiosensors.

KEYWORDS

- **sensing technology**
- **nanobiosensors**
- **noncarbon-based nanobiosensors**
- **environmental applications**

REFERENCES

Abdel-Karim, R.; Reda, Y.; Abdel-Fattah, A. Nanostructured Materials-Based Nanosensors. *J. Electrochem. Soc.* **2020,** *29*, *167* (3), 037554.

Afsharan, H. Highly Sensitive Electro Chemiluminescence Detection of p53 Protein Using Functionalized Ru–Silica Nanoporous@Gold Nanocomposite. *Biosens. Bioelectron.* **2016,** *80*, 146–53.

Afzal, A.; Cioffi, N.; Sabbatini, L.; Torsi, L. NOx Sensors Based on Semiconducting Metal Oxide Nanostructures: Progress and Perspectives. *Sens. Actuat. B Chem.* **2012,** *171*, 25–42.

Ahmad, R.; Ahn, M. S.; Hahn, Y. B. ZnO Nanorods Array Based field-Effect Transistor Biosensor for Phosphate Detection. *J. Colloid Interf. Sci.* **2017A,** *498*, 292–297.

Ahmad, R.; Tripathy, N.; Ahn, M. S.; Hahn, Y. B.. Solution Process Synthesis of High Aspect Ratio ZnO Nanorods on Electrode Surface for Sensitive Electrochemical Detection of Uric Acid. *Sci. Rep.* **2017B**, *7*, 46475.

Alharbi, O. M.; Khattab, R. A.; Ali, I. Health and Environmental Effects of Persistent Organic Pollutants. *J. Mol. Liquids* **2018,** *1, 263*, 442–453.

Ansari, A. A.; Solanki,P. R.; Malhotra, B. D. Hydrogen Peroxide Sensor Based on HRP Immobilized Nanostructured Cerium Oxide Film. *J. Biotechnol.* **2009A**, *142*, 179–184.

Ansari, A. A.; Sumana, G.; Pandey, M. K.; Malhotra, B. D. Sol-Gel Derived Titanium Oxide-Cerium Oxide Biocompatible Nanocomposite Film for Urea Sensor. *J. Mater. Res.* **2009B,** *24* (5), 1667–1673.

Arvand, M.; Mirroshandel, A. A. Highly-Sensitive Aptasensor Based on Fluorescence Resonance Energy Transfer between L-Cysteine Capped ZnS Quantum Dots and Graphene Oxide Sheets for the Determination of Edifenphos Fungicide. *Biosens. Bioelectron.* **2017,** *96*, 324–331,

Aziz, M. A. Amperometric Immuno Sensing Using an Indium Tin Oxide Electrode Modified with Multi-Walled Carbon Nanotube and Poly (Ethylene Glycol)–Silane Copolymer. *Chem. Commun.* **2007,** 2610–2612.

Badshah, M. A.; Koh, N. Y.; Zia, A. W.; Abbas, N.; Zahra, Z.; Saleem, M. W. Recent Developments in Plasmonic Nanostructures for Metal Enhanced Fluorescence-Based Biosensing. *Nanomaterials* **2020,** *10* (9), 1749.

Bahadir, E. B.; Sezgintürk, M. K. Label-Free, ITO-Based Immunosensor for the Detection of a Cancer Biomarker: Receptor for Activated C Kinase 1. *Analyst* **2016,** *141*, 5618–5626.

Bala, R.; Dhingra, S.; Kumar, M.; Bansal, K.; Mittal, S.; Sharma, R. K.; Wangoo, N. Detection of Organophosphorus Pesticide: Malathion in Environmental Samples Using Peptide and Aptamer Based Nanoprobes. *Chem. Eng. J.* **2017,** *311*, 111–116.

Baruah, S.; Dutta, J. Nanotechnology Applications in Pollution Sensing and Degradation in Agriculture: A Review. *Environ. Chem. Lett.* **2009,** *1, 7* (3), 191–204.

Bhalla, N.; Pan, Y.; Yang, Z.; Payam, A. F. Opportunities and Challenges for Biosensors and Nanoscale Analytical Tools for Pandemics: COVID-19. *ACS Nano* **2020,** *18, 14* (7), 7783–7807.

Bisen, P. S.; Debnath, M.; Prasad, G. B. *Molecular Diagnostics: Promises and Possibilities*; Springer: The Netherlands, 2010.

Buckley, D.; Black, N. C. G.; Castanon, E.; Melios, C.; Hardman, M.; Kazakova, O. Frontiers of Graphene and 2D Material-Based Gas Sensors for Environmental Monitoring. *2D Materials* **2020,** *7* (3), 032002.

Burdușel, A. C.; Gherasim, O.; Grumezescu, A. M.; Mogoantă, L.; Ficai, A.; Andronescu, E. Biomedical Applications of Silver Nanoparticles: An Up-to-Date Overview. *Nanomaterials. Basel: Switzerland*, **2018,** *8* (9), 681.

Cai, X. Ratiometric Electrochemical Immunoassay Based on Internal Reference Value for Reproducible and Sensitive Detection of Tumor Marker. *Biosens. Bioelectron.* **2016,** *81*, 173–180.

Cakiroglu, B.; Ozacar, M. A Self-Powered Photoelectrochemical Biosensor for H_2O_2, and Xanthine Oxidase Activity Based on Enhanced Chemiluminescence Resonance Energy Transfer through Slow Light Effect in Inverse Opal TiO_2. *Biosens. Bioelectron.* **2019,** *141*, 111385.

Cao, X. Silver Nanowire-Based Electrochemical Immunoassay for Sensing Immunoglobulin G with Signal Amplification Using Strawberry-Like ZnO Nanostructures as Labels. *Biosens. Bioelectron.* **2013,** *49*, 256–262.

Chandra, A. Synthesis and Physicochemical Profile of Transition and Lanthanide Nanoparticles and Their Transduced Molecular Interaction with Proteins, PhD thesis. Submitted to School of Chemical Sciences, Central University of Gujarat, Gandhinagar, India, 2017.

Chandra, A.; Bhattarai, A.; Yadav, A. K.; Adhikari, J.; Singh, M.; Giri, B. Green Synthesis of Silver Nanoparticles Using Tea Leaves from Three Different Elevations. *Chem. Select* **2020,** *16, 5* (14), 4239–4246.

Chandra, A.; Singh, M. Biosynthesis of Amino Acid Functionalized Silver Nanoparticles for Potential Catalytic and Oxygen Sensing Applications. *Inorg. Chem. Front.* **2018,** *5* (1), 233–257.

Chang, H. et al. Pt NPs and DNAzyme Functionalized Polymer Nanospheres as Triple Signal Amplification Strategy for Highly Sensitive Electrochemical Immunosensor of Tumor Marker. *Biosens Bioelectron.* **2016,** *86*, 156–163.

Chen, K.; Zhang, W.; Zhang, Y.; Huang, L.; Wang, R.; Yue, X.; Zhu, W.; Zhang, D.; Zhang, X.; Zhang, Y.; Wang, J. Label-Free Fluorescence Aptasensor for Sensitive Determination of Bisphenol S by the Salt-Adjusted FRET between CQDs and MoS_2. *Sens. Actuat. B Chem.* **2018,** *259*, 717–724.

Chen, Q.; Wu, X.; Wang, D.; Tang, W.; Li, N.; Liu, F. Oligonucleotide-Functionalized Gold Nanoparticles-Enhanced QCM-D Sensor for Mercury(ii) Ions with High Sensitivity and Tunable Dynamic Range. *Analyst* **2011,** *136*, 2572–2577.

Chen, X.; Yang, Z.; Si, S. Potentiometric Urea Biosensor Based on Immobilization of Urease Onto Molecularly Imprinted TiO_2 Film. *J. Electroanaly. Chem.* **2009,** *635* (1), 1–6.

Cheng, N.; Song, Y.; Fu, Q.; Du, D.; Luo, Y.; Wang, Y.; Lin, Y. Aptasensor Based on Fluorophore-Quencher Nano-Pair and Smartphone Spectrum Reader for on-Site Quantification of Multi-Pesticides. *Biosens. Bioelectron.* **2018,** *117*, 75–83.

Cho, I. H. Current Technologies of Electrochemical Immunosensors: Perspective on Signal Amplification. *Sensors.* **2018,** *18*, 207.

D'Souza, A. A.; Kumari, D.; Banerjee, R. Nanocomposite Biosensors for Point-of-Care—Evaluation of Food Quality and Safety. *Nanobiosensors* **2017,** 629–676.

Dai, Z.; Guo, J.; Xu, J.; Liu, C.; Gao, Z.; Song, Y. Y. Target-Driven Nanozyme Growth in TiO_2 Nano Channels for Improving Selectivity in Electrochemical Biosensing. *Analy. Chem.* **2020,** *92* (14), 10033–10041.

Deng H-H.; Hong G-L.; Lin F-L.; Liu A-L.; Xia X-H.; Chen, W. Colorimetric Detection of Urea, Urease, and Urease Inhibitor Based on the Peroxidase-Like Activity of Gold Nanoparticles. *Anal. Chim. Acta* **2016,** *915*, 74–80.

Deng, Z.; Gong, Y.; Luo, Y.; Tian, Y. WO_3 Nanostructure Facilitate Electron Transfer of Enzyme: Application to Detection of H_2O_2 with High Sensitivity. *Biosens. Bioelectron.* **2009,** *24*, 2465–2469.

Dey, A. Semiconductor Metal Oxide Gas Sensors: A Review. *Mater. Sci. Eng. B* **2018,** *229*, 206–217.

Díez, N. Highly Packed Graphene–CNT Films as Electrodes for Aqueous Super Capacitors with High Volumetric Performance. *J. Mater. Chem. A* **2018,** *6*, 3667–3673.

Dong, S.; Li, N.; Huang, T.; Tang, H.; Zheng, J. Myoglobin Immobilized on LaF_3 Doped CeO_2 and Ionic Liquid Composite Film for Nitrite Biosensor. *Sens. Actuat. B* **2012,** *173*, 704–709.

Dong, Y.; Zheng, J. A Nonenzymatic L-Cysteine Sensor Based on SnO_2-MWCNTs Nanocomposites. *J. Mol. Liq.* **2014,** *196*, 280–284.

Doria, G.; Conde, J.; Veigas, B.; Giestas, L.; Almeida, C.; Assunção, M.; Rosa, J.; Baptista, P. V. Noble Metal Nanoparticles for Biosensing Applications. *Sensors*. **2012,** *12* (2), 1657–1687.

Dou, X.; Chu, X.; Kong, W.; Luo, J.; Yang, M. A Gold-Based Nanobeacon Probe for Fluorescence Sensing of Organophosphorus Pesticides. *Anal. Chim. Acta* **2015,** *891*, 291–297.

El-Said, W. A.; Choi, J. W. Electrochemical Biosensor Consisted of Conducting Polymer Layer on Gold Nano Dots Patterned Indium Tin Oxide Electrode for Rapid and Simultaneous Determination of Purine Bases. *Electrochim. Acta* **2014,** *123*, 51–57.

Erban, S. I.; Enesca, A. Metal Oxides-Based Semiconductors for Biosensors Applications. *Front. Chem.* **2020,** *8*, 354.

Erickson, M. D. Environmental PCB Forensics: Processes and Issues. *Environ. Sci. Pollut. Res.* **2020,** *27* (9), 8926–8937.

Fan, D. Electro Chemical Immunosensor for Detection of Prostate Specific Antigen Based on an Acid Cleavable Linker into MSN-Based Controlled Release System. *Biosens. Bioelectron.* **2016,** *85*, 580–586.

Fan, L.; Zhao, G.; Shi, H.; Liu, M.; Wang, Y.; Ke, H. A Femtomolar Level and Highly Selective 17 β-Estradiol Photoelectrochemical Aptasensor Applied in Environmental Water Samples Analysis. *Environ. Sci. Technol* **2014,** *48*, 5754–5761.

Fang, C. S. An Ultrasensitive and Incubation-Free Electrochemical Immunosensor Using a Gold-Nanocatalyst Label Mediating Outer-Spherereaction- Philic and Inner-Sphere-Reaction-Philic Species. *Chem. Commun.* **2016,** *52*, 5884–5887.

Feng, J.; Li, Y.; Gao, Z.; Lv, H.; Zhang, H.; Dong, Y.; Wang, P.; Fan, D.; Wei, Q. A Competitive Type Photoelectrochemical Immunosensor for Aflatox in B1 Detection Based on flower-Like WO_3 as Matrix and Ag_2S-Enhanced $BiVO_4$ for Signal Amplification. *Sens. Actuat. B Chem.* **2018,** *270*, 104–111.

Fortunati, S. Single-Walled Carbon Nanotubes as Enhancing Substrates for PNA-Based Amperometric Geno Sensors. *Sensors* **2019,** *19*, 588.

Goldman, E. R.; Clapp, A. R.; Anderson, G. P.; Uyeda, H. T.; Mauro, J. M.; Medintz, I. L.; Mattoussi, H. Multiplexed Toxin Analysis Using Four Colors of Quantum Dot Fluororeagents. *Analy. Chem.* **2004,** *76* (3), 684–688.

Gong, N. C.; Le LI, Y.; Jiang, X.; Zheng, X. F.; Wang, Y. Y.; Huan, S. Y. Fluorescence Resonance Energy Transfer-Based Biosensor Composed of Nitrogen-Doped Carbon Dots and Gold Nanoparticles for the Highly Sensitive Detection of Organophosphorus Pesticides. *Analy. Sci.* **2016,** *32* (9), 951–956.

Goswami, N.; Yao, Q.; Luo, Z.; Li, J.; Chen, T.; Xie, J. Luminescent Metal Nanoclusters with Aggregation-Induced Emission. *J. Phys. Chem. Lett.* **2016,** *7* (6), 962–975.

Guan, H.; Liu, X.; Wang, W.; Liang, J. Direct Colorimetric Biosensing of Mercury (II) Ion Based on Aggregation of Poly-(γ-Glutamic Acid)-Functionalized Gold Nanoparticles. *Spectrochim. Acta A* **2014,** *121*, 527–532.

Guo, H.; Li, J.; Li, Y.; Wu, D.; Ma, H.; Wei, Q.; Du, B. Exciton Energy Transfer-Based Fluorescent Sensor for the Detection of Hg^{2+} through Aptamer-Programmed Self-Assembly of QDs, *Anal. Chim. Acta* **2019,** *1048*, 161–167.

Hahn, Y. B.; Ahmad, R.; Tripathy, N. Chemical and Biological Sensors Based on Metal Oxide Nanostructures. *Chem. Commun.* **2012,** *48* (84), 10369–10385.

Han, L. Enhanced Conductivity of rGO/Ag NPs Composites for Electrochemical Immunoassay of Prostate-Specific Antigen. *Biosens. Bioelectron.* **2017,** *87*, 466–472.
Han, T.; Nag, A.; Mukhopadhyay, S. C.; Xu, Y. Carbon Nanotubes and Its Gas-Sensing Applications: A Review. *Sens. Actuators A Phys.* **2019,** *291*, 107–143.
Harihran, V.; Radhakrishnan, S.; Parthibavarman, M.; Dilipkumar, R.; Sekar, C. Synthesis of Polyethylene Glycol (PEG) assisted Tungsten Oxide (WO_3) Nanoparticles for l-Dopa Bio-Sensing Application, *Talanta* **2011,** *85*, 2166–2174.
He, Y.; Ma, L.; Zhou, L.; Liu, G.; Jiang, Y.; Gao, J. Preparation and Application of Bismuth/MXene Nano-Composite as Electrochemical Sensor for Heavy Metal Ions Detection. *Nanomaterials*, **2020,** *10* (5), 866.
He, Z. J.; Kang, T. F.; Lu, L. P.; Cheng, S. Y. An Electrochemiluminescence Sensor Based on CdSe@CdS-Functionalized MoS_2 and a GOD-Labeled DNA Probe for the Sensitive Detection of Hg(ii). *Anal. Methods* **2020,** *12*, 491–498.
Henry, F. A.; Koleangan, H.; Wuntu, A. D. Synthesis of Silver Nanoparticles Using Aqueous Extract of Medicinal Plants' (*Impatiens balsamina* and *Lantana camara*) Fresh Leaves and Analysis of Antimicrobial Activity. *Int. J. Microbiol.* **2019,** *8*.
Hill, M. K. Understanding Environmental Pollution; Cambridge University Press, 2020; pp 1–15.
Holzinger, M.; Le Goff, A.; Cosnier, S. Nanomaterials for Biosensing Applications: A Review. *Front. Chem.* **2014,** *27*, 2–63.
Hoon, C.; Dong, H. K.; Sangsoo, P. Electrochemical Biosensors: Perspective on Functional Nanomaterials for On-Site Analysis. *Biomater. Res.* **2020,** *24*, 1–12.
Hrbac, J.; Halouzka, V.; Zboril, R.; Papadopoulos, K.; Triantis, T. Carbon Electrodes Modified by Nanoscopic Iron (III) Oxides to Assemble Chemical Sensors for the Hydrogen Peroxide Amperometric Detection. *Electroanalysis* **2007,** *19* (17), 1850–1854.
Hu, T.; Xu, J.; Ye, Y.; Han, Y.; Li, X.; Wang, Z.; Ni, Z. Visual Detection of Mixed Organophosphorous Pesticide Using QD-AChE Aerogel Based Microfluidic Arrays Sensor. *Biosens. Bioelectron.* **2019A**, *136*, 112–117.
Hu, W.; Zhang, Z.; Li, L.; Ding, Y.; An, J. Preparation of Electrospun SnO_2 Carbon Nanofiber Composite for Ultrasensitive Detection of APAP and p-Hydroxyacetophenone. *Sens. Actuat. B* **2019B**, *299*, 127003.
Hu, Y.; Xue, Z.; He, H.; Ai, R.; Liu, X.; Lu, X. Photoelectrochemical Sensing for Hydroquinone Based on Porphyrin-Functionalized Au Nanoparticles on Graphene. *Biosens. Bioelectron.* **2013,** *47*, 45–49.
Ibrahim, H.; Temerk, Y.; Farhan, N. A Novel Biosensor Based on Nanocomposite Au-In_2O_3-Chitosan Modifide Acetyl Black Paste Electrode for Sensitive Detection of Antimycotic-ciclopirox Olamine. *Talanta* **2018,** *179*, 75–85.
Jafari, S.; Mahyad, B.; Hashemzadeh, H.; Janfaza, S.; Gholikhani, T.; Tayebi, L. Biomedical Applications of TiO_2 Nanostructures: Recent Advances. *Int. J. Nanomed.* **2020,** *15*, 3447.
Jaroenapibal, P.; Boonma, P.; Saksilaporn, N.; Horprathum, M.; Amornkitbamrung, V.; Triroj, N. Improved NO_2 Sensing Performance of Electrospun WO_3 Nanofibers with Silver Doping. *Sens. Actuat. B Chem.* **2018,** *255*, 1831–1840.
Jeyapragasam, T.; Saraswathi, R. Electrochemical Biosensing of Carbofuran Based on Acetylcholinesterase Immobilized onto Iron Oxide—Chitosan Nanocomposite. *Sens. Actuat. B Chem.* **2014,** *191*, 681–687.
Ji, X.; Zheng, J.; Xu, J.;Rastogi, V. K.; Cheng, T-C.; DeFrank, J. J.; Leblanc, R. M. (CdSe)ZnS Quantum Dots and Organophosphorus Hydrolase Bioconjugate as Biosensors for Detection of Paraoxon. *J. Phys. Chem. B* **2005,** *109*, 3793–3799.

Jiang, D.; Du, X.; Liu, Q.; Zhou, L.; Dai, L.; Qian, J.; Wang, K. Silver Nanoparticles Anchored on Nitrogen-Doped Graphene as a Novel Electrochemical Biosensing Platform with Enhanced Sensitivity for Aptamer-Based Pesticide Assay. *Analyst* **2015,** *140*, 6404–6411.

Jiao, Y.; Hou, W.; Fu, J.; Guo, Y.; Sun, X.; Wang, X.; Zhao, J. A Nanostructured Electrochemical Aptasensor for Highly Sensitive Detection of Chlorpyrifos. *Sens. Actuat. B Chem* **2017,** *243*, 1164–1170.

Jindal, K.; Tomar, M.; Gupta, V. CuO Thin Film Based Uric Acid Biosensor with Enhanced Response Characteristic. *Biosens. Bioelectron.* **2012,** *38*, 11–18.

Kafi, A. K. M.; Wu, G.; Benvenuto, P.; Chen, A. Highly Sensitive Amperometric H_2O_2 Biosensor Based on Hemoglobin Modified TiO_2 Nanotubes. *J. Electroanal. Chem.* **2011,** *662*, 64–69.

Kanan, S. M.; El-Kadri, O. M.; Abu-Yousef, I. A.; Kanan, M. C. Semiconducting Metal Oxide Based Sensors for Selective Gas Pollutant Detection. *Sensors*, **2009,** *9* (10), 8158–8196.

Kaushik, A.; Ahmad, S. Iron Oxide-Chotosan Hybrid Nanobiocomposite Based Nucleic Acid Sensor for Pyrethroid Detection. *Biochem. Eng. J.* **2009,** *46* (2), 132–140.

Kayhomayun, Z.; Ghani, K.; Zargoosh, K. Surfactant-Assisted Synthesis of Fluorescent $SmCrO_3$ Nanopowder and Its Application for Fast Detection of Nitroaromatic and Nitramine Explosives in Solution. *Mater. Chem. Phys.* **2020,** 122899.

Koedrith, P.; Thasiphu, T.; Weon, J. I.; Boonprasert, R.; Tuitemwong, K.; Tuitemwong, P. Recent Trends in Rapid Environmental Monitoring of Pathogens and Toxicants: Potential of Nanoparticle-Based Biosensor and Applications. *Sci. World J.* **2015,** *2015*.

Korotcenkov, G. Gas Response Control through Structural and Chemical Modification of Metal Oxide Films: State of the Art and Approaches. *Sens. Actuat. B Chem.* **2005,** *107* (1), 209–232.

Korotcenkov, G. Metal Oxides for Solid-State Gas Sensors: What Determines Our Choice? *Mater. Sci. Eng. B* **2007,** *139* (1), 1–23.

Korotcenkov, G.; Cho, B. K. Metal Oxide Composites in Conductometric Gas Sensors: Achievements and Challenges. *Sens. Actuat. B Chem.* **2017,** *244*, 182–210.

Kumar, V. V.; Anbarasan, S.; Christina, L. R.; SaiSubramanian, N.; Anthony, S. P. Bio-Functionalized Silver Nanoparticles for Selective Colorimetric Sensing of Toxic Metal Ions and Antimicrobial Studies. *Spectrochim. Acta A* **2014,** *129*, 35–42.

Kumar, V.; Guleria, P. Application of DNA-Nanosensor for Environmental Monitoring: Recent Advances and Perspectives. *Curr. Pollution. Rep.* **2020.**

Kwon, Y. J.; Kang, S. Y.; Mirzaei, A.; Choi, M. S.; Bang, J. H.; Kim, S. S.; Kim, H. W. Enhancement of Gas Sensing Properties by the Functionalization of ZnO-Branched SnO_2 Nanowires with Cr_2O_3 Nanoparticles. *Sens. Actuat. B Chem.* **2017,** *249*, 656–666.

Lai, G. Amplified Inhibition of the Electrochemical Signal of Ferrocene by Enzyme-Functionalized Graphene Oxide Nanoprobe for Ultrasensitive Immunoassay. *Anal. Chim. Acta.* **2016,** *902*, 189–195.

Lang, Q.; Han, L.; Hou, C.; Wang, F.; Liu, A. A Sensitive Acetylcholinesterase Biosensor Based on Gold Nanorods Modified Electrode for Detection of Organophosphate Pesticide. *Talanta* **2016,** *156–157*, 34–41.

Lavanya, N.; Radhakrishnan, S.; Sekar, C. 2012. Fabrication of Hydrogen Peroxide Biosensor Based on Ni Doped SnO_2 Nanoparticles. *Biosens. Bioelectron.* **2012,** *36*, 41–47.

Lee, S. W.; Lee, W.; Hong, Y.; Lee, G.; Yoon, D. S. Recent Advances in Carbon Material-Based NO_2 Gas Sensors. *Sens. Actuat. B Chem.* **2018,** *255*, 1788–1804.

Lega, M.; Teta, R. Environmental Forensics: Where Techniques and Technologies Enforce Safety and Security Programs. *Int. J. Saf. Sec. Eng.* **2016,** *6* (4), 709–719.

Lei, Y.; Yan, X.; Zhao, J.; Liu, X.; Song, Y.; Luo, N. Improved Glucose Electrochemical Biosensor by Appropriate Immobilization of Nano-ZnO. *Colloid. Surf. B* **2011,** *82*, 168–172.

Li, C.; Wei, D.; DNA-Templated Silver Nanocluster as a Label-Free Fluorescent Probe for the Highly Sensitive and Selective Detection of Mercury Ions. *Sens. Actuat. B Chem.* **2017,** *242*, 563–568.

Li, D.; Yuan, X.; Li, C.; Luo, Y.; Jiang, Z. A Novel Fluorescence Aptamer Biosensor for Trace Pb (II) Based on Gold-Doped Carbon Dots and DNAzyme Synergetic Catalytic Amplification. *J. Luminescence* **2020,** *221*, 117056.

Li, J.; Chen, L.; Lou, T.; Wang, Y. Highly Sensitive SERS Detection of As^{3+} Ions in Aqueous Media Using Glutathione Functionalized Silver Nanoparticles. *ACS Appl. Mater. Interf.* **2011,** *3*, 3936–3941.

Li, J.; Zhao, T., Chen, T.; Liu, Y.; Ong, C. N.; Xie, J. Engineering Noble Metal Nanomaterials for Environmental Applications. *Nanoscale* **2015,** *7* (17), 7502–7519.

Li, L.; Huang, J.; Wang, Y.; Zhang, H.; Liu, Y.; Li, J. An Excellent Enzyme Biosensor Based on Sb-Doped SnO_2 Nanowires. *Biosens. Bioelectron.* **2010,** *25*, 2436–2441.

Li, M. An Ultrasensitive Sandwich-Type Electrochemical Immunosensor Based on the Signal Amplification Strategy of Mesoporous Core–Shell Pd@Pt Nanoparticles/Amino Group Functionalized Graphene Nanocomposite. *Biosens. Bioelectron.* **2017,** *87*, 752–759.

Li, M.; Kong, Q.; Bian, Z.; Ma, C.; Ge, S.; Zhang, Y.; Yu, J.; Yan, M. Ultrasensitive Detection of Lead Ion Sensor Based on Gold Nanodendrites Modified Electrode and Electrochemiluminescent Quenching of Quantum Dots by Electrocatalytic Silver/Zinc Oxide Coupled Structures. *Biosens. Bioelectron.* **2015,** *65*, 176–182.

Li, Y.; Hsu, P.; Chen, S. Multi-Functionalized Biosensor at WO_3–TiO_2 Modified Electrode for Photoelectrocatalysis of Norepinephrine and Riboflavin. *Sens. Actuat. B Chem.* **2012,** *174*, 427.

Li, Z.; Liu, X.; Liang, X. H.; Zhong, J.; Guo, L.; Fu, F. Colorimetric Determination of Xanthine in Urine Based on Peroxidase-Like Activity of WO_3 Nanosheets. *Talanta* **2019,** *1, 204*, 278–284.

Li, Z.; Miao, X.; Xing, K.; Peng, X.; Zhu, A.; Ling, L. Ultrasensitive Electrochemical Sensor for Hg^{2+} by Using Hybridization Chain Reaction Coupled with Ag@Au Core—Shell Nanoparticles. *Biosens. Bioelectron.* **2016,** *80*, 339–343.

Liang, Y. R. A Highly Sensitive Signal-Amplified Gold Nanoparticle-Based Electrochemical Immunosensor for Di-Butyl Phthalate Detection. *Biosens. Bioelectron.* **2017,** *91*, 199–202.

Lim, D. J.; Sim, M.; Oh, L. Carbon-Based Drug Delivery Carriers for Cancer Therapy. *Arch. Pharm. Res.* **2014,** *37*, 43–52.

Lin, B.; Yu, Y.; Li, R.; Cao, Y.; Guo, M. Turn-On Sensor for Quantification and Imaging of Acetamiprid Residues Based on Quantum Dots Functionalized with Aptamer. *Sens. Actuat. B Chem.* **2016,** *229*, 100–109.

Lin, D. Triple Signal Amplification of Graphene Film, Polybead Carried Gold Nanoparticles as Tracing Tag and Silver Deposition for Ultrasensitive Electrochemical Immunosensing. *Anal. Chem.* **2012,** *84*, 3662–3668.

Lin, H-Y.; Huang, C-H.; Lu, S-H.; Kuo, I. T.; Chau, L-K. Direct Detection of Orchid Viruses Using Nanorod-Based Fiber Optic Particle Plasmon Resonance Immunosensor. *Biosens. Bioelectron.* **2014,** *51*, 371–378.

Lin, T-J.; Chung, M-F. Detection of Cadmium by a Fiber-Optic Biosensor Based on Localized Surface Plasmon Resonance. *Biosens. Bioelectron.* **2009,** *24*, 1213–1218.

Lin, T-J.; Huang, K-T.; Liu C-Y. Determination of Organophosphorous Pesticides by a Novel Biosensor Based on Localized Surface Plasmon Resonance. *Biosens. Bioelectron.* **2006,** *22*, 513–518.

Lisa, M.; Chouhan, R. S.; Vinayaka, A. C.; Manonmani, H. K.; Thakur, M. S. Gold Nanoparticles Based Dipstick Immunoassay for the Rapid Detection of Dichlorodipheny ltrichloroethane: An Organochlorine Pesticide. *Biosens. Bioelectron.* **2009,** *25*, 224–227.

Liu, B.; Cao, Y.; Chen, D.; Kong, J.; Deng, J. Amperometric Biosensor Based on a Nanoporous ZrO2 Matrix. *Anal. Chim. Acta* **2003,** *478*, 59–66.

Liu, B.; Zhuang, J.; Wei, G. Recent Advances in the Design of Colorimetric Sensors for Environmental Monitoring. *Environ. Sci. Nano* **2020,** *7* (8), 2195–2213.

Liu, D.; Guo, Q.; Zhang, X.; Hou, H.; You, T. PdCo Alloy Nanoparticle-Embedded Carbon Nanofiber for Ultrasensitive Nonenzymatic Detection of Hydrogen Peroxide and Nitrite. *J. Colloid Interf. Sci.* **2015,** *450*, 168–173.

Liu, H.; Duan, C.; Yang, C.; Chen, X.; Shen, W.; Zhu, Z. A Novel Nitrite Biosensor Based on the Direct Electron Transfer Hemoglobin Immobilized in the WO_3 Nanowires with High Length–Diameter Ratio. *Mater. Sci. Eng.* **2015,** 43–49.

Liu, H.; Guo, K.; Duan, C.; Dong, X.; Gao, J. Hollow TiO_2 Modified Reduced Graphene Oxide Microspheres Encapsulating Hemoglobin for a Mediator-Free Biosensor. *Biosens. Bioelectron.* **2017,** *87*, 473–479.

Liu, J.; Li, Y.; Hung, X.; Zhu, Z. Tin Oxide Nanorod Array-Based Electrochemical Hydrogen Peroxide Biosensor. *Nanosc. Res. Lett.* **2017,** *5*, 1177.

Liu, J.; Ren, S.; Cao, J.; Tsang, D. C.; Beiyuan, J.; Peng. Y.; Fang, F.; She, J.; Yin, M.; Shen, N.; Wang, J. Highly Efficient Removal of Thallium in Wastewater by $MnFe_2O_4$-Biochar Composite. *J. Hazard. Mater.* **2020,** *25*, *401*, 123311.

Liu, S.; Dai, Z.; Chen, H.; Ju, H. Immobilization of hemoglobin on zirconium dioxide nanoparticles for preparation of a novel hydrogen peroxide biosensor. *Biosens. Bioelectron.* **2004,** *19*, 963–969.

Liu, X.; Li, W-J.; Li, L.; Yang, Y.; Mao, L-G.; Peng, Z. A Label-Free Electrochemical Immunosensor Based on Gold Nanoparticles for Direct Detection of Atrazine. *Sens. Actuat. B Chem.* **2014,** *191*, 408–414.

Liu, X.; Zhang, J.; Liu, S.; Zhang, Q.; Liu, X.; Wong, D. K. Gold Nanopar-Ticle Encapsulated-Tubular TiO_2 Nanocluster as a Scaffold for Development of Thiolated Enzyme Biosensors. *Anal. Chem.* **2013,** *85*, 4350–4356.

Lu, X.; Wen, Z.; Li, J. Hydroxyl-Containing Antimony Oxide Bromide Nanorods Combined with Chitosan for Biosensors. *Biomaterials* **2006,** *27*, 5740–5747.

Luo, X.; Yang, J. A Survey on Pollution Monitoring Using Sensor Networks in Environment Protection. *J. Sens.* **2019,** *2019*.

Ma, H.; Li, X.; Yan, T.; Li, Y.; Zhang, Y.; Wu, D.; Du, B. Electro Chemi Luminescent Immunosensing of Prostate-Specific Antigen Based on Silver Nanoparticles-Doped Pb (II) Metal-Organic Framework. *Biosens. Bioelectron.* **2016,** *79*, 379–385.

Madianos, L.; Tsekenis, G.; Skotadis, E.; Patsiouras, L.; Tsoukalas, D. A Highly Sensitive Impedimetric Aptasensor for the Selective Detection of Acetamiprid and Atrazine Based on Microwires Formed by Platinum Nanoparticles. *Biosens. Bioelectron.* **2018,** *101*, 268–274.

Maduraiveeran, G.; Jin, W. Nanomaterials Based Electrochemical Sensor and Biosensor Platforms for Environmental Applications. *Trends Environ. Analy. Chem.* **2017,** *1* (13), 10–23.

Malhotra, B. D.; Ali, M. A. Nanomaterials in Biosensors: Fundamentals and Applications. *Nanomater. Biosens.* **2018,** *1*.

Malik, P.; Katyal, V.; Malik, V.; Asatkar, A.; Inwati, G.; Mukherjee, T. K. Nanobiosensors: Concepts and Variations. *Int. Sch. Res. Notices*. **2013,** *2013*, 1–9.

Manam, D. V. Pharmacological Potential of Silver Nano Particles Bio-Synthesized by the Marine Algae *Colpomenia sinuosa* (Mertens ex Roth) Derbes and Solier and *Halymenia porphyroides* Boergesen (Crypton) with Reference to Their Antimicrobial Efficacy, PhD Thesis. Submitted to University of Madras, India, 2015.

Manam, D. V.; Murugesan, S. Biogenic Silver Nanoparticles by *Halymenia poryphyroides* and Its in Vitro Anti-Diabetic Efficacy. *J. Chem. Pharm. Res.* **2013,** *5* (12), 1001–1008.

Manam, D. V.; Murugesan, S. Biological Synthesis of Silver Nanoparticles from Marine Alga *Colpomenia sinuosa* and Its in Vitro Anti-Diabetic Activity. *Am. J. Bio-Pharmacol. Biochem. Lifesci.* **2014,** *3* (1), 1–7.

Manam, D. V.; Murugesan, S. Biosynthesis and Characterization of Silver Nanoparticles from Marine Macroscopic Red Seaweed *Halymenia porphyroides* Boergesen (Crypton). *J. Nanosci. Technol.* **2020**A, *6* (2), 886–890.

Manam, D. V.; Murugesan, S. Biosynthesis and Characterization of Silver Nanoparticles from Marine Macroscopic Brown Seaweed *Colpomenia sinuosa* (Mertens ex Roth) Derbes and Solier. *J. Adv. Chem. Sci.* **2020**B, *6* (1), 663–666.

Manam, D. V.; Murugesan, S. Investigation of Biosynthesized Silver Nano Particles Interaction from *Halymenia porphyroides* with the E7 Protein Using Bioinformatics Tool. *Biomed. J. Sci. Techn. Res.* **2017,** *1* (4), 1–4.

Martin-Gracia, B.; Martin-Barreiro, A.; Cuestas-Ayllon, C.; Grazu, V.; Line, A.; Llorente, A.; de la Fuente, J. M.; Moros, M. Nanoparticle-Based Biosensors for Detection of Extracellular Vesicles in Liquid Biopsies. *J. Mater. Chem. B* **2020.**

Mayorga-Martinez, C. C.; Pino, F.; Kurbanoglu, S.; Rivas, L.; Ozkan, S. A.; Merkoçi, A. Iridium Oxide Nanoparticle Induced Dual Catalytic/Inhibition Based Detection of Phenol and Pesticide Compounds. *J. Mater. Chem. B* **2014,** *2*, 2233–2239.

Mehrotra, P. Biosensors and Their Applications—A Review. *J. Oral Biol. Craniofac. Res.* **2016,** *6* (2), 153–159.

Miller, D. R.; Akbar, S. A.; Morris, P. A. Nanoscale Metal Oxide-Based Heterojunctions for Gas Sensing: A Review. *Sens. Actuat. B Chem* **2014,** *204*, 250–272.

Mogha, N. K.; Sahu, V.; Sharma, M.; Sharma, R. K.; Masram, D. T. Biocompatible ZrO_2—Reduced Graphene Oxide Immobilized AChE Biosensor for Chlorpyrifos Detection. *Mater. Des.* **2016,** *111*, 312–320.

Munawar, A.; Ong, Y.; Schirhagl, R.; Tahir, M. A.; Khan, W. S.; Bajwa, S. Z. Nanosensors for Diagnosis with Optical, Electric and Mechanical Transducers. *RSC Adv.* **2019,** *9* (12), 6793–6803.

Mura, S.; Greppi, G.; Roggero, P. P.; Musu, E.; Pittalis, D.; Carletti, A.; Ghiglieri, G.; Irudayaraj, J. Functionalized Gold Nanoparticles for the Detection of Nitrates in Water. *Int. J. Environ. Sci. Technol.* **2015,** *12*, 1021–1028.

Nadzirah, S.; Gopinath, S. C.; Parmin, N. A.; Hamzah, A. A.; Mohamed, M. A.; Chang, E. Y.; Dee, C. F. State-of-the-Art on Functional Titanium Dioxide-Integrated Nano-Hybrids in Electrical Biosensors. *Crit. Rev. Analy. Chem.* **2020,** 1–12.

Nandita, D.; Shivendu, R.; Eric, L. *Environmental Nanotechnology*, Vol. 4; Springer Nature, Technology & Engineering, 2020; p 410.

Njagi, J.; Ispas, C.; Andreescu, S. Mixed Ceria-Based Metal Oxides Biosensor for Operation in Oxygen Restrictive Environments. *Anal. Chem.* **2008,** *80*, 7266.

Noah, N. M. Design and Synthesis of Nanostructured Materials for Sensor Applications. *J. Nanomater.* **2020,** *2020*.

Norizan, M. N.; Moklis, M. H.; Demon, S. Z.; Halim, N. A.; Samsuri, A.; Mohamad, I. S.; Knight, V. F.; Abdullah, N. Carbon Nanotubes: Functionalisation and Their Application in Chemical Sensors. *RSC Adv.* **2020,** *10* (71), 43704–43732.

Okoth, O. K.; Yan, K.; Feng, J.; Zhang, J. Label-Free Photoelectrochemical Aptasensing of Diclofenac Based on Gold Nanoparticles and Graphene-Doped CdS. *Sens. Actuat. B* **2018,** *256*, 334–341.

Pabbi, M.; Mittal, S. K. An Electrochemical Algal Biosensor Based on Silica Coated ZnO Quantum Dots for Selective Determination of Acephate. *Anal. Methods* **2017,** *9*, 1672–1680.

Pandey, P.; Datta, M.; Malhotra, B. D. Prospects of Nanomaterials in Biosensors. *Analy. Lett.* **2008,** *1*, *41* (2), 159–209.

Parth, M.; Varun, K.; Vibhuti, M.; Archana, A.; Gajendra, I.; Tapan, K. M. Nanobiosensors: Concepts and Variations. *Int. Sch. Res. Notices* **2013,** *9.*

Peng, D.; Hu, B.; Kang, M.; Wang, M.; He, L.; Zhang, Z.; Fang, S. Electrochemical Sensors Based on Gold Nanoparticles Modified with Rhodamine B Hydrazide to Sensitively Detect Cu (II). *Appl. Surf. Sci.* **2016,** *390*, 422–429.

Penza, M.; Rossi, R.; Alvisi, M.; Cassano, G.; Signore, M. A.; Serra, E.; Giorgi, R. Pt-and Pd-Nanoclusters Functionalized Carbon Nanotubes Networked Films for Sub-ppm Gas Sensors. *Sens. Actuat. B Chem.* **2008,** *135* (1), 289–297.

Piplode, S. Solar Photocatalytic Degradation of Methomyl, Oxamyl, Carbofuran, Carbaryl and Propoxur Pesticides Using Flower Like Nano BiOCl, PhD thesis. Submitted to Vikram University, Ujjain, MP, India, 2017.

Piplode, S.; Dawane, V. BiOX (X=F, Cl, Br, I) Promising Nanomaterial for Environmental Remediation. In *Role of Chemical Sciences in Research and Development for Sustainability*; Dr. Aruna, Ed.; Immortal Publications, 2020; pp 136–147.

Piplode, S.; Dawane, V., Joshi, V.; Pare, B. BiOCl Nano Pallets Preparation and Their White/Solar Light Mediated Photo Catalytic Activities Evaluation on Carbamate Pesticide Oxamyl and Synthetic Dye Azure B. *Biosci. Biotech. Res. Comm.* **2020,** *13* (2), 676–782.

Ponzoni, A.; Comini, E.; Ferroni, M.; Sberveglieri, G. Nanostructured WO_3 Deposited by Modified Thermal Evaporation for Gas-Sensing Applications. *Thin Solid Films* **2005,** *490* (1), 81–85.

Rackauskas. S.; Barbero, N.; Barolo, C.; Viscardi, G. ZnO Nanowire Application in Chemoresistive Sensing: A Review. *Nanomaterials* **2017,** *7* (11), 381.

Rahman, M. A. Functionalized Conducting Polymer as an Enzyme Immobilizing Substrate: An Amperometric Glutamate Micro Biosensor for in Vivo Measurements. *Anal. Chem.* **2005,** *77*, 4854–4860.

Rahmanian, R.; Mozaffari, S. A. Electrochemical Fabrication of ZnO-Polyvinyl Alcohol Nanostuctured Hybrid Film for Application to Urra Biosensor. *Sens. Actuat. B Chem.* **2015,** *207*, Part A, 772–781.

Rajendran, S.; Manoj, D.; Raju, K.; Dionysiou, D. D.; Naushad, M.; Gracia, F. Influence of Mesoporous Defect Induced Mixed- Valent NiO (Ni_2+/Ni_3+)-TiO_2 Nanocomposite for Non-Enzymatic Glucose Biosensors. *Sens. Actuat. B* **2018,** *264*, 27–37.

Ravikumar, P.; Panneerselvam, K.; Radhakrishnan, A. A. B.; Christus, S.; Sivanesan, MoS_2 Nanosheets as an Effective Fluorescent Quencher for Successive Detection of Arsenic Ions in Aqueous System. *Appl. Surf. Sci.* **2018,** *449*, 31–38.

Rim, Y. S.; Chen, H.; Zhu, B.; Bae, S. H.; Zhu, S.; Li, P. J.; Yang, Y. Interface Engineering of Metal Oxide Semiconductors for Biosensing Applications. *Adv. Mater. Interf.* **2017,** *4* (10), 1700020.

Rizwan, M.; Mohd-Naim, N. F.; Ahmed, M. U. Trends and Advances in Electrochemiluminescence Nanobiosensors. *Sensors* **2018,** *18* (1), 166.

Ryu, K. R.; Ha, J. W. Influence of Shell Thickness on the Refractive Index Sensitivity of Localized Surface Plasmon Resonance Inflection Points in Silver-Coated Gold Nanorods. *RSC Adv.* **2020,** *10* (29), 16827–16831.

Salimi, A.; Hallaj, R.; Mamkhezri, H.; Mohamad, S.; Hosiani, T. Electrochemical Propretis and Electrocataltyic Activity of FAD Immobilized Onto Cobalt Oxide Nanoparricles: Application to Nitrate Detection. *J. Electroanaly. Chem.* **2008,** *619*, 31–38.

Santos, L.; Silveira, C. M.; Elangovan, E.; Neto, J. P.; Nunes, D.; Pereira, L. Synthesis of WO_3 Nanoparticles for Biosensing Applications. *Sens. Actuat. B Chem.* **2016,** *223*, 186–194.

Sarika, S.; Shivakumar, M. S.; Shivakumara, C.; Krishnamurthy, G.; Narsimha Murthy, B.; Lekshami, I. C. A Novel Amperometric Catechol Biosensor Based on Alpha-Fe_2O_3 Nanocrystals-Modified Carbon Paste Electrode. *Artif. Cells, Nanomed. Biotechnol.* **2017,** *45* (3), 625–634.

Sayago, I.; Santos, H.; Horrillo, M. C.; Aleixandre, M.; Fernández, M. J.; Terrado, E.; Martínez, M. T. Carbon Nanotube Networks as Gas Sensors for NO_2 Detection. *Talanta* **2008,** *77* (2), 758–764.

Scida, K.; Stege, P. W.; Haby, G.; Messina, G. A.; García, C. D. Recent Applications of Carbon-Based Nanomaterials in Analytical Chemistry: Critical Review. *Analy. Chim. Acta.* **2011,** *8, 691* (1–2), 6–17.

Şensoy, K. G.; Muti, M. The Novel Nanomaterials Based Biosensors and Their Applications. In *Novel Nanomaterials*; IntechOpen, 2020; p 7.

Şerban, I.; Enesca, A. Metal Oxides-Based Semiconductors for Biosensors Applications. *Front. Chem.* **2020,** *8*.

Shahdost-fard, F.; Roushani, M. Designing an ultra-sensitive aptasensor based on an AgNPs/thiol-GQD Nanocomposite for TNT Detection at Femtomolar Levels Using the Electrochemical Oxidation of Rutin as a Redox Probe. *Biosens. Bioelectron.* **2017,** *87*, 724–731.

Shaheen, S. M.; Antoniadis, V.; Kwon, E.; Song, H.; Wang, S. L.; Hseu, Z. Y.; Rinklebe, J. Soil Contamination by Potentially Toxic Elements and the Associated Human Health Risk in Geo-and Anthropogenic Contaminated Soils: A Case Study from the Temperate Region (Germany) and the Arid Regions (Egypt). *Environ. Pollut.* **2020,** *5*, 114312.

Sharma, A. K.; Mahajan, A.; Saini, R.; Bedi, R. K.; Kumar, S.; Debnath, A. K.; Aswal, D. K. Reversible and Fast Responding ppb Level Cl2 Sensor Based on Noncovalent Modified Carbon Nanotubes with Hexadecafluorinated Copper Phthalocyanine. *Sens. Actuat. B Chem.* **2018,** *255*, 87–99.

Shi,H.;Zhao,G.; Liu,M.; Fan,L.; Cao,T. Aptamer-Based Colorimetric Sensing of Acetamiprid in Soil Samples: Sensitivity, Selectivity and Mechanism. *J. Hazard. Mater.* **2013,** *260*, 754–761.

Shirsat, M.; Too, C. O.; Wallace, G. Amperometric Glucose Biosensor on Layer by Layer Assembled Carbon Nanotube and Polypyrrole Multilayer Film. *Electroanalysis*. **2007,** *20*, 150–156.

Shtenberg, G.; Massad-Ivanir, N.; Segal, E. Detection of Trace Heavy Metal Ions in Water by Nanostructured Porous Si Biosensors. *Analyst* **2015,** *140*, 4507–4514.

Skotadis, E.; Tsekenis, G.; Chatzipetrou, M.; Patsiouras, L.; Madianos, L.; Bousoulas, P.; Zergioti, I.; Tsoukalas, D. Heavy Metal Ion Detection Using DNAzyme-Modified Platinum Nanoparticle Networks. *Sens. Actuat. B* **2017,** *239*, 962–969.

Song, B. Y.; Zhang, M.; Teng, Y.; Zhang, X. F.; Deng, Z. P.; Huo, L. H.; Gao, S. Highly Selective ppb-Level H_2S Sensor for Spendable Detection of Exhaled Biomarker and Pork Freshness at Low Temperature: Mesoporous SnO_2 Hierarchical Architectures Derived from Waste Scallion Root. *Sens. Actuat. B Chem.*, **2020,** *307*, 127662.

Song, L.; Mao, K.; Zhou, X. Hu, J. A Novel Biosensor Based on Au@Ag Core—Shell Nanoparticles for SERS Detection of Arsenic (III). *Talanta* **2016,** *146*, 285–290.

Srivastava, M.; Srivastava, N.; Mishra, P. K.; Malhotra, B. D. Prospects of Nanomaterials-Enabled Biosensors for COVID-19 Detection. *Sci. Total Environ.* **2020,** *16*, *754*, 142363.

Srivastava, S.; Ali, M. A.; Solanki, P. R.; Chavhan, P. M.; Pandey, M. K.; Mulchandani, A.; Srivastava, A.; Malhotra, B. D. Mediator-Free Microfluidics Biosensor Based on Titania–Zirconia Nanocomposite for Urea Detection. *RSC Adv.* **2013,** *3*, 228.

Su, Y-T.; Lan, G-Y.; Chen, W-Y.; Chang, H-T. Detection of Copper Ions through Recovery of the Fluorescence of DNA-Templated Copper/Silver Nanoclusters in the Presence of Mercaptopropionic Acid. *Anal. Chem.* **2010,** *82*, 8566–8572.

Sugunan, A.; Dutta, J. Pollution Treatment, Remediation and Sensing. *Nanotechnol. Online* **2010,** *15*, 125–146.

Sundarmurugasan, R.; Gumpu, M. B.; Ramachandra, B. L.; Nesakumar, N.; Sethuraman, S.; Krishnan, U. M.; Rayappan, J. B. B. Simultaneous Detection of Monocrotophos and Dichlorvos in Orange Samples Using Acetylcholinesterase-Zincoxide Modified Platinume Lectrode with Linear Regression Calibration. *Sens. Actuat. B Chem.* **2016,** *230*, 306–313.

Tan, L.; Chen, Z.; Zhao, Y.; Wei, X.; Li, Y.; Zhang, C.; Wei, X.; Hu, X. Dual Channel Sensor for Detection and Discrimination of Heavy Metal Ions Based on Colorimetric and Fluorescence Response of the AuNPs-DNA Conjugates. *Biosens. Bioelectron.* **2016,** *85*, 414–421.

Taylor, I. M. Enhanced Dopamine Detection Sensitivity by PEDOT/Graphene Oxide Coating on in Vivo Carbon Fiber Electrodes. *Biosens. Bioelectron.* **2017,** *89*, 400–410.

Tekaya, N.; Saiapina, O.; Ben Ouada, H.; Lagarde, F.; Namour, P.; Ben Ouada, H.; Jaffrezic-Renault, N. Bi-Enzymatic Conductometric Biosensor for Detection of Heavy Metal Ions and Pesticides in Water Samples Based on Enzymatic Inhibition in Arthrospira Platensis. *J. Environ. Prot.* **2014,** *5*, 441–453.

Tertiş, M. Highly Selective Electrochemical Detection of Serotonin on Polypyrrole and Gold Nanoparticles-Based 3D Architecture. *Electrochem. Commun.* **2017,** *75*, 43–47.

Theerthagiri, J.; Salla, S.; Senthil, R. A.; Nithyadharseni, P.; Madankumar, A.; Arunachalam, P.; Kim, H. S. A Review on ZnO Nanostructured Materials: Energy, Environmental and Biological Applications. *Nanotechnology* **2019,** *30* (39), 392001.

Trindade, E. K. G. A Probe Less and Label-Free Electrochemical Immunosensor for Cystatin C detection Based on Ferrocene Functionalized Graphene Platform. *Biosens. Bioelectron.* **2019,** *138*, 111311.

Tuitemwong, P.; Songvorawit, N.; Tuitemwong, K. Facile and Sensitive Epifluorescent Silica Nanoparticles for the Rapid Screening of EHEC. *J. Nanomater.* **2013,** *2013*.

Ueda, T.; Katsuki, S.; Takahashi, K.; Narges, H. A.; Ikegami, T.; Mitsugi, F. Fabrication and Characterization of Carbon Nanotube Based High Sensitive Gas Sensors Operable at Room Temperature. *Diamond Relat. Mater.* **2008,** *17* (7–10), 1586–1589.

Upadhyay, S.; Rao, G. R.; Sharma, M. K.; Bhattacharya, B. K.; Rao, V. K.; Vijayaraghavan, R. Immobilization of Acetylcholineesterase—Choline Oxidase on a Gold—Platinum Bimetallic Nanoparticles Modified Glassy Carbon Electrode for the Sensitive Detection of Organophosphate Pesticides, Carbamates and Nerve Agents. *Biosens. Bioelectron.* **2009,** *25*, 832–838.

Valera, E.; Ramón-Azcón, J.; Sanchez, F. J.; Marco, M. P.; Rodriguez, Á. Conductimetric Immunosensor for Atrazine Detection Based on Antibodies Labelled with Gold Nanoparticles. *Sens. Actuat. B Chem* **2008,** *134*, 95–103.

Verma, M. L.; Rani, V. Biosensors for Toxic Metals, Polychlorinated Biphenyls, Biological Oxygen Demand, Endocrine Disruptors, Hormones, Dioxin, Phenolic and Organophosphorus Compounds: A Review. *Environ. Chem. Lett.* **2020,** *19*, 1–0.

Vinayak, A. C.; Basheer, S.; Thakur, M. S. Bioconjugation of CdTe Quantum Dot for the Detection of 2,4-Dichlorophenoxyacetic Acid by Competitive Fluoroimmunoassay Based Biosensor. *Biosens. Bioelectron.* **2009,** *24*, 1615–1620.

Viswanathan, S. Disposable Electrochemical Immunosensor for Arcino Embryonic Antigen Using Ferrocene Liposomes and MWCNT Screen Printed Electrode. *Biosens. Bioelectron.* **2009,** *24*, 1984–1989.

Viswanathan, S.; Radecka, H.; Radecki, J. Electrochemical Biosensors for Food Analysis. *Monatsh. Chem.* **2009,** *140*, 891.

Wang, C.; Sun, R.; Li, X.; Sun, Y.; Sun, P.; Liu, F.n Lu, G. Hierarchical flower-like WO_3 Nanostructures and Their Gas Sensing Properties. *Sens. Actuat. B Chem.* **2014,** *204*, 224–230.

Wang, D.; Saleh, N. B.; Sun, W.; Park, C. M.; Shen, C.; Aich, N.; Peijnenburg, W. J.; Zhang, W.; Jin, Y.; Su, C. Next-Generation Multifunctional Carbon–Metal Nanohybrids for Energy and Environmental Applications. *Environ. Sci. Technol.* **2019,** *14, 53* (13), 7265–7287.

Wang, H. Facile Synthesis of Cuprous Oxide Nanowires Decorated Graphene Oxide Nanosheets Nanocomposites and Its Application in Label Free Electrochemical Immunosensor. *Biosens. Bioelectron.* **2017,** *87*, 745–751.

Wang, J.; Wu, Y.; Zhou, P.; Yang, W.; Tao, H.; Qiu, S.; Feng, C. A Novel Fluorescent Aptasensor for Ultrasensitive and Selective Detection of Acetamiprid Pesticide Based on the Inner Filter Effect between Gold Nanoparticles and Carbon Dots, *Analyst* **2018,***143*, 5151–5160.

Wang, M.; Yin, H.; Zhou, Y.; Sui, C.; Wang, Y.; Meng, X. Photo Electro Chemical Biosensor for microRNA Detection Based on a MoS_2/g-C_3N_4/Black TiO_2 Heterojunction with Histostar@AuNPs for Signal Amplification. *Biosens. Bioelectron.* **2019,** *128*, 137–143.

Wang, P. An Ultrasensitive Sandwich-Type Electrochemical Immunosensor Based on the Signal Amplification System of Double-Deck Gold Film and Thionine Unite with Platinum Nanowire Inlaid Globular SBA-15 Microsphere. *Biosens. Bioelectron.* **2017,** *91*, 424–430.

Wang, P.; Zhao, L.; Shou, H.; Wang, J.; Zheng, P.; Jia, K.; Liu, X. Dual-Emitting Fluorescent Chemosensor Based on Resonance Energy Transfer from Poly(Arylene Ether Nitrile) to Gold Nanoclusters for Mercury Detection. *Sens. Actuat. B Chem.* **2016,** *230*, 337–344.

Wang, S. F.; Tan, Y. M. A Novel Amperometric Immunosensor Based on Fe_3O_4 Magnetic Nanoparticles/Chitosan Composite Film for Determination of Ferritin. *Anal. Bioanal. Chem.* **2007,** *387*, 703–8.

Wang, Y.; Liu, J.; Cui, X.; Gao, Y.; Ma, J.; Sun, Y.; Lu, G. NH3 Gas Sensing Performance Enhanced by Pt-Loaded on Mesoporous WO_3. *Sens. Actuat. B Chem.*, **2017,** *238*, 473–481.

Wang, Y.; Zu, M.; Zhou, X.; Lin, H.; Peng, F.; Zhang, S. Designing Efficient TiO_2-Based Photo Electro Catalysis Systems for Chemical Engineering and Sensing. *Chem. Eng. J.* **2020,** *381*, 122605.

Wei, M.; Wang, J. A Novel Acetylcholinesterase Biosensor Based on Ionic Liquids-AuNPs-Porous Carbon Composite Matrix for Detection of Organophosphate Pesticides. *Sens. Actuat. B Chem.* **2015,** *211*, 290–296.

Wei, M.; Zeng, G.; Lu, Q. Determination of Organophosphate Pesticides Using an Acetylcholinesterase- Based Biosensor Based on a Boron-Doped Diamond Electrode

Modified with Gold Nanoparticles and Carbon Spheres. *Microchim. Acta* **2014,** *181*, 121–127.

Wen, N.; Zhang, L.; Jiang, D.; Li, B.; Sun, C.; Guo, Z. Emerging Flexible Sensors Based on Nanomaterials: Recent Status and Applications. *J. Mater. Chem A* **2020**.

Wu, C. M.; Naseem, S.; Chou, M. H.; Wang, J. H.; Jian, Y. Q. Recent Advances in Tungsten-Oxide-Based Materials and Their Applications. *Front. Mater.* **2019,** *6*, 49.

Wu, C-S.; Khaing Oo, M. K.; Fan, X. Highly Sensitive Multiplexed Heavy Metal Detection Using Quantum-Dot-Labeled DNAzymes. *ACS Nano* **2010,** *4*, 5897–5904.

Wu, P.; Wang, S.; Hou, P.; Wu, J.; Fluorescence Sensor for Facile and Visual Detection of Organophosphorus Pesticides Using AIE Fluorogens-SiO_2-MnO_2 Sandwich Nanocomposites. *Talanta* **2019,** *198*, 8–14.

Wu, X.; Zhang, H.; Huang, K.; Zeng, Y.; Zhu, Z. Rose Petal and P123 Dual-Templated Macro-Mesoporous TiO_2 for a Hydrogen Peroxide Biosensor. *Bioelectrochemistry* **2018,** *120*, 150–156.

Wu, Y.; Liu, L.; Zhan, S.; Wang, F.; Zhou, P. Ultrasensitive Aptamer Biosensor for Arsenic(III) Detection in Aqueous Solution Based on Surfactant-Induced Aggregation of Gold Nanoparticles. *Analyst* **2012,** *137*, 4171–4178.

Wu, Z.; Yao, Q.; Zang, S.; Xie, J. Directed Self-Assembly of Ultrasmall Metal Nanoclusters. *ACS Mater. Lett.* **2019,** *1* (2), 237–248.

Xiangqian, L.; Huizhong, X.; Zhe-Sheng, C.; Guofang, C. Biosynthesis of Nanoparticles by Microorganisms and Their Applications. *J. Nanomater*. **2011,** *16*.

Yagati, A. K.; Lee, T.; Min, J.; Choi, J. An Enzymatic Biosensor for Hydrogen Peroxide Based on CeO_2 Nanostructure Electrodeposited on ITO Surface. *Biosens. Bioelectron.* **2013,** *47*, 385–390.

Yang, L.; Wang, G.; Liu, Y.; Wang, M. Development of a Biosensor Based on Immobilization of Acetylcholinesterase on NiO Nanoparticles–Carboxylic Graphene–Nafion Modified Electrode for Detection of Pesticides. *Talanta* **2013** *113*, 135–141.

Yang, Q.; Zhou, L.; Wu, Y. X.; Zhang, K.; Cao, Y.; Zhou, Y.; Wu, D.; Hu, F.; Gan, N. A Two Dimensional Metal–Organic Framework Nanosheets-Based Fluorescence Resonance Energy Transfer Aptasensor with Circular Strand-Replacement DNA Polymerization Target-Triggered Amplification Strategy for Homogenous Detection of Antibiotics, *Anal. Chim. Acta* **2018,** *1020*, 1–8.

Yang, S.; Jiang, C.; Wei, S. H. Gas sensing in 2D materials. *Appl. Phys. Rev.*, **2017,** *4* (2), 021304.

Yang, Y.; Guo, M.; Yang, M.; Wang, Z.; Shen, G.; Yu, R. Determination of Pesticides in Vegetable Samples Using an Acetylcholinesterase Biosensor Based on Nanoparticles ZrO2/ Chitosan Composite Film. *Int. J. Environ. Anal.Chem.* **2005,** *85* (3) 163–175.

Yang, Z. Hollow Platinum Decorated Fe_3O_4 Nanoparticles as Peroxidase Mimetic Couple with Glucose Oxidase for Pseudobienzyme Electrochemical Immunosensor. *Sens. Actuat. B Chem.* **2014,** *193*, 461–466.

Yao, S.; Xu, J.; Wang, Y.; Chen, X.; Xu, Y.; Hu, S. A Highly Sensitive Hydrogen Peroxide Amperometric Sensor Based on MnO_2 Nanoparticles and Dihexadecyl Hydrogen Phosphate Composite Film. *Anal. Chim. Acta* **2006,** *557*, 78–84.

Yaraki, M. T.; Rezaei, S. D.; Tan, Y. N. Simulation Guided Design of Silver Nanostructures for Plasmon-Enhanced Fluorescence, Singlet Oxygen Generation and SERS Applications. *Phys. Chem. Chem. Phys* **2020,** *22* (10), 5673–5687.

Yaraki, M. T.; Tan, Y. N. Recent Advances in Metallic Nanobiosensors Development: Colorimetric, Dynamic Light Scattering and Fluorescence Detection. *Sens. Int.* **2020,** *2020*, 100049.

Yin, H. Electrochemical Immunosensor for N6-Methyladenosine Detection in Human Cell Lines Based on Biotin-Streptavidin System and Silver-SiO_2 Signal Amplification. *Biosens. Bioelectron.* **2017,** *90,* 494–500.

Yu, J.; Liu, S.; Ju, H. Mediator-Free Phenol Sensor Based on Titania Sol–Gel Encapsulation Matrix for Immobilization of Tyrosinase by a Vapor Deposition Method. *Biosens. Bioelectron.* **2003,** *19,* 509.

Yu, L. Ultrasmall Silver Nanoparticles Supported on Silica and Their Catalytic Performances for carbon Monoxide Oxidation. *Catal. Commun.* **2011,** *12,* 616–620.

Yuea, H. Y.; Zhanga, H. J.; Huanga, S.; Lua, X. X; Gaoa, X.; Songa, S. S. Highly Sensitive and Selective Dopamine Biosensor Using Au Nanoparticles- ZnO Nanocone Arrays/Graphene Foam Electrode. *Mater. Sci. Eng. C.* **2020,** *108,* 110490.

Zhang, B.; Gao, P. X Metal Oxide Nanoarrays for Chemical Sensing: A Review of Fabrication Methods, Sensing Modes, and Their Inter-Correlations. *Front. Mater.* **2019,** *6,* 55.

Zhang, B.; Wang, H.; Xi, J.; Zhao, F.; Zeng, B. In Situ Formation of Inorganic/Organic Heterojunction Photocatalyst of WO_3/Au/ Polydopamine for Immunoassay of Human Epididymal Protein 4. *Electrochim. Acta* **2020,** *331,* 135350.

Zhang, H. A Novel Electrochemical Immunosensor Based on Nonenzymatic Ag@Au-Fe_3O_4 Nanoelectrocatalyst for Protein Biomarker Detection. *Biosens. Bioelectron.* **2016,** *85,* 343–350.

Zhang, R.; Belwal, T.; Li, L.; Lin, X.; Xu, Y.; Luo, Z. Nanomaterial-Based Biosensors for Sensing Key Foodborne Pathogens: Advances from Recent Decades. *Comprehen. Rev. Food Sci. Food Saf.* **2020,** *19* (4), 1465–1487.

Zhang, X.; Guo, Q.; Cui, D. Recent Advances in Nanotechnology Applied to Biosensors. *Sensors* **2009,** *9* (2), 1033–1053.

Zhang, X.; MA, W.; Nan, H.; Wang, G. Ultrathin Zinc Oxide Nanofilm on Zinc Substrate for High Performance Electrochemical Sensors. *Electrochim. Acta* **2014,** *144,* 186–193.

Zhang, X.; Wang, H.; Yang, C.; Du, D.; Lin, Y. Preparation, Characterization of Fe_3O_4 at TiO_2 Magnetic Nanoparticles and Their Application for Immunoassay of Biomarker of Exposure to Organophosphorus Pesticides. *Biosens. Bioelectron.* **2013,** *41,* 669.

Zhang, Y.; Zeng, G. M.; Tang, L.; Huang, D. L.; Jiang, X. Y.; Chen, Y. N. A Hydroquinone Biosensor Using Modified Core—Shell Magneticnanoparticles Supported on Carbon Paste Electrode. *Biosens. Bioelectron.* **2007,** *22,* 2121–2126.

Zhao, G.; Lei, Y.; Zhang, Y.; Li, H.; Liu, M. Growth and Favorable Bioelectrocatalysis of Multishaped Nanocrystal Au in Vertically Aligned TiO_2 Nanotubes for Hemoprotein. *J. Phys. Chem. C* **2008,** *112,* 14786–14795.

Zhao, J.; Wu, D.; Zhi, J. A Novel Tyrosinse Biosensor Based on Biofunctional ZnO Nanorod Microarray on the Nanocrystalline Diamind Electrode for Detection of Phenolic Compounds. *Bioelectrochemistry* **2009,** *5* (1), 44–49.

Zhao, M.; Fan, G. C.; Chen, J. J.; Shi, J. J.; Zhu, J. J. Highly Sensitive and Selective Photoelectrochemical Biosensor for Hg^{2+} Detection Based on Dual Signal Amplification by Exciton Energy Transfer Coupled with Sensitization Effect. *Analy. Chem* **2015,** *87* (24), 12340–12347.

Zhao, M.; Fan, G-C.; Chen, J-J.; Shi, J-J.; Zhu, J-J. Highly Sensitive and Selective Photo Electrochemical Biosensor for Hg^{2+} Detection Based on Dual Signal Amplification by Excite on Energy Transfer Coupled with Sensitization Effect. *Anal. Chem.* **2015,** *87,* 12340–12347.

Zhao, Y.; Zhang, W.; Lin, Y.; Du, D. The Vital Function of Fe_3O_4@Au Nanocomposites for Hydrolase Biosensor Design and Its Application in Detection of Methyl Parathion. *Nanoscale* **2013,** *5,* 1121–1126.

Zhou, Q.; Yang, L.; Wang, G.; Yang, Y. Acetylcholinesterase Biosensor Based on SnO_2 Nanoparticles–Carboxylic Graphene–Nafion Modified Electrode for Detection of Pesticides. *Biosens. Bioelectron.* **2013,** *49*, 25–31.

Zhou, S.; Wang, Y.; Zhu, J. J. Simultaneous Detection of Tumor Cell Apoptosis Regulators Bcl-2 and Bax through a Dual-Signal-Marked Electrochemical Immunosensor. *ACS. Appl. Mater. Interf.* **2016,** *8*, 7674–7682.

Zhou, Y.; Yang, L.; Li, S.; Dang, Y. A Novel Electrochemical Sensor for Highly Sensitive Detection of Bisphenol A Based on the Hydrothermal Synthesized Na-Doped WO_3 Nanorods. *Sens. Actuat. B Chem.* **2017,** *245*, 238–246.

Zong, X.; Zhu, R. ZnO Nanorod-Based FET Biosensor for Continuous Glucose Monitoring. *Sens. Actuat. B Chem.* **2018,** *255*, 2448–2453.

CHAPTER 9

Carbon-Based Nanomaterials: Nanobiosensors Detecting Environmental Pollution

SWAPNALI JADHAV[1], EKTA B. JADHAV[1], SWAROOP S. SONONE[1], MAHIPAL SINGH SANKHLA[2*], and RAJEEV KUMAR[2]

[1]*MSc Forensic Science, Government Institute of Forensic Science, Aurangabad, Maharashtra, India*

[2]*Department of Forensic Science, School of Basic and Applied Sciences, Galgotias University, Greater Noida, India*

[*]*Corresponding author. E-mail: mahipal4n6@gmail.com*

ABSTRACT

Environmental pollution is one of the major threats across the globe due to the rapid growth of industries, technologies, and urbanization. Development in the manufacturing and production sector has increased pollution momentously. Hazardous pollutant sources, such as heavy metals, pesticides, pathogens, and other contaminants released in the surroundings are primarily responsible for the pollution of the environment and can cause significant influences on the well-being of humans. Nanotech is the upcoming area which needs to be integrated with environmental pollution. The convergence of nanotechnology and biosensors is an effective alternative to the conventional analytical technique to analyze minute details in the environment. Biosensors are considered as potential devices to recognize the analyte. In recent years, nanomaterials have been implied for biosensor applications due

Nanotechnology for Environmental Pollution Decontamination: Tools, Methods, and Approaches for Detection and Remediation. Fernanda Maria Policarpo Tonelli, Rouf Ahmad Bhat, & Gowhar Hamid Dar (Eds.)

to their characteristic features. Integration of carbon-based nanomaterials (CBNs) such as carbon nanotubes, graphene, graphene oxide, and carbon nanowires can be used with biosensors for monitoring, sensing, and rapid detection of environmental pollutants.

9.1 INTRODUCTION

In the last few decades, various environmental issues of concern, such as contamination of water, air, soil, and other ecosystems are continuously increasing due to globally growing anthropogenic activities, industrialization, and urbanization, especially in rapidly developing nations. Anthropogenic industrial and agricultural activities are highly contributing this issue of contamination to the environment by disposing hazardous solid and liquid waste into water, releasing noxious gases into atmosphere, and excessive use of pesticides, synthetic fertilizers, pharmaceuticals, which eventually bioaccumulated in the tissues of consuming organisms (Ramnani et al., 2016). Environmental pollutants include inorganic pollutants (heavy metal ions, nitrite, etc.), gaseous analytes (SO_2, CO, NH3, methanol vapor, etc.), organic pollutants (phenolic compound, aromatic amine, nitroaromatic explosives, and pesticides), endocrine-disrupting chemicals (estrogen, bisphenol, and nonylphenol), and biological substances, such as pathogens and antibiotics. The regular consumption of such hazardous water, food, and air contaminated with such toxic chemicals may affect human health by causing severe or even fatal effects. Therefore, it is important to develop a constantly environment monitoring tool, which is sensitive, well-grounded, cost-effective, user as well as eco-friendly for the detection of contaminants in different ecosystems (Wang and Hu, 2016). The most commonly used analytical techniques for environmental pollutant detection includes gas chromatography–mass spectrometry (GC–MS), high-performance liquid chromatography (HPLC), surface-enhanced Raman spectroscopy (SERS), and atomic absorption spectroscopy (AAS) (Auroux et al., 2002; Padrón-Sanz et al., 2005; Van Loon et al., 2012; Li et al., 2013). Although these analytical methods are quick to detect, costly, time-consuming, not useful for on-site detection and require trained users or experts for operating such devices. Biosensors itself contain incorporated chemical sensors, which are able to analyze the analytes qualitatively or semiquantitatively by incorporating a bioreceptor that is indirectly connected with a transducer (Thévenot et al., 2001). A biosensor is a detection tool used specially for qualitative sensing of a chemical substance that interacts with some biorecognition component. This interaction ismeasured by a transducer which outputs in readable form which shows the presence of particular target analyte. Thus, the biosensor consists of three major components: the biological receptor, signal transducer, and the detector. The biosensors are applied for sensing materials, such as antibodies, hormones, gas and immunological molecules, pharmaceuticals, proteins, enzymes, etc. (Kissinger, 2005). An extensive range of biosensors detection tool with nanomaterials have been investigated in a number of studies all over the world. Several studies are conducted regarding the application of nanotechnology and nanomaterials for designing these biosensing tools, and such devices are known as nanobiosensors (Malik et al., 2013). The nanosize of such nanomaterials (1–100 nm) makes them highly effective because most of their atoms placed at or close to surface and having all physicochemical characteristics which are greatly different from the same material at the bulk range which enhanced signaling and transduction. The nanomaterials can be efficiently applied in the biosensors for the detection of analyte. The nanomaterials, majorly in use include nanorods, nanotubes, nanoparticles, and nanowires (Jianrong et al., 2004). The carbon-based nanomaterials, such as CNT, graphene, graphene oxide (GO), have been widely used in biosensors for advanced detection of sample, and they have become a novel interdisciplinary frontier between biosensing and material science (Prasad et al., 2017). Merging of nanotechnology with biosensors can be used to monitor the environmental contamination at high speed. In this chapter, we emphasize on the various carbon-based nanomaterials which are used for biosensing, their synthesis and application for the detection of environmental pollutants.

9.2 NANOBIOSENSORS

The nanobiosensors are biosensors modified with nanomaterials. They can be actively employed for the identification of biomolecules, and these highly sensitive nanobiosensors can rapidly sense high levels of environmental pollutants, and therefore, they can play a major role in potential environmental biomonitoring (Mohammadi Aloucheh et al., 2018). In the 21st century, along with the progress of science, miniature biosensors based on nanotechnology were developed for the detection of analyte (Rai et al., 2012). Due to their high sensitivity, cost-effectiveness, user-friendliness, reliability, and detection potential for specific molecules of these nanobiosensors, their development has been quite interesting and fruitful with a wide range of applications, including determination of emerging pollutants without any isolation, concentration, or pre-sampling procedures (Steffens et al., 2017).

9.2.1 COMPONENTS OF NANOBIOSENSORS

Nanobiosensor is an analytical device. It is the advanced version of a biosensor device containing a biological recognition element in contact with the sensor for measuring the biological or biochemical signal and its conversion into electrical signal (Kanjana, 2017). The immobilized layer may consist of proteins, virus, cellular lipid bilayers, DNA/RNA, microbial cells, etc. The ideal nanobiosensor should possess special features such as potential of real-time monitoring of the molecule of interest, versatility specificity and selectivity toward particular target analyte, rapid, on-site detection, accuracy and repeatability over the useful analytical scale for monitoring the intended environment (Kwak et al., 2017; Thakur and Ragavan, 2013).

Based on the mode of action, the nanobiosensors consist of (1) the bioreceptor, such as antibodies, immunological molecules, proteins, enzymes, etc., which act as a template for the target analyte; (b) the transducer, which converts bioreceptor–bioanalyte interaction into electronic signal; and (c) the detector system that detects the electrical signal and amplify it properly (Malik et al., 2013).

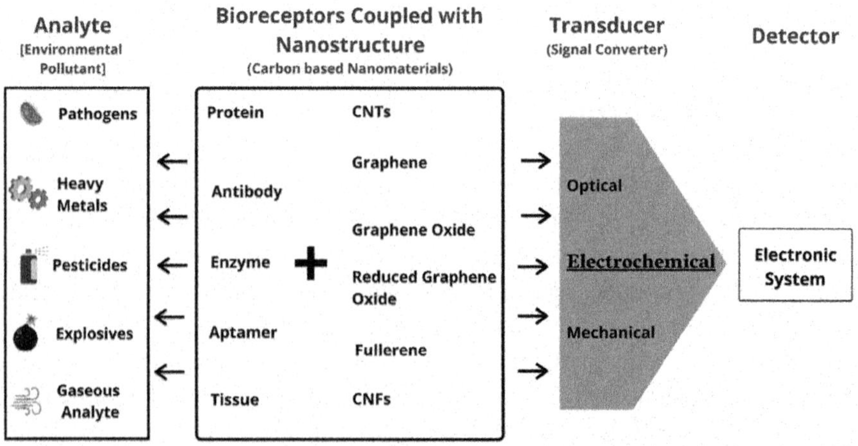

FIGURE 9.1 Components of carbon-based nanomaterials nanobiosensors.

9.2.2 TYPES OF NANOBIOSENSOR

Nanobiosensors are categorized either by nanomaterials used for enhancing the biosensor properties, or by biorecognition element utilized for molecular identification, or by the kind of transducer utilized for identification.

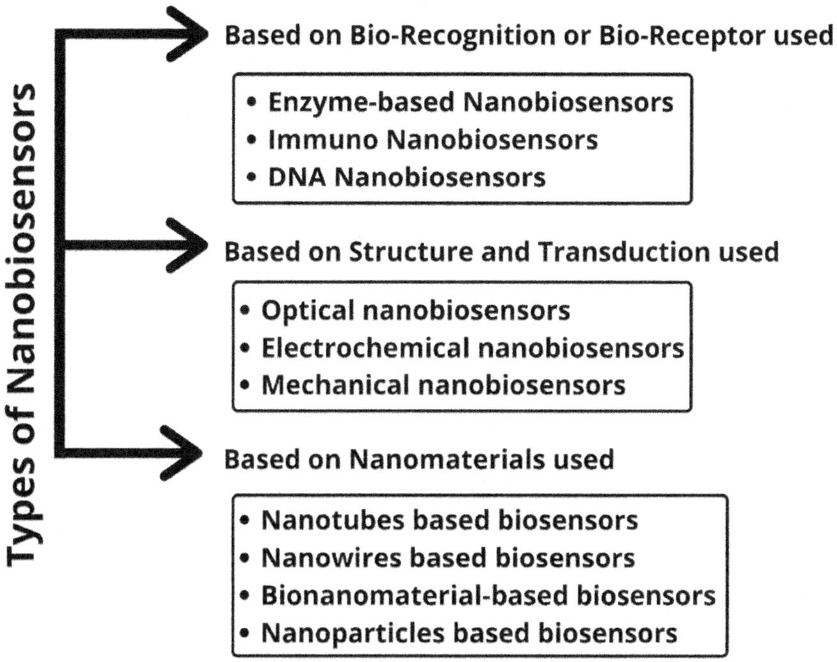

FIGURE 9.2 Types of nanobiosensors.

9.2.2.1 NANOBIOSENSORS BASED ON NANOMATERIALS USED

9.2.2.1.1 *Nanoparticles-Based Biosensors*

Nanoparticle-based biosensors are more effective due to their nanoscale size and consequently, they exhibit high surface-to-volume ratio and are highly attractive and cost-effective due to easy and bulk synthesis using standard chemical methods. They need not be fabricated or modified with advanced approaches. Several metal nanoparticles are used to label different kinds of biomolecules by maintaining the integrity of their biological activities (Hrapovic et al., 2004).

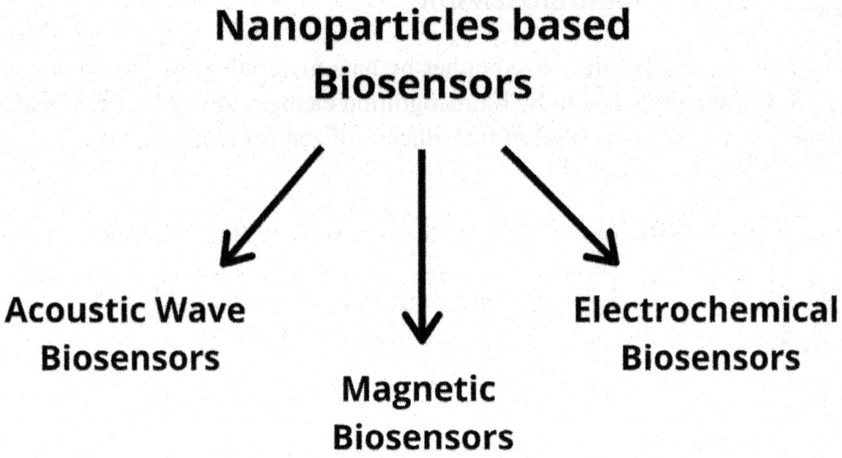

FIGURE 9.3 Nanoparticles-based biosensors.

a) Acoustic Wave Biosensors

Acoustic wave biosensors were advanced to importantly progress the responsiveness and restrictions of recognition (Ward and Ebersole, 1996). There can be various stimuli-dependent impacts in such type of sensor. Mass-dependent types of such sensors comprise the coupling of antibody-altered sol elements that fix themselves on the exterior of the electrode which has been mixed with the particle of analytes combined in such a way that antibody particles are halted on the electrode. The great amount of mass of bound sol elements of the antibody outcomes in an alteration in the vibrational frequency of the quartz-dependent detecting platforms, and this alteration acts as the basis of identification. In common, the accepted diameter of sol-dependent antibody particle is in the range of 5–100 nm. Elements of platinum, gold, titanium dioxide, and cadmium sulfide are most commonly used (Su et al., 2000; Liu et al., 2004).

b) Magnetic Biosensors

Magnetic biosensor uses the particularly intended magnetic nanoparticle. Such magnetic nanoparticles are commonly ferrite-based material, whether used separately or in a shared system. The conventionally used biodetectors and diagnostic tools become more sensitive, powerful, and versatile due to the incorporation of magnetic nanomaterials (Richardson et al., 2001). Magnetic nanoparticle is a commanding and adaptable analytic instrument in ecology and medication. These nanoparticles can be frequently synthesized in the arrangement of either single domain or greigite (Fe_3S_4), superparamagnetic (Fe_3O_4), maghemite (g-Fe_2O_3), different kinds of ferrites like MeO_Fe_2O_3, where Me = Mn, Ni, Mg, Co, Zn, etc. Destined to biorecognitive particles, magnetic nanoparticles can be utilized to disperse or to improve the analytes to be recognized (Šafařík and Šafaříková, 1999). Distinct gadgets like SQUID (superconducting quantum interference devices) have been utilized for fast discovery of biotic targets utilizing the super paramagnetic nature of magnetic nanoparticles. Superconducting quantum interference devices are utilized to curtain the exact antigens from the combinations by using antibodies destined to the magnetic nanoparticle (Chemla et al., 2000).

c) Electrochemical Biosensors

Nanoparticles are utilized as labels in the electrochemical detection of analyte. Often, electrochemical sensors are used for potential environmental monitoring due to the formation of electrochemically detectable ions by the oxidation of metal nanoparticles (Koedrith et al., 2015). Electrochemical biosensors are popular because of their low-priced, comparatively quick response time, user-friendliness, and small dimensions. Such sensors fundamentally work to enhance or examine the biological responses with the help of enhanced electric means. These gadgets are frequently dependent on the metallic nanoparticles. Chemical reactions among the biomolecules can be effortlessly and professionally carried out with the assistance of metallic nanoparticles, which significantly help in attaining immobilization of one of the reactants. This capability makes these responses very precise and removes any chance of receiving unwanted by-products. In the present condition, colloidal gold-dependent nanoparticles have been utilized to increase the immobilization of DNA on gold electrode that has significantly amplified the efficacy of a complete biosensor by additionally letting down the recognition boundary (Cai et al., 2001).

9.2.2.1.2 Nanotubes-Based Biosensors

Carbon nanotubes (CNTs) are nanosized materials with best electrochemical and electrical properties. Carbon nanotubes are suitable for transport of electrical indications produced upon identification of a subject, and thus, it plays a significant part in the current growth of enzyme-dependent biosensor (Balasubramanian and Burghard, 2006). CNTs are hopeful resources for detection purposes because of their fascinating characteristics. Particularly, the greater length-to-diameter part proportions offer for larger surface-to-volume proportions (Sotiropoulou et al., 2003). Furthermore, CNTs have an excellent capability to facilitate quick electron transfer kinetics for a diverse variety of electroactive types, such as H_2O_2 or hydrogen peroxide or NADH. Additionally, CNTs' chemical utilization can be done to assign nearly every chosen chemical type to themselves, this permits us to improve the biocompatibility and solubility of the nanotubes (Wang et al., 2003). CNTs, along with its exciting electrical and electrochemical features, are perfect together as an electrode and as a transducer component into a biosensor. Separate SWCNTs (single-walled CNTs) are tremendously subtle to their neighboring situation. Both chemoresistors and chemically delicate field-effect transistors (FETs) integrating unspoiled or precisely working CNTs have been revealed to be capable of sensing biomolecule (Balasubramanian and Burghard, 2006).

9.2.2.1.3 Nanowires-Based Biosensors

Nanowires are tube-shaped structures similar to that of CNTs, consisting of measurements in the order of a few microns to cm and widths in the nanoseries. Nanowires are the 1D structures with great electron-transport effects. Specifically, the movement of current transporters in the nanowire is dynamically enhanced and are dissimilar from those in majority of the supplies. The tiny dimensions and ability of nanowires make them suitable option to be utilized for biodetection of pathogen and numerous real-time monitoring of a widespread variety of biotic and biochemical classes, therefore, massively extemporizing the carriage precisions of currently used in vivo analytical methods (Cui and Lieber, 2001). Nanowire biosensor can be ornamented with practically any possible biochemical or biotic molecular detection units, by suitable superficial features. The nanomaterial transduces the biochemical binding action on its exterior into an alteration in the conductance of the nanowires in an tremendously subtle, realistic, and measurable manner (Wang et al., 2009; Wang, 2002).

9.2.2.1.4 Bionanomaterial-Based Biosensors

The special features of nanomaterials additionally when combined with catalytic function of biomolecules form various efficient nanosensors. The self-organization of biological molecules within sensors are led to obtain a well-defined nanostructure with biomaterials (Abdel-Karim et al., 2020). Inside the biosensor, the bioreceptor is united with an appropriate transducer that creates an indication after collaboration with the particle of attention. The occurrence of the biotic component forms the system tremendously precise and very subtle, offering an advantage to the traditional techniques. Throughout the time, various naturally occurring and man-made biotic sources have been utilized in the biosensor. The utmost significant botic sources are the dendrimers, enzymes, tinny films, etc. In an enzyme-dependent biosensor, the biotic material is an enzyme that reacts distinctively with its substratum (Guilbault et al., 2004).

9.2.2.2 BASED ON BIORECOGNITION OR BIORECEPTOR

Nanobiosensors can be categorized into the following:

a. Enzyme-dependent nanobiosensors
b. Immuno-nanobiosensors
c. DNA-nanobiosensors (Fu et al., 2017)

9.2.2.3 BASED ON STRUCTURE AND TRANSDUCTION

a) Optical nanobiosensors

Optical nanobiosensor works on fiber optic by means of absorbance, total internal reflection, or fluorescence, bioluminescence, chemiluminescence, evanescent wave, or SPR (surface plasmon resonance). Optical nanobiosensors are usually handled for the recognition of pathogens depending on the fluorescence and SPR (surface plasmon resonance). Usually, this approach is depending on the tracking of the shift in an optical sign among the working pathogens and nanomaterials. The major advantage of this technique is that with tiny cell disturbance, the device can mix into the inner parts of the cells. Optical transducers are promising devices to design a potential, low cost, user-friendly and on-site nanobiosensing tool (Fu et al., 2017; Kabariya and Ramani., 2017).

b) Electrochemical nanobiosensors

Electrochemical transducer is potentiometric, amperometric, conductometric, and impedimetric systems. The structure of carbon nanomaterial includes two types, which are 2-dimensional, that is, planar in the case of graphene and 1-dimensional, that is, tubular in shape for CNTs. These form them perfect transduction constituents. Moreover, its electrochemical features like fast electron transfer kinetic, lesser residual charge, available reusable surfaces, widespread probable window and improved current with noteworthy decrease in excess potentials create carbon nanomaterial a suitable option for biosensing applications when compared with other carbon-dependent electrode material like glassy carbon (Wang et al., 2009).

i) Carbon nanomaterials used for electrochemical Biosensors

Nanomaterial are utilized for the manufacture of electrochemical biosensor and are established for the ecological examination. Electrochemical biosensor can be categorized based on the source of nanomaterials and transducer utilized for the electrode alterations (Mazzei et al., 2015). The carbon nanomaterials have been widely utilized for an electrochemical biosensor due to their larger external region, because of this, numerous recognition actions can occur concurrently on its external area and the binding of biomolecule is easily possible too. Carbon nanomaterials exhibit the electrical, physical, mechanical, and optical characteristics that make them ideal option to be utilized in the biosensor. Their charge storing and electron transfer property can be put together for the electrochemical uses The electrochemical performances of the biosensors could be improved by adjusting the structure of their molecules to engineer their electrochemical activity. Carbon nanomaterials do own a greater surface-to-volume ratio, electrical conductance, and motorized power which make thema suiable option for use in the electrochemical biosensor (Royea et al., 2006; Goyal et al., 2010; Revin and John, 2012).

b) Mechanical nanobiosensors

Usually, nanomaterial is used in the biorecognition of materials (1) as subsistence for the load of many pointers, for example, fluorescent dyes, biomolecules, or Raman reporters to intensify the detection action by its high surface-volume proportion or (2) as the pointer which is formed with the help of chemical reaction to attain numerous signal intensification (Fu et al., 2017). Biomolecular collaboration is calculated through motorized nanoscale biosensor. Depending upon the alterations in the external stress created by the interaction among the target and probe particles on its surface, chemical vapor at tiny level can be recognized. The extent of the shift in the surface stress depends on the type of contact which happened, consisting of the forces of electrostatic, van der Waals, hydrogen bonding, etc. There are three ways for transforming detection of the targeted analytes into the cantilever micromechanical bending's: (1) bending in response to a surface stress, (b) bending in response to a mass loadings, and (c) bending as a consequence of a temperature alteration (Choudhary et al., 2015).

9.3 CARBON-BASED NANOMATERIALS FOR BIOSENSING

Nanomaterial structures are commonly recognized as exciting gadgets with precise chemical and physical features because of their smaller dimension impacts and larger surface which provide the specific and varied characteristic as compared with majority of the substances. The study of such varied features offers the chance to advance biosensor characteristics and enhance the strength of identification through dimension and structure regulation (Pandit et al., 2016). Presently, along with nanotechnology advancements, the investigations about the features of nanomaterials and nanobiosensors are too increasing. The curiosity and significance of nanomaterials in the manufacture of nanobiosensors lie in their unusual features which makes them an optimal choice for detection devices. (Malik et al., 2013). Materials containing carbon show superior characteristics like greater conductance, higher stability, lesser price, wider usable window, and easier surface working (Zhang et al., 2016). Thus, these have been widely considered and used for different electroanalytical usages, especially the upcoming nanoscale carbon material groups of CNTs, graphene, and nano/mesoporous carbon. This nanostructure enables promising revelation of surface groups for the association among analytes and transducing materials, forwarding to notable identification action for ecological contaminants (Ramnani et al., 2016; Brett, 2001). Integration of nanotechnology with biosensors became faster, cleverer, cheap, and easy to use. The transducing mechanism has been notably enhanced with the use nanostructures and nanomaterials (Pandit et al., 2016).

9.3.1 PROPERTIES OF CBNS

9.3.1.1 CARBON NANOTUBES

CNTs initially defined in the year 1991 by Sumio Iijima (S. Iijima, 1991), In carbon nanotubes, each carbon atom having three electrons makes a trigonal coordinated *s* bond to three carbon atoms by using sp2 hybridization (Iijima, 1991; Yang et al., 2019; Kumar et al., 2017). Carbon nanotubes are fundamentally one sheet of graphene trundled in the manner of a hollow tube flawlessly. CNTs have the characteristic feature of rolled graphene sheets piled in the tubular/cylindrical structure with a diameter of several nanometers. The CNTs can have diverse lengths, diameters, layers and chirality vectors (Iijima, 2002; Iijima and Ichihashi, 1993). CNTs generally become the option for the manufacture of biosensor due to their greater power and their notable physicochemical features. CNTs are connected with superior conductance, higher sensitivity, great biocompatibility, and excellent chemical stability (Hu and Hu, 2009). Based upon its chirality, CNTs can have properties of metal, semiconductors, and have vigorous optical absorption in the near-infrared range. 1-D nature, charge transfers in CNTs were discovered to happen by quantum impacts and single blowout along with the tube axis, while causing no scattering. Such exclusive electric and physical features of CNTs make them suitable to be used in biosensors so that some alterations in the electron transfer can signify the incidence of a biotic marvel whose drive is to identify it. The surface of CNTs can be changed with varied chemical groups, or biomolecule can be controlled on the total area of surface, mechanical, uniform distribution in solution, optical absorption, ability of conduct electrical current, and strength abilities to attain enhanced functionality of the electrochemical biosensors. For such motives, CNTs are widely utilized as a surface transformer to plan the electrochemical biosensor (Heydari-Bafrooei and Ensafi, 2019). Furthermore, the end and sidewall of the CNT can be effortlessly altered by assigning any wanted chemical groups. CNTs can be outstanding transducer in the nanoscale sensor because of their noteworthy sensitivity. Through literature, it is described that CNTs can increase the electrochemical reactivity of significant biomolecule, and arbitrate quicker electron transfer kinetics for a wide variety of electroactive types (Wang, 2005). Generally, CNTs can be classified on the basis of structure as into two types: single-walled carbon nanotubes SWCNTs and multiwalled CNTs (MWCNTs). SWCNTs consist of only a sheet of graphene containing *sp*2 hybridized carbon structured into a hexagonal honeycomb networks with diameters around 0.4–3 nm and a length of up to several μm (Balasubramanian and Burghard, 2005). MWCNTs are made of concentrical cylinder of trundled graphene sheets, having variable diameters of several tens' of nanometers relying on the number of graphene sheets, and the distance between sheets and interlayer spacing is nearly 0.34 nm (Ajayan, 1999). SWCNTs can be of semiconducting or metallic nature relying on their chirality and diameters of the tube (Wilder et al., 1998). There are three chiralities for SMCNTs: armchair, zigzag and chiral armchair. SWCNTs are metallic and remaining chirality are semiconducting in nature. MWCNTs consist of multilayers of graphene, and every layer can have a diverse chirality (Gao et al., 2012). SWCNTs show superior chemical stability, greater mechanical power, and a variety of electrical conductance characteristics. MWCNTs show metallic electric features which are parallel to metallic SWCNT (Hu et al., 2010), which in few aspects make them an optimal for the electrochemical application. Optical characteristics shown by MWCNTs are lesser prominent than that of the SWCNTs, so they are utilized as a delivery system for greater biomolecule consisting of plasmid (DNA) into cell. Both these types of CNTs consist of 1-D arrangement and show superior features, such as conductance, adsorptive capability, and higher bioconsistency. Such features allow carbon nanotubes to transport high fluxes with slight heating (Wei et al., 2001). The less width and specific physical characteristics consisting of superior electrical conductivity, ampere density, modulus, and stability of SWCNTs make them suitable option for making electrochemical detecting gadgets (Sarkar and Srinives, 2018).

9.3.1.2 GRAPHENE

Graphene (GR) structure consists of 2-dimensional sheets of carbon atoms in a form of hexagonal structure like a honeycomb arrangement. In the structure of graphene, carbon atoms are sp2 hybridized, and one carbon is bonded with three other carbon atoms at a 120 degree bond angle, all of them are arranged in the same plane. For such reasons, a grid of hexagon is done. Various functional groups, along with hydrogen atoms can be arranged into the free space (Erol et al., 2018). Graphene, as the most popular 2-D nanomaterial, has widespread concern from its discovery in the year 2004. This astonishing nanostructure discloses excellent electric conductance, higher electron transfer rate, mechanical power, and larger surface area, which makes it capable for numerous potential usages in the area of tangible electronics, energy-storing gadgets, detection gadgets, catalyst, membrane, and nanocomposite. Specifically, in detecting instruments, graphene has showcased huge hopes in the aspects of its outstanding conductance, larger surface region, wider electrochemical window, and electrochemical stability (Ping et al., 2011; Yao and Ping, 2018). Graphene is being utilized as a transducing nanomaterial in conductometric, impedimetric, and amperometric detectors for sensing the pathogen and bacteria's (Xu et al., 2017). Graphene can be exfoliated and reacted for the formation of GO to create stable spreading in water because of the existence of the epoxide, hydroxyl, and carboxyl functional group (Chang et al., 2014). After reducing, graphene is altered into rGO or reduced GO with low oxygen content and organizational defects giving higher thermal conductance. GO, rGO or reduced graphene oxide is utilized with superior outcomes in the electrochemical biosensors: graphene showcases certainly a greater surface region and thus it could be effortlessly immobilizing the enzyme by amidic bonds compared with rGO but rGO, being similar to pristine graphene, should showcase, in principles, much greater conducting property than GO (Carbone et al., 2015; Pumera, 2011). Generally, the derivative of graphene-dependent material consists of GO, graphene quantum dots, single and multilayer form of graphene, and rGO. GO has an oxygen reliant functional group, that makes it simple to spread in an aqueous and organic solvent. This is an extreme benefit of graphene that enables it to syndicate with remaining compounds, such as ceramics and polymers. Though, GO is cloistered concerning the electrical conductance. To restore the electrical conductance, it is necessary to recover the hexagonal honeycomb arrangement of graphene again by reducing GO into rGO. rGO will not be handily spread as it has a propensity of combination. Like CNTs, graphene could also be chemically working, refining its features for utilization in various applications (Heydari-Bafrooei and Ensafi, 2019). GO has a higher density of oxygen-carrying functional group, and is not very conductive because of disrupted coupled π–π bonds. Its conductance can be restored by reducing, performing thermally, chemically, or electro chemically, and these materials are labeled as rGO. Though graphene sheets do not contain any oxygen, it can spread up to 30% in GO; yet, the quantity of oxygen can be reduced to about 5–10% in rGO through reduction (Filip et al., 2015). Graphene is being used in a variety of biosensors, and specifically for affinity-dependent biosensors like immunosensors or DNA sensors for detecting higher molecular mass analyte, like DNA or protein (Akhavan et al., 2012). Graphene reliant material with higher surface area and varied workings enables the immobilization of enzymes and antibodies that can intensely increase the electrochemical readout by signal enhancement. As GO or rGO is much less costly when compared with the remaining nanomaterials, then the strategy can result in the cheaper manufacture of ultrasensitive affinity-dependent electrochemical biosensor (Filip et al., 2015).

9.3.1.3 CARBON NANOFIBERS

CNFs or carbon nanofibers are the cylinder-shaped cable type nanostructure in which graphene panes are loaded in dissimilar structures, such as ribbon, herringbone, or platelet. The extent of CNFs differs in the direction of microns and can be up to 10 μm while its widths vary in the range 10–500 nm. Its machine-driven power and electronic features are similar to CNTs. CNFs also have properties like the virtuous electric conduction, great superficial region, biocompatibility, and effortless manufacturing procedure which are important for electrochemical detection application. Furthermore, CNFs can be effortlessly utilized to ensemble a specific sensing process (Endo et al., 2002).

9.3.1.4 FULLERENE

The allotropic alteration of carbon, commonly recognized as fullerene, was revealed in 1985 by H. W. Kroto, R. F. Curl, and R. F. Smalley (Kroto et al., 1985). It is considered as the first nanomaterial to be magnificently separated. The typical characteristic of fullerene is the creation of an amount of atomic Cn groups ($n > 20$) of carbon on a sphere-shaped exterior. Carbon atoms make the covalent bond among themselves in the sp2 hybridized state in fullerene. These carbon atoms are frequently existing on the superficial layer of the sphere shape at the vertexes of hexagons and pentagons. C60 is fullerene which has been extensively researched and examined. It has extremely symmetrical sphere-shaped particles containing 60 C-atoms, existing at the apexes of 20 hexagons and 12 pentagons or 60 carbon atoms including 20-six member rings and 12-five member rings (Paredes et al., 2008). Carbon atoms in fullerene are organized in a sphere-shaped arrangement commonly known as buckyball structure, and hence they do not have any size. Fullerene can create unswervingly from graphene the formation of pentagons on the sides of graphene, that is trailed by the deforming of the graphene plane to a spherical shape, which is called wrapped-up graphene (Afreen et al., 2015). Fullerene has substantial electrical, chemical, and physical features and is thus considered to be a perfect solid for constructing nano assemblages for a number of applications. Fullerene has sphere-shaped assembly that makes it likely to quicken the transfer of control and electron in numerous tenders. Fullerene could be transformed into a multipurpose material that is completed by functionalization with noncovalent and covalent communications. Fullerenes are automatically robust and can withstand great compressions (Heydari-Bafrooei and Ensafi, 2015).

9.3.1.5 CARBON NANO HORNS

CNHs or carbon nanohorns are pointed crates that are alike SWCNTs. They are sp2 hybridized, semiconductive, extremely resilient to oxidation, and with equal response rates. Though, they consist of greater thickness of flaws that enable functionalization, they are typically broader, letting unrestricted drive of captured particles (Russell et al., 2016), and can be produced in higher amounts at room temperature in the nonappearance of possibly deadly metal substances. The application of CNHs has grown at a considerable sluggish rate than CNTs due to their combination into sphere-shaped group, which produces dispersal and superficial alteration problem. A novel method for unravelling groups into separate CNHs was described lately that could hasten its expansion (Karousis et al., 2016).

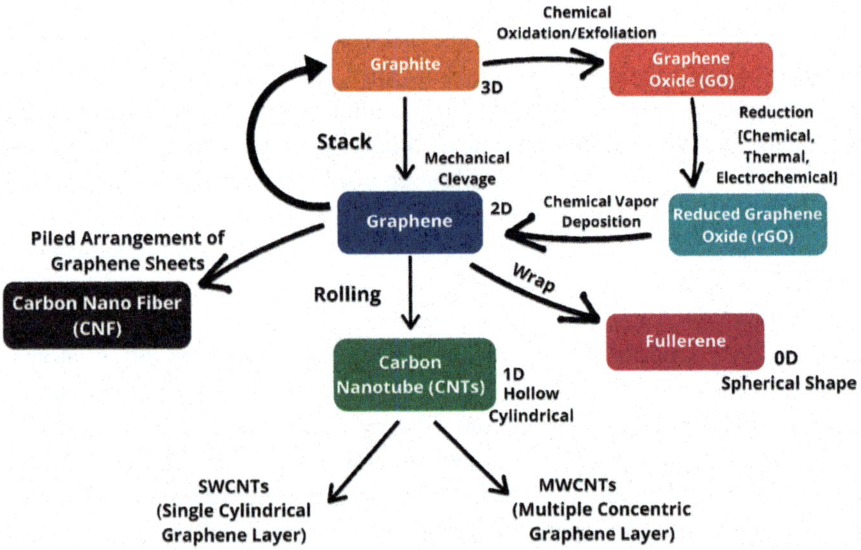

FIGURE 9.4 Various ways of preparation of carbon-based nanomaterials.

9.3.2 SYNTHESIS OF CARBON NANOMATERIALS FOR BIOSENSING

9.3.2.1 CARBON NANOTUBES

The most widely used approaches to the synthesis of CNTs include arc discharge, laser ablation, and chemical vapor deposition (CVD). Primarily SWCNT and MWCNTs are fabricated with arc-discharge approach. SWCNT and MWCNT were obtained by Iijima of the NEC Corporation. Moreover, Since 1985, laser ablation method was used for synthesizing the fullerene as well as CNTs. High power laser beam radiation applied on the graphite sheet in the presence of an Ar or an N_2, evaporated carbon transformed at 1200°C to MWCNT in an Ar or an N_2 atmosphere. Impregnation method is used for the production of SWCNT bundles. Graphite can be impregnated with certain catalysts, such as nickel(Ni) and cobalt (Co) to enhance the property (Xu et al., 2016).

MWCNT is produced with arc-discharge method. In this method, arc is placed between two carbon rods under some conditions whereas suitable metal catalysts, such as Co, Fe, Ni, Pt, and Rh are utilized for the fabrication of SWCNT. CVD is another adaptable and cost-effective approach for the production of CNTs on a diverse scale based on the morphology and crystalline structure. In this method, carbon sources (generally gaseous) are heated under high temperature in the temperature furnace. Carbon sources with high temperature are split into carbon atoms in the presence of transition metal catalysts, such as Co, Ni, or Fe which initiate the development of CNTs. CNT production is very high with the CVD approach and the nanotubes produced are structurally defectivecompared with other approaches. CNT produced with the processes mentioned above is purified through refluxing with some strong acid, sonication-assisted surfactants dispersion in aqueous solution, or by the process of air oxidation (Kroto et al., 1985).

9.3.2.2 GRAPHENE

Various methods were intended for graphene synthesis, such methods are mechanical exfoliation, pyrolysis, chemical reduction of GO, chemical exfoliation, epitaxial growth, CVD approach, and plasma synthesis (Xu et al., 2016). Peel-off or micromechanical exfoliation method depends on the separation p-stacked layers in graphite through repetitive exfoliation with scotch tape. Peel-off method of graphene synthesis is highly superior, though the sample size is very small or is in micron size (Hernandez et al., 2008).

In the chemical reduction approach, graphene is produced with the reduction of GO. In this approach, Hummer's method is used for the oxidation of graphite and to form graphite oxide, and then exfoliated through sonication for the generation of GO flakes (Paredes et al., 2008). Graphite is treated with potassium permanganate, sodium nitrate, and concentrated sulfuric acid for a short time to bring polar oxygen-containing functional groups, such as epoxide, carboxyl, carbonyl, and hydroxyl to introduce hydrophilic nature in the layers (Hummers et al., 2007); To make stable dispersion of single layers and few layers of GO in a various organic solvent or the water, oxygen-containing layers in GO can be mechanically exfoliated with the sonication process. An exfoliated solution of graphene is stabilized in the surface-active agent, then the flakes of GO can be applied on a suitable substrate with spray coating and drop-casting.

Chemical defects and crystalline damages are in greater compromise in the electrical properties of exfoliated GO than pristine graphene. The chemical oxide group and sp3 defects must be reduced to remove such imperfection to restore the electrical properties (Stankovich et al., 2007). GO is reduced electrochemically with the application of reducing potential on GO layers or with reducing agents (e.g., hydrazine), or by using high temperature (Mao et al., 2011; Ramnani et al., 2016). To synthesize pure graphene at a manufacturing level, the frequent method which is used to produce graphene sheets is the method of decomposition of SiC at high temperature (De Heer et al., 2007), or the graphene is epitaxially grown on transition metals, such as Ni, Pd, Ru, Ir, and Cu by CVD of only carbon and hydrogen (Hass et al., 2008; Kim et al., 2009). Graphene can be utilized for the electrode modification and sensing mechanism due to its distinctive characteristics that are rapid conductivity, elevated elasticity, mechanical strength, surface area, and fast heterogeneous transfer rate (Akhgari et al., 2015).

9.3.2.3 CARBON NANOFIBERS

The CNFs can be obtained by laser ablation (Guo et al., 1995) approach. Other methods, such as thermal chemical vapor deposition, plasma-enhanced chemical vapor deposition (PECVD), and electrospinning methods are also used for the production of CNFs (Sharma et al., 2011). In the thermal chemical vapor deposition method, an organic compound is decomposed with a metal catalyst at a stable temperature (Zahid et al., 2018). The CVD fabricates CNFs along with impurities which need an additional complex decontamination process whereas the electrospinning method produces CNFS through a very simple process with high purity (Liu et al., 2008; Tang et al., 2011).

9.3.2.4 FULLERENE

Fullerenes are synthesized simply with the primary laser ablation of graphite, followed by arc-discharge method, and thermal decomposition of an aromatic organic compound. Fullerene was first discovered by Kroto et al. (1985). It was obtained by sublimation of graphite through a pulsed laser beam (Kroto et al., 1985).

9.4 APPLICATION OF CARBON NANOMATERIAL-BASED BIOSENSORS FOR MONITORING ENVIRONMENTAL POLLUTION

The rapidly growing global urbanization and industrialization are directly impacting over environmental safety and eventually human health. The hazardous pollutants, which are challenging the environmental quality by imparting issues of concern related to water and soil pollution include various organic and inorganic toxicants, heavy metals, steroids, hormones, pesticides, industrial chemicals, and infectious microbes, etc. Even these toxicants are found in very small quantities, such as micrograms or nanograms per liter, they have strong potential to damage human health (Salouti et al., 2020). Since commonly found heavy metals contaminant are cadmium, lead, mercury, arsenic, and nickel are nonbiodegradable, carcinogenic, and have the ability of bioaccumulation within the organism. Their small dosage may also cause severe health hazards (Sankhla et al., 2016; Sonone et al., 2020; Steffens et al., 2017). The natural and synthetic hormones toxicity may lead to testicular and prostate cancer, low sperm count, endometriosis, and breast enhancement (Moraes et al., 2015; Salouti et al., 2020). Moreover, pesticide pollutants are carcinogenic, teratogenic, and mutagenic and may cause hematopoietic cancers, immunologic abnormalities, and adverse reproductive, genotoxic, and neurotoxic effects (Eskenazi et al., 1999; Ünal et al., 2011; Gulia et al., 2020; Sankhla et al., 2018; Yadav et al., 2019).

To overcome such severe environmental pollutants, it is important to develop a good environmental quality monitoring and biosensing tool. Carbon nanomaterial-based biosensors can be used for the assessment of the biological and chemical quality by monitoring major biological and chemical environmental pollutants (Salouti et al., 2020). These extremely sensitive, rapid, portable and low-cost biosensors target and detect the significant pollutants pathogens, pesticides, explosives, hormones, antibiotics, noxious gases, and heavy metal ions. The better utilization of these biosensors is beneficial and they are considered to be excellent tools to improve the environmental quality (Ramnani et al., 2016; Lin et al., 2019; Salouti et al., 2020). To increase the sensitivity of the biosensing platform, it is modified with highly specific recognition elements, such as a molecularly imprinted polymer (MIP), biological entities including different antibodies and aptamers (Torrinha et al., 2020).

9.4.1 SENSING OF CHEMICAL CONTAMINANTS

9.4.1.1 SMALL ORGANIC MOLECULES

The mostly occurring organic pollutants of concern include natural as well as synthetic organic matter as humus substances, nitroaromatic explosives, dyes, toxins, pharmaceuticals, pesticides, personal care, and domestic waste compounds and their sensing is very important for maintaining the environmental quality (Smith and Rodrigues, 2015). Conventional detection methods are not able to accurately sense due to the minute size of such pollutants. But these compounds usually include some electroactive functional group which can be oxidized or reduced with the application of potential and this feature is broadly used to sense the pollutants using carbon-based electrochemical nanobiosensors (Ramnani et al., 2016).

9.4.1.1.1 Nitro-Based Explosives

Advancement in the development of nanomaterials shows strong potential to create electrochemical sensors for detecting explosives due to very high surface area-to-volume ratio, convergent rather than linear diffusion, enhanced selectivity, catalytic activity, and distinctive electrical and optical properties that can be exploited for highly sensitive molecular adsorption detection (O'Mahony and Wang, 2013). Since 2,4,6-trinitrotoluene (TNT) is a best known explosive material and is also used for the synthesis of various dyes, pesticides, and plasticizers. The occurrence of its residues in the soil as well as in water sources as persistent pollutants is a matter of environmental concern. The highly sensitive, selective, and rapid sensing of small molecules of TNT can be achieved by using portable SWCNTs-based chemiresistive affinity biosensor (Pennington and Brannon, 2002; Park et al., 2010). Graphene containing sensing device constructed by electrochemical reduction of GO onto a glassy carbon electrode (GCE) is employed for the detection of residues of TNT and other nitroaromatic explosives, such as dinitrotoluene (DNT), trinitrobenzene (TNB), and dinitrobenzene (DNB) (Chen et al., 2011). The CNTs and graphene nanoribbons are considered as potential biosensors for tracing of nitro-based explosives in seawater (O'Mahony and Wang, 2013).

9.4.1.1.2 Pharmaceutical Pollutants

The growing medicinal needs and eventually increasing pharmaceutical production and consumption has become a global issue of environmental concern. The pharmaceutical pollutants may include various chemical products, such as antibiotics, hormones, β-blockers, anti-inflammatories, anticonvulsants, antacids, antidiabetics, analgesics, CNS agents, which are highly contaminating the ecosystems, especially water systems and causing severe impacts on human health. As antibiotics, hormones, and anti-inflammatories are highly consuming pharmaceuticals and consequently major occurring environmental pollutants, they were mostly studied for environmental biosensing (Torrinha et al., 2020). Here also, carbon nanoparticle-based biosensors are useful in the detection of these pollutants because of their extra properties such as high chemical toxicity, low cost, and commercial availability (Torrinha et al., 2020).

9.4.1.1.2.1 Antibiotics

Antibiotics are one of those types of pharmaceutical pollutants which are released not only from industrial or hospital waste but also through domestic waste and commercial livestock farming (Ramnani et al., 2015). Bueno et al. showed individual amperometric detection of six types of antibiotic residues belonging to a class of sulfonamide antibiotics or sulfa drugs using GCE/MWCNT-Nafion® sensor (Bueno et al., 2013). While recently, Silva and Cesarino developed an advanced electrochemical biosensor, that is, GCE/rGO–AuNPs electrode for selective detection of sulfamethazine antibiotic from polluted water, even in the occurrence of additional potential organic interferences. This nanocomposite was prepared by immobilizing gold nanoparticles (AuNPs) and rGO on GCE and then employed for biosensing under differential pulse voltammetry (DPV) optimized conditions (Silva and Cesarino, 2019). Tetracycline is also one of the most consumed classes of antibiotics. Lorenzetti et al. employed disposable and portable biosensor, that is, SPE/rGO with adsorptive stripping DPV (AdS-DPV) technique of in situ detection for sensing three types of natural tetracyclines (tetracycline, chlortetracycline, and oxytetracycline) and one semisynthetic tetracycline, that is, doxycycline. But this method has limited applicability due to short linear range and relatively low interelectrode precision. The tetracycline antibiotic can be analyzed by basically graphene-based (GCE/graphene/L-cysteine, MWCNT-based (GCE/MWCNT-Nafion®), and CB-based (GCE/CB-PS) nanobiosensors (Vega et al., 2007; Sun et al., 2017; Delgado et al., 2018).

9.4.1.1.2.2 *Endocrine-Disrupting Chemicals*

Endocrine-disrupting chemicals (EDCs) are synthetic hormones, such as bisphenol-A (BPA), 17α-ethinylestradiol, and diethylstilbestrol, which causes adverse impacts over human as well as animal health mainly by affecting reproductive characteristics (Diamanti-Kandarakis et al., 2009; Vandenberg et al., 2012). Numerous studies have been conducted for sensitive detection of 17β-estradiol under DPV optimized conditions using graphene-based electrochemical sensors, such as GCE/rGO-MIP-Fe_3O_4, GCE/rGO-CuTthP, GCE/exfoliated graphene as well as with biochar-based nanobiosensors, such as GCE/biocharNPs (Li et al., 2015; Moraes et al., 2015; Hu et al., 2015; Dong et al., 2018). Recently, in 2019, Duan et al. proposed potential 17β-estradiol sensing mechanism GCE/MOF(Al)–CNT/PB/MIP(PPy) using CNT-based biosensor with the highest sensitivity (21,000 μA/μmolL). This advanced mechanism was modified as the double sensitization material with MIP for highly sensitive recognition and electrodeposition of a Prussian blue layer and polypyrrole (PPy) for better imprinting effect and rapid sensing (Duan et al., 2019). Several studies have been carried out for the potential detection of another disrupting hormone, that is, diethylstilbestrol using graphene quantum dots (GQD) of modified screen-printed electrodes as a nanobiosensing platform, that is, SPE/GQD, GCE modified with MWCNT, gold nanoparticles, and cobalt phthalocyanine film electrode as a nanobiosensing platform, that is, GCE/AuNPs/MWCNT-CoPc, and carbon black paste electrode as nanobiosensing platform, that is, CPE(CB) (Qu et al., 2012; Aragão et al., 2017; Gevaerd et al., 2019).

9.4.1.2 *HEAVY METALS*

Heavy metal pollutants are mainly released from metallurgical industries and are hazardous to human health as they are highly persistent in nature and bioaccumulative in living organisms. These heavy metals are mercury (Hg), copper (Cu), lead (Pb), nickel (Ni), cadmium (Cd), and zinc (Zn). Thus, their potential, rapid, and on-site sensing and real-time monitoring is a vital need to maintain environmental quality as well as individual well-being (Yilmaz et al., 2007; Ramnani et al., 2015). The detection of these elements is also possible with regular spectroscopic techniques, such as AAS, inductively coupled plasma mass spectroscopy (ICP-MS), etc. But these techniques are not suitable for real-time and on-site detection, as well as that are expensive and relatively time-consuming also. Alternatively, enzyme-based detection methods can also be employed for sensitive sensing of metals, which is achieved by the inhibition of enzyme due to their binding with metal ions. But this method also has some limitations such as poor stability of enzymes at physiological pH and reduced selectivity due to other interference. Therefore, for potential sensing of heavy metals, the sensing tool must overcome such limitations (Burlingame et al., 1996; Wanekaya et al., 2008).

Several studies have been conducted to develop a suitable carbon-based nanobiosensing tool for highly sensitive, selectivity, real-time and anti-interference detection of heavy metals (Ramnani et al., 2015). The label-free and sensitive detection of Hg (II) can be conducted by using a chemiresistor based on SWNTs within the linear range 100 nM–1 μM (Gong et al., 2013). Li et al. (2009) proposed an electrochemical sensor GCE modified with graphene for the detection of Cd (II). Morton et al. (2009) conducted another study to develop the sensor GCE modified with carboxylated CNT for the detection of Cu (II) with cyclic voltammetry (CV). The sensing of Arsenic (III) was also performed on GCE modified with DNA-functionalized, single-walled carbon nanotube (DNA–SWNT) hybrid using linear sweep voltammetry (LSV) as a detection technique (Liu and Wei, 2008). The electrochemical sensor with graphene-modified GCE, DNA-wrapped metallic SWNTs, carboxylated CNT-modified GCE as potential biosensors for Pb (II) (Li et al., 2009; Morton et al., 2009; Lian et al., 2014).

9.4.1.3 PESTICIDES

Pesticides used for agricultural purposes are increasingly polluting the surrounding ecosystems and the consumption may cause hazardous effects on human health. Carbon-nanoparticle-based biosensors can be successfully employed for their detection and simultaneously environmental monitoring, because conventional chromatographic or spectrometric methods are not suitable for rapid, low cost and real-time monitoring. Carbamate (CBMs) and organophosphorus (OPs) pesticides are extensively used due to their desired and extensive applications. But both these pesticides are highly neurotoxic and may cause cholinergic dysfunction, neurological, and motor level complications (da Silva et al., 2018; Gulia et al., 2020).

Carbaryl is the widely used pesticides from the CBMs class. The graphite-based nanobiosensor, poly-p phenylenediamine/ionic liquid/CPE having linear range within 0.5–200 μM and limit of detection (LOD) 0.09 μM is developed for the detection of carbaryl residues in water (Salih et al., 2018). Several phenyl-CBMs, such as carbaryl, carbofuran, and isoprocarb can be detected in water ecosystems using single carbon black nanoparticles-based (CBNs) biosensing tools (Della Pelle et al., 2016). Another pesticide of the same class, carbendazim can be sensitively sensed in river water, rice field water, and soil using graphene-modified CPE having linear range 2.61–52.3 μM and LOD of 0.58 μM (Noyrod et al., 2014). Moreover, many studies have been conducted to develop graphene-, CNT-, CB-, GCE- and boron-doped diamond (BDD)-based potential environmental biosensors for the detection of CBM residues present in freshwater, sea water, paddy water, field water, and soil (Oliveira et al., 2020).

Both OPs and CBMs have neurotoxic action and they inhibit the catalytic activity of acetylcholinesterase (AChE) which prevents the breakdown of the transmitter choline and consequently blocks the nerve transmission. Therefore, the characteristics of AChE inhibition have been widely used in developing electrochemical-enzymatic biosensors for sensitive, more selective, and noninterference sensing of CBM and OP pesticides. An amperometric biosensor based on the modification of the graphene surface by epoxy resins and immobilized AChE (graphite epoxy composite/AChE) having range within 6–20 μM and LOD of 0.001 μM can be employed for the detection of CBMs and OPs in water environments (Montes et al., 2018).

9.4.1.4 NOXIOUS GASES

The increasing industrialization leads to a release of toxic gases, such as NOx, NH_3 and CO in the air. As an electronic property of graphene is strongly influenced by the adsorption of such noxious gas molecules, graphene and grapheme-derivative-based biosensors are considered as most promising gas sensing tools for environmental biomonitoring. The hybrid structure formed by blending of nanoparticles of metal oxides with graphene or its derivatives have been investigated, which showed a more selectivity and sensitivity with high gas sensing potential (Chatterjee et al., 2015).

9.4.2 SENSING OF BIOLOGICAL CONTAMINANTS

Various pathogenic bacteria and viruses are increasingly polluting the environment and these pathogenic microbes are causing major infectious diseases, which may lead to fatal effects on human health. Therefore, the need for the development of potential environmental monitoring tools for rapid and cost-effective sensing microbial pathogens is also increasing. The commonly used detection techniques, such as enzyme-linked immunosorbent assay (ELISA) and polymerase chain reaction (PCR) are not suitable for real-time and on-site detection because such techniques require sample pretreatment and long experimental procedures. Due to high sensitivity, rapidness and selectivity of carbon nanomaterial-based electrochemical biosensors can detect a broad spectrum of microbial outbreaks in drinking water ecosystems and surrounding environment with easy and cost-effective manner (Rodríguez-Lázaro et al., 2005; Ramnani et al., 2015).

9.4.2.1 PATHOGENIC VIRUS

Viruses are the smallest intracellular pathogens or infectious agents which affect the human body with various severe diseases. The carbon nanoparticles, such as CNT, QD, and GO can be employed to detect these pathogenic viruses present in the environment. Carbon nanomaterial-based aptamer and DNA sensors are mostly applied for sensing viruses (Mokhtarzadeh et al., 2017). Wang et al. (2013)developed a sensor, that is, ssDNA-SWCNTs/AuNPs for sensing hepatitis B and papilloma virus, where the DNAs of these viruses were captured by probe ssDNA and immobilized on CNT-based platform coated with gold nanoparticles. A target-specific and sensitive rGO-based aptamer sensor developed to detect HIV (human immunodeficiency virus) (Bi et al., 2012; Kim et al., 2014). Furthermore, various carbon nanomaterials have been developed to detect commonly found viruses in the environment, such as influenza virus, hepatitis B virus, rotavirus, etc. (Mokhtarzadeh et al., 2017).

9.4.2.2 PATHOGENIC BACTERIA

The potential and immediate biosensing of bacteria can be achieved with carbon-based nanomaterials-modified apt sensors. The selective and sensitive sensing of very commonly found bacteria, that is, *Escherichia coli* can be achieved with an aptamer-functionalized SWNT-based field-effect transistor (So et al., 2008). Graphene-containing biosensor can sense the metabolic activities of live bacterial cells in real-time, which has the potential to differentiate live and dead bacterial cells (Huang et al., 2011). Similarly, carbon nanotube-based potentiometric apt sensor can be used for immediate recognition of *Salmonella typhi* from an intricate sample (Zelada-Guillén et al., 2009). Hernández et al. reported a simply constructed rGO and GO-modified aptameric sensors for sensing challenging microbe *Staphylococcus aureus* with ultra-low detection limits (Hernández et al., 2014).

9.5 CONCLUSION

Excess population, rapid urbanization, and wrongful usage of technologies have increased Environmental pollution. There is an urgent need to use methods that detect such pollutants which affect the environment, humans, and overcome the load of conventional analytical techniques. Biosensors development play a vital role in the recognition of contaminants with few techniques, while the inclusion of nanostructures with biosensor enhances their sensitivity, reproducibility, and speed of operation. Nanomaterial applications are being increasingly used to design the novel biosensors. Nanomaterials have excellent properties, which improve the signal conversion and mechanism of biosensors. Carbon nanomaterials, such as CNTs, graphene, GOs, rGOs, may improve the sensitivity and accuracy of biosensors. Excellent physical and electric properties of CNMs improves the functionality of biosensors. These carbon-based nanomaterials nanobiosensors are widely used because of their low cost, compact size, robustness, and portability. Integration of carbon-based nanomaterials in biosensors increased the functionality of sensing devices and can easily monitor and detect various environmental pollutants. Sustaining life on earth will be the responsibility of everyone. Carbon-based nanomaterials nanobiosensors will open up a whole new dimension for the detection of environmental pollution.

KEYWORDS

- **nanomaterials**
- **biosensors**
- **nanotechnology**
- **pollutants**
- **environment**
- **pollution**
- **carbon**

REFERENCES

Abdel-Karim, R.; Reda, Y.; Abdel-Fattah, A. Nanostructured Materials-Based Nanosensors. *J. Electrochem. Soc.* **2020,** *167* (3), 037554.

Afreen, S.; Muthoosamy, K.; Manickam, S.; Hashim, U. Functionalized Fullerene (C60) as a Potential Nanomediator in the Fabrication of Highly Sensitive Biosensors. *Biosens. Bioelectron.* **2015,** *63*, 354–364. https://doi.org/10.1016/j.bios.2014.07.044

Ajayan, P. M. Nanotubes from Carbon. *Chem. Rev.* **1999,** *99* (7), 1787–1800.

Akhavan, O.; Ghaderi, E.; Rahighi, R. Toward Single-DNA Electrochemical Biosensing by Graphene Nanowalls. *ACS Nano* **2012,** *6* (4), 2904–2916.

Akhgari, F.; Fattahi, H.; Oskoei, Y. M. Recent Advances in Nanomaterial-Based Sensors for Detection of Trace Nitroaromatic Explosives. *Sens. Actuat. B Chem.* **2015,** *221*, 867–878. https://doi.org/10.1016/j.snb.2015.06.146

Aragão, J. S.; Ribeiro, F. W.; Portela, R. R.; Santos, V. N.; Sousa, C. P.; Becker, H.; de Lima-Neto, P. Electrochemical Determination Diethylstilbestrol by a Multi-Walled Carbon Nanotube/Cobalt Phthalocyanine Film Electrode. *Sens. Actuat. B Chem.* **2017,** *239*, 933–942. https://doi.org/10.1016/j.snb.2016.08.097

Auroux, P. A.; Iossifidis, D.; Reyes, D. R.; Manz, A. Micro Total Analysis Systems. 2. Analytical Standard Operations and Applications. *Analy. Chem.* **2002,** *74* (12), 2637–2652.

Balasubramanian, K.; Burghard, M. Biosensors Based on Carbon Nanotubes. *Analy. Bioanaly. Chem.* **2006,** *385* (3), 452–468. https://doi.org/10.1007/s00216-006-0314-8

Balasubramanian, K.; Burghard, M. Chemically Functionalized Carbon Nanotubes. *Small* **2005,** *1* (2), 180–192. https://doi.org/10.1002/smll.200400118

Bi, S.; Zhao, T.; Luo, B. A Graphene Oxide Platform for the Assay of Biomolecules Based on Chemiluminescence Resonance Energy Transfer. *Chem. Commun.* **2012,** *48* (1), 106–108. https://doi.org/10.1039/c1cc15443e

Brett, C. M. Electrochemical Sensors for Environmental Monitoring. Strategy and Examples. *Pure Appl. Chem.* **2001,** *73* (12), 1969–1977. https://doi.org/10.1351/pac200173121969

Bueno, A. M.; Contento, A. M.; Ríos, Á. Validation of a Screening Method for the Rapid Control of Sulfonamide Residues Based on Electrochemical Detection Using Multiwalled Carbon Nanotubes-Glassy Carbon Electrodes. *Analy. Methods* **2013,** *5* (23), 6821–6829. https://doi.org/10.1039/c3ay41437j

Burlingame, A. L.; Boyd, R. K.; Gaskell, S. J. Mass Spectrometry. *Analy. Chem. Am. Chem. Soc.* **1996,** *68* (12). https://doi.org/10.1021/a1960021u

Cai, H.; Xu, C.; He, P.; Fang, Y. Colloid Au-enhanced DNA Immobilization for the Electrochemical Detection of Sequence-Specific DNA. *J. Electroanaly. Chem.* **2001,** *510* (1–2), 78–85. https://doi.org/10.1016/S0022-0728(01)00548-4

Carbone, M.; Gorton, L.; Antiochia, R. An Overview of the Latest Graphene-Based Sensors for Glucose Detection: The Effects of Graphene Defects. *Electroanalysis* **2015,** *27* (1), 16–31. https://doi.org/10.1002/elan.201400409

Chang, J.; Zhou, G.; Christensen, E. R.; Heideman, R.; Chen, J. Graphene-Based Sensors for Detection of Heavy Metals in Water: A Review. *Analy. Bioanaly. Chem.* **2014,** *406* (16), 3957–3975. https://doi.org/10.1007/s00216-014-7804-x

Chatterjee, S. G.; Chatterjee, S.; Ray, A. K.; Chakraborty, A. K. Graphene–Metal Oxide Nanohybrids for Toxic Gas Sensor: A Review. *Sens. Actuat. B Chem.* **2015,** *221*, 1170–1181. https://doi.org/10.1016/j.snb.2015.07.070

Chemla, Y. R.; Grossman, H. L.; Poon, Y.; McDermott, R.; Stevens, R.; Alper, M. D.; Clarke, J. Ultrasensitive Magnetic Biosensor for Homogeneous Immunoassay. *Proc. Natl. Acad. Sci.* **2000,** *97* (26), 14268–14272. https://doi.org/10.1073/pnas.97.26.14268

Chen, T. W.; Sheng, Z. H.; Wang, K.; Wang, F. B.; Xia, X. H. Determination of Explosives Using Electrochemically Reduced Graphene. *Chem.–Asian J.* **2011,** *6* (5), 1210–1216. https://doi.org/10.1002/asia.201000836

Choudhary, M. K.; Singh, M.; Saharan, V. Application of Nanobiosensors in Agriculture. *Popular Kheti* **2015,** *3*, 130–135.

Cui, Y.; Lieber, C. M. Functional Nanoscale Electronic Devices Assembled Using Silicon Nanowire Building Blocks. *Science* **2001,** *291* (5505), 851–853. https://doi.org/10.1126/science.291.5505.851

da Silva, M. K.; Vanzela, H. C.; Defavari, L. M.; Cesarino, I. Determination of Carbamate Pesticide in Food Using a Biosensor Based on Reduced Graphene Oxide and Acetylcholinesterase Enzyme. *Sens. Actuat. B Chem.* **2018,** *277*, 555–561. https://doi.org/10.1016/j.snb.2018.09.051

Davis, J. J.; Coleman, K. S.; Azamian, B. R.; Bagshaw, C. B.; Green, M. L. Chemical and Biochemical Sensing with Modified Single Walled Carbon Nanotubes. *Chem. Eur. J.* **2003,** *9* (16), 3732–3739. https://doi.org/10.1002/chem.200304872

De Heer, W. A.; Berger, C.; Wu, X.; First, P. N.; Conrad, E. H.; Li, X.; Potemski, M. Epitaxial Graphene. *Solid State Commun.* **2007,** *143* (1–2), 92–100, 154.

Delgado, K. P.; Raymundo-Pereira, P. A.; Campos, A. M.; Oliveira Jr, O. N.; Janegitz, B. C. Ultralow Cost Electrochemical Sensor Made of Potato Starch and Carbon Black Nanoballs to Detect Tetracycline in Waters and Milk. *Electroanalysis* **2018,** *30* (9), 2153–2159. https://doi.org/10.1002/elan.201800294

Della Pelle, F.; Del Carlo, M.; Sergi, M.; Compagnone, D.; Escarpa, A. Press-Transferred Carbon Black Nanoparticles on Board of Microfluidic Chips for Rapid and Sensitive Amperometric Determination of Phenyl Carbamate Pesticides in Environmental Samples. *Microchim. Acta* **2016,** *183* (12), 3143–3149. https://doi.org/10.1007/s00604-016-1964-7

Diamanti-Kandarakis, E.; Bourguignon, J. P.; Giudice, L. C.; Hauser, R.; Prins, G. S.; Soto, A. M.; Gore, A. C. Endocrine-Disrupting Chemicals: an Endocrine Society Scientific Statement. *Endocr. Rev.* **2009,** *30* (4), 293–342. https://doi.org/10.1210/er.2009-0002

Dong, X.; He, L.; Liu, Y.; Piao, Y. Preparation of Highly Conductive Biochar Nanoparticles for Rapid and Sensitive Detection of 17β-Estradiol in Water. *Electrochim. Acta* **2018,** *292*, 55–62. https://doi.org/10.1016/j.electacta.2018.09.129

Duan, D.; Si, X.; Ding, Y.; Li, L.; Ma, G.; Zhang, L.; Jian, B. A Novel Molecularly Imprinted Electrochemical Sensor Based on Double Sensitization by MOF/CNTs and Prussian Blue for Detection of 17β-Estradiol. *Bioelectrochemistry* **2019,** *129*, 211–217. https://doi.org/10.1016/j.bioelechem.2019.04.014

Endo, M.; Kim, Y. A.; Hayashi, T.; Fukai, Y.; Oshida, K.; Terrones, M.; Dresselhaus, M. S. Structural Characterization of Cup-Stacked-Type Nanofibers with an Entirely Hollow Core. *Appl. Phys. Lett.* **2002,** *80* (7), 1267–1269. https://doi.org/10.1063/1.1450264

Erol, O.; Uyan, I.; Hatip, M.; Yilmaz, C.; Tekinay, A. B.; Guler, M. O. Recent Advances in Bioactive 1D and 2D Carbon Nanomaterials for Biomedical Applications. *Nanomed. Nanotechnol. Biol. Med.* **2018,** *14* (7), 2433–2454. https://doi.org/10.1016/j.nano.2017.03.021

Eskenazi, B.; Bradman, A.; Castorina, R. Exposures of Children to Organophosphate Pesticides and Their Potential Adverse Health Effects. *Environ. Health Perspect.* **1999,** *107* (Suppl 3), 409–419. https://doi.org/10.1289/ehp.99107s3409

Filip, J.; Kasák, P.; Tkac, J. Graphene as Signal Amplifier for Preparation of Ultrasensitive Electrochemical Biosensors. *Chem. Papers* **2015,** *69* (1), 112–133. https://doi.org/10.1515/chempap-2015-0051

Fu, X.; Chen, L.; Choo, J. Optical Nanoprobes for Ultrasensitive Immunoassay. *Analy. Chem.*, **2017,** *89* (1), 124–137. https://doi.org/10.1021/acs.analchem.6b02251

Gao, C.; Guo, Z.; Liu, J. H.; Huang, X. J. The New Age of Carbon Nanotubes: An Updated Review of Functionalized Carbon Nanotubes in Electrochemical Sensors. *Nanoscale* **2012,** *4* (6), 1948–1963. https://doi.org/10.1039/C2NR11757F

Gevaerd, A.; Banks, C. E.; Bergamini, M. F.; Marcolino-Junior, L. H. Graphene Quantum Dots Modified Screen-printed Electrodes as Electroanalytical Sensing Platform for Diethylstilbestrol. *Electroanalysis* **2019,** *31* (5), 838–843. https://doi.org/10.1002/elan.201800838

Gong, J. L.; Sarkar, T.; Badhulika, S.; Mulchandani, A. Label-Free Chemiresistive Biosensor for Mercury (II) Based on Single-Walled Carbon Nanotubes and Structure-Switching DNA. *Appl. Phys. Lett.* **2013,** *102* (1), 013701. https://doi.org/10.1063/1.4773569

Goyal, R. N.; Chatterjee, S.; Rana, A. R. S. The Effect of Modifying an Edge-Plane Pyrolytic Graphite Electrode with Single-Wall Carbon Nanotubes on Its Use for Sensing Diclofenac. *Carbon*, **2010,** *48* (14), 4136–4144. https://doi.org/10.1016/j.carbon.2010.07.024

Guilbault, G. G.; Pravda, M.; Kreuzer, M.; O'sullivan, C. K. Biosensors—42 Years and Counting. *Analy. Lett.* **2004,** *37* (8), 1481–1496. https://doi.org/10.1081/AL-120037582

Gulia, S.; Rohilla, R. K.; Sankhla, M. S.; Kumar, R.; Sonone, S. S.; Impact of Pesticide Toxicity in Aquatic Environment. *Biointerf. Res. Appl. Chem.* **2020,** *11* (3), 10131–10140. https://dx.doi.org/10.33263/BRIAC113.1013110140

Guo, T.; Nikolaev, P.; Rinzler, A. G.; Tomanek, D.; Colbert, D. T.; Smalley, R. E. Self-Assembly of Tubular Fullerenes. *J. Phys. Chem.* **1995,** *99* (27), 10694–10697. https://doi.org/10.1021/j100027a002

Hass, J.; De Heer, W. A.; Conrad, E. H. The Growth and Morphology of Epitaxial Multilayer Graphene. *J. Phys. Condens. Matter* **2008,** *20* (32), 323202. https://doi.org/10.1088/0953-8984/20/32/323202

Hernández, R.; Vallés, C.; Benito, A. M.; Maser, W. K.; Rius, F. X.; Riu, J. Graphene-Based Potentiometric Biosensor for the Immediate Detection of Living Bacteria. *Biosens. Bioelectron.* **2014,** *54*, 553–557. https://doi.org/10.1016/j.bios.2013.11.053

Hernandez, Y.; Nicolosi, V.; Lotya, M.; Blighe, F. M.; Sun, Z.; De, S.; Boland, J. J. High-Yield Production of Graphene by Liquid-Phase Exfoliation of Graphite. *Nat. Nanotechnol.*, **2008,** *3* (9), 563–568. https://doi.org/10.1038/nnano.2008.215

Heydari-Bafrooei, E.; Ensafi, A. A. Typically Used Carbon-Based Nanomaterials in the Fabrication of Biosensors. In *Electrochemical Biosensors*; Elsevier, 2019; pp 77–98. https://doi.org/10.1016/B978-0-12-816491-4.00004-8

Hrapovic, S.; Liu, Y.; Male, K. B.; Luong, J. H. Electrochemical Biosensing Platforms Using Platinum Nanoparticles and Carbon Nanotubes. *Analy. Chem.* **2004,** *76*(4), 1083–1088. https://doi.org/10.1021/ac035143t

Hu, C.; Hu, S. Carbon Nanotube-Based Electrochemical Sensors: Principles and Applications in Biomedical Systems. *J. Sens.* **2009**. https://doi.org/10.1155/2009/187615

Hu, L.; Cheng, Q.; Chen, D.; Ma, M.; Wu, K. Liquid-Phase Exfoliated Graphene as Highly-Sensitive Sensor for Simultaneous Determination of Endocrine Disruptors: Diethylstilbestrol and Estradiol. *J. Hazardous Mater.* **2015,** *283*, 157–163. https://doi.org/10.1016/j.jhazmat.2014.08.067

Hu, P.; Zhang, J.; Li, L. E.; Wang, Z.; O'Neill, W.; Estrela, P. Carbon Nanostructure-Based Field-Effect Transistors for Label-Free Chemical/Biological Sensors. *Sensors* **2010,** *10* (5), 5133–5159. https://doi.org/10.3390/s100505133

Huang, Y.; Dong, X.; Liu, Y.; Li, L. J.; Chen, P. Graphene-Based Biosensors for Detection of Bacteria and Their Metabolic Activities. *J. Mater. Chem.* **2011,** *21* (33), 12358–12362. https://doi.org/10.1039/c1jm11436k

Hummers Jr, W. S.; Offeman, R. E. Preparation of Graphitic Oxide. *J. Am. Chem. Soc.* **1958,** *80* (6), 1339–1339.

Iijima, S. Carbon Nanotubes: Past, Present, and Future. *Phys. B Condens. Matter* **2002,** *323* (1–4), 1–5. https://doi.org/10.1016/S0921-4526(02)00869-4

Iijima, S. Helical Microtubules of Graphitic Carbon. *Nature* **1991,** *354* (6348), 56–58. https://doi.org/10.1038/354056a0

Iijima, S.; Ichihashi, T. Single-Shell Carbon Nanotubes of 1-nm Diameter. *Nature* **1993,** *363* (6430), 603–605. https://doi.org/10.1038/363603a0

Jianrong, C.; Yuqing, M.; Nongyue, H.; Xiaohua, W.; Sijiao, L. Nanotechnology and Biosensors. *Biotechnol. Adv.* **2004,** *22* (7), 505–518. https://doi.org/10.1016/j.biotechadv.2004.03.004

Kabariya, J. H.; Ramani, V. M. Nanobiosensors, as a Next-Generation Diagnostic Device for Quality & Safety of Food and Dairy Product. In *Nanotechnology*; Springer: Singapore, 2017; pp 115–129. https://doi.org/10.1007/978-981-10-4678-0_7

Kanjana, D. Advancement of Nanotechnology Applications on Plant Nutrients Management and Soil Improvement. In *Nanotechnology*; Springer: Singapore, 2017; pp 209–234. https://doi.org/10.1007/978-981-10-4678-0_12

Karousis, N.; Suarez-Martinez, I.; Ewels, C. P.; Tagmatarchis, N. Structure, Properties, Functionalization, and Applications of Carbon Nanohorns. *Chem. Rev.* **2016,** *116* (8), 4850–4883. https://doi.org/10.1021/acs.chemrev.5b00611

Kim, K. S.; Zhao, Y.; Jang, H.; Lee, S. Y.; Kim, J. M.; Kim, K. S.; Hong, B. H. Large-Scale Pattern Growth of Graphene Films for Stretchable Transparent Electrodes. *Nature* **2009,** *457* (7230), 706–710. https://doi.org/10.1038/nature07719

Kim, M. G.; Shon, Y.; Lee, J.; Byun, Y.; Choi, B. S.; Kim, Y. B.; Oh, Y. K. Double Stranded Aptamer-Anchored Reduced Graphene Oxide as Target-Specific Nano Detector. *Biomaterials* **2014,** *35* (9), 2999–3004. https://doi.org/10.1016/j.biomaterials.2013.12.058

Kissinger, P. T. Biosensors—a Perspective. *Biosens. Bioelectron.* **2005,** *20* (12), 2512–2516. https://doi.org/10.1016/j.bios.2004.10.004

Koedrith, P.; Thasiphu, T.; Weon, J. I.; Boonprasert, R.; Tuitemwong, K.; Tuitemwong, P. Recent Trends in Rapid Environmental Monitoring of Pathogens and Toxicants: Potential of Nanoparticle-Based Biosensor and Applications. *Sci. World J.* **2015**. https://doi.org/10.1155/2015/510982

Kroto, H. W.; Heath, J. R.; O'Brien, S. C.; Curl, R. F.; Smalley, R. E. C60: Buckminsterfullerene. *Nature* **1985,** *318* (6042), 162–163.

Kumar, S.; Rani, R.; Dilbaghi, N.; Tankeshwar, K.; Kim, K. H. Carbon Nanotubes: A Novel Material for Multifaceted Applications in Human Healthcare. *Chem. Soc. Rev.* **2017,** *46* (1), 158–196. https://doi.org/10.1039/C6CS00517A

Kwak, S. Y.; Wong, M. H.; Lew, T. T. S.; Bisker, G.; Lee, M. A.; Kaplan, A.; Hamann, C. Nanosensor Technology Applied to Living Plant Systems. *Annu. Rev. Analy. Chem.* **2017,** *10*, 113–140.

Li, J.; Guo, S.; Zhai, Y.; Wang, E. High-Sensitivity Determination of Lead and Cadmium Based on the Nafion-Graphene Composite Film. *Analy. Chim. Acta* **2009,** *649* (2), 196–201. https://doi.org/10.1016/j.aca.2009.07.030

Li, Y. T.; Qu, L. L.; Li, D. W.; Song, Q. X.; Fathi, F.; Long, Y. T. Rapid and Sensitive in-Situ Detection of Polar Antibiotics in Water Using a Disposable Ag–Graphene Sensor Based on Electrophoretic Preconcentration and Surface-Enhanced Raman Spectroscopy. *Biosens. Bioelectron.*, **2013,** *43*, 94–100. https://doi.org/10.1016/j.bios.2012.12.005

Li, Y.; Zhao, X.; Li, P.; Huang, Y.; Wang, J.; Zhang, J. Highly Sensitive Fe_3O_4 Nanobeads/Graphene-Based Molecularly Imprinted Electrochemical Sensor for 17β-Estradiol in Water. *Analy. Chim. Acta* **2015,** *884*, 106–113. https://doi.org/10.1016/j.aca.2015.05.022

Lian, Y.; Yuan, M.; Zhao, H. DNA Wrapped Metallic Single-Walled Carbon Nanotube Sensor for Pb (II) Detection. *Fullerenes Nanotubes Carbon Nanostruct.* **2014,** *22* (5), 510–518. https://doi.org/10.1080/1536383X.2012.690462

Lin, Z.; Wu, G.; Zhao, L.; Lai, K. W. C. Carbon Nanomaterial-Based Biosensors: A Review of Design and Applications. *IEEE Nanotechnol. Magaz.* **2019,** *13* (5), 4–14. https://doi.org/10.1109/MNANO.2019.2927774

Liu, T.; Tang, J. A.; Jiang, L. The Enhancement Effect of Gold Nanoparticles as a Surface Modifier on DNA Sensor Sensitivity. *Biochem. Biophys. Res. Commun.* **2004,** *313* (1), 3–7. https://doi.org/10.1016/j.bbrc.2003.11.098

Liu, Y.; Hou, H.; You, T. Synthesis of Carbon Nanofibers for Mediatorless Sensitive Detection of NADH. *Electroanaly. Int. J. Devoted Fundament. Pract. Aspects Electroanaly.* **2008,** *20* (15), 1708–1713. https://doi.org/10.1002/elan.200804242

Liu, Y; Wei, W. Layer-by-Layer Assembled DNA Functionalized Single-Walled Carbon Nanotube Hybrids for Arsenic (III) Detection. *Electrochem. Commun.* **2008,** *10* (6), 872–875. https://doi.org/10.1016/j.elecom.2008.03.013

Malik, P.; Katyal, V.; Malik, V.; Asatkar, A.; Inwati, G.; Mukherjee, T. K. Nanobiosensors: Concepts and Variations. *Int. Sch. Res. Notices* **2013.**

Mao, S.; Yu, K.; Lu, G.; Chen, J. Highly Sensitive Protein Sensor Based on Thermally-Reduced Graphene Oxide Field-Effect Transistor. *Nano Res.* **2011,** *4* (10), 921. https://doi.org/10.1007/s12274-011-0148-3

Mazzei, F.; Favero, G.; Bollella, P.; Tortolini, C.; Mannina, L.; Conti, M. E.; Antiochia, R. Recent Trends in Electrochemical Nanobiosensors for Environmental Analysis. *Int. J. Environ. Health* **2015,** *7* (3), 267–291. https://doi.org/10.1504/IJENVH.2015.073210

Mohammadi Aloucheh, R.; Alaee Mollabashi, Y.; Asadi, A.; Baris, O.; Gholamzadeh, S. The Role of Nanobiosensors in Identifying Pathogens and Environmental Hazards. *Anthropogenic Pollut. J.* **2018,** *2* (2), 16–25. https://dx.doi.org/10.22034/ap.2018.572812.1024

Mokhtarzadeh, A.; Eivazzadeh-Keihan, R.; Pashazadeh, P.; Hejazi, M.; Gharaatifar, N.; Hasanzadeh, M.; de la Guardia, M. Nanomaterial-Based Biosensors for Detection of Pathogenic Virus. *TrAC Trends Analy. Chem.* **2017,** *97*, 445–457. https://doi.org/10.1016/j.trac.2017.10.005

Montes, R.; Céspedes, F.; Gabriel, D.; Baeza, M. Electrochemical Biosensor Based on Optimized Biocomposite for Organophosphorus and Carbamates Pesticides Detection. *J. Nanomater.* **2018.** https://doi.org/10.1155/2018/7093606

Moraes, F. C.; Rossi, B.; Donatoni, M. C.; de Oliveira, K. T.; Pereira, E. C. Sensitive Determination of 17β-Estradiol in River Water Using a Graphene Based Electrochemical Sensor. *Analy. Chim. Acta* **2015,** *881*, 37–43. https://doi.org/10.1016/j.aca.2015.04.043

Noyrod, P.; Chailapakul, O.; Wonsawat, W.; Chuanuwatanakul, S. The Simultaneous Determination of Isoproturon and Carbendazim Pesticides by Single Drop Analysis Using a Graphene-Based Electrochemical Sensor. *J. Electroanaly. Chem.* **2014,** *719*, 54–59. https://doi.org/10.1016/j.jelechem.2014.02.001

Oliveira, T. M.; Ribeiro, F. W.; Sousa, C. P.; Salazar-Banda, G. R.; de Lima-Neto, P.; Correia, A. N.; Morais, S. Current Overview and Perspectives on Carbon-Based (Bio) Sensors for Carbamate Pesticides Electroanalysis. *TrAC Trends Analy. Chem.* **2020,** *124,* 115779. https://doi.org/10.1016/j.trac.2019.115779

O'Mahony, A. M.; Wang, J. Nanomaterial-Based Electrochemical Detection of Explosives: A Review of Recent Developments. *Analy. Methods* **2013,** *5* (17), 4296–4309. https://doi.org/10.1039/c3ay40636a

Padrón-Sanz, C.; Halko, R.; Sosa-Ferrera, Z.; Santana-Rodríguez, J. J. Combination of Microwave Assisted Micellar Extraction and Liquid Chromatography for the Determination of Organophosphorus Pesticides in Soil Samples. *J. Chromatogr. A* **2005,** *1078* (1–2), 13–21. https://doi.org/10.1016/j.chroma.2005.05.005

Pandit, S.; Dasgupta, D.; Dewan, N.; Prince, A. Nanotechnology Based Biosensors and Its Application. *Pharma Innov.* **2016,** *5* (6, Part A), 18.

Paredes, J. I.; Villar-Rodil, S.; Martínez-Alonso, A.; Tascon, J. M. D. Graphene Oxide Dispersions in Organic Solvents. *Langmuir* **2008,** *24* (19), 10560–10564. https://doi.org/10.1021/la801744a

Park, M.; Cella, L. N.; Chen, W.; Myung, N. V.; Mulchandani, A. Carbon Nanotubes-Based Chemiresistive Immunosensor for Small Molecules: Detection of Nitroaromatic Explosives. *Biosens. Bioelectron.* **2010,** *26* (4), 1297–1301. https://doi.org/10.1016/j.bios.2010.07.017

Pennington, J. C.; Brannon, J. M. Environmental Fate of Explosives. *Thermochim. Acta* **2002,** *384* (1–2), 163–172. https://doi.org/10.1016/S0040-6031(01)00801-2

Ping, J.; Wang, Y.; Fan, K.; Wu, J.; Ying, Y. Direct Electrochemical Reduction of Graphene Oxide on Ionic Liquid Doped Screen-Printed Electrode and Its Electrochemical Biosensing Application. *Biosens. Bioelectron.* **2011,** *28* (1), 204–209. https://doi.org/10.1016/j.bios.2011.07.018

Prasad, R.; Bhattacharyya, A.; Nguyen, Q. D. Nanotechnology in Sustainable Agriculture: Recent Developments, Challenges, and Perspectives. *Front. Microbiol.*, **2017,** *8,* 1014. https://doi.org/10.3389/fmicb.2017.01014

Pumera, M. Graphene in Biosensing. *Mater. Today*, **2011,** *14* (7–8), 308–315. https://doi.org/10.1016/S1369-7021(11)70160-2

Qu, K.; Zhang, X.; Lv, Z.; Li, M.; Cui, Z.; Zhang, Y.; Kong, Q. Simultaneous Detection of Diethylstilbestrol and Malachite Green Using Conductive Carbon Black Paste Electrode. *Int. J. Electrochem. Sci*, **2012,** *7,* 1827–1839.

Rai, V.; Acharya, S.; Dey, N. Implications of Nanobiosensors in Agriculture. *J. Biomater. Nanobiotechnol.* **2012,** *3,* 315–324.

Ramnani, P.; Saucedo, N. M.; Mulchandani, A. Carbon Nanomaterial-Based Electrochemical Biosensors for Label-Free Sensing Of Environmental Pollutants. *Chemosphere* **2016,** *143,* 85–98. https://doi.org/10.1016/j.chemosphere.2015.04.063

Revin, S. B.; John, S. A. Electrochemical Sensor for Neurotransmitters at Physiological pH Using a Heterocyclic Conducting Polymer Modified Electrode. *Analyst* **2012,** *137* (1), 209–215. https://doi.org/10.1039/C1AN15746A

Richardson, J.; Hawkins, P.; Luxton, R. The Use of Coated Paramagnetic Particles as a Physical Label in a Magneto-Immunoassay. *Biosens. Bioelectron.* **2001,** *16* (9–12), 989–993. https://doi.org/10.1016/S0956-5663(01)00201-9

Rodríguez-Lázaro, D.; D'Agostino, M.; Herrewegh, A.; Pla, M.; Cook, N.; Ikonomopoulos, J. Real-Time PCR-Based Methods for Detection of *Mycobacterium avium* Subsp. Paratuberculosis in Water and Milk. *Int. J. Food Microbiol.* **2005,** *101* (1), 93–104. https://doi.org/10.1016/j.ijfoodmicro.2004.09.005

Royea, W. J.; Hamann, T. W.; Brunschwig, B. S.; Lewis, N. S. A Comparison between Interfacial Electron-Transfer Rate Constants at Metallic and Graphite Electrodes. *J. Phys. Chem. B* **2006,** *110* (39), 19433–19442. https://doi.org/10.1021/jp062141e

Russell, B. A.; Khanal, P.; Calbi, M. M.; Yudasaka, M.; Iijima, S.; Migone, A. D. Sorption Kinetics on Open Carbon Nanohorn Aggregates: The Effect of Molecular Diameter. *Molecules* **2016,** *21* (4), 521. https://doi.org/10.3390/molecules21040521

Šafařík, I.; Šafaříková, M. Use of Magnetic Techniques for the Isolation of Cells. *J. Chromatogr. B Biomed. Sci. App.* **1999,** *722* (1–2), 33–53. https://doi.org/10.1016/S0378-4347 (98)00338-7

Salih, F. E.; Oularbi, L.; Halim, E.; Elbasri, M.; Ouarzane, A.; El Rhazi, M. Conducting Polymer/Ionic Liquid Composite Modified Carbon Paste Electrode for the Determination of Carbaryl in Real Samples. *Electroanalysis* **2018,** *30* (8), 1855–1864. https://doi.org/10.1002/elan.201800152

Salouti, M.; Derakhshan, F. K. Biosensors and Nanobiosensors in Environmental Applications. In *Biogenic Nano-Particles and Their Use in Agro-Ecosystems*; Springer: Singapore, 2020; pp 515–591. https://doi.org/10.1007/978-981-15-2985-6_26

Sankhla, M. S.; Kumari, M.; Nandan, M.; Kumar, R.; Agrawal, P. Heavy Metals Contamination in Water and Their Hazardous Effect on Human Health-A Review. *Int. J. Curr. Microbiol. App. Sci.* **2016**, *5* (10), 759–766. https://doi.org/10.20546/ijcmas.2016.510.082

Sankhla, M. S.; Kumari, M.; Sharma, K.; Kushwah, R. S.; Kumar, R. Water Contamination through Pesticide & Their Toxic Effect on Human Health. *Int. J. Res. Appl. Sci. Eng. Technol. (IJRASET)* **2018,** *6* (1), 967–970. https://doi.org/10.22214/ijraset.2018.1146

Sarkar, T.; Srinives, S. Single-Walled Carbon Nanotubes-Calixarene Hybrid for Sub-ppm Detection of NO_2. *Microelectron. Eng.* **2018,** *197*, 28–32. https://doi.org/10.1016/j.mee.2018.05.004

Sharma, C. S.; Katepalli, H.; Sharma, A.; Madou, M. Fabrication and Electrical Conductivity of Suspended Carbon Nanofiber Arrays. *Carbon* **2011,** *49* (5), 1727–1732. https://doi.org/10.1016/j.carbon.2010.12.058

Silva, M.; Cesarino, I. Evaluation of a Nanocomposite Based on Reduced Graphene Oxide and Gold Nanoparticles as an Electrochemical Platform for Detection of Sulfamethazine. *J. Compos. Sci.* **2019,** *3* (2), 59. https://doi.org/10.3390/jcs3020059

Smith, S. C.; Rodrigues, D. F. Carbon-Based Nanomaterials for Removal of Chemical and Biological Contaminants from Water: A Review of Mechanisms and Applications. *Carbon* **2015,** *91*, 122–143. https://doi.org/10.1016/j.carbon.2015.04.043

So, H. M.; Park, D. W.; Jeon, E. K.; Kim, Y. H.; Kim, B. S.; Lee, C. K.; Lee, J. O. Detection and Titer Estimation of *Escherichia coli* Using Aptamer-Functionalized Single-Walled Carbon-Nanotube Field-Effect Transistors. *Small* **2008,** *4* (2), 197–201. https://doi.org/10.1002/smll.200700664

Sonone, S. S.; Jadhav, S.; Sankhla, M. S.; Kumar, R. Water Contamination by Heavy Metals and Their Toxic Effect on Aquaculture and Human Health through Food Chain. *Lett. Appl. NanoBioSci.* **2020,** *10* (2), 2148–2166. https://doi.org/10.33263/LIANBS102.21482166

Sotiropoulou, S.; Gavalas, V.; Vamvakaki, V.; Chaniotakis, N. A. Novel Carbon Materials in Biosensor Systems. *Biosens. Bioelectron.* **2003,** *18* (2–3), 211–215. https://doi.org/10.1016/S0956-5663(02)00183-5

Stankovich, S.; Dikin, D. A.; Piner, R. D.; Kohlhaas, K. A.; Kleinhammes, A.; Jia, Y.; Ruoff, R. S. Synthesis of Graphene-Based Nanosheets via Chemical Reduction of Exfoliated Graphite Oxide. *Carbon* **2007,** *45* (7), 1558–1565. https://doi.org/10.1016/j.carbon.2007.02.034

Steffens, C.; Steffens, J.; Graboski, A. M.; Manzoli, A.; Leite, F. L. Nanosensors for Detection of Pesticides in Water. In *New Pesticides and Soil Sensors*; Academic Press, 2017; pp 595–635. https://doi.org/10.1016/B978-0-12-804299-1.00017-5

Su, X.; Chew, F. T.; Li, S. F. Design and Application of Piezoelectric Quartz Crystal-Based Immunoassay. *Analy. Sci.* **2000,** *16* (2), 107–114. https://doi.org/10.2116/analsci.16.107

Sun, X. M.; Ji, Z.; Xiong, M. X.; Chen, W. The Electrochemical Sensor for the Determination of Tetracycline Based on Graphene/L-Cysteine Composite Film. *J. Electrochem. Soc.* **2017,** *164* (4), B107. https://doi.org/10.1149/2.0831704jes

Tang, X.; Liu, Y.; Hou, H.; You, T. A nonenzymatic sensor for Xanthine Based on Electrospun Carbon Nanofibers Modified Electrode. *Talanta 83* (5), 1410–1414. https://doi.org/10.1016/j.talanta.2010.11.019

Thakur, M. S.; Ragavan, K. V. Biosensors in Food Processing. *J. Food Sci. Technol.* **2013,** *50* (4), 625–641. https://doi.org/10.1007/s13197-012-0783-z

Thévenot, D. R.; Toth, K.; Durst, R. A.; Wilson, G. S. Electrochemical Biosensors: Recommended Definitions and Classification. *Biosens. Bioelectron.* **2001,** *16* (1–2), 121–131. https://doi.org/10.1016/S0956-5663(01)00115-4

Torrinha, Á.; Oliveira, T. M.; Ribeiro, F. W.; Correia, A. N.; Lima-Neto, P.; Morais, S. Application of Nanostructured Carbon-Based Electrochemical (Bio) Sensors for Screening of Emerging Pharmaceutical Pollutants in Waters and Aquatic Species: A Review. *Nanomater.* **2020,** *10* (7), 1268. https://doi.org/10.3390/nano10071268

Ünal, F.; Yüzbaşıoğlu, D.; Yılmaz, S.; Akıncı, N.; Aksoy, H. Genotoxic Effects of Chlorophenoxy Herbicide Diclofop-Methyl in Mice in Vivo and in Human Lymphocytes in Vitro. *Drug Chem. Toxicol.* **2011,** *34* (4), 390–395. https://doi.org/10.3109/01480545.2010.538695

Vandenberg, L. N.; Colborn, T.; Hayes, T. B.; Heindel, J. J.; Jacobs Jr, D. R.; Lee, D. H.; Zoeller, R. T. Hormones and Endocrine-Disrupting Chemicals: Low-Dose Effects and Nonmonotonic Dose Responses. *Endocr. Rev.* **2012,** *33* (3), 378–455. https://doi.org/10.1210/er.2011-1050

Vega, D.; Agüí, L.; González-Cortés, A.; Yáñez-Sedeño, P.; Pingarrón, J. M. Voltammetry and Amperometric Detection of Tetracyclines at Multi-Wall Carbon Nanotube Modified Electrodes. *Analy. Bioanaly. Chem.* **2007,** *389* (3), 951–958. https://doi.org/10.1007/s00216-007-1505-7

Wanekaya, A. K.; Chen, W.; Mulchandani, A. Recent Biosensing Developments in Environmental Security. *J. Environ. Monitor.* **2008,** *10* (6), 703–712. https://doi.org/10.1039/b806830p

Wang, J. Carbon-Nanotube Based Electrochemical Biosensors: A Review. *Electroanaly. Int. J. Devoted o Fundament. Pract. Aspects Electroanaly.* **2005,** *17* (1), 7–14. https://doi.org/10.1002/elan.200403113

Wang, J. Electrochemical Nucleic Acid Biosensors. *Analy. Chim.Acta* **2002,** *469* (1), 63–71. https://doi.org/10.1016/S0003-2670(01)01399-X

Wang, J.; Musameh, M.; Lin, Y. Solubilization of Carbon Nanotubes by Nafion toward the Preparation of Amperometric Biosensors. *J. Ame. Chem. Soc.* **2003,** *125* (9), 2408–2409. https://doi.org/10.1021/ja028951v

Wang, J.; Yang, S.; Guo, D.; Yu, P.; Li, D.; Ye, J.; Mao, L. Comparative Studies on Electrochemical Activity of Graphene Nanosheets and Carbon Nanotubes. *Electrochem. Commun.* **2009a,** *11* (10), 1892–1895. https://doi.org/10.1016/j.elecom.2009.08.019

Wang, L.; Chen, W.; Xu, D.; Shim, B. S.; Zhu, Y.; Sun, F.; Kotov, N. A. Simple, Rapid, Sensitive, and Versatile SWNT–Paper Sensor for Environmental Toxin Detection Competitive with ELISA. *Nano Lett.* **2009b,** *9* (12), 4147–4152. https://doi.org/10.1021/nl902368r

Wang, S.; Li, L.; Jin, H.; Yang, T.; Bao, W.; Huang, S.; Wang, J. Electrochemical Detection of Hepatitis B and Papilloma Virus DNAs Using SWCNT Array Coated with Gold Nanoparticles. *Biosens. Bioelectron.*, **2013,** *41*, 205–210. https://doi.org/10.1016/j.bios.2012.08.021

Wang, Y.; Hu, S. Applications of Carbon Nanotubes and Graphene for Electrochemical Sensing of Environmental Pollutants. *J. Nanosci. Nanotechnol.* **2016,** *16* (8), 7852–7872. https://doi.org/10.1166/jnn.2016.12762

Ward, M. D.; Ebersole, R. C. US Pat. 5,501,986 **1996.**

Wei, B. Q.; Vajtai, R.; Ajayan, P. M. Reliability and Current Carrying Capacity of Carbon Nanotubes. *Appl. Phys. Lett.* **2001,** *79* (8), 1172–1174. https://doi.org/10.1063/1.1396632

Wilder, J. W.; Venema, L. C.; Rinzler, A. G.; Smalley, R. E.; Dekker, C. Electronic Structure of Atomically Resolved Carbon Nanotubes. *Nature* **1998,** *391* (6662), 59–62. https://doi.org/10.1038/34139

Xu, J.; Lu, X.; Li, B. Synthesis, Functionalization, and Characterization. In *Biomedical Applications and Toxicology of Carbon Nanomaterials*; Chen, C., Wang, H., Eds.; Wiley-VCH Verlag GmbH & Co. KGaA: Weinheim, Germany, 2016; pp 1e28.

Xu, M.; Wang, R.; Li, Y. Electrochemical Biosensors for Rapid Detection of *Escherichia coli* O157: H7. *Talanta*, **2017,** *162*, 511–522. https://doi.org/10.1016/j.talanta.2016.10.050

Yadav, H.; Sankhla, M. S.; Kumar, R. Pesticides-Induced Carcinogenic & Neurotoxic Effect on Human. *Forensic Res. Criminol. Int. J.* **2019,** *7* (5), 243–245. https://doi.org/10.15406/frcij.2019.07.00288

Yang, N.; Yu, S.; Macpherson, J. V.; Einaga, Y.; Zhao, H.; Zhao, G.; Jiang, X. Conductive Diamond: Synthesis, Properties, and Electrochemical Applications. *Chem. Soc. Rev.* **2019,** *48* (1), 157–204. https://doi.org/10.1039/C7CS00757D

Yao, Y.; Ping, J. Recent Advances in Graphene-Based Freestanding Paper-Like Materials for Sensing Applications. *TrAC Trends Analy. Chem.* **2018,** *105*, 75–88. https://doi.org/10.1016/j.trac.2018.04.014

Yılmaz, F.; Özdemir, N.; Demirak, A.; Tuna, A. L. Heavy Metal Levels in Two Fish Species *Leuciscus cephalus* and *Lepomis gibbosus*. *Food Chem.* **2007,** *100* (2), 830–835. https://doi.org/10.1016/j.foodchem.2005.09.020

Zahid, M. U.; Pervaiz, E.; Hussain, A.; Shahzad, M. I.; Niazi, M. B. K. Synthesis of Carbon Nanomaterials from Different Pyrolysis Techniques: A Review. *Mater. Res. Express* **2018,** *5* (5), 052002. https://doi.org/10.1088/2053-1591/aac05b

Zelada-Guillén, G. A.; Riu, J.; Düzgün, A.; Rius, F. X. Immediate Detection of Living Bacteria at Ultralow Concentrations Using a Carbon Nanotube Based Potentiometric Aptasensor. *Angew. Chem. Int. Ed.* **2009,** *48* (40), 7334–7337. https://doi.org/10.1002/anie.200902090

Zhang, W.; Zhu, S.; Luque, R.; Han, S.; Hu, L.; Xu, G. Recent Development of Carbon Electrode Materials and Their Bioanalytical and Environmental Applications. *Chem. Soc. Rev.* **2016,** *45* (3), 715–752. https://doi.org/10.1039/C5CS00297D

PART IV
Nanoremediation

CHAPTER 10

Nanomaterials to Remediate Water Pollution

FERNANDA MARIA POLICARPO TONELLI[1],
FLÁVIA CRISTINA POLICARPO TONELLI[2*], MOLINE SEVERINO LEMOS[1],
and DANILO ROBERTO CARVALHO FERREIRA[3]

[1]*Department of Cell Biology, ICB/UFMG, Belo Horizonte, Brazil*

[2]*Department of Pharmacy, UFSJ/CCO, Divinópolis, Brazil*

[3]*Department of Materials, CDTN/CNEN, Belo Horizonte, Brazil*

[*]*Corresponding author. E-mail: flacristinaptonelli@gmail.com*

ABSTRACT

Environmental pollution is a concern worldwide as it may cause serious damage to living beings' health, threatening the existence of live on Earth. Activities performed by humans such as industrial ones are important sources of pollution. Regarding water, the increasing world's population needs to receive from water treatment plants, a product with considerable quality to survive in a healthy way. However, with the enormous array of organic and inorganic contaminants that can pollute groundwater and surface water is becoming even more difficult to regulatory agencies to set quality standards for each of them and to assure if a water sample is secure to be used. Strategies to deal with pollution in order to reduce the risk it represents to ecosystems are highly necessary. When it comes to water pollution, nanoscience offers interesting alternatives. This chapter will address nanomaterials in protocols aiming to remediate water pollution.

Nanotechnology for Environmental Pollution Decontamination: Tools, Methods, and Approaches for Detection and Remediation. Fernanda Maria Policarpo Tonelli, Rouf Ahmad Bhat, & Gowhar Hamid Dar (Eds.)

10.1 INTRODUCTION

Environmental pollution is an old problem that has become even more risky as human actions are increasingly performed ignoring sustainability polices (Peuke and Rennenberg, 2005). Since Bronze Age, for example, the heavy metal lead (Longman et al., 2018), from which poisoning can impair cognitive ability and cause developmental disorders or at high levels may damage circulatory and nervous systems leading to death, is a harmful environmental contaminant (Yamada et al., 2020). However, the rapid industrialization and urbanization that happened and/or has been happening in several countries worldwide, increases the negative impacts over ecosystems threatening life on Earth (Chaoua et al., 2019; Zheng et al., 2020).

Organic (e.g., pesticides and dyes (Tonelli and Tonelli, 2020)) and inorganic (e.g., heavy metals and radionuclides (Salt et al., 1995)) pollutants that can be accumulated by living forms passing through food chain increasing their harmful potential and/or persist on polluted environs for a long period of time are of special concern (Guerra et al., 2018; Varjani et al., 2019; Zhang et al., 2019). They, at low levels and without prolonged exposure, can cause acute intoxication and can damage DNA and as the dose and time of exposure increase, severe diseases, such as chronic kidney disease, cancer or even death may occur (Gavrilescu et al., 2015; Bagazgoïtia et al., 2018; Fatima et al., 2018; Singh et al., 2018; Ali et al., 2019; Farkhondeh et al., 2020; Fulekar and Pathak, 2020).

When it comes to water pollution, two important contaminant's sources deserve to be highlighted: landfills and wastewaters.

Landfills are highly dangerous waste disposal practice, they pollute not only water but also soil and air. Leachate production can end up contaminating the surface water and also groundwater (Kim and Owens, 2010).

Domestic wastewater possesses large concentration of organic material (Boutin and Eme, 2016), which can disrupt the ecosystem where the discharge occurs, and also causes hazard to living being in contact with this water prior to treatment (Moretti et al., 2015; Newton et al., 2015), and in great urban centers also a high level of drugs and human estrogenic hormones that are dangerous to ecosystem's balance (Manickum et al., 2011, Gimiliani et al., 2016; Deng et al., 2020). Mixed pollutants are present in domestic wastewater and this composition changes as it reflects population's lifestyle (Gray, 2004). Emerging contaminants such as estrogenic endocrine-disrupting chemicals (e.g., bisphenol A and polychlorinated biphenyls) (Roy et al., 2009) and pharmaceuticals are serious threats to living beings and that can even be found in treated wastewater (Klatte et al., 2017; Zhou et al., 2019; Yeung et al, 2020). The presence of pharmaceuticals in wastewaters is a consequence of the increase in medicines availability to treat humans and animals and as bioactive compounds that can induce effects even at small amounts. They can damage not only the species directly in contact with them but also to species that will consume contaminated water or organisms (Bebianno and Gonzalez-Rey, 2015; Sangion and Gramatica, 2016; Mezzelani et al., 2018).

Industrial wastewater presents an enormous nocive potential as it may contains toxic heavy metal, pharmaceuticals, veterinary products, personal care products, dyes, pesticides, food additives, industrial compounds/by-products, among others dangerous substances, mainly persistent ones, to a greater or lesser extent depending on the type of industry from which it comes (Tayeb et al., 2015; Yang et al., 2017; Iloms et al., 2020). Textile industry's residues are rich in synthetic dyes, such as azo dyes (El-Sikaily et al., 2012) that commonly present mutagenic and/or carcinogenic potential that can be enhanced depending on the water treatment employed to treat them. If aromatic amines are generated, the mutagenic/carcinogenic potential enhances (Bhaskar et al., 2003; Pinheiro et al., 2004; Khan and Malik, 2018).

Some pollutants, such as phosphorus and organics are not eliminated by commonly applied treatments such as chemically enhanced primary sedimentation. Industries sometimes perform treatment before throwing the pollutants on rivers (where they modify chemical composition affecting biological equilibrium) (Kazlauskienė et al., 2012; Zhou et al., 2020). And together with waters' movement, the pollution may spread, thus contaminating a large area and also groundwater (Kim et al., 2017; Sposito et al., 2018; Ziadi et al., 2019).

Water is a vital resource for humans and animals and its pollution should receive attention. Remediation strategies are necessary once water resources present limited self-recovery capacity and groundwater resources are being depleted (Sharma et al., 2016; Kurwadkar, 2019). Scarcity threatens societies and conventional purification methods that are generally not sustainable procedures, receive an increasing demand to deal with the most different kinds of pollutant once most surface water before treatment presents low quality containing a large amount of pollutants (Schwarzenbach et al., 2006; Ying et al., 2017).

Remediation strategies capable of dealing with water pollution in a cost-effective, sustainable, and efficient way are urgently required as same as human's adoption of practices to reduce pollution generation. Water treatment technologies' development alone proved to be unable to avoid reduction in biodiversity and other negative effects of environmental contaminants (Melián, 2020). Wastewater treatment plants commonly require large energy input and the use of chemicals resulting in a relevant source of air pollution (Hao et al., 2019).

10.2 THE NANOTECHNOLOGY FIELD AND THE NANOMATERIALS

Nanotechnology field has been offering materials in nanoscale since late 1950s to a large variety of uses. The term was introduced by Norio Taniguchi in 1974 (Taniguchi, 1974) and different protocols to synthesize and chemically modify these materials to perform specific task have been proposed since then (Drexler, 2004; Weiss et al., 2006; Zhang et al., 2008).

The field has been in a continuous development process to advance to offer efficient and low-cost solutions to environmental science field (Lofrano et al., 2017). Green-fabricated nanomaterials have emerged as eco-friendly tools and green synthesis a strategic alternative to conventional physico-chemical strategies (that commonly apply hazardous substances) to fabricate nanomaterials (Bolade et al., 2020; Nasrollahzadeh et al., 2021).

Nanomaterials (NMs) can be classified as nanoparticles (NPs), nanotubes (NTs), nanofibers (NFs), nanowires (NWs), and nanomembranes (NMBs) according to shape or morphology. NMs can be simply divided into inorganic NMs and organic NMs (Lu and Astruc, 2020).

Nanomaterials are versatile in function (such as delivery of drugs and genes, vaccine development, imaging/diagnosis, energy storage, tissue engineering (Tonelli et al., 2015, 2016, 2020)) due to unique properties. At least in one dimension, these materials present a size ranging from 1 to 100 nm, displaying as a consequence quantum confinement and high surface-to-volume ratio (Bethi and Sonawane, 2018; Werkneh et al., 2020). Other relevant physicochemical properties to nanomaterial's different functions include chemical reactivity, high stability (thermal and physical/chemical), and catalytic activity (Pradhan et al., 2001; Zhang et al., 2014; Xu et al., 2019).

The majority of released nanoparticles aggregate as soon as hydrated, thus become the sediment in different rates. The extent of aggregation depends on the nanoparticle's surface charge and charge magnitude. The coverage of the nanoparticle's surface is likely made by mono- and divalent cations, by natural organic matter (NOM) or other organic molecules (Keller et al, 2010).

It is also possible to improve some desirable characteristic and design a nanomaterial specifically to perform a task. Protocols to synthesize and functionalize (perform chemical modification on nanomaterial's surface) nanomaterials can be developed to originate a material that will meet researchers' necessities. Nanosensors efficient to water pollution (Priyadarshni et al., 2018; Das et al., 2019; Rao et al., 2019; Xie et al., 2019) as same as nanostructures to remediate water removing contaminants (Santhosh et al., 2016; Khan and Malik, 2019) can be produced.

The main concern involving nanomaterials' use to remediate polluted water is the safety of these materials to the living beings. It is necessary to keep the acceptable security level on guaranteeing that the final product of treatment will not threat the health of who is consuming the water or dealing with it in treatment plants. The risks are considered low by research community from nanoscience field. However, studies on nanomaterials biocompatibility and safety must continue to be carried out to ensure that they are safe to be used in large scale (Pulizzi and Sun, 2018).

Regarding chemical constitution, nanomaterials can be divided into carbon-based (e.g., graphene, graphene oxide, fullerene, carbon nanotube, carbon quantum dots) or NOM-carbon-based (e.g., magnetic nanoparticles and silver quantum dots) (Drexler, 2004; Weiss et al., 2006; Zhang et al., 2008; Wicki et al., 2015; Werkneh and Rene, 2019; Trivedi et al., 2020).

Many nanoadsorbents and nanomembranes are now known to be in the development stage for the sake of large-scale production and industrialization. Given the fact that water recycling is considered an important aspect of sustainable development in human communities, particularly remembering the expansion of the water scarcity crisis around the world. Discovering new technologies that improve water quality are even more important in this scenario (Ghadimi et al., 2020).

10.2.1 IMPORTANT NONCARBON-BASED NANOMATERIALS FOR WATER REMEDIATION

Noncarbon-based nanomaterial, different from the carbon-based ones does not possess their structure based on carbon atoms. There are noncarbon-based nanomaterials that are structurally based on metallic atoms, for example, that have proved to be able to outcompete activated carbon in the capacity to deal with metal's pollution in water samples (Table 10.1) (Sharma et al., 2009).

Recently, nanoscale zero-valent iron (nZVI) has been receiving increasing attention as an efficient manner to reduce the availability of inorganic pollutants in water samples (Gil-Díaz et al., 2020). Uranium present in acid mine water could be remediated by nZVI as same as Al, sulfates, As, Be, Cd, Cr, Cu, Ni, V, and Zn (Klimkova et al., 2011). High

TABLE 10.1 Examples of Noncarbon-Based Nanomaterial to Remediate Polluted Water.

Nanomaterial	Pollutant(s)	References
Nanoscale zero-valent iron (nZVI)	U, Al, sulfates, As, Be, Cd, Cr, Cu, Ni, V, and Zn	Klimkova et al. (2011)
	Cr^{6+}	Gheju, 2011; Guan et al. (2015)
	Cu^{2+}	Li et al. (2013)
	Cu and Ni	Gil-Díaz et al. (2020)
	Phosphate, ammonia, Pb, nitrate, chloride	Shad et al. (2020)
Nanoscale zero-valent iron nanoparticles coated with 4% bentonite	Cr^{6+}	Wang et al. (2018)
Sulfide-modified nZVI associated to graphene aerogel composite	Trichloroethylene	Bin et al. (2020)
Nanoscale zero-valent iron conjugated to biochar	Pb^{2+}	Li, Wang et al. (2020)
	Trichloroethylene	Li, Z. et al. (2020)
Iron oxide nanoparticles	Nitrate, nitrite, phosphate, ammonium	Hesni et al. (2020)
Iron oxide nanoparticles associated to biochar	Arsenite and arsenate	Priyadarshni et al. (2020)
Coated magnetite nanoparticles	Organic dyes and toxic metal ions	Das et al. (2020)
Iron magnetic nanoparticles	As^{+5}	Zeng et al. (2020)
TiO_2 nanoparticles	As	Deliyanni et al. (2003; Mayo et al. (2007)
TiO_2 decorated with gold nanoparticles	Methylene blue	Perera et al. (2020)
Hydrated Ti oxide nanoparticles on a support of agricultural waste rice straw	Cu	Chen, Y. et al. (2020)

TABLE 10.1 *(Continued)*

Nanomaterial	Pollutant(s)	References
TiO_2 nanoparticles decorated with SnO_2 quantum dots	Rhodamine B	Lee et al. (2014)
ZnO quantum dots over SiO_2 nanotubes	Rhodamine B	Zhang et al. (2012)
Cadmium telluride quantum dot associated to europium-metal organic framework and	Rhodamine 6G	Kaur et al. (2016)
Silver quantum dots	Organic contaminants	Kumar et al. (2019)
Silver nanoparticles coated with nanomagnetic biochar dots	Dyes	Zahedifar et al. (2020)
Polyvinylpyrrolidone-coated magnetite nanoparticles	Oil	Mirshahghassemi and Lead, 2015)
	Cd, Cr, Ni, and Pb	Hong et al. (2020)

level of Cu^{2+} could be remediated with efficiency greater than 96% by nZVI from wastewater in field experiment, and the nanomaterial could be easily separated and recycled enhancing cost-effectiveness and up-scalability (Li et al., 2013). When compared with the capacity to remediate soil samples, nZVI's capacity to remediate water samples dealing with Cu and Ni pollution, by immobilizing them, is higher (Gil-Díaz et al., 2020). This nanomaterial presents high reactivity, low toxicity, low cost, the possibility of coated versions and bimetalic versions development and the capacity to also deal with organic pollutants in water samples (O'Carroll et al., 2013; Jiang et al., 2018). nZVI nanoparticles (nZVI-NPs) synthesized through a green method using the leaf extract of *Mentha piperita* as a reducing agent proved to be able to remove organic and inorganic contaminants from the canal at the University of Birmingham, UK (Shad et al., 2020). nZVI-NPs could efficiently remediate hexavalent chromium-contaminated wastewater (Gheju, 2011; Guan et al., 2015), but when coated with 4% bentonite, nZVI-NPs were even more efficient (Wang et al., 2018). When conjugated with biochar, the nZVI's efficiency to deal with pollution increases. The hydrophilic biochar bound to nZVI presented increased binding force to deal with Pb^{2+}'s pollution through reduction, complexation, and coprecipitation mechanisms (Li, Wang et al., 2020). Maize derived biochars bound to nZVIs could promote the total removal of trichloroethylene from groundwater condition in 20 min through reactive oxygen species (ROS) that promoted pollutant's dechlorination (Li, Z. et al., 2020). nZVI's functionalized version (sulfide-modified) associated with graphene aerogel composite proved to be an efficient option to remediate water contaminated by organic pollutants such as trichloroethylene (Bin et al., 2020).

Magnetic nanoparticles constituted by iron oxide are biocompatible and biodegradable, present electronic properties of interest, can be found as ferrimagnetic maghemite (γ-Fe_2O_3) or ferrimagnetic magnetite ($Fe_3O_4 \equiv FeO \cdot Fe_2O_3$), and their functionalization allows their optimization to perform desirable tasks (Lesiak et al, 2019). Iron oxide nanoparticles reduced the discharge burden of the effluent generated in fish farms, with a capacity that increased in first 6 h. The treatment in reactor had reduced nitrate, nitrite, phosphate, ammonium, total suspend, and dissolved solid levels (Hesni et al., 2020). Oil could be removed from oil–water mixture due to the use of PVP (polyvinylpyrrolidone)-coated magnetite nanoparticles (Mirshahghassemi and Lead, 2015). However, these coated nanoparticles can also remediate metals. Synthetic soft water and sea water polluted by Cd, Cr, Ni, and Pb at 0.1 mg/L could be remediated (near to 100%) by 167 mg/L of PVP– magnetite nanoparticles (Hong et al., 2020). Iron magnetic nanoparticles prepared in a friendly way from iron-containing sludge could co-precipitate arsenate in contaminated water removing at 0.2 g/L 400 μg/L of As^{+5} in 1 h (Zeng et al., 2020). Magnetite nanoparticles can also be generated in coated version through low-cost green synthesis protocol using crude latex of *Jatropha curcas* or leaf extract of *Cinnamomum tamala* offering a noncytotoxic material capable of remediating organic dyes and toxic metal ions' pollution in water, and also presenting antibacterial agents' action as same as antioxidant potential (Das et al., 2020). Iron oxide nanoparticles, as same as copper oxide nanoparticles were able to remediate arsenic pollution (arsenite and arsenate) and eliminate microbial contaminants from water samples, and to solve irreversible aggregation problem association to rice-husk biochar helped to stabilize the structure (Priyadarshni et al., 2020).

Not only nanosized magnetite but also TiO_2 nanoparticles could deal with arsenic pollution through adsorption in a way more efficient than activated carbon (Deliyanni et al., 2003; Mayo et al., 2007; Zahra et al., 2020). In fact, the skeleton of activated carbon, as same as other porous matrix can be used with metal (hydr)oxide nanoparticles to deal with not only pollution promoted by arsenic but also by organic cocontaminants (Hristovski et al., 2009a, 2009b). Clay immobilizing TiO_2 and ZnO nanoparticles present the ability to remediate wastewater being suitable for the generation of filters to treat water (Mustapha et al., 2020). Hydrated titanium oxide nanoparticles on a support of rice straw, an agricultural waste could efficiently adsorb copper from contaminated wastewater (Chen, Y. et al., 2020). Nano-TiO_2 proved to be phototoxic being an interesting option to remediate water especially through protocols involving solar energy or UV (Brunet et al., 2009). A low cost and eco-friendly protocol using green tea to generate a nanocomposite containing gold nanoparticles decorating TiO_2 resulted in a material able to promote the adsorption of methylene blue in contaminated water samples at a rate of 8185cmg/g (Perera et al., 2020).

Silver nanoparticles coated with nanomagnetic biochar dots generated through green synthesis using leaf extract of the plant *Pistacia atlantica* could remediate dyes from aqueous samples besides presenting the option of being reused and separated through magnetic process (Zahedifar et al., 2020).

When it comes to quantum dots, silver quantum dots are also efficient tools to deal with organic contaminants present in polluted water through photocatalysis (Kumar et al., 2019). SnO_2 quantum dots attached to the surface of TiO_2 nanoparticles contributed to charge separation and ROS generation promoting more efficient Rhodamine B degradation (Lee et al., 2014). Rhodamine B degradation was also efficiently promoted by ZnO quantum dots over SiO_2 nanotubes (Zhang et al., 2012). Complete degradation of Rhodamine 6G could be achieved in less than an hour by the nanocomposite generated by europium–metal organic framework and cadmium telluride quantum dot (Kaur et al., 2016).

10.2.2 IMPORTANT CARBON-BASED NANOMATERIALS FOR WATER REMEDIATION

The versatility of carbon atoms allows the use of this tetravalent element, capable of hybridizing in sp, sp^2 and sp^3 configurations to generate a large array of different substances by binding to other carbon atoms and/or to other elements. It presents not only the naturally occurring allotropes (e.g., diamond and graphite) but also the synthetic ones (e.g., the carbon-based nanomaterials) (Silva, 2008; Rauti et al., 2019). Carbon-based nanomaterials are also capable of remediating polluted water (Fig. 10.1; Table 10.2).

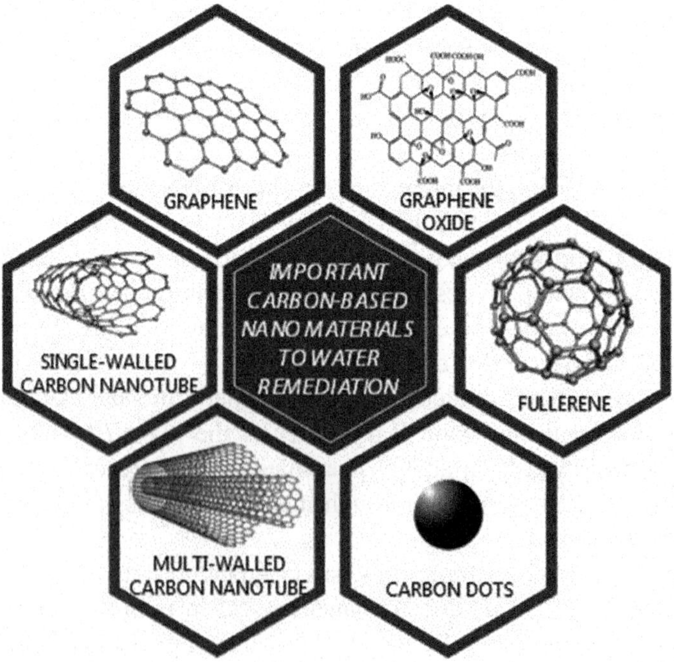

FIGURE 10.1 Carbon-based nanomaterial useful to water remediation protocols.

TABLE 10.2 Examples of Carbon-Based Nanomaterial to Remediate Polluted Water.

Nanomaterial	Pollutant(s)	References
Amino-functionalized C_{60} immobilized on 3-(2-succinic anhydride)propyl functionalized silica	Cimetidine and ranitidine	Lee et al. (2010)
Ag-embedded C_{60}	4-Nitrophenol, orange G dye	Liu, R. et al. (2020)
C_{60} modified by Ag_3PO_4	Dye acid red 18	Xu et al. (2016)
Fullerene coating nano-titanium dioxide	Mesotrione (herbicide)	Djordjevic et al. (2018)
Polyhydroxy fullerene		
C_{60}-modified graphite phase carbonitride	Methylene blue and phenol	Bai et al. (2014)
Anatase-TiO_2 nanoparticles modified by fullerene	Methylene blue	Qi et al. (2016)
Carbon dots impregnated in defect-rich g-C_3N_4	Hexavalent chromium, ofloxacin, bisphenol A and ciprofloxacin	Liu H. et al. (2020)
Carbon dots decorating hollow carbon nitride nanospheres	Naproxen	Wu, Y. et al. (2020)
Nitrogen-doped carbon quantum dots hybridized with g-C_3N_4	Methylene	Seng et al. (2020)
Carbon dots attached to tungsten oxide	Cd^{2+}, crystal violet	Smrithi et al. (2020)
Graphene associated with iron oxide	Methylene blue, rhodamine B, acid orange 7, and phenol	Hammad et al. (2020)
Graphene oxide in magnetic aerogel	Congo red, methylene blue, Cu^{2+}, Pb^{2+}, Cd^{2+}, Cr^{3+}	Xiong et al. (2020)
Reduced graphene oxide functionalized and associated to MnO_2	Methylene blue, arsenite	Tara et al. (2020)
Graphene oxide platform containing nickel–benzene dicarboxylate nanosized nickel metal organic framework	Methylene blue	Ahsan et al. (2020)
Carbon nanotube platform containing nickel–benzene dicarboxylate nanosized nickel metal organic framework		
Carbon nanotubes grafted antifouling layer of polyacryloyl hydrazide in poly(vinylidene fluoride) membrane	Micropollutants and heavy metals	Chen, Z. et al. (2020)
Chitosan–carbon nanotube supporting palladium nanoparticles	2-Nitroaniline, 4-nitrophenol, 4-nitro-*o*-phenylenediamine and 2,4-dinitrophenol, methyl orange, Congo red, methyl red, and methylene blue dyes	Sargin et al. (2020)

The fullerene allotropic family's most explored member is the C_{60}: from which the discovery was recognized by the Nobel Prize in Chemistry in 1996 (Biglova and Mustafina, 2019). It is generated by carbon atoms hybridized in sp^2 disposed rings of five members (twelve) or six members (twenty) to form a hollow sphere. The bonds located on the borders of hexagons and pentagons present 1.45 Å in length and the bonds located on the borders of two hexagons present 1.38 Å in length (being considered as double bonds) (Hirsch, 1995). Functionalized versions of fullerene proved to be useful to treat polluted water samples. Amino-functionalized C_{60} (photoactive structure) could be immobilized on 3-(2-succinic anhydride) propyl functionalized silica to efficiently remediate the contamination caused by pharmaceuticals pollutants (e.g., Cimetidine and Ranitidine) (Lee et al., 2010). In a version encapsulated in poly (N-vinylpyrrolidone, C_{60} proved to be more efficient than its hydroxylated version to generate ROS to remediate pollutants and induce the death of microorganisms sensible to superoxide and singlet oxygen (Brunet et al., 2009). Ag-embedded C_{60} was able to enhance the catalytic reduction of 4-nitrophenol. Photodegradation of orange G dye was improved making it an interesting tool to wastewater remediation (Liu, R. et al., 2020). C_{60} modified by Ag_3PO_4 photodegraded with an efficiency of 90% the dye Acid Red 18 after 1 h presenting more than 3.0 times of degradation rate improvement when compared with unmodified version (Xu et al., 2016). Besides being modified by other structures, fullerene can also be used as a coat to other nanomaterials.Nano titanium dioxide coated by fullerene and polyhydroxy fullerene could remediate the herbicide mesotrione (Djordjevic et al., 2018). Graphite phase carbonitride modified by C_{60} showed the efficient photocatalytic degradation of the dye methylene blue and phenol from contaminated samples (Bai et al., 2014). Anatase–TiO_2 nanoparticles after being modified by fullerene exhibited enhanced photocatalytic activity to degrade the dye methylene blue (Qi et al., 2016).

Carbon dots are considered zero-dimensional nanomaterials due to their low size, and are highly useful in chemical sensor/bioimaging, it is a field capable of remediating polluted water (Rani et al., 2020). These quantum dots, besides presenting high catalytic activity and being easy to functionalize through different protocols, possess superconductivity, good dispersibility, strong fluorescence, and high crystallization (Semeniuk et al., 2019; Wang et al., 2019a). Defect-rich g-C_3N_4 impregnated by N-doped carbon quantum dots presented high performance to deal with mixed inorganic (hexavalent chromium) and organic (ofloxacin, bisphenol A and ciprofloxacin) water pollution (Liu, H. et al., 2020). Sustainable nitrogen-doped carbon quantum dots derived from *Bombyx mori* silk fibroin could improve photocatalytic capability of TiO_2, thus converting it into a more suitable material to remediate water contamination (Wang et al., 2020). Hollow carbon nitride nanospheres decorated with carbon dots originated a metal-free photocatalytic nanoreactor that could efficiently degrade 10 mg/L of naproxen after 5 min of natural solar irradiation (Wu, Y. et al., 2020). Nitrogen-doped carbon quantum dots were hybridized with g-C_3N_4 to offer an efficient pollution remediator of methylene blue presenting 2.6-fold increase in activity when compared with g-C_3N_4 alone (Seng et al., 2020). Carbon dots obtained through green synthesis protocol using the peels of *Trichosanthes cucumerina* were attached to tungsten oxide originating a nanosystem capable of Cd^{2+} removal and crystal violet degradation (Smrithi et al., 2020).

Graphene is a bi-dimensional nanomaterial that consists of sp^2 carbon atoms organized in hexagonal rings with interesting mechanical and optical properties and a high thermal conductivity to be developed for different desirable roles (Novoselov et al, 2004). Graphene and the oxide generated from this nanomaterial can be used chemically designed/modified to be suitable for desirable roles, such as water remediation (Tiwari et al., 2020; Wang et al., 2019b). Graphene-containing biochar, for example, is an interesting tool to water remediation (Fang et al., 2020). Iron oxide's capacity to degrade organic pollutants (such as methylene blue, rhodamine B, acid orange 7, and phenol) could be enhanced by graphene that presented high photo-Fenton activity. The nanosystem was stable at pH from 3 to 9 being suitable to serve as feasible water treatment material (Hammad et al., 2020). By oxidizing graphene flakes and then exfoliating, graphene oxide can be generated (Aliyev *et al.*, 2019). This nanomaterial does not depend upon the surfactant's use to be dispersed in water and present lower toxicity and higher biocompatibility when compared with graphene (Pinto et al., 2013; Kiew et al., 2016). An aerogel framework consisted on amphiprotic microcrystalline cellulose enabled self-gelation of Fe_3O_4 and graphene oxide to generate a magnetic aerogel that presented good adsorption capacity for dyes (Congo red and methylene blue) and metal ions (Cu^{2+}, Pb^{2+}, Cd^{2+}, and Cr^{3+}) (Xiong et al., 2020). Reduced graphene oxide functionalized with functional groups of black cumin seeds and associated with MnO_2 particles could remediate water samples contaminated by methylene blue and arsenide exhibiting Langmuir sorption capacity of 232.5 and 14.7 mg/g (Tara et al., 2020). Graphene oxide, as same as carbon nanotube, served as platforms to nickel–benzene dicarboxylate nanosized nickel metal organic framework. The system was able to perform the adsorption of the water pollutant methylene blue (Ahsan et al., 2020).

Carbon atoms in hexagonal ring can also originate tube-like nanomaterials known as carbon nanotubes, product of graphite sheet rolled that presents high thermal conductivity and elasticity. The main methods used to generate these materials include chemical vapor deposition, electric arc, and laser deposition. The single-walled carbon nanotube possesses only one layer, the double-walled carbon nanotube possesses two and multiwalled carbon nanotubes present more layers of carbon atoms (Dresselhaus et al., 2000; Anzar et al., 2020). Carbon nanotubes are applied in different protocols of sensor's development and can also as previously mentioned, be applied in water remediation procedures, especially after functionalization to optimize desirable characteristics such as capacity to promote efficient attachment to contaminants (Ahsan et al., 2020; Sousa-Moura et al., 2019; Verma and Balomajumder, 2020; Wang et al., 2019b). An antifouling layer of polyacryloyl hydrazide (PAH)-grafted-carbon nanotubes in poly (vinylidene fluoride) membrane revealed to be able to remediate trace micropollutants and heavy metals from water samples (Chen, Z. et al., 2020). Chitosan–carbon nanotube-supported palladium nanoparticles were efficient to deal with organic pollution promoting the reduction of 2-nitroaniline, 4-nitrophenol, 4-nitro-*o*-phenylenediamine and 2,4-dinitrophenol, and the degradation methyl orange, Congo red, methyl red, and methylene blue dyes (Sargin et al., 2020).

10.3 KEY MECHANISMS FOR WATER REMEDIATION THROUGH NANOMATERIALS

The technologies of water remediation using nanomaterials are advantageous especially when compared with classic advantages, classic physicochemical protocols that are applied to perform this role. Physicochemical protocols present high costs when performing large-scale remediation as same as some bioremediation protocols based on enzymes, and may also negatively impact the environment generating harmful waste (Ye et al., 2017; Baldissarelli et al., 2019; Crini and Lichtfouse, 2019). Nanotechnologies, on their turn, can be performed at inexpensive manner (e.g., nano-TiO_2 are stable to corrosion and inexpensive nanoparticles that present high photocatalytic activity are capable of efficiently degrading the pollutant dye methylene blue (Jain and Vaya, 2017)) and it is also possible to develop green synthesis protocols to generate eco-friendly nanomaterials to be used in water remediation in an ecologically sustainable way. There are nanotechnologies that are compatible to existing water treatment set-ups being easily integrated to make the process costs less. The enzyme Protocatechuate 3,4-dioxygenase from *Rhizobium* sp. LMB-1, for example, is capable of remediating water containing organic pollutants that could be immobilized to modified Fe_3O_4 nanoparticles making reusability possible and increasing the stability of protein (Zhang, L. et al., 2017).

Nanomaterials suitable for water remediation can be catalysts, nanoadsorbents or constituents of nanomembranes and the main mechanisms through which nanomaterials can remedy contaminated water include filtration, adsorption, and degradation. They will be addressed separately.

10.3.1 FILTRATION TO DEAL WITH WATER POLLUTION: THE NANOMEMBRANES

Filtration (regular filtration, nanofiltration, ultrafiltration), as same as reverse osmosis, is performed when there is necessity to separate solid pollutants from water. At the end of filtration, treated water is obtained and the membrane retains pollutants that were eliminated from aqueous environment (Fig. 10.2) (Singh, R. et al., 2020). For example, membranes constituted by reduced GO (rGO)@MoO_2 and rGO@WO_3 presented a high separation efficiency to organic water pollutants rhodamine B, Evans blue, methylene blue (Thebo et al., 2018).

Nanomembranes present nanofibers and nanopores among them, being able to retain solid contaminants present in aqueous samples (e.g., metallic ions) and possessing higher size than the one from nanopores. In membranes of molybdenum disulfide in nanosheets, hydration state severely impacts pore structure, reflecting on water permeability and filtration capacity. Fully hydrated membranes present moderate-to-high molecular and ionic rejection and high water permeability. Dry membranes are almost impermeable to water (Wang et al., 2017).

Chemical modification and the addition of functional nanomaterials into nanomembranes also can allow improvements in thermal and mechanical stability, exhibition of multifunction (e.g., degradation capacity to be exhibited by the membrane) and synergism, adjustments in permeability (Han et al., 2013; Qu et al., 2013). Mechanical stability of thin film composites membranes could be enhanced by zirconia nanoparticles grown over polydopamine/polyetheylenimine that also provided high salt retention and water flux (Lv et al., 2016). Through chemical modification, protocols have been developed to allow nanomembranes to exhibit selective permeability to deal with specific pollutants of interest (Yin et al., 2013; Turchanin and Gölzhäuser, 2016; Yaqoob et al., 2020). Functionalization can also be performed to prevent membrane fouling and membrane clogging. Fouling layers can be prevented to build up by adding titanium and silver-based nanomaterials to the surface of nanomembranes (where they can oxidize organic contaminants) avoiding clogging (Mollahosseini and Rahimpour, 2014; Gehrke et al., 2015; Sumisha et al., 2015; Li, Liu et al., 2016). To solve the low water permeance problem from membranes of pure graphene oxide and total rGO, reduction degree was adjusted to a weak one, avoiding small interlayer spacing from total rGO and the lack of pristine graphitic sp^2 domains from pure graphene oxide. As a consequence, stability and separation performance were optimized (Zhang, Q. et al., 2018).

Characteristics of materials used to generate nanomembrane directly impact membrane's ability to filtrate different pollutants. Polyamide layer generated through embedding polysulfone support with zeolite nanoparticles offered a membrane with high rejection to small molecular weight negatively charged pharmaceuticals (Dong et al., 2016). ZnO nanospheres offered a similar effect by being added to inner-skinned hollow fiber composite nanofiltration membrane (Li, Shi et al., 2016), as same as silica nanospheres added to trimesoyl chloride and piperazine distributed over polysulfone support (Li, Wang, Song et al., 2015). Superior separation performance was observed on nanostrand-channelled graphene oxide ultrafiltration membrane that could deal with water polluted by rhodamine B through a fast separation (Huang et al., 2013). Despite being able to efficiently deal with positively charged dyes (such as methyl viologen and methylene blue), neutral ones (such as rhodamine B) and negatively charges ones (such as methyl orange, orange G, brilliant blue, methyl blue, and rose bengal), the graphene-based membranes generated by shear-induced alignment of liquid crystals of GO presented higher retention to negatively charges molecules (Akbari et al., 2016).

Polymeric membranes can receive nanomaterials to become able to degrade pollutants present in polluted water. Catalyst nanomaterials such as metallic ones are interesting examples. Nano zero-valent iron (nZVI), alone or supporting noble metals, can offer efficient reductive degradation of chlorinated organic pollutants (Wu et al., 2005; Wu and Ritchie, 2008). Nanomembranes containing TiO_2 can also perform photocatalysis to degrade organic pollutants (Leong et al., 2014). Copper mining water could be efficiently treated by using a ceramic nanofiltration membrane that possess α-Al_2O_3 flat-sheet supporter containing Fe_2O_3, TiO_2, or ZrO_2 nanoparticles. This membrane also served to waste salt recycling and purification of seawater (Wang and Wang, 2020).

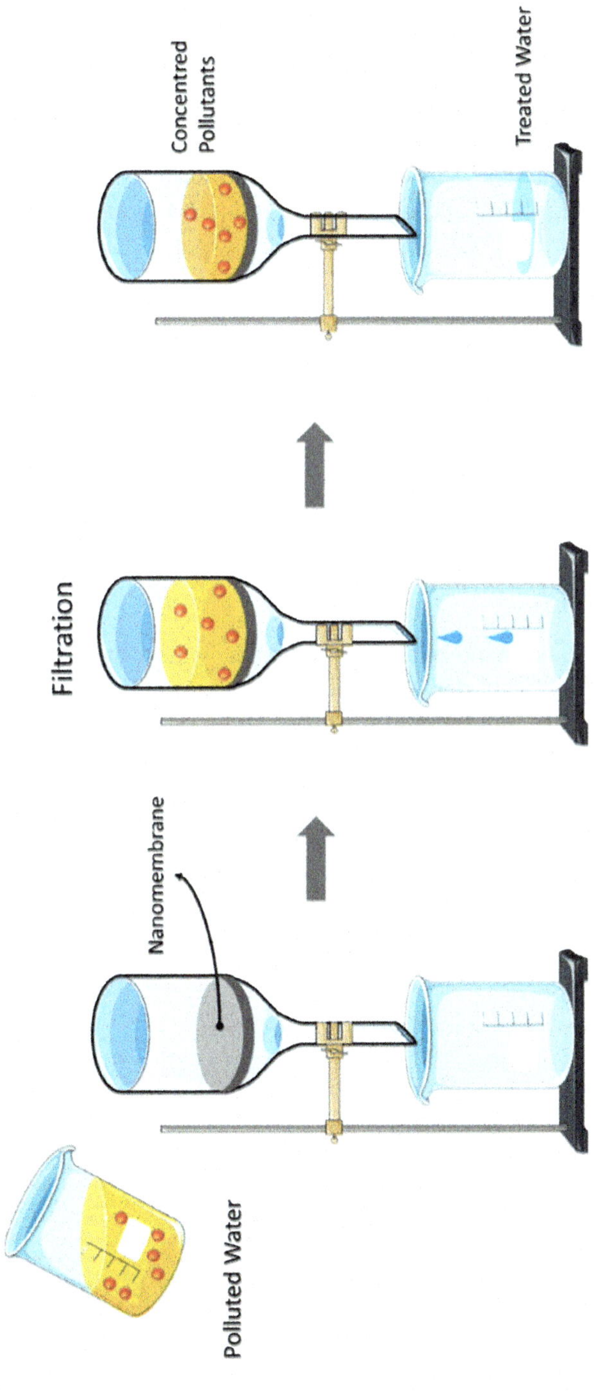

FIGURE 10.2 Nanomembranes to treat polluted water through filtration.

In order to prevent bacterial attachment to the membranes and biofilm formation, nanomaterials exhibiting antimicrobial properties, such as carbon nanotubes and silver nanoparticles can be used in association with filtration membranes (Andrade et al., 2015; Liu et al., 2018; Qing et al., 2018). There are also nanomaterials functionalized to present antimicrobial properties that can serve as nanomembranes to treat contaminated water, such as graphene and graphene oxide (Li, Liu et al., 2016; Bodzek et al., 2020).

10.3.2 ADSORPTION OF WATER POLLUTANTS PROMOTED BY NANOMATERIALS

Adsorption of water pollutants can be performed by regular nanoadsorbents or by nanomotors offering adsorption capacity.

10.3.2.1 NANOADSORBENTS

Nanoadsorbents or nanosorbents remediate water by interacting with pollutants through van der Walls interactions or polar ones, restraining pollutants mobility and attaching to them (the coagulation process). If the complex possess higher density when compared with water, it precipitates (sedimentation process) to the bottom of the recipient. If the complex is less dense than water, it floats (flotation process) to the surface. So, depending on the polluted being remediated, the nanosorbent–pollutant complex will suffer sedimentation or flotation (Fig. 10.3) (Boutilier et al., 2009; Medvedeva et al., 2015; Kyzas and Matis, 2018; Yaqoob et al., 2020).

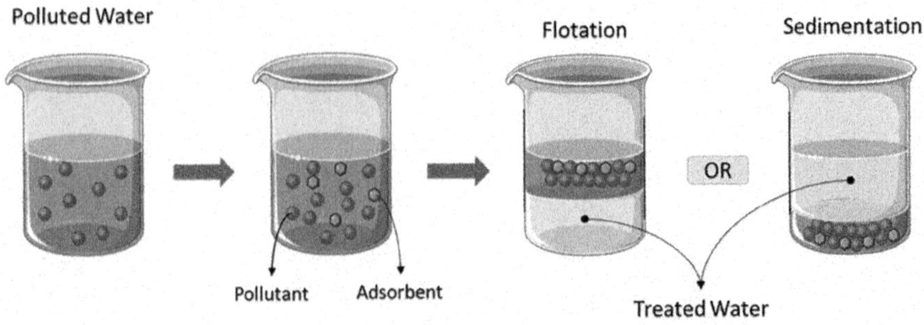

FIGURE 10.3 Nanoadsorbents remediating polluted water.

Prussian blue-based nanoparticles immobilized in a platform can be used to remediate water polluted by Cs ions and this insoluble dye can also be useful to treat patients that were exposed to Cs radionuclides as in Brazil in 1987 (IAEA, 1988). The dye efficiently performs adsorption of heavy metals and on its crystal form, the face-centered cubic Bravais lattices have dimensions similar to Cs hydration radius (Kim et al., 2017). Polyacrylonitrile nanofibers with Prussian blue nanoparticles could promote the adsorption of ^{137}Cs from polluted water (Kim et al., 2018). Activated charcoal–Prussian blue nanoparticles could adsorb at pH 6.8 the elements Cs^{+}, Sr^{+2}, and Co^{+2}, only saturating at 67.9, 29.6 and 15.3 mg/g, respectively (Ali et al., 2020). Polyacrylonitrile nanofibers decorated with Prussian blue nanoparticles offered stable structure, able to remove Cs ions efficiently from contaminated samples (Gwon et al., 2020). Mesoporous silica gel doped by Prussian blue nanoparticles could rapidly promote the removal of rubidium and cesium, even at low concentration, from complex matrixes (Yuan et al., 2020). The Prussian blue-based nanoparticles also present magnetism, favoring the removal of the complex nanosorbent–pollutant from water by applying a magnetic field (Wang and Huang, 2011) and photocatalytic activity (Li, X. et al., 2015); γ-Al_2O_3 spheres with Prussian Blue nanoparticles could deal with orange G dye's pollution though nanoadsorption and nanophotocalysis (Doumic et al., 2015).

Pollution provoked by pharmaceuticals (such as fluoxetine hydrochloride) can be remediated by carbon nanotubes I association to activated carbon by adsorption (Sousa-Moura et al., 2019). Single-walled carbon nanotubes could efficiently show adsorption for benzoic acid offering a high adsorption rate to its water contaminant (Li, De Silva et al., 2020). Multiwalled carbon nanotube is a highly effective adsorbent to deal with water pollution caused by ciprofloxacin hydrochloride (Avcı et al., 2020). Ion-imprinted carbon paste electrodes developed based on multiwalled carbon nanotube composites presented excellent selective adsorption for Pb^{2+} being suitable to remediate water contaminated by the ion (Wang, H. et al., 2020). Batch adsorption of pollutants like trichloroethylene and Congo red was performed involving nanomaterials, including carbon nanotubes, by insoluble nanosponge β-cyclodextrin polyurethane modified with phosphorylated multiwalled carbon nanotubes decorated with titanium dioxide and silver nanoparticles (Taka et al., 2020). Dyes removal from polluted water could also be remediated by carbon nanotube attached to laccase enzyme, with adsorption capacity being regenerated by enzymatic degradation of contaminants (Zhang et al., 2020).

Organic pollutants and heavy metals can be remediated through the use of the tailored adsorbents known as dendrimers through adsorption based on electrostatic interactions, complexation, hydrogen bonding, or hydrophobic effect. Dendrimers can promote the adsorption of pollutants and also be designed to present catalytic activity (Crooks et al., 2001; Wazir et al., 2020). Its external branches can be terminated in functional groups, such as amine and hydroxyl, displaying affinity to heavy metals adsorption and the interior shells, by being hydrophobic environments, can deal with organic pollutants. Dendrimer-ultrafiltration system could be designed to recover Cu^{2+} ions from aqueous solutions (Diallo et al., 2005). Poly(propyleneimine) dendrimer was used to functionalize a magnetic zirconium-based metal–organic framework nanocomposite that presented good adsorption capacity for the dyes Direct red 31 and Acid blue 92, respectively 173.7 and 122.5 mg/g (Far et al., 2020). To promote the adsorption of Mn^{2+} and Co^{2+} present in polluted water, Schiff base functionalized polyamidoamine dendrimer/silica can be used (Qiao, W. et al., 2020).

Graphene oxide can also promote the adsorption of pollutants, and presents a high water dispersion due to functional groups, such as phenol, hydroxyl, carbonyl, carboxyl, and epoxy (Aliyev et al., 2019). Organic dyes and Pb^{2+} and Cd^{2+}'s remediation could be performed by a sustainable nanomaterial: date syrup-based graphene sand hybrid (Khan et al., 2019). The radionuclides U^{6+}, Pu^{2+}, and Th^{4+} could be adsorbed and coagulated by graphene oxide, being easily removed from polluted water after centrifugation (Li et al, 2012; Romanchuk et al., 2013; Laver, 2020). Cs^{+} could be removed at a rate of 220 mg/g by graphene oxide fibers functionalized with sodium. After a wash step using NaOH, the nanomaterial is ready to a new removal cycle (Lee et al., 2019). Adsorption followed by catalytic advanced oxidation processes of personal care products could be performed by nanohybrids of reduced graphene oxide and nZVI, remediating 82–99% of the pollutants (Masud et al., 2020). The nanocomposite composed of magnetic graphene oxide and ZnO was able not only to adsorb but also to degrade the pharmaceutical pollutant tetracycline (Qiao, D. et al., 2020). Graphene oxide, after being modified by bis(2-pyridylmethyl) amino groups could act as adsorbent to remediate the pollution caused by Cu^{2+}, Ni^{2+}, and Co^{2+} with desorption/regeneration capacities higher than 10 cycles (Chaabane et al., 2020). The reusable chitosan-reinforced graphene oxide–hydroxyapatite composite matrix exhibited high adsorption to remediate water contamination promoted by the dyes Congo red, Acid red 1 and Reactive Red (Sirajudheen et al., 2020). Graphene oxide nanoribbons are interesting options to remediate As^{5+}

and Hg^{2+}'s pollution (Sadeghi et al., 2020). Graphene oxide after suffering functionalization by carbon disulfide improved graphene oxide's capacity to adsorpt the pollutant Pb^{2+} in 83.2% (Lian et al., 2020). A network of nanomaterials containing graphene oxide (carbon nanotubes/graphene oxide/sodium alginate, triple-network nanocomposite hydrogel) could be successfully used to remediate antibiotics' pollution (Ma, J. et al., 2020).

Pb^{2+} ions' pollution can be remediated through nanoadsorption up to 125 mg/g by oval CuO nanostructures (Farghali et al., 2013), as same as methylene blue dye from which water pollution can be efficiently remediated by cupric oxide nanoparticles (Mustafa et al., 2013). Copper atoms can be converted into Cu nanoparticles by green synthesis protocols using *Cynomorium coccineum* extract, and the nanostructure was used to promote adsorption of methylene blue dye (Sebeia et al., 2020). Anion exchangers doped with binary Cu^{2+}–Fe^{3+} oxide proved to be able to promote the adsorption of As^{3+} water remediating arsenate pollution (Jacukowicz-Sobala et al., 2020).

The industrial dye Acidic dye 36 could be efficiently adsorbed by nZVI in 90 s at an optimum environment at pH = 5.5, nZVI concentration of 0.5 and dye's concentration of 30 mg/L (Delnavaz and Kazemimofrad, 2020). The commercial nZVI Nanofer 25 could not only deal efficiently with organic and inorganic pollution but also eliminated pathogens from contaminated water (Oprčkal et al., 2017). Core-shell nanoscale zero-valent iron showed efficient adsorption of Cu^{2+} being suitable to remediate water polluted by this ion (Dada et al., 2020). Additional function can be conferred to nZVI depending on the type of chemical modification performed on the nanomaterial. For example, phosphate modification favors not only Cr^{6+} adsorption but also ion's reduction making remediation more efficient (Li, M. et al., 2020). The same occurs with the composite composed by flax straw biochar and nanoscale zero-valent iron that presented higher capacity to promote Cr^{6+} adsorption and reduction than flax straw biochar alone (Ma, F. et al., 2020). The system containing magnetic composite microbial extracellular polymeric substances in Fe_3O_4 could present enhanced Sb^{+5} reducibility/adsorbability after association with nZVI (Yang et al., 2020).

Magnetic nanoparticles of Fe_3O_4 can also adsorb pollutants from water. Hg^{2+} (at a maximum rate of 141.57 mg/g) could be remediated by this nanoparticle's humic acid functionalized version attached to oyster shell (He et al., 2019). In a version coated with citric acid, magnetite nanoparticles could remediate Cd^{2+}'s pollution (Singh et al., 2014). Hg^{2+} and Pb^{2+} from industrial wastewater could be remediated by reusable microspheres of silica and Fe_3O_4 (Hu et al., 2010). Metal oxide nanoadsorbents generally can be regenerated by pH change and after various reuse cycles, they commonly maintain adsorption capacity, but in some cases, it may decrease (Deliyanni et al., 2003; Hu et al., 2006; Sharma et al., 2009). Green synthesis of these particles could be performed by using tea waste template to generate a nanomaterial able to promote As^{3+} and As^{5+}'s adsorption (Lunge et al., 2014).

10.3.2.2 NANOMOTORS OFFERING ADSORBENT CAPACITY

Nanomotors are synthetic nanomaterials that can promote pollutants adsorption and/or catalysis (Ying and Pumera, 2018) (Fig. 10.4). They are complex structures capable of converting different sources of energy into motion, moving by themselves, eliminating the need for the mixing step during remediation (Sánchez et al., 2015). Inorganic and biological/organic materials can be used to generate them (Gao and Wang, 2014).

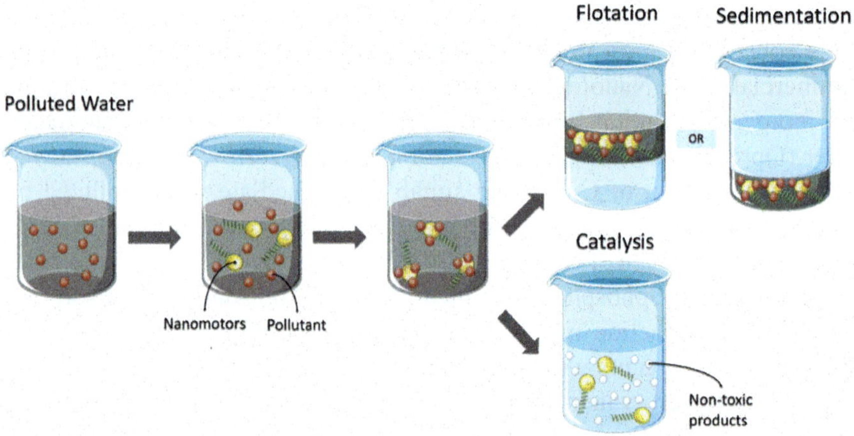

FIGURE 10.4 Nanomotors to remediate polluted water.

When performing adsorption of pollutants, they work in a way similar to regular nanosorbents, forming a complex with pollutants and precipitating or floating depending on the density of the complex.

Nonomotors with noncatalytic properties are not so common to treat water pollution. Generally, nanomotors are able to promote not only the adsorption but also contaminant's degradation (Yaqoob et al., 2020). A popular application for noncatalytic nanomotors is to treat Pb intoxication. Nanomotors capable of promoting excessive blood level of Pb^{2+} ions' adsorption inside red blood cells are useful to treat lead poisoning. A magnetic mesoporous silica/ε-polylysine nanomotor-based remover could perform this task presenting low cytotoxicity and blood compatibility (Liu, Z. et al., 2020). A magnetic nanomotor adsorbent containing Fe_3O_4 nanoparticle modified with meso-2, 3-dimercaptosuccinic acid proved to be also efficient to perform the Pb^{2+} adsorption in pig models (Wang et al., 2021).

10.3.3 DEGRADATION OF WATER POLLUTANTS PROMOTED BY NANOMATERIALS

Degradation of water pollutants can be performed by regular nanophoto-catalysts or by nanomotors offering catalysis capacity.

10.3.3.1 NANOMOTORS OFFERING CATALYSIS CAPACITY

Nanomotors can also promote pollutants' degradation, converting them into nontoxic or less toxic subproducts by catalysis (Ying and Pumera, 2018).

Fresh water and sea water, for example, contaminated by mixed pollution (organic per- and polyfluoroalkyl substances and inorganic heavy metal ions), could be remediated by a metal–organic framework (MOF) nanomotor system (Guo, Z. et al., 2020). Azo dyes could be degraded by bubble-propelled Fe^0 Janus nanomotors, low-cost and biocompatible nanostructures stabilized by an ultrathin iron oxide shell (Teo et al., 2016). Green cellulose nanocrystals (obtained from renewable biomass) decorated with Fe_2O_3/Pd nanoparticles (magnetoresponsive and catalytically active structure), in a self-propelled way not requiring surfactants, could in situ degrade pollutants, such as dyes, being interesting tool for water remediation (Dhar et al., 2020). A rotating magnetic field and the presence of $NaBH_4$ allowed mesoporous magnetic nanorods containing a core of CoNi and a Pt shell to be propelled and degrade methylene blue, 4-nitrophenol, and rhodamine B (organic pollutants) (García-Torres et al., 2017). Self-propelled multimotion modes helix carbon nanocoil/TiO_2 nanomotors could degrade more than 50% of the pollutant phenol in 30 min (Li, X. et al., 2019).

Nanomaterials can also be used to generate micromotors to be applied in water remediation (Yaqoob et al., 2020). Colloidal carbon WO_3 nanoparticle composite spheres could be used to generate a light-driven Au-WO_3-C Janus micromotor to degrade sodium-2,6-dichloroindophenol and rhodamine B even in low concentration (Zhang, Q. et al., 2017). MnO_2 wrapped carbon nanotubes could be used to generate a bubble-propelled micromotor which can degrade the organic pollutant dye Congo red (Wu, X. et al., 2020).

10.3.3.2 NANOPHOTOCATALYSTS

Nanophotocatalysts can convert light radiation (e.g., sun light) into chemical energy to promote pollutant's degradation (Fig. 10.5) (Gómez-Pastora et al., 2017). The photocatalysis can happen through advanced oxidative processes (AOPs) or by electron–hole pair formation. In the first scenario, ROS are generated in the surrounding (Werkneh and Rene, 2019), and in the second scenario, the electron–hole pair formation in a semiconductor produces two redox polos: that can catalyze the formation of ROS or act directly in contaminants' degradation (Li, Yu et al., 2016).

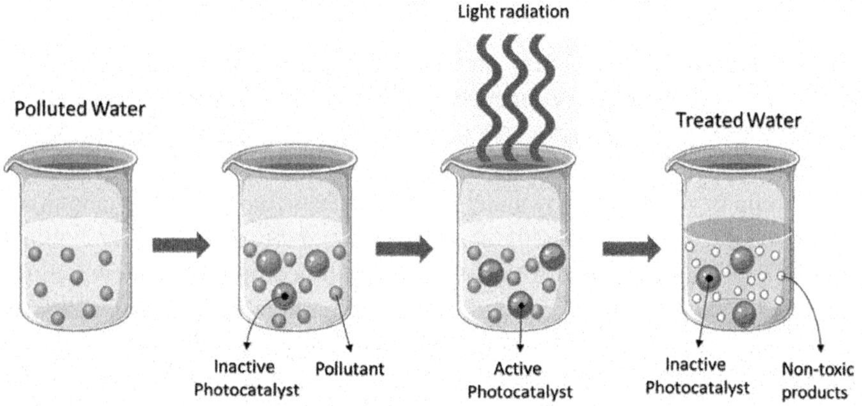

FIGURE 10.5 Nanophotocatalysts promoting water remediation.

Metallic oxides, especially iron and zinc ones, generally perform AOP, being indirectly responsible for pollutants degradation (the direct responsible are ROS). The energy from a luminous photon is absorbed, hydroxyl ROS (OH^- and $OH^{\bullet}$) are generated through photo-Fenton reaction from water or hydrogen peroxide, and finally ROS can react with contaminants and degrade them (Singh and Pandey, 2020; Vaiano et al., 2020).

Semiconductor nanomaterials generally perform electron–hole pair formation, being able to degrade pollutants through ROS or directly (Lam et al., 2016). These nanomaterials are capable of, after suffering stimulation, performing electron jumping from the valence band to conduction band (Lam et al., 2016; Li, Yu et al., 2016). The energy from a luminous photon is absorbed stimulating band gap and a negative charge is arriving on conduction band. The valence band, from where the electron came from, suffers from an electron's absence: a "hole" is generated creating a positive electric charge (Horikoshi and Serpone, 2020). As previously discussed, nanomaterials possess a high ratio superficial area/volume, and after the described process, the negative polo in surface (caused by the electron in the conductive band) will be able to reduce pollutants nearby, and the positive polo (due to "hole" in valence band) will be able to oxidize contaminants (Kuang et al., 2020).

There are strategies that use only one nanomaterial to perform photocatalysis of pollutants (such as ZnO or TiO_2 or iron oxide nanoparticles or Prussian blue nanoparticles) (El-Kemary et al., 2010; Das et al., 2014; Pastrana-Martinez et al., 2015; Singh and Pandey, 2020). However, association of two or more photocatalysts nanomaterials may offer better results benefiting from synergism (Bhanvase et al., 2017).

Due to fast electrical conductivity and large surface area and other unique properties (thermal, electric and mechanical), graphene and its oxide are popular support to be used as photocatalyst for nanocomposites (And and Jimmy, 2011; Li, Yu et al, 2016). Associated with ZnO, for example, graphene could participate in rhodamine B's photocatalytic degradation better than the substances separately (Li and Cao, 2011).

Heavy metals and organic compounds could be remediated by carbon nanotubes containing TiO_2/SiO_2 in the hollow structure (Rasheed et al., 2019).

Prussian blue nanoparticles can not only act as nanoadsorbent but also can act as nanophotocatalyst (Wang et al., 2011; Mai et al., 2017). The term "nanozyme" is used to refer to this nanomaterial once its ascorbic acid oxidase-like end peroxidase- activities can sometimes exceed the efficiency of the enzymes themselves (Zhang, W. et al., 2018; Komkova et al., 2018). Prussian blue nanoparticles can be generated in an eco-friendly way through green synthesis, for example, using β-cyclodextrin as surfactant and ethyl alcohol as a green solvent. The synthetized nanoparticle was stable and could degrade rhodamine B under visible exhibiting high catalytic efficiency (Yang et al., 2021). In the presence of light and in association with TiO_2, Prussian blue (PB) can suffer reduction, induced by TiO_2 photocatalysis, enhancing PB's capacity to promote Cs^+ adsorption and water remediation (Park et al., 2020). The nanocomposite composed of zinc oxide and Prussian blue promoted the efficient photocatalytic degradation of pollutants phenol, phenol, 3-aminophenol, and 2,4-dinitrophenol (Rachna et al., 2020).

Carbon dots can also act as photocatalysts (Cailotto et al., 2020). Methylene blue, for example, could be degraded by green-emissive version of the nanomaterial presenting photocatalytic activity (Das et al., 2019). They can also be used to modify other structures conferring them a high catalytic activity. $ZnSnO_3$ cubes modified by carbon dots generated a composite that when compared with pure $ZnSnO_3$, present 21 times higher photocatalytic degradation kinetic toward tetracycline under visible light (Guo, F. et al., 2020). g-C_3N_4/SnO_2 photocatalyst modified with carbon dots presented enhanced capacity to degrade indomethacin (Li et al., 2021).

Silver-based nanomaterials can also be used to remediate polluted water. rhodamine B, for example, could be degraded by silver quantum dots with 2D SnO_2 nanoflakes through photocatalysis (Kumar et al., 2019). A BiOI nanocomposite co-doped with metal silver and carbon dots could efficiently show the photocatalytic degradation of the pollutant 4-Chlorophenol (Guo, Y. et al., 2020). The composite involving titanium dioxide–silver could be successfully applied to promote the photocatalytic degradation of methylene blue (Ibukun and Jeong, 2020).

Fullerene-based nanomaterials are also interesting option to remediate contaminated water. Through photocatalytic activity, the pollutants phenol and methylene blue could be remediated by functionalized fullerene (using zinc porphyrin) associated with TiO_2 (Regulska et al., 2019). Carboxylated zinc phthalocyanine–carboxylated C_{60}-titanium dioxide nanosheets could promote photocatalytic reduction of nitrobenzene to aniline (Liu, L. et al., 2020). Metal-free P-doped g-C3N4 photocatalyst was decorated by fullerene and could degrade efficiently the pesticide imidacloprid through photocatalysis (Sudhaik et al., 2020). Ag(I)-fullerene (C_{60}) composite could promote the visible light-driven photodegradation of organic pollutants (Yi et al., 2020).

$ZnWO_4$ nanoparticles have proved to be able to convert into less toxic forms the pollutants V^{5+} and Cr^{6+} making it possible to recover V^{6+} and Cr^{3+} (Zhao et al., 2016). $ZnWO_4$ (ethylene glycol 62.5) nanoparticles could promote methylene blue, rhodamine B, methyl orange, and Cr^{6+} photocatalytic removal (He et al., 2020). High stability of $ZnWO_4$ nanoparticles also have been proved to efficiently degrade para-aminobenzoic acid (Faka et al., 2021). $Fe_3O_4/ZnWO_4/CeVO_4$ nanoparticles could promote the degradation of 90% of methyl violet and 70% of methylene blue, respectively. This degradation was observed after visible light irradiation (Marsooli et al., 2020).

Dye-contaminated industrial wastewater could be successfully remediated by mixed-phase bismuth ferrite nanoparticles (Kalikeri and Kodialbail,

2018). Bi_2WO_6 exhibits the capacity to promote the photocatalytic degradation of rhodamine B, however, after loading of Ag and graphene, the degradation efficiency enhanced considerably (Low et al., 2014). Bismuth ferrite nanoparticles promoted the photocatalysis of the dye Acid Yellow-17 efficiently (Kalikeri and Kodialbail, 2020). Nitrogen-doped graphene oxide-supported gadolinium-doped bismuth ferrite was efficient in remediating rhodamine B (Dixit et al., 2020).

CuO nanoparticles can be generated through green synthesis using *Citrofortunella microcarpa*, and the nanomaterial proved to be useful to perform efficient rhodamine B's photocatalytic degradation under UV light (Rafique et al., 2020). In fact, green synthesis of different nanomaterials to perform water remediation have been proposed recently (Bessa et al., 2020; Bolade et al., 2020; Nasrollahzadeh et al., 2021).

10.4 CONCLUSIONS

The environment has been suffering with human actions ignoring sustainability principles, and environmental pollution has become a serious problem worldwide. Live on Earth is threaten and urgent actions need to be performed in order to try to attenuate or even solve the risks. As water is a vital resource, water remediation protocols possessing low-cost and being eco-friendly, and able to be performed at large scale are highly necessary. Nanotechnology field has been presenting different kinds of nanomaterials, chemically modified or not, to efficiently treat contaminated water and restore its quality.

10.5 FUTURE PERSPECTIVES

Nanotechnology field has been advancing in a fast pace and is expected to continue to rapidly develop specially when it comes to protocols to generate biodegradable materials, preferentially through green synthesis, to perform desirable roles such as remediating polluted water. It is also expected that field experiments involving these materials to restore contaminated environs increase in number as same as research works on nanomaterials safety, biocompatibility and feasibility to be used in large scale. However, the success of implementing nanomaterials to remediate environmental pollution in aquatic environments is not sufficient to solve the problems generated by harmful anthropic actions. Environmental awareness and concern for sustainable development must be widely adopted worldwide.

KEYWORDS

- **nanomaterials**
- **water pollution remediation**
- **nanotechnology**
- **environmental pollution**

REFERENCES

Ahsan, M. A.; Jabbari, V.; Imam, M. A.; Castro, E.; Kim, H.; Curry, M. L.; Valles-Rosales, D. J.; Noveron, J. C. Nanoscale Nickel Metal Organic Framework Decorated Over Graphene Oxide and Carbon Nanotubes for Water Remediation. *Sci. Total Environ.* **2020,** *698*, 134214.

Akbari, A.; Sheath, P.; Martin, S. T.; Shinde, D. B.; Shaibani, M.; Banerjee, P. C.; Tkacz, R.; Bhattacharyya, D.; Majumder, M. Large-Area Graphene-Based Nanofiltration Membranes by Shear Alignment of Discotic Nematic Liquid Crystals of Graphene Oxide. *Nat. Commun.* **2016,** *7*, 10891.

Ali, H.; Khan, E.; Ilahi, I. Environmental Chemistry and Ecotoxicology of Hazardous Heavy Metals: Environmental Persistence, Toxicity, and Bioaccumulation. *J. Chem.* **2019,** 1–14.

Ali, M. M. S.; Sami, N. M.; El-Sayed, A. A. Removal of Cs+, Sr2+ and Co2+ by Activated Charcoal Modified with Prussian Blue Nanoparticle (PBNP) from Aqueous Media: Kinetics and Equilibrium Studies. *J. Radioanal. Nucl. Chem.* **2020,** *324*, 189–201.

Aliyev, E.; Filiz, V.; Khan, M. M.; Lee, Y. J.; Abetz, C.; Abetz, V. Structural Characterization of Graphene Oxide: Surface Functional Groups and Fractionated Oxidative Debris. *Nanomaterials* **2019,** *9*, 1180.

And, X.; Jimmy, C. Y. Graphene-Based Photocatalytic Composites. *Rsc. Adv.* **2011,** *1*, 1426–1434.

Andrade, P. F.; de Faria, A. F.; Oliveira, S. R.; Arruda, M. A. Z.; Gonçalves, M. C. Improved Antibacterial Activity of Nanofiltration Polysulfone Membranes Modified with Silver Nanoparticles. *Water Res.* **2015,** *81*, 333–342.

Anzar, N.; Hasan, R.; Tyagi, M.; Yadav, N.; Narang, J. Carbon Nanotube—A Review on Synthesis, Properties and Plethora of Applications in the Field of Biomedical Science. *Sensors. Int.* **2020,** *1*, 100003.

Avcı, A.; İnci, I.; Baylan, N. Adsorption of Ciprofloxacin Hydrochloride on Multiwall Carbon Nanotube. *J. Mol. Struct.* **2020,** *1206*, 127711.

Bagazgoïtia, N. V. E.; Bailey, H. D.; Orsi, L.; Lacour, B.; Guerrini-Rousseau, L.; Bertozzi, A. I.; Leblond, P.; Faure-Conter, C.; Pellier, I.; Freycon, C.; Doz, F.; Puget, S.; Ducassou, S.; Clavel, J. Maternal Residential Pesticide use During Pregnancy and Risk of Malignant Childhood Brain Tumors: A Pooled Analysis of the ESCALE and ESTELLE Studies (SFCE). *Int. J. Cancer* **2018,** *142*, 489–497.

Bai, X.; Wang, L.; Wang, Y.; Yao, W.; Zhu, Y. Enhanced Oxidation Ability of g-C3N4 Photocatalyst via C60 Modification. *Appl. Catal. B Environ.* **2014,** *152–153*, 262–270.

Baldissarelli, D. P.; Vargas, G. D. L. P.; Korf, E. P.; Galon, L.; Kaufmann, C.; Santos, J. B. Remediation of Soils Contaminated by Pesticides Using Physicochemical Processes: A Brief Review. *Planta Daninha* **2019,** *37*, e019184975.

Bebianno, M. J.; Gonzalez-Rey, M. Ecotoxicological Risk of Personal Care Products and Pharmaceuticals. In *Aquatic Ecotoxicology: Advancing Tools for Dealing with Emerging Risks*; Academic Press: Cambridge, USA, 2015; p 518.

Bessa, A.; Gonçalves, G.; Henriques, B.; Domingues, E. M.; Pereira, E.; Marques, P. A. A. P. Green Graphene–Chitosan Sorbent Materials for Mercury Water Remediation. *Nanomaterials* **2020,** *10*, 1474.

Bethi, B.; Sonawane, S. H. Nanomaterials and Its Application for Clean Environment. In *Nanomaterials for Green Energy*, 2018; pp 385–409.

Bhanvase, B. A.; Shende, T. P.; Sonawane, S. H. A Review on Graphene–TiO_2 and Doped Graphene–TiO_2 Nanocomposite Photocatalyst for Water and Wastewater Treatment. *Environ. Technol. Rev.* **2017,** *6*, 1–14.

Bhaskar, M.; Gnanamani, A.; Ganeshjeevan, R. J.; Chandrasekar, R.; Sadulla, S.; Radhakrishnan, G. Analyses of Carcinogenic Aromatic Amines Released from Harmful Azo Colorants by *Streptomyces* sp. SS07. *J. Chromatogr. A* **2003,** *1018*, 117–123.

Biglova, Y. N.; Mustafina, A. G. Nucleophilic Cyclopropanation of [60]Fullerene by the Addition–Elimination Mechanism. *RSC Adv.* **2019,** *9*, 22428–22498.

Bin, Q.; Lin, B.; Zhu, K.; Shen, Y.; Man, Y.; Wang, B.; Lai, C.; Chen, W. Superior Trichloroethylene Removal from Water by Sulfide-Modified Nanoscale Zero-Valent Iron/Graphene Aerogel Composite. *J. Environ. Sci.* **2020,** *88*, 90–102.

Bodzek, M.; Konieczny, K.; Kwiecińska-Mydlak, A. Nanotechnology in Water and Wastewater Treatment. Graphene—The Nanomaterial for Next Generation of Semipermeable Membranes. *Crit. Rev. Environ. Sci. Technol.* **2020,** *50*, 1515–1579.

Bolade, O. P.; Williams, A. B.; Benson, N. U. Green Synthesis of Iron-Based Nanomaterials for Environmental Remediation: A Review. *Environ. Nanotechnol. Monit. Manag.* **2020,** *13*, 100279.

Boutilier, L.; Jamieson, R.; Gordon, R.; Lake, C.; Hart, W. Adsorption, Sedimentation, and Inactivation of *E. coli* within Wastewater Treatment Wetlands. *Water Res.* **2009,** *43*, 4370–4380.

Boutin, C.; Eme, C. *Domestic Wastewater Characterization by Emission Source*; 13eme Congres Spécialisé IWA on Small Water and Wastewater Systems, Athènes, Greece, 2016; p 8 https://hal.archives-ouvertes.fr/hal-01469077/document

Brunet, L.; Lyon, D. Y.; Hotze, E. M.; Alvarez, P. J. J.; Wiesner, M. R. Comparative Photoactivity and Antibacterial Properties of C60 Fullerenes and Titanium Dioxide Nanoparticles. *Environ. Sci. Technol.* **2009,** *43*, 4355–4360.

Cailotto, S.; Negrato, M.; Daniele, S.; Luque, R.; Selva, M.; Emanuele Amadio, E.; Perosa, A. Carbon Dots as Photocatalysts for Organic Synthesis: Metal-Free Methylene–Oxygen-Bond Photocleavage. *Green Chem.* **2020,** *22*, 1145–1149.

Chaabane, L.; Beyou, E.; El Ghali, A.; Baouab, M. H. V. Comparative Studies on the Adsorption of Metal Ions from Aqueous Solutions Using Various Functionalized Graphene Oxide Sheets as Supported Adsorbents. *J. Hazard. Mater.* **2020,** *389*, 121839.

Chaoua, S.; Boussaa, S.; El Gharmali, A.; Boumezzough, A. Impact of Irrigation with Wastewater on Accumulation of Heavy Metals in Soil and Crops in the Region of Marrakech in Morocco. *J. Saudi Soc. Agric. Sci.* **2019,** *18*, 429–436.

Chen, Y.; Shi, H.; Guo, H.; Ling, C.; Yuan, X.; Li, P. Hydrated Titanium Oxide Nanoparticles Supported on Natural Rice Straw for Cu (II) Removal from Water. *Environ. Technol. Inno.* **2020,** *20*, 101143.

Chen, Z.; Mahmud, S.; Cai, L.; He, Z.; Yang, Y.; Zhang, L. Zhao, S.; Xiong, Z. Hierarchical Poly(Vinylidene Fluoride)/Active Carbon Composite Membrane with Self-Confining Functional Carbon Nanotube Layer for Intractable Wastewater Remediation. *J. Membr. Sci.* **2020,** *603,* 118041.

Crini, G.; Lichtfouse, E. Advantages and Disadvantages of Techniques Used for Wastewater Treatment. *Environ. Chem. Lett.* **2019,** *17,* 145–155.

Crooks, R. M.; Zhao, M.; Sun, L.; Chechik, V.; Yeung, L. K. Dendrimer-Encapsulated Metal Nanoparticles: Synthesis, Characterization, and Applications to Catalysis. *Acc. Chem. Res.* **2001,** *34,* 181–190.

Dada, A. O.; Adekola, F. A.; Odebunmi, E. O.; Dada, F. E.; Bello, O. S.; Ogunlaja, A. S. Bottom-up Approach Synthesis of Core-Shell Nanoscale Zerovalent Iron (CS-nZVI): Physicochemical and Spectroscopic Characterization with Cu(II) Ions Adsorption Application. *MethodsX* **2020,** *7,* 100976.

Das, C.; Sen, S.; Singh, T.; Ghosh, T.; Paul, S. S.; Kim, T. W.; Jeon, S.; Maiti, D. K.; Im, J.; Biswas, G. Green Synthesis, Characterization and Application of Natural Product Coated Magnetite Nanoparticles for Wastewater Treatment. *Nanomaterials* **2020,** *10,* 1615.

Das, G. S.; Shim, J. P.; Bhatnagar, A.; Tripathi, K. M.; Kim, T. Biomass-Derived Carbon Quantum Dots for Visible-Light-Induced Photocatalysis and Label-Free Detection of Fe(III) and Ascorbic Acid. *Sci. Rep.* **2019,** *9,* 15084.

Das, R.; Sarkar, S.; Chakraborty, S.; Choi, H.; Bhattacharjee, C. Remediation of Antiseptic Components in Wastewater by Photocatalysis Using TiO_2 Nanoparticles. *Ind. Eng. Chem. Res.* **2014,** *53,* 3012–3020.

Deliyanni, E. A.; Bakoyannakis, D. N.; Zouboulis, A. I.; Matis, K. A. Sorption of As(V) Ions by Akaganeite-Type Nanocrystals. *Chemosphere* **2003,** *50,* 155–163.

Delnavaz, M.; Kazemimofrad, Z. Nano Zerovalent Iron (NZVI) Adsorption Performance on Acidic Dye 36 Removal: Optimization of Effective Factors, Isotherm and Kinetic Study. *Environ. Prog. Sustain. Energy* **2020,** *39,* e13349.

Deng, Y.; Guo, C.; Zhang, H.; Yin, X.; Chen, L.; Wu, D.; Xu, J. Occurrence and Removal of Illicit Drugs in Different Wastewater Treatment Plants with Different Treatment Techniques. *Environ. Sci. Eur.* **2020,** *32,* 28.

Dhar, P.; Narendren, S.; Gaur, S. S.; Sharma, S.; Kumar, A.; Katiyar, V. Self-Propelled Cellulose Nanocrystal Based Catalytic Nanomotors for Targeted Hyperthermia and Pollutant Remediation Applications. *Int. J. Biol. Macromol.* **2020,** *158,* 1020–1036.

Diallo, M. S.; Christie, S.; Swaminathan, P.; Johnson, J. H.; Goddard, W. A. Dendrimer Enhanced Ultrafiltration. 1. Recovery of Cu(II) from Aqueous Solutions Using PAMAM Dendrimers with Ethylene Diamine Core and Terminal NH2 Groups. *Environ. Sci. Technol.* **2005,** *39* (5), 1366–1377.

Dixit, T. K.; Sharma, S.; Sinha, A. K. Synergistic Effect of N-rGO Supported Gd Doped Bismuth Ferrite Heterojunction on Enhanced Photocatalytic Degradation of Rhodamine B. *Mat. Sci. Semicon. Proc.* **2020,** In press.

Djordjevic, A.; Merkulov, D. S.; Lazarevic, M.; Borisev, I.; Medic, I.; Pavlovic, V.; Miljevic, B.; Abramovic, B. Enhancement of Nano Titanium Dioxide Coatings by Fullerene and Polyhydroxy Fullerene in the Photocatalytic Degradation of the Herbicide Mesotrione. *Chemosphere* **2018,** *196,* 145–152.

Dong, L. X.; Huang, X. C.; Wang, Z.; Yang, Z.; Wang, X. M.; Tang, C. Y. Separation and Purification Technology A Thin-Film Nanocomposite Nanofiltration Membrane Prepared on a Support with in Situ Embedded Zeolite Nanoparticles. *Sep. Purif. Technol.* **2016,** *166,* 230–239.

Doumic, L.; Salierno, G.; Cassanello, M.; Haure, P.; Ayude, M. Efficient Removal of Orange G Using Prussian Blue Nanoparticles Supported Over Alumina. *Catal. Today* **2015,** *240,* 67–72.

Dresselhaus, M. S.; Dresselhaus, G.; Eklund, P. C.; Rao, A. M. Carbon Nanotubes. In *The Physics of Fullerene-Based and Fullerene-Related Materials*; Springer: Dordrecht, 2000; pp 331–379.

Drexler, K. E. Nanotechnology: From Feynman to Funding. *B. Sci. Technol. Soc.* **2004,** *24,* 21–27.

El-Kemary, M.; El-Shamy, H.; El-Mehasseb, I. Photocatalytic Degradation of Ciprofloxacin Drug in Water Using ZnO Nanoparticles. *J. Lumin.* **2010,** *130,* 2327–2331.

El-Sikaily, A.; Khaled, A.; El-Nemr, A. Textile Dyes Xenobiotic and Their Harmful Effect. In *Non-Conventional Textile Waste Water Treatment*; Nova Science Publishers, Inc., 2012; pp 31–64.

Faka, V.; Tsoumachidou, S.; Moschogiannaki, M.; Kiriakidis, G.; Poulios, I.; Binas, V. $ZnWO_4$ Nanoparticles as Efficient Photocatalyst for Degradation of Para-Aminobenzoic Acid: Impact of Annealing Temperature on Photocatalytic Performance. *J. Photochem. Photobiol. A: Chem.* **2021,** *406,* 113002.

Fang, Z.; Gao, Y.; Bolan, N.; Shaheen, S. M.; Xu, S.; Wu, X.; Xu, X.; Hu, H.; Lin, J.; Zhang, F.; Li, J.; Rinklebe, J.; Wang, H. Conversion of Biological Solid Waste to Graphene-Containing Biochar for Water Remediation: A Critical Review. *Chem. Eng. J.* **2020,** *390,* 124611.

Far, H. S.; i Hasanzadeh, M.; Nashtaei, M. S.; Rabbani, M.; Haji, A.; Moghadam, B. H. PPI-Dendrimer-Functionalized Magnetic Metal–Organic Framework (Fe3O4@MOF@PPI) with High Adsorption Capacity for Sustainable Wastewater Treatment. *ACS Appl. Mater. Interf.* **2020,** *12,* 25294–25303.

Farghali, A. A.; Bahgat, M.; Allah, A. E.; Khedr, M. H. (2013) Adsorption of Pb (II) Ions from Aqueous Solutions Using Copper Oxide Nanostructures, Beni-Seuf Univ. *J. Appl. Sci.* **2013,** *2,* 61–71.

Farkhondeh, T.; Naseri, K.; Esform, A.; Aramjoo, H.; Naghizadeh, A. Drinking Water Heavy Metal Toxicity and Chronic Kidney Diseases: A Systematic Review. *Rev. Environ. Health* **2020,** 0110.

Fatima, S. A.; Hamid, A.; Yaqub, G.; Javed, A.; Akram, H. Detection of Volatile Organic Compounds in Blood of Farmers and Their General Health and Safety Profile. *Nat. Environ. Pollut. Technol.* **2018,** *17,* 657–660.

Fulekar, M. H.; Pathak, B. *Bioremediation Technology: Hazardous Waste Management*; CRC Press, 2020; p 366.

Gao, W.; Wang, J. The Environmental Impact of Micro/Nanomachines: A Review. *ACS Nano* **2014,** *8,* 3170–3180.

García-Torres, J.; Serra, A.; Tierno, P.; Alcobe, X.; Valles, E. Magnetic Propulsion of Recyclable Catalytic Nanocleaners for Pollutant Degradation. *ACS Appl. Mater. Interf.* **2017,** *9,* 23859–23868.

Gavrilescu, M.; Demnerová, K.; Aamand, J.; Agathos, S.; Fava, F. N. Emerging Pollutants in the Environment: Present and Future Challenges in Biomonitoring, Ecological Risks and Bioremediation. *Biotechnol.* **2015,** *32,* 147–156.

Gehrke, I.; Geiser, A.; Somborn-Schulz, A. Innovations in Nanotechnology for Water Treatment. *Nanotechnol. Sci. Appl.* **2015,** *8,* 1–17.

Ghadimi, M.; Zangenehtabar, S.; Homaeigohar, S. An Overview of the Water Remediation Potential of Nanomaterials and Their Ecotoxicological Impacts. *Water*, **2020,** *12* (4), 1–23.

Gheju, M. Hexavalent Chromium Reduction with Zero-Valent Iron (ZVI) in Aquatic Systems. *Water Air Soil Pollut.* **2011,** *222,* 103–148.

Gil-Díaz, M.; Álvarez, M. A.; Alonso, J.; Lobo, M. C. Effectiveness of Nanoscale Zero-Valent Iron for the Immobilization of Cu and/or Ni in Water and Soil Samples. *Sci. Rep.* **2020,** *10,* 15927.

Gimiliani, G. T.; Fontes, R. F. C.; Abessa, D. M. S. Modeling the Dispersion of Endocrine Disruptors in the Santos Estuarine System (Sao Paulo State, Brazil). *Braz. J. Oceanogr.* **2016,** *64,* 1–8.

Gómez-Pastora, J.; Dominguez, S.; Bringas, E.; Rivero, M. J.; Ortiz, I.; Dionysiou, D. D. Review and Perspectives on the Use of Magnetic Nanophotocatalysts (MNPCs) in Water Treatment. *Chem. Eng. J.* **2017,** *310,* 407–427.

Gray, N. F. *Biology of Wastewater Treatment*; Imperial College: London, 2004; p 1444.

Guan, X.; Sun, Y.; Qin, H.; Li, J.; Lo, I. M.; He, D.; Dong, H. The Limitations of Applying Zero-Valent Iron Technology in Contaminants Sequestration and the Corresponding Countermeasures: The Development in Zero-Valent Iron Technology in the Last Two Decades (1994–2014). *Water Res.* **2015,** *75,* 224–248.

Guerra, F. D.; Attia, M. F.; Whitehead, D. C.; Alexis, F. Nanotechnology for Environmental Remediation: Materials and Applications. *Molecules* **2018,** *23,* 1760.

Guo, F.; Huang, X.; Chen, Z.; Sun, H.; Shi, W. Investigation of Visible-Light-Driven Photocatalytic Tetracycline Degradation via Carbon Dots Modified Porous $ZnSnO_3$ Cubes: Mechanism and Degradation Pathway. *Sep. Purif. Technol.* **2020,** *253,* 117518

Guo, Y.; Lay, C. H.; Zhou, D.; Dong, S.; Zhang, J.; Ren, N. Enhanced Photocatalytic Performance of Metal Silver and Carbon Dots Co-Doped BiOI Photocatalysts and Mechanism Investigation. *Environ. Sci. Pollut. Res.* **2020,** *27,* 17516–17529.

Guo, Z.; Liu, J.; Li, Y.; McDonald, J. A.; Zulkifli, M. Y. B.; Khan, S. J.; Xie, L.; Gu,Z.; Kong, B.; Liang, K. Chemical Communications Biocatalytic Metal–Organic Framework Nanomotors for Active Water Decontamination. *Chem. Comm.* **2020,** *56,* 14837–14840.

Gwon, Y. J.; Lee, J. J.; Lee, K. W.; Ogden, M. D.; Harwood, L. W.; Lee, T. S. Prussian Blue Decoration on Polyacrylonitrile Nanofibers Using Polydopamine for Effective Cs Ion Removal. *Ind. Eng. Chem. Res.* **2020,** *59,* 4872–4880.

Hammad, M.; Fortugno, P.; Hardt, S.; Kim, C.; Salamon, S.; Schmidt, T. C.; Wende, H.; Schulz, C.; Wiggers, H. Large-Scale Synthesis of Iron Oxide/Graphene Hybrid Materials as Highly Efficient Photo-Fenton Catalyst for Water Remediation. *Environ. Technol. Innov.* **2020,** In Press.

Han, Y.; Xu, Z.; Gao, C. Ultrathin Graphene Nanofiltration Membrane for Water Purification. *Adv. Funct. Mater.* **2013,** *23,* 3693–3700.

Hao, X.; Wang, X.; Liu, R.; Li, S.; van Loosdrecht, M. C. M.; Jiang, H. Environmental Impacts of Resource Recovery from Wastewater Treatment Plants. *Water Res.* **2019,** *160,* 268–277.

He, H.; Luo, Z.; Yu, C. Multifunctional $ZnWO_4$ Nanoparticles for Photocatalytic Removal of Pollutants and Disinfection of Bacteria. *J. Photochem. Photobiol. A Chem.* **2020,** *401,* 112735.

He, C.; Qu, J.; Yu, Z.; Chen, D.; Su, T.; He, L.; Zhao, Z.; Zhou, C.; Hong, P.; Li, Y.; Sun, S.; Li, C. Preparation of Micro-Nano Material Composed of Oyster Shell/Fe3O4 Nanoparticles/Humic Acid and Its Application in Selective Removal of Hg(II). *Nanomaterials* **2019,** *9,* 953.

Hesni, M.A; Hedayati, S. A.; Qadermarzi, A.; Pouladi, M.; ZangiAbadi, S.; Naghshbandi, N. Application of Iron Oxide Nanoparticles in the Reactor for Treatment of Effluent from Fish Farms. *Iran. J. Fish. Sci.* **2020,** *19,* 1319–1328.

Hirsch, A. Addition Reactions of Buckminsterfullerene (C60). *Synthesis* **1995,** *8*, 895–913.

Hong, J.; Xie, J.; Mirshahghassemi, S.; Lead, J. Metal (Cd, Cr, Ni, Pb) Removal from Environmentally Relevant Waters Using Polyvinylpyrrolidone-Coated Magnetite Nanoparticles. *RSC Adv.* **2020,** *10*, 3266–3276.

Horikoshi, S.; Serpone, N. Can the Photocatalyst TiO_2 Be Incorporated Into a Wastewater Treatment Method? Background and Prospects. *Catal. Today* **2020,** *340*, 334–346.

Hristovski, K. D.; Nguyen, H.; Westerhoff, P. K. Removal of Arsenate and 17-Ethinyl Estradiol (EE2) by Iron (Hydr)Oxide Modified Activated Carbon Fibers. *J. Environ. Sci. Heal. A* **2009a,** *44*, 354–361.

Hristovski, K. D.; Westerhoff, P. K.; Moller, T.; Sylvester, P. Effect of Synthesis Conditions on Nano-Iron (Hydr)Oxide Impregnated Granulated Activated Carbon. *Chem. Eng. J.* **2009b,** *146*, 237–243.

Hu, H.; Wang, Z.; Pan, L. Synthesis of Monodisperse Fe_3O_4@ Silica Core–Shell Microspheres and Their Application for Removal Of Heavy Metal Ions from Water. *J. Alloys. Compd.* **2010,** *492*, 656–661.

Hu, J.; Chen, G.; Lo, I. M. C. Selective Removal of Heavy Metals from Industrial Wastewater Using Maghemite Nanoparticle: Performance and Mechanisms. *J. Environ. Eng.* **2006,** *132*, 7.

Huang, H.; Song, Z.; Wei,N.; Shi, L.; Mao, Y.; Ying, Y.; Sun, L.; Xu, Z.; Peng, X. Ultrafast Viscous Water Flow through Nanostrand-Channelled Graphene Oxide Membranes. *Nat. Commun.* **2013,** *4*, 2979.

Ibukun, O.; Jeong, H. K. Tailoring Titanium Dioxide by Silver Particles for Photocatalysis. *Curr. Appl. Phys.* **2020,** *20*, 23–28.

Iloms, E.; Ololade, O. O.; Ogola, H. J. O.; Selvarajan R. Investigating Industrial Effluent Impact on Municipal Wastewater Treatment Plant in Vaal, South Africa. *Int. J. Environ. Res. Public Health* **2020,** *17*, 1096.

International Atomic Energy Agency. *The Radiological Accident in Goiânia*; IAEA: Vienna, 1988.

Jacukowicz-Sobala, I.; Ociński, D.; Mazur, P.; Stanisławska, E.; Kociołek-Balawejder, E. Cu(II)-Fe(III) Oxide Doped Anion Exchangers—Multifunctional Composites for Arsenite Removal from Water via As(III) Adsorption and Oxidation. *J. Hazard. Mater.* **2020,** *394*, 122527.

Jain, A.; Vaya, D. Photocatalytic Activity of TiO_2 Nanomaterial. *J. Chil. Chem. Soc.* **2017,** *62*, 3683–3690.

Jiang, D.; Zeng, G.; Huang, D.; Chen, M.; Zhang, C.; Huang, C.; Wan, J. Remediation of Contaminated Soils by Enhanced Nanoscale Zero Valent Iron. *Environ. Res.* **2018,** *163*, 217–227.

Kalikeri, S.; Kodialbail, V. S. Auto-Combustion Synthesis of Narrow Band-Gap Bismuth Ferrite Nanoparticles for Solar Photocatalysis to Remediate Azo Dye Containing Water. *Environ. Sci. Pollut. Res.* **2020,** *56*, 2020.

Kalikeri, S.; Shetty Kodialbail, V. Solar Light-Driven Photocatalysis Using Mixed-Phase Bismuth Ferrite (BiFeO(3)/Bi(25)FeO(40)) Nanoparticles for Remediation of Dye-Contaminated Water: Kinetics and Comparison with Artificial UV and Visible Light-Mediated Photocatalysis. *Environ. Sci. Pollut. Res. Int.* **2018,** *25*, 13881–13893.

Kaur, R.; Vellingiri, K.; Kim, K. H.; Paul, A. K.; Deep, A. Efficient Photocatalytic Degradation of Rhodamine 6G with a Quantum Dot-Metal Organic Framework Nanocomposite. *Chemosphere* **2016,** *154*, 620–627.

Kazlauskienė, N.; Svecevičius, G.; Marčiulionienė, D.; Montvydienė, D.; Kesminas, V.; Staponkus, R.; Taujanskis, E.; Slučkaitė, A. The Effect of Persistent Pollutants on Aquatic Ecosystem: A Complex Study. *IEEE/OES Baltic Int. Symp.* **2012,** *2012*, 1–6.

Keller, A. A.; Wang, H.; Zhou, D.; Lenihan, H. S.; Cherr, G.; Cardinale, B. J.; Miller, R.; Ji, Z. Stability and Aggregation of Metal Oxide Nanoparticles in Natural Aqueous Matrices. *Environ. Sci. Technol.* **2010,** *44*, 1962–1967.

Khan, S.; Edathil, A. A.; Banat, F. Sustainable Synthesis of Graphene-Based Adsorbent Using Date Syrup. *Sci. Rep.* **2019,** *9*, 18106.

Khan, S. T.; Malik, A. Engineered Nanomaterials for Water Decontamination and Purification: From Lab to Products. *J. Hazard Mater.* **2019,** *363*, 295–308.

Khan, S.; Malik, A. Toxicity Evaluation of Textile Effluents and Role of Native Soil Bacterium in Biodegradation of a Textile Dye. *Environ. Sci. Pollut. Res. Int.* **2018,** *25*, 4446–4458.

Kiew, S. F.; Kiew, L. V.; Lee, H. B.; Imae, T.; Chung, L. Y. Assessing Biocompatibility of Graphene Oxide-Based Nanocarriers: A Review. *J. Control Release* **2016,** *226*, 217–228.

Kim, K. H.; Kabir, E.; Jahan, S. A. Exposure to Pesticides and the Associated Human Health Effects. *Sci. Total Environ.* **2017,** *575*, 525–529.

Kim, H.; Kim, M.; Lee, W.; Kim, S. Rapid Removal of Radioactive Cesium by Polyacrylonitrile Nanofibers Containing Prussian Blue. *J. Hazard. Mater.* **2018,** *347*, 106–113.

Kim, K. R.; Owens, G. Potential for Enhanced Phytoremediation of Landfills Using Biosolids—A Review. *J. Environ. Manage.* **2010,** *91*, 791–797.

Klatte, S.; Schaefer, H. C.; Hempel, M. Pharmaceuticals in the Environment: A Short Review on Options to Minimize the Exposure of Humans, Animals and Ecosystems. *Sustain. Chem. Pharm.* **2017,** *5*, 61–66.

Klimkova, S.; Cernik, M.; Lacinova, L.; Filip, J.; Jancik, D.; Zboril, R. Zero-Valent Iron Nanoparticles in Treatment of Acid Mine Water from in Situ Uranium Leaching. *Chemosphere* **2011,** *82*, 1178–1184.

Komkova, M. A.; Karyakina, E. E.; Karyakin, A. A. Catalytically Synthesized Prussian Blue Nanoparticles Defeating Natural Enzyme Peroxidase. *J. Am. Chem. Soc.* **2018,** *140*, 11302–11307.

Kuang, P.; Sayed, M.; Fan, J.; Cheng, B.; Yu, J. 3D Graphene-Based H2-Production Photocatalyst and Electrocatalyst. *Adv. Energy Mater.* **2020,** *10*, 1903802.

Kumar, N. S.; Asif, M.; Reddy, T. R. K.; Shanmugam, G.; Ajbar, A. Silver Quantum Dot Decorated 2D-SnO_2 Nanoflakes for Photocatalytic Degradation of the Water Pollutant Rhodamine B. *Nanomaterials* **2019,** *9*, 1536.

Kurwadkar, S. Occurrence and Distribution of Organic and Inorganic Pollutants in Groundwater. *Water Environ. Res.* **2019,** *91*, 1001–1008.

Kyzas, G. Z.; Matis, K. A. Flotation in Water and Wastewater Treatment. *Processes* **2018,** *6*, 116.

Lam, S. M.; Sin, J. C.; Mohamed, A. R. A Review on Photocatalytic Application of g-C3N4/Semiconductor (CNS) Nanocomposites towards the Erasure of Dyeing Wastewater. *Mater. Sci. Semicond. Process* **2016,** *47*, 62–84.

Laver, N. R. The Removal of Radionuclides from Contaminated Water Samples Using Graphene Oxide Nano-Flakes, Doctoral dissertation, UCL (University College London), 2020.

Lee, H.; Lee, K.; Kim, S. O.; Lee, J. S.; Oh, Y. Effective and Sustainable Cs+ Remediation via Exchangeable Sodium-Ion Sites in Graphene Oxide Fibers. *J. Mater. Chem. A* **2019,** *7*, 17754–17760.

Lee, J.; Mackeyev, Y.; Cho, M.; Wilson, L. J.; Kim, J. H.; Alvarez, P. J. J. C60 Aminofullerene Immobilized on Silica as a Visible-Light-Activated Photocatalyst. *Environ. Sci. Technol.* **2010,** *44*, 9488–9495.

Lee, K. T.; Lin, C. H.; Lu, S. Y. SnO2 Quantum Dots Synthesized with a Carrier Solvent Assisted Interfacial Reaction for Band-Structure Engineering of TiO_2 Photocatalysts. *J. Phys. Chem. C* **2014,** *118*, 14457–14463.

Leong, S.; Razmjou, A.; Wang, K.; Hapgood, K.; Zhang, X.; Wang, H. TiO_2 Based Photocatalytic Membranes: A Review. *J. Membr. Sci.* **2014,** *472*, 167–184.

Lesiak, B.; Rangam, N.; Jiricek, P.; Gordeev, I.; Tóth, J.; Kövér, L.; Mohai, M.; Borowicz, P. Surface Study of Fe_3O_4 Nanoparticles Functionalized with Biocompatible Adsorbed Molecules. *Front. Chem.* **2019,** *7*, 642.

Li, B.; Cao, H. ZnO@ Graphene Composite with Enhanced Performance for the Removal of Dye from Water. *J. Mater. Chem.* **2011,** *21*, 3346–3349.

Li, D.; Huang, J.; Li, R.; Chen, P.; Chen, D.; Cai, M.; Liu, H.; Feng, Y.; Lv, W.; Liu, G. Synthesis of a Carbon Dots Modified g-C3N4/SnO2 Z-Scheme Photocatalyst with Superior Photocatalytic Activity for PPCPs Degradation Under Visible Light Irradiation. *J. Hazard. Mater.* **2021,** *401*, 123257.

Li, H.; Shi, W.; Zhu, H.; Zhang, Y.; Du, Q.; Qin, X. Effects of Zinc Oxide Nanospheres on the Separation Performance of Hollow Fiber Poly(Piperazine-Amide) Composite Nanofiltration Membranes. *Fiber Polym.* **2016,** *17*, 836–846.

Li, J.; Liu, X.; Lu, J.; Wang, Y.; Li, G.; Zhao, F. Anti-Bacterial Properties of Ultrafiltration Membrane Modified by Graphene Oxide with Nano-Silver Particles. *J. Colloid Interf. Sci.* **2016,** *484*, 107–115.

Li, M.; Mu, Y.; Shang, H.; Mao, C.; Cao, S.; Ai, Z.; Zhang, L. Phosphate Modification Enables High Efficiency and Electron Selectivity of nZVI toward Cr(VI) Removal. *Appl. Catal. B: Environ.* **2020,** *263*, 118364.

Li, Q.; Wang, Y.; Song, J.; Guan, Y.; Yu, H.; Pan, X.; Wu, F.; Zhang, M. Influence of Silica Nanospheres on the Separation Performance of Thinfilm Composite Poly(Piperazineamide) Nanofiltration Membranes. *Appl. Surf. Sci.* **2015,** *324*, 757–764.

Li, S.; Wang, W.; Yanb, W.; Zhang, W. X. Nanoscale Zero-Valent Iron (nZVI) for the Treatment of Concentrated Cu(ii) Wastewater: A Field Demonstration. *Environ. Sci.: Processes Impacts* **2014,** *16*, 524–533.

Li, S.; De Silva, T.; Arsano, I.; Gallaba, D.; Karunanithy, R.; Wasala, M.; Zhang, X.; Sivakumar, P.; Migone, A.; Tsige, M.; Ma, X.; Talapatra, S. High Adsorption of Benzoic Acid on Single Walled Carbon Nanotube Bundles. *Sci. Rep.* **2020,** *10*, 10013.

Li, S.; Yang, F.; Li, J.; Cheng, K. Porous Biochar-Nanoscale Zero-Valent Iron Composites: Synthesis, Characterization and Application for Lead Ion Removal. *Sci. Total Environ.* **2020,** *746*, 141037.

Li, X.; Sun, Y. M.; Zhang, Z. Y.; Feng,N. X.; Song, H.; Liu, Y. L.; Hai, L.; Cao, J. M.; Wang, G. P. Visible Light-Driven Multi-Motion Modes CNC/TiO2 Nanomotors for Highly Efficient Degradation of Emerging Contaminants. *Carbon* **2019,** *155*, 195–203.

Li, X.; Wang, J.; Rykov, A. I.; Sharma, V. K.; Wei, H.; Jin, C.; Dionysiou, D. D. Prussian Blue/TiO_2 Nanocomposites as a Heterogeneous Photo-Fenton Catalyst for Degradation of Organic Pollutants in Water. *Catal. Sci. Technol.* **2015,** *5*, 504–514.

Li, X.; Yu, J.; Wageh, S.; Al-Ghamdi, A. A.; Xie, J. Graphene in Photocatalysis: A Review. *Small* **2016,** *12*, 6640–6696.

Li, Z.; Chen, F.; Yuan, L.; Liu, Y.; Zhao, Y.; Chai, Z.; Shi, W. Uranium (VI) Adsorption on Graphene Oxide Nanosheets from Aqueous Solutions. *Chem. Eng. J.* **2012,** *210*, 539–546.

Li, Z.; Sun, Y.; Yang, Y.; Han, Y.; Wang, T.; Chen, J.; Tsang, D. C. W. Biochar-Supported Nanoscale Zero-Valent Iron as an Efficient Catalyst for Organic Degradation in Groundwater. *J. Hazard. Mater.* **2020,** *383*, 121240.

Lian, Q.; Ahmad, Z. U.; Gang, D. D.; Zappi, M. E.; Fortela, D. L. B.; Hernandez, R. The Effects of Carbon Disulfide Driven Functionalization on Graphene Oxide for Enhanced Pb(II) Adsorption: Investigation of Adsorption Mechanism. *Chemosphere* **2020,** *248*, 126078.

Liu, D.; Mao, Y.; Ding, L. Carbon Nanotubes as Antimicrobial Agents for Water Disinfection and Pathogen Control. *J. Water Health* **2018,** *16*, 171–180.

Liu, H.; Liang, J.; Fu, S.; Li, L.; Cui, J.; Gao, P.; Zhao, F.; Zhou, J. N Doped Carbon Quantum Dots Modified Defect-Rich g-C3N4 for Enhanced Photocatalytic Combined Pollutions Degradation and Hydrogen Evolution. *Colloids Surf A: Physicochem Eng Asp* **2020,** *591*, 124552.

Liu, L.; Liu, X.; Chai, Y.; Wu, B.; Wang, C. Surface Modification of TiO_2 Nanosheets with Fullerene and Zinc-Phthalocyanine for Enhanced Photocatalytic Reduction Under Solar-Light Irradiation. *Sci. China Mater.* **2020,** *63*, 2251–2260.

Liu, R.; Hou, Y.; Jiang, S.; Nie, B. Ag(I)-Hived Fullerene Microcube as an Enhanced Catalytic Substrate for the Reduction of 4-Nitrophenol and the Photodegradation of Orange G Dye. *Langmuir* **2020,** *36*, 5236–5242.

Liu, Z.; Xu,T.; Wang, M.; Mao, C.; Chi, B. Magnetic Mesoporous Silica/ε-Polylysine Nanomotor-Based Removers of Blood Pb2+. *J. Mater. Chem. B* **2020,** In press.

Longman, J.; Veres, D.; Finsinger, W.; Ersek, V. Exceptionally High Levels of Lead Pollution in the Balkans from the Early Bronze Age to the Industrial Revolution. *Proc. Natl. Acad. Sci. USA* **2018,** *115*, E5661–E5668.

Low, J.; Yu, J.; Li, Q.; Cheng, B. Enhanced Visible-Light Photocatalytic Activity of Plasmonic Ag and Graphene Co-Modified Bi_2WO_6 Nanosheets. *Phys. Chem. Chem. Phys.* **2014,** *16*, 1111–1120.

Lunge, S.; Singh, S.; Sinha, A. Magnetic Iron Oxide (Fe_3O_4) Nanoparticles from Tea Waste for Arsenic Removal. *J. Magn. Magn. Mater.* **2014,** *356*, 21–31.

Lu F, Astruc D, Nanocatalysts and Other Nanomaterials for Water Remediation from Organic Pollutants. *Coord. Chem. Rev.* **2020,** 408.

Lv, Y.; Yang, H. C.; Liang, H. Q.; Wan, L. S.; Xu, Z. K. Novel Nanofiltration Membrane R with Ultrathin Zirconia Film as Selective Layer. *J. Membr. Sci.* **2016,** *500*, 265–271.

Ma, F.; Philippe, B.; Zhao, B.; Diao, J.; Li, J. Simultaneous Adsorption and Reduction of Hexavalent Chromium on Biochar-Supported Nanoscale Zero-Valent Iron (nZVI) in Aqueous Solution. *Water Sci. Technol.* **2020,** *82*, 1339–1349.

Ma, J.; Jiang, Z.; Cao, J.; Yu, F. Enhanced Adsorption for the Removal of Antibiotics by Carbon Nanotubes/Graphene Oxide/Sodium Alginate Triple-Network Nanocomposite Hydrogels in Aqueous Solutions. *Chemosphere* **2020,** *242*, 125188.

Mai, N. X. D.; Yoon, J.; Kim, J. H.; Kim, I. T.; Son, H. B.; Bae, J.; Hur, J. Hybrid Hydrogel and Aerogel Membranes Based on Chitosan/Prussian Blue for Photo-Fenton-Based Wastewater Treatment Using Sunlight. *Sci. Adv. Mater.* **2017,** *9*, 1484–1487.

Manickum, T.; John, W.; Terry, S. Determination of Selected Steroid Estrogens in Treated Sewage Effluent in the Umsunduzi (Duzi) River Water Catchment Area. *Hydrol. Current Res.* **2011,** *2*, 117.

Marsooli, M. A.; Nasrabadi, M. R.; Fasihi-Ramandi, M.; Adib, K.; Pourmasoud, S.; Ahmadi, F.; Eghbali, M.; Sobhani, A.; Nasab, S.; Tomczykowa, M.; Plonska-Brzezinska, M. E. Synthesis of Magnetic $Fe_3O_4/ZnWO_4$ and $Fe_3O_4/ZnWO_4/CeVO_4$ Nanoparticles: The Photocatalytic Effects on Organic Pollutants upon Irradiation with UV-Vis Light. *Catalysts* **2020,** *10,* 494.

Masud, A.; Soria, N. G. C.; Aga, D. S.; Aich, N. Adsorption and Advanced Oxidation of Diverse Pharmaceuticals and Personal Care Products (PPCPs) from Water Using Highly Efficient rGO–nZVI Nanohybrids. *Environ. Sci.: Water Res. Technol.* **2020,** *6*, 2223–2238.

Mayo, J. T.; Yavuz, C.; Yean, S.; Cong, L.; Shipley, H.; Yu, W.; Falkner, J.; Kan, A.; Tomson, M.; Colvin, V. L. The Effect of Nanocrystalline Magnetite Size on Arsenic Removal. *Sci. Technol. Adv. Mater.* **2007,** *8*, 71–75.

Medvedeva, I.; Bakhteeva, I.; Zhakov, S.; Revvo, A.; Uimin, M.; Yermakov, A.; Shchegoleva, N. Separation of Fe_3O_4 Nanoparticles from Water by Sedimentation in a Gradient Magnetic Field. *JWARP* **2015,** *7*, 111.

Melián, J. A. H. Sustainable Wastewater Treatment Systems (2018–2019). *Sustainability* **2020,** *12*, 1940.

Mezzelani, M.; Gorbi, S.; Regoli, F. Pharmaceuticals in the Aquatic Environments: Evidence of Emerged Threat and Future Challenges for Marine Organisms. *Mar. Environ. Res.* **2018,** *140*, 41–60.

Mirshahghassemi, S.; Lead, J. R. Oil Recovery from Water under Environmentally Relevant Conditions Using Magnetic Nanoparticles. *Environ. Sci. Technol.* **2015,** *49*, 11729–11736.

Mollahosseini, A.; Rahimpour, A. Interfacially Polymerized Thin Film Nanofiltration Membranes on TiO_2 Coated Polysulfone Substrate. *J. Ind. Eng. Chem.* **2014,** *20*, 1261–1268.

Moretti, S. M. L.; Bertoncini, E. I.; Abreu-Junior, C. H. Composting Sewage Sludge with Green Waste from Tree Pruning. *Sci. Agric.* **2015,** *72*, 432–439.

Mustafa, G.; Tahir. H.; Sultan, M.; Akhtar, N. Synthesis and Characterization of Cupric Oxide (CuO) Nanoparticles and Their Application for the Removal of Dyes. *Afr. J. Biotechnol.* **2013,** *12*, 6650–6660.

Mustapha, S.; Ndamitso, M. M.; Abdulkareem, A. S.; Tijani, J. O., Shuaib, D. T.; Ajala, A. O.; Mohammed, A. K. Application of TiO_2 and ZnO Nanoparticles Immobilized on Clay in Wastewater Treatment: A Review. *Appl. Water Sci.* **2020,** *10*, 49.

Nasrollahzadeh, M.; Sajjadi, M.; Iravani, S.; Varma, R. S. Green-Synthesized Nanocatalysts and Nanomaterials for Water Treatment: Current Challenges and Future Perspectives. *J. Hazard. Mater.* **2021,** *401*, 123401.

Newton, R. J.; McLellan, S. L.; Dila, D. K.; Vineis, J. H.; Morrison, H. G.; Eren, A. M.; Sogin, M. L. Sewage Reflects the Microbiomes of Human Populations. *mBio* **2015,** *6*, e02574-14.

Novoselov, K. S.; Geim, A. K.; Morozov, S. V.; Jiang, D.; Zhang, Y.; Dubonos, S. V.; Firsov, A. A. Electric Field Effect in Atomically Thin Carbon Films. *Science* **2004,** *306*, 666–669.

O'Carroll, D.; Sleep, B.; Krol, M.; Boparai, H.; Kocur, C. Nanoscale Zero Valent Iron and Bimetallic Particles for Contaminated Site Remediation. *Adv. Water Resour.* **2013,** *51*, 104–122.

Oprčkal, P.; Mladenovič, A.; Vidmar, J.; Pranjić, A. M.; Milačič, R.; Ščančar, J. Critical Evaluation of the Use of Different Nanoscale Zero-Valent Iron Particles for the Treatment of Effluent Water from a Small Biological Wastewater Treatment Plant. *Chem. Eng. J.* **2017,** *321*, 20–30.

Park, J. H.; Kim, M.; Kim, H.; Kim, W.; Ryu, J.; Lim, J. M.; Kim, S. Enhancement of Cesium Adsorption on Prussian Blue by TiO_2 Photocatalysis: Effect of the TiO_2/PB Ratio. *J. Water Process Eng.* **2020,** *38*, 101571.

Pastrana-Martinez, L. M.; Pereira, N.; Lima, R.; Faria, J. L.; Gomes, H. T.; Silva, A. M. Degradation of Diphenhydramine by Photo-Fenton Using Magnetically Recoverable Iron Oxide Nanoparticles as Catalyst. *Chem. Eng. J.* **2015,** *261*, 45–52.

Perera, M.; Wijenayaka, L. A.; Siriwardana, K.; Dahanayake, D.; de Silva, K. M. N. Gold Nanoparticle Decorated Titania for Sustainable Environmental Remediation: Green Synthesis, Enhanced Surface Adsorption and Synergistic Photocatalysis. *RSC Adv.* **2020,** *10*, 29594–29602.

Peuke, A. D.; Rennenberg, H. Phytoremediation. *EMBO Reports* **2005,** *6*, 497–501.

Pinheiro, H. M.; Thomas, O.; Touraud, E. Aromatic Amines from Azo Dye Reduction: Status Review with Emphasis on Direct UV Spectrophotometric Detection in Textile Industry Wastewaters. *Dyes Pigm.* **2004,** *61*, 121–139.

Pinto, A. M.; Gonçalves, I. C.; Magalhães, F. D. Graphene-Based Materials Biocompatibility: A Review. *Colloid Surf. B* **2013,** *111*, 188–202.

Pradhan, N.; Pal, A.; Pal, T. Catalytic Reduction of Aromatic Nitro Compounds by Coinage Metal Nanoparticles. *Langmuir* **2001,** *17*, 1800–1802.

Priyadarshni, N.; Nath, P.; Nagahanumaiah; Chanda, N. Sustainable Removal of Arsenate, Arsenite and Bacterial Contamination from Water Using Biochar Stabilized Iron and Copper Oxide Nanoparticles and Associated Mechanism of the Remediation Process. *J. Water Process Eng.* **2020,** *37*, 101495.

Priyadarshni, N.; Nath, P.; Nagahanumaiah; Chanda, N. DMSA-Functionalized Gold Nanorod on Paper for Colorimetric Detection and Estimation of Arsenic (III and V) Contamination in Groundwater. *ACS Sustain. Chem. Eng.* **2018,** *6*, 6264–6272.

Pulizzi, F.; Sun, W. Treating Water with Nano. *Nature Nanotech.* **2018,** *13*, 633.

Qi, K.; Selvaraj, R.; Al Fahdi, T.; Al-Kindy, S.; Kim, Y.; Wang, G.; Tai, C. W.; Sillanpaa, M. Enhanced Photocatalytic Activity of Anatase-TiO_2 Nanoparticles by Fullerene Modification: A Theoretical and Experimental Study. *Appl. Surf. Sci.* **2016,** *387*, 750–758.

Qiao, D.; Li, Z.; Duan, J.; He, X. Adsorption and Photocatalytic Degradation Mechanism of Magnetic Graphene Oxide/ZnO Nanocomposites for Tetracycline Contaminants. *Chem. Eng. J.* **2020,** *400*, 125952.

Qiao, W.; Zhang, P.; Sun, L.; Ma, S.; Xu, W.; Xu, S.; Niu, Y. Adsorption Performance and Mechanism of Schiff Base Functionalized Polyamidoamine Dendrimer/Silica for Aqueous Mn(II) and Co(II). *Chin. Chem. Lett.* **2020,** *31*, 2742–2746.

Qing, Y.; Cheng, L.; Li, R.; Liu, G.; Zhang, Y.; Tang, X.; Wang, J.; Liu, H.; Qin, Y. Potential Antibacterial Mechanism of Silver Nanoparticles and the Optimization of Orthopedic Implants by Advanced Modification Technologies. *Int. J. Nanomed.* **2018,** *13*, 3311–3327.

Qu, X.; Alvarez, P. J.; Li, Q. Applications of Nanotechnology in Water and Wastewater Treatment. *Water Res.* **2013,** *47*, 3931–3946.

Rachna, Rani, M.; Shanker, U. Sunlight Assisted Degradation of Toxic Phenols by Zinc Oxide Doped Prussian Blue Nanocomposite. *J. Environ. Chem. Eng.* **2020,** *8*, 104040.

Rafique, M.; Shafiq, F.; Gillani, S. S. A.; Shakil, M.; Tahir, M. B.; Sadaf, I. Eco-Friendly Green and Biosynthesis of Copper Oxide Nanoparticles Using Citrofortunella Microcarpa Leaves Extract for Efficient Photocatalytic Degradation of Rhodamin B Dye form Textile Wastewater. *Optik* **2020,** *208*, 164053.

Rani, U. A.; Ng, L. Y.; Ng, C. Y.; Mahmoudi, E. A Review of Carbon Quantum Dots and Their Applications in Wastewater Treatment. *Adv. Colloid Interf. Sci.* **2020,** *278*, 102124.

Rao, H.; Xue, X.; Wang, H.; Xue, Z. Gold Nanorod Etching-Based Multicolorimetric Sensors: Strategies and Applications. *J. Mater. Chem. C* **2019,** *7*, 4610–4621.

Rasheed, T.; Adeel, M.; Nabeel, F.; Bilal, M.; Iqbal, H. M. N. TiO_2/SiO_2 Decorated Carbon Nanostructured Materials as a Multifunctional Platform for Emerging Pollutants Removal. *Sci. Total Environ.* **2019,** *688*, 299–311.

Rauti, R.; Musto, M.; Bosi, S.; Prato, M.; Ballerini, L. Carbon Review Article Properties and Behavior of Carbon Nanomaterials When Interfacing Neuronal Cells: How Far Have We Come? *Carbon* **2019,** *143*, 430–446.

Regulska, E.; Rivera-Nazario, D. M.; Karpinska, J.; Plonska-Brzezinska, M. E.; Echegoyen, L. Zinc Porphyrin-Functionalized Fullerenes for the Sensitization of Titania as a Visible-Light Active Photocatalyst: River Waters and Wastewaters Remediation. *Molecules* **2019,** *24*, 1118.

Romanchuk, A. Y.; Slesarev, A. S.; Kalmykov, S. N.; Kosynkin, D. V.; Tour, J. M. Graphene Oxide for Effective Radionuclide Removal. *Phys. Chem. Chem. Phys.* **2013,** *15*, 2321–2327.

Roy, J. R.; Chakraborty, S.; Chakraborty, T. R. Estrogen-like Endocrine Disrupting Chemicals Affecting Puberty in Humans—A Review. *Med. Sci. Monit.* **2009,** *15*, 137–145.

Sadeghi, M. H.; Tofighy, M. A.; Mohammadi, T. One-Dimensional Graphene for Efficient Aqueous Heavy Metal Adsorption: Rapid Removal of Arsenic and Mercury Ions by Graphene Oxide Nanoribbons (GONRs). *Chemosphere* **2020,** *253*, 126647.

Salt, D. E.; Blaylock, M.; Kumar, N. P. B. A.; Dushenkov, V.; Ensley, B. D.; Chet, I., Raskin, I. Phytoremediation: A Novel Strategy for the removal of Toxic Metals from the Environment Using Plants. *Nat. Biotechnol.* **1995,** *13*, 468–474.

Sánchez, S.; Soler, L.; Katuri, J. Chemically Powered Micro-and Nanomotors. *Angew. Chem. Int. Ed.* **2015,** *54*, 1414–1444.

Sangion, A.; Gramatica, P. PBT Assessment and Prioritization of Contaminants of Emerging Concern: Pharmaceuticals. *Environ. Res.* **2016,** *147*, 297–306.

Santhosh, C.; Velmurugan, V.; Jacob, G.; Jeong, S. K.; Grace, A. N.; Bhatnagar, A. Role of Nanomaterials in Water Treatment Applications: A Review. *Chem. Eng. J.* **2016,** *306*, 1116–1137.

Sargin, I.; Baran, T.; Arslan, G. Environmental Remediation by Chitosan-Carbon Nanotube Supported Palladium Nanoparticles: Conversion of Toxic Nitroarenes Into Aromatic Amines, Degradation of Dye Pollutants and Green Synthesis of Biaryls. *Sep. Purif. Technol.* **2020,** *247*, 116987.

Schwarzenbach, R. P.; Escher, B. I.; Fenner, K.; Hofstetter, T. B.; Johnson, C. A.; von Gunten, U.; Wehrli, B. The Challenge of Micropollutants in Aquatic Systems. *Science* **2006,** *313*, 1072–1077.

Sebeia, N.; Jabli, M.; Ghith, A.; Saleh, T. A. Eco-friendly synthesis of Cynomorium Coccineum Extract for Controlled Production of Copper Nanoparticles for Sorption of Methylene Blue Dye. *Arab. J. Chem.* **2020,** *13*, 4263–4274.

Semeniuk, M.; Yi, Z.; Poursorkhabi, V.; Tjong, J.; Jaffer, S.; Lu, Z. H.; Sain, M. Future Perspectives and Review on Organic Carbon Dots in Electronic Applications. *ACS Nano* **2019,** *13*, 6224–6255.

Seng, R. X.; Tan, L. L.; Lee, W. P. C.; Ong, W. J.; Chai, C. P. Nitrogen-Doped Carbon Quantum Dots-Decorated 2D Graphitic Carbon Nitride as a Promising Photocatalyst for Environmental Remediation: A Study on the Importance of Hybridization Approach. *J. Environ. Manag.* **2020,** *255*, 109936.

Shad, S.; Belinga-Desaunay-Nault, M. F. A.; Sohail; Bashir, N.; Lynch, I. Removal of Contaminants from Canal Water Using Microwave Synthesized Zero Valent Iron Nanoparticles. *Water Res. Technol.* **2020,** *6*, 3057–3065.

Sharma, A. K.; Tiwari, R. K.; Gaur, M. S. Nanophotocatalytic UV Degradation System for Organophosphorus Pesticides in Water Samples and Analysis by Kubista Model. *Arab. J. Chem.* **2016,** *9*, 1755–1764.

Sharma, Y. C.; Srivastava, V.; Singh, V. K.; Kaul, S. N.; Weng, C. H. (2009). Nano-Adsorbents for the Removal of Metallic Pollutants from Water and Wastewater. *Environ. Technol.* **2009,** *30*, 583–609.

Silva, G. A. Nanotechnology Approaches to Crossing the Blood-Brain Barrier and Drug Delivery to the CNS. *BMC Neurosci.* **2008,** *9*, S4.

Singh, D.; Gautam, R. K.; Kumar, R.; Shukla, B. K.; Shankar, V.; Krishna, V. Citric Acid Coated Magnetic Nanoparticles: Synthesis, Characterization and Application in Removal of Cd (II) Ions from Aqueous Solution. *J. Water Process Eng.* **2014,** *4*, 233–241.

Singh, N. S., Sharma, R., Parween, T., and Patanjali, P. K. Pesticide Contamination and Human Health Risk Factor. In *Modern Age Environmental Problems and Their Remediation*; Iqbal Cham: Springer, 2018; pp 49–68.

Singh, R.; Bhadouria, R.; Singh, P.; Kumar, A.; Pandey, S.; Singh, V. K. Nanofiltration Technology for Removal of Pathogens Present in Drinking Water. Waterborne Pathogens **2020,** *2020*, 463–489.

Singh, S.; Pandey, P. C. Synthesis and Application of Functional Prussian Blue Nanoparticles for Toxic Dye Degradation. *J. Environ. Chem. Eng.* **2020,** *8*, 103753.

Sirajudheen, P.; Karthikeyan, P.; Ramkumar, K.; Meenakshi, S. Effective Removal of Organic Pollutants by Adsorption Onto Chitosan Supported Graphene Oxide-Hydroxyapatite Composite: A Novel Reusable Adsorbent. *J. Mol. Liq.* **2020,** *318*, 114200.

Smrithi, S. P.; Kottam, N.; Arpitha, V.; Narula, A.; Anilkumar, G. N.; Subramanian, K. R. V. Tungsten Oxide Modified with Carbon Nanodots: Integrating Adsorptive and Photocatalytic Functionalities for Water Remediation. *J. Sci. Adv. Mater. Dev.* **2020,** *5*, 73–83.

Sousa-Moura, D.; Matsubara, E. Y.; Machado Ferraz, I. B.; Oliveira, R.; Szlachetka, Ĺ. O.; da Silva, S. W.; Camargo, N. S.; Rosolen, J. M.; Grisolia, C. K.; da Rocha, M. C. O. CNTs Coated Charcoal as a Hybrid Composite Material: Adsorption of Fluoxetine Probed by Zebrafish Embryos and Its Potential for Environmental Remediation. *Chemosphere* **2019,** *230*, 369–376.

Sposito, J. C. V.; Montagner, C. C.; Casado, M.; Navarro-Martín, L.; Jut Solórzano, J. C.; Piña, B.; Grisolia, A. B. Emerging Contaminants in Brazilian Rivers: Occurrence and Effects on Gene Expression in Zebrafish (*Danio rerio*) Embryos. *Chemosphere* **2018,** *209*, 696–705.

Sudhaik, A.; Raizada, P.; Singh, P.; Hosseini-Bandegharaei, A.; Thakur, V. K.; Nguyen, V. H. Highly Effective Degradation of Imidacloprid by H_2O_2/Fullerene Decorated P-Doped g-C_3N_4 Photocatalyst. *J. Environ. Chem. Eng.* **2020,** *8*, 104483.

Sumisha, A.; Arthanareeswaran, G.; Ismail, A. F.; Kumar, D. P.; Shankar, M. V. Functionalized Titanate Nanotube–Polyetherimide Nanocomposite Membrane for Improved Salt Rejection under Low Pressure Nanofiltration. *RSC Adv.* **2015,** *5*, 39464–39473.

Taka, A. L.; Fosso-Kankeu, E.; Pillay, K.; Mbianda, X. Y. Metal Nanoparticles Decorated Phosphorylated Carbon Nanotube/Cyclodextrin Nanosponge for Trichloroethylene and Congo Red Dye Adsorption from Wastewater. *J. Environ. Chem. Eng.* **2020,** *8*, 103602.

Tara, N.; Siddiquia, S. I.; Bach, Q. V.; Chaudhry, S. A. Reduce Graphene Oxide-Manganese Oxide-Black Cumin Based Hybrid Composite (rGO-MnO_2/BC): A Novel Material for Water Remediation. *Mater. Today Commun.* **2020,** *25*, 101560.

Tayeb, A.; Chellali, M. R.; Hamou, A.; Debbah, S. Impact of Urban and Industrial Effluents on the Coastal Marine Environment in Oran, Algeria. *Mar. Pollut. Bull.* **2015,** *98*, 281–288.

Teo, W. Z.; Zboril, R.; Medrik, I.; Pumera, M. Fe^0 Nanomotors in Ton Quantities (10^{20} Units) for Environmental Remediation. *Chem. Eur. J.* 2016, *22*, 4789–4793.

Thebo, K. H.; Qian, X.; Wei, Q.; Zhang, Q.; Cheng, H. M.; Ren, W. Reduced Graphene Oxide/Metal Oxide Nanoparticles Composite Membranes for Highly Efficient Molecular Separation. *J. Mater. Sci. Technol.* **2018,** *34*, 1481–1486.

Tonelli, F. M. P.; Lacerda, S. M. S. N.; Paiva, N. C. O.; Lemos, M. S.; Jesus, A. C.; Pacheco, F. G.; Correa Junior, J. D.; Ladeira, L. O.; Furtado, C. A.; Franca, L. R.; Resende, R.

R. Efficiently and Safely in Gene Transfection in Fish Spermatonial Stem Cells Using Nanomaterials. *RSC Adv.* **2016,** *58*, 1058.

Tonelli, F. M. P.; Lacerda, S. M. S. N.; Paiva, N. C. O.; Pacheco, F. G.; Scalzo Junior, S. R. A.; Macedo, F. H. P.; Cruz, J. S.; Pinto, M. C. X.; Correa Junior, J. D.; Ladeira, L. O.; França, L. R.; Guatimosim, S. C.; Resende, R. R. Functionalized Nanomaterials: Are They Effective to Perform Gene Delivery to Difficult-to-Transfect Cells with No Cytotoxicity? *Nanoscale* **2015,** *7*, 18036–18043.

Tonelli, F. C. P.; Tonelli, F. M. P. Concerns and Threats of Xenobiotics on Aquatic Ecosystems. In *Bioremediation and Biotechnology Vol 3: Persistent and Recalcitrant Toxic Substances*; Springer: New York, USA, 2020; p 360.

Tonelli, F. M. P.; Tonelli, F. C. P.; Ferreira, D. R. C.; Silva, K. E.; Cordeiro, H. G.; Ouchida, A. T.; Nunes, N. A. M. Biocompatibility and Functionalization of Nanomaterials. Intelligent Nanomaterials for Drug Delivery Applications. In *Intelligent Nanomaterials for Drug Delivery Applications*; Elsevier: New York, 2020; p 212.

Tiwari, S. K.; Sahoo, S.; Wang, N.; Huczko, A. Graphene Research and Their Outputs: Status and Prospect. *J. Sci. Adv. Mat. Dev.* **2020,** *5*, 10–29.

Trivedi M, Johri P, Singh A, Singh R, Tiwari RK. Latest Tools in Fight Against Cancer: Nanomedicines. In *NanoBioMedicine*; Springer: Singapore, 2020; pp 139–164.

Turchanin, A.; Gölzhäuser, A. Carbon Nanomembranes. *Adv. Mater.* **2016,** *28*, 6075–6103.

Vaiano V, Sannino D, Sacco O. The Use of Nanocatalysts (and Nanoparticles) for Water and Wastewater Treatment by Means of Advanced Oxidation Processes. In *Nanotechnology in the Beverage Industry*; Elsevier, 2020; pp 241–264.

Varjani, S.; Kumar, G.; Rene, E. R. Developments in Biochar Application for Pesticide Remediation: Current Knowledge and Future Research Directions. *J. Environ. Manage.* **2019,** *232*, 505–513.

Verma, B.; Balomajumder, C. Surface Modification of One-Dimensional Carbon Nanotubes: A Review for the Management of Heavy Metals in Wastewater. *Environ. Technol. Inno.* **2020,** *17*, 100596.

Wang, F.; Yang, W.; Zheng, F.; Sun, Y. Removal of Cr (VI) from Simulated and Leachate Wastewaters by Bentonite-Supported Zero-Valent Iron Nanoparticles. *Int. J. Environ. Res. Public Health,* **2018,** *15*, 2162.

Wang, H.; Huang, Y. Prussian-Blue-Modified Iron Oxide Magnetic Nanoparticles as Effective Peroxidase-Like Catalysts to Degrade Methylene Blue with H_2O_2. *J. Hazard Mater.* **2011,** *191*, 163–169.

Wang, H.; Shang, H.; Sun, X.; Hou, L.; Wen, M.; Qiao, Y. Preparation of Thermo-Sensitive Surface Ion-Imprinted Polymers Based on Multi-Walled Carbon Nanotube Composites for Selective Adsorption of Lead(II) Ion. *Colloids Surf. Physicochem. Eng. Aspects* **2020,** *585*, 124139.

Wang, M.; Bao, T.; Yan, W.; Fang, D.; Yu, Y.; Liu, Z.; Yin, G.; Wan, M.; Mao, C.; Shi, D. Bioactive Materials Nanomotor-Based Adsorbent for Blood Lead(II) Removal in Vitro and in Pig Models. *Bioact. Mater.* **2021,** *6*, 1140–1149.

Wang, Q.; Cai, J.; Biesold-McGee, G. V.; Huang, J.; Ng, Y. H.; Sun, H.; Wang, J.; Lai, Y.; Lin, Z. Silk Fibroin-Derived Nitrogen-Doped Carbon Quantum Dots Anchored on TiO_2 Nanotube Arrays for Heterogeneous Photocatalytic Degradation and Water Splitting. *Nano Energy* **2020,** *78*, 105313.

Wang, X.; Feng, Y.; Dong, P.; Huang, J. A Mini Review on Carbon Quantum Dots: Preparation, Properties, and Electrocatalytic Application. *Front. Chem.* **2019a,** *7*, 671.

Wang, Y.; Pan, C.; Chu, W.; Vipin, A. K.; Sun, L. Environmental Remediation Applications of Carbon Nanotubes and Graphene Oxide: Adsorption and Catalysis. *Nanomaterials* **2019b,** *9*, 439.

Wang, Z.; Wang, T. C. Quantum Dots Ceramic Nano Membrane for Copper Mining Water Treatment, Turning Na2SO4 to NaOH and Pretreatment for Seawater Desalination. *ChemRxiv* **2020,** Pre-print.

Wang, Z.; Tu, Q.; Zheng, S.; Urban, J. J.; Li, S.; Mi, B. Understanding the Aqueous Stability and Filtration Capability of MoS_2 Membranes. *Nano Lett.* **2017,** *17*, 7289–7298.

Wazir, M. B.; Daud, M.; Ali, F.; Al-Harthi, M. A. Dendrimer Assisted Dye-Removal: A Critical Review of Adsorption and Catalytic Degradation for Wastewater Treatment. *J. Mol. Liq.* **2020,** *315*, 113775.

Weiss, J.; Takhistov, P.; McClements, D. J. Functional Materials in Food Nanotechnology. *J. Food Sci.* **2006,** *71*, R107–R116.

Werkneh, A. A.; Rene, R. R.; Lens, P. N. L. Application of Nanomaterials in Food, Cosmetics and Other Process Industries. In *Nanotoxicology: Toxicity Evaluation, Risk Assessment and Management*; CRC Press Taylor and Francis Group, 2020; pp 63–79.

Werkneh, A. A.; Rene, E. R. Applications of Nanotechnology and Biotechnology for Sustainable Water and Wastewater Treatment. In *Water and Wastewater Treatment Technologies*; Springer: Singapore, 2019; pp 405–430.

Wicki, A.; Witzigmann, D.; Balasubramanian, V.; Huwyler, J. Nanomedicine in Cancer Therapy: Challenges, Opportunities, and Clinical Applications. *J. Control Release* **2015,** *200*, 138–157.

Wu, X.; Chen, L.; Zheng, C.; Yan, X.; Dai, P.; Wang, Q.; Li, W.; Chen, W. Bubble-Propelled Micromotors Based on Hierarchical MnO_2 Wrapped Carbon Nanotube Aggregates for Dynamic Removal of Pollutants. *RSC Adv.* **2020a,** *10*, 14846–14855.

Wu, Y.; Wang, F.; Jin, X.; Zheng, X.; Wang, Y.; Wei, D.; Zhang, Q.; Feng, Y.; Xie, Z.; Chen, P.; Liu, H.; Liu, G. Highly Active Metal-Free Carbon Dots/g-C_3N_4 Hollow Porous Nanospheres for Solar-Light-Driven PPCPs Remediation: Mechanism Insights, Kinetics and Effects of Natural Water Matrices. *Water Res.* **2020b,** *172*, 115492.

Xie, H.; Li, P.; Shao, J.; Huang, H.; Chen, Y.; Jiang, Z.; Chu, P. K.; Yu, X. F. Electrostatic Self-Assembly of $Ti_3C_2T_x$ MXene and Gold Nanorods as an Efficient Surface-Enhanced Raman Scattering Platform for Reliable and High-Sensitivity Determination of Organic Pollutants. *ACS Sens.* **2019,** *4*, 2303–2310.

Xiong, J.; Zhang, D.; Lin, H.; Chen, Y. Amphiprotic Cellulose Mediated Graphene Oxide Magnetic Aerogels for Water Remediation. *Chem. Eng. J.* **2020,** *400*, 125890.

Xu, C.; Nasrollahzadeh, M.; Sajjadi, M.; Maham, M.; Luque, R.; Puente-Santiago, A. R. Benign-by-Design Nature-Inspired Nanosystems in Biofuels Production and Catalytic Applications. *Renew. Sustain. Energy Rev.* **2019,** *112*, 195–252.

Xu, T. Y.; Zhu, R. L.; Zhu, J. X.; Liang, X. L.; Zhu, G. Q.; Liu, Y.; Xu, Y.; He, H. P. Fullerene Modification of Ag_3PO_4 for the Visible-Light-Driven Degradation of Acid Red 18. *RSC Adv.* **2016,** *6*, 85962–85969.

Yamada, D.; Hiwatari, M.; Hangoma, P.; Narita, D.; Mphuka, C.; Chitah, B.; Yabe, J.; Nakayama, S. M. M.; Nakata, H.; Choongo, K.; Ishizuka, M. Assessing the Population-Wide Exposure to Lead Pollution in Kabwe, Zambia: An Econometric Estimation Based on Survey Data. *Sci. Rep.* **2020,** *10*, 15092.

Yang, J.; Huang, Z.; Yang, B.; Lin, H.; Qin, L.; Nie, M.; Li, Q. Green Route to Prussian Blue Nanoparticles with High Degradation Efficiency of RhB under Visible Light. *J. Mater. Sci.* **2021,** *56*, 3268–3279.

Yang, J.; Zhou, L.; Ma, F.; Zhao, H.; Deng, F.; Pi, S.; Tang, A.; Li, A. Environmental Research Magnetic Nanocomposite Microbial Extracellular Polymeric Substances@ Fe3O4 Supported nZVI for Sb(V) Reduction and Adsorption under Aerobic and Anaerobic Conditions. *Environ. Res.* **2020,** *189*, 109950.

Yang, Y.; Ok, Y. S.; Kim, K. H.; Kwon, E. E.; Tsang, Y. F. Occurrences and Removal of Pharmaceuticals and Personal Care Products (PPCPs) in Drinking Water and Water/ Sewage Treatment Plants: A Review. *Sci. Total Environ.* **2017,** *596*, 303–320.

Yaqoob, A. A.; Parveen, T.; Umar, K.; Nasir, M.; Ibrahim, M. N. Role of Nanomaterials in the Treatment of Wastewater: A Review. *Water* **2020,** *12*, 495.

Ye, S.; Zeng, G.; Wu, H.; Zhang, C.; Dai, J.; Liang, J.; Yu, J.; Ren, X.; Yi, H.; Cheng, M.; Zhang, C. Biological Technologies for the Remediation of Co-Contaminated Soil. *Crit. Rev. Biotechnol.* **2017,** *37*, 1062–1076.

Yeung, K. W. Y.; Zhou, G. J.; Hilscherová, K.; Giesy, J. P.; Leung, K. M. Y. Current Understanding of Potential Ecological Risks of Retinoic Acids and Their Metabolites in Aquatic Environments. *Environ. Int.* **2020,** *136*, 105464.

Yi, H.; Liu, R.; Chen, Z.; Nie, B. Chemosphere Visible-Light Driven Photodegradation on Ag Nanoparticle-Embedded Fullerene (C60) Heterostructural Microcubes. *Chemosphere* **2020,** *258*, 127355.

Yin, J.; Yang, Y.; Hu, Z.; Deng, B. Attachment of Silver Nanoparticles (AgNPs) Onto Thin-Film Composite (TFC) Membranes through Covalent Bonding to Reduce Membrane Biofouling. *J. Membr. Sci.* **2013,** *441*, 73–82.

Ying, Y. L.; Ying, W.; Li, Q. C.; Meng, D. H.; Ren, G. H.; Yan, R. X.; Peng, X. S. Recent Advances of Nanomaterial-Based Membrane for Water Purification. *Appl. Mater. Today* **2017,** *7*, 144–158.

Ying, Y.; Pumera, M. Micro/Nanomotors for Water Purification. *Chem. Eur. J.* **2018,** *25*, 106–121.

Yuan, T.; Chen, Q.; Shen, X. Chinese Chemical Letters Communication Adsorption of Cesium Using Mesoporous Silica Gel Evenly Doped by Prussian Blue Nanoparticles. *Chin. Chem. Lett.* **2020,** *31*, 2835–2838.

Zahedifar, M.; Seyedi, N.; Salajeghe, M.; Shafiei, S. Nanomagnetic Biochar Dots Coated Silver NPs (BCDs-Ag/MNPs): A Highly Efficient Catalyst for Reduction of Organic Dyes. *Mater. Chem. Phys.* **2020,** *246*, 122789.

Zahra, Z.; Habib, Z.; Chung, S.; Badshah, M. A. Exposure Route of TiO2 NPs from Industrial Applications to Wastewater Treatment and Their Impacts on the Agro-Environment. *Nanomaterials* **2020,** *10*, 1469.

Zeng, H.; Zhai, L.; Qiao, T.; Yu, Y.; Zhang, J.; Li, D. Efficient Removal of As(V) from Aqueous Media by Magnetic Nanoparticles Prepared with Iron-Containing Water Treatment Residuals. *Sci. Rep.* **2020,** *10*, 9335.

Zhang, L. S.; Fang, Y.; Zhou, Y.; Ye, B. C. Improvement of the Stabilization and Activity of Protocatechuate 3,4-Dioxygenase Isolated from *Rhizobium sp.* LMB-1 and Immobilized on Fe_3O_4 Nanoparticles. *Appl. Biochem. Biotechnol.* **2017,** *183*, 1035–1048.

Zhang, Q.; Dong, R.; Wu, Y.; Gao, W.; He, Z.; Ren, B. Light-Driven Au-WO3@ C Janus Micromotors for Rapid Photodegradation of Dye Pollutants. *ACS App. Mater. Interf.* **2017,** *9*, 4674–4683.

Zhang, Q. Qian, X.; Thebo, K. H.; Cheng, H. M.; Ren, W. Controlling Reduction Degree of Graphene Oxide Membranes for Improved Water Permeance. *Sci. Bull.* **2018,** *63*, 788–794.

Zhang, W.; Yang, Q.; Luo, Q.; Shi, L.; Meng, S. Laccase-Carbon Nanotube Nanocomposites for Enhancing Dyes Removal. *J. Clean. Prod.* **2020,** *242*, 118425.

Zhang, W.; Wu, Y.; Dong, H. J.; Yin, J. J.; Zhang, H.; Wu, H. A.; Zhang, Y. Sparks fly between Ascorbic Acid and Iron-Based Nanozymes: A Study on Prussian Blue Nanoparticles. *Colloid Surf. B* **2018,** *163*, 379–384.

Zhang, X.; Lin, M.; Lin, X.; Zhang, C.; Wei, H.; Zhang, H.; Yang, B. Polypyrrole-Enveloped Pd and Fe3O4 Nanoparticle Binary Hollow and Bowl-Like Superstructures as Recyclable Catalysts for Industrial Wastewater Treatment *ACS Appl. Mater. Interf.* **2014,** *6*, 450–458.

Zhang, X.; Shao, C.; Zhang, Z.; Li, J.; Zhang, P.; Zhang, M.; Mu, J.; Guo, Z.; Liang, P.; Liu, Y. In Situ Generation of Well-Dispersed ZnO Quantum Dots on Electrospun Silica Nanotubes with High Photocatalytic Activity. *ACS Appl. Mater. Interfaces* **2012,** *4*, 785–790.

Zhang, Y.; Yang, M.; Portney, N. G.; Cui, D.; Budak, G.; Ozbay, E.; Ozkan, M.; Ozkan, C. S. Zeta Potential: A Surface Electrical Characteristic to Probe the Interaction of Nanoparticles with Normal and Cancer Human Breast Epithelial Cells. *Biomed. Microdevices* **2008,** *10*, 321–328.

Zhang, Z.; Zhu, Z.; Shen, B.; Liu, L. Insights Into Biochar and Hydrochar Production and Applications: A Review. *Energy* **2019,** *171*, 581–598.

Zhao, Z.; Zhang, B.; Chen, D.; Guo, Z.; Peng, Z. Simultaneous Reduction of Vanadium (V) and Chromium (VI) in Wastewater by Nanosized ZnWO4 Photocatalysis. *J. Nanosci. Nanotechnol.* **2016,** *16*, 2847–2852.

Zheng H, Zhang C, Liu B, Liu G, Zhao M, Xu G, Luo X, Li F, Xing B. Biochar for Water and Soil Remediation: Production, Characterization, and Application. In *A New Paradigm for Environmental Chemistry and Toxicology*; Springer: Singapore, 2020; pp. 153–196.

Zhou, G. J.; Lin, L.; Li, X. Y.; Leung, K. M. Y. Removal of Emerging Contaminants from Wastewater during Chemically Enhanced Primary Sedimentation and Acidogenic Sludge Fermentation. *Water Res.* **2020,** *175*, 115646.

Zhou, G. J.; Li, X. Y.; Leung, K. M. Y. Retinoids and Oestrogenic Endocrine Disrupting Chemicals in Saline Sewage Treatment Plants: Removal Efficiencies Cological Risks to Marine Organisms. *Environ. Int.* **2019,** *127*, 103–113.

Ziadi, A.; Hariga, N. T.; Tarhouni, J. Mineralization and Pollution Sources in the Costal Aquifer of Lebna, Cap Bom, Tunisia. *J. Afr. Earth Sci.* **2019,** *151*, 391–402.

CHAPTER 11

Carbonaceous Materials for Nanoremediation of Polluted and Nutrient-Depleted Soils

GUILHERME MAX DIAS FERREIRA[1], GABRIEL MAX DIAS FERREIRA[2], JOSÉ ROMÃO FRANCA[3], and JENAINA RIBEIRO-SOARES[3*]

[1]*Department of Chemistry, Institute of Natural Science, Federal University of Lavras, 37200-900, Lavras, Minas Gerais, Brazil*

[2]*Department of Chemistry, Federal University of Ouro Preto, Campus Morro do Cruzeiro, 35400-000, Ouro Preto, Minas Gerais, Brazil*

[3]*Department of Physics, Institute of Natural Science, Federal University of Lavras, 37200-900, Lavras, Minas Gerais, Brazil*

[*]*Corresponding author. E-mail: jenaina.soares@ufla.br*

ABSTRACT

New strategies for remediation of polluted and nutrient-depleted soils are in high demand due to the increasing pressure of the growth of worldwide population and environmental issues generated by industrial activities. Approximately, three-fourth of the world population lives in tropical regions, where the heavy rains promote the fast lixiviation of soil nutrients and contaminants, and the high temperatures are related to the fast degradation of soil organic matter. Carbonaceous nanomaterials present novel properties arising from their reduced dimensionality that are valuable to elaborate new effective strategies to solve these problems. For example, these materials present high specific surface area, mechanical strength, versatility to be tuned

Nanotechnology for Environmental Pollution Decontamination: Tools, Methods, and Approaches for Detection and Remediation. Fernanda Maria Policarpo Tonelli, Rouf Ahmad Bhat, & Gowhar Hamid Dar (Eds.)

from stable to reactive forms, and superior adsorption capacity for a large variety of contaminants. These properties are valuable to formulate solutions to immobilize contaminants, avoiding the contamination of groundwater, to promote carbon sequestration in soil, as well as to formulate bioinspired slow-release fertilizers. In this chapter the structural, chemical, adsorption, and application properties of carbon nanotubes, graphene, graphene oxide, biochar (artificially synthesized or from anthropogenic soils), and nanomaterials-based fertilizers are discussed. The adsorption aspects in soil are studied for the different materials, showing the benefits of their use in diverse scenarios. Scanning electron microscopy, transmission electron microscopy, atomic force microscopy, Raman spectroscopy, X-ray photoelectron spectroscopy, Fourier transform infrared spectroscopy, and energy-dispersive X-ray analysis are presented as tools for the comprehension of the structural and chemical properties of these materials and their use in soil. The consequence of their application is also discussed in different contexts. Structural disorder in nanoscale and the presence of surface functional groups are shown to be related to the superior performance of these materials in soil applications, being promising to the elaboration of sustainable soil use practices.

11.1 INTRODUCTION

The scarcity of foodstuff products in some regions of the planet has been the focus of studies by world organizations regarding their availability. According to the Food and Agriculture Organization of the United Nations (FAO), one of the aspects that inspire this study theme is related to the population increase, which suggests the increase in the necessity of production of foodstuffs (FAO, 2017). Soil degradation, scarcity, and unavailability of some nutrients are the main challenges found for food production. According to Achari and Kowshik (2018), studies have shown that the soils available for agricultural production are increasingly deficient in macro (N, P, and K) and micronutrients (Zn, Cu, B, Mn, among others), due to biological transformations. Also, environmental concerns are being leveraged in recent years regarding soil contamination with toxic metals (As, Cd, Ni, Pb, Cr, etc.), pesticides, pharmaceuticals, among others (Gavrilescu et al., 2015; Gong et al., 2018; Iavicoli et al., 2017; Mattina et al., 2003) and their implication for food security. From these observations, new strategies aimed to treat polluted or nutrient-depleted soils are in high demand, and new materials presenting enhanced properties emerging from their reduced dimensionality, such as nanomaterials, are becoming valuable options. Among different nanomaterials used for that purpose, carbonaceous nanomaterials are on the rise and are the focus in this chapter.

In the next sections, the structural, chemical, and adsorption properties of several carbonaceous nanomaterials are discussed in greater detail, as well as the results and perspectives of their use in soil remediation.

11.1.1 CARBON NANOMATERIALS AS TOOLS FOR SOIL REMEDIATION

In an attempt to increase the fertility of agricultural soils, fertilizers have become essential agents of correction. However, conventional fertilizers are characterized by low absorption efficiency by plants and quick release, requiring several applications to obtain better results (Raliya et al., 2017). As a strategy to overcome such difficulties, the literature has suggested the use of nanoparticles as fertilizing agents, being called nanofertilizers (NF), or even intelligent fertilizers (Achari and Kowshik, 2018). NF are differentiated from conventional fertilizers because they have a nanostructured formulation, allowing nutrients to be delivered more efficiently, enabling better absorption by plants and even slow release of their active compounds, beneficial for crop development (Raliya et al., 2017). Due to their dimensions in the order of nanometers (1–100 nm), the NF have characteristics, such as high specific superficial area, porous structure, and crystallinity, which act as reservoirs of functional groups that are responsible for the properties of the nutrient release rate, reducing material waste when applied (Achari and Kowshik, 2018; Dimkpa and Bindraban, 2017). In addition, their adsorption capacity is improved due to these properties, being widely used for the removal (or immobilization) of several contaminants.

Carbon nanomaterials (CNM) are promising in the perspective of the development of several cultures (Mukherjee et al., 2016; Vithanage et al., 2017; Zaytseva and Neumann, 2016) and contaminated soil remediation (Baragano et al., 2020; Dai et al., 2019; Gong et al., 2019; Lei et al., 2019; Yan et al., 2017). According to Vithanage et al. (2017), carbon nanotubes have been studied due to the ease of being absorbed by plants, influencing their development, increasing root growth, and mainly stimulating the increase in plant biomass and seed germination. Carbon nanotubes are also effective in immobilizing contaminants, such as chlorinated hydrocarbons (Ma et al., 2010), antibiotics (Su et al., 2016), paracetamol (Yan et al., 2017), among several others. Graphene (GN) and graphene oxide (GO) present high mechanical resistance and functional groups on their surface, which allow their adaptation on several other surfaces (Alves, 2013; Mukherjee et al., 2016). These materials are being used in the transport and slow release of micronutrients for plants (Kabiri et al., 2018), lead (Zhang et al., 2018b) and cadmium (Liu et al., 2016) adsorption, and mainly for improvements in the physical and chemical stability of biocarbons and fertilizers (Kabiri et al., 2017).

An emerging important class of carbonaceous materials for soil remediation is the biochar (BC) and biochar-based fertilizer (BBF). These materials are obtained by the pyrolysis of different feedstocks and are mixed with conventional fertilizers, being bioinspired in the occurrence and properties of the "Terras Pretas de Índios" (TPI) or Indian Dark Earths. TPI are anthropogenic soils found in the Amazonian region with remarkable fertility and resilience (Glaser, 2007; Glaser et al., 2000; Neves et al., 2004). They are nanostructured with a mixture of nanographitic and amorphous carbon (Archanjo et al., 2014; Archanjo et al., 2015; Jório et al., 2012; Ribeiro-Soares et al., 2013), with beneficial uses reported as carbon retention in soil, improvement of chemical and physical properties of soil, nutrient retention, improvement of crop yields according to the soil properties, and adsorption of a broad number of contaminants (Chew et al., 2020; Purakayastha et al., 2019; Ye et al., 2020; Zhang et al., 2013).

The next sections will treat the general structural properties of these materials, and then their adsorption, structural, chemical, and application aspects will be related to their use for soil remediation from both polluted and nutrient-depleted soil remediation.

11.1.2 CARBON NANOTUBES, GRAPHENE, AND GRAPHENE OXIDE: GENERAL STRUCTURAL MODELS

Carbon is a chemical element presenting three conventional hybridization states (sp^3, sp^2, or sp), which allows the formation of complex structures with different properties (Zarbin and Oliveira, 2013). Thinking only of three-dimensional (or bulk) structures formed purely for carbon atoms, graphite and diamond are two representative materials. The graphite carbon atoms present sp^2 hybridization state, assuming the hexagonal structure, as shown in Figure 11.1a. The diamond is formed by carbon atoms in the sp^3 state of hybridization, enabling the formation of a tetrahedral structure, as illustrated in Figure 11.1b.

Although diamond and graphite are formed purely by carbon atoms, both have different properties. For example, diamond is a shiny, insulating, rare, hard, and expensive material. Graphite, on the other hand, is an abundant material in nature, opaque, electricity conductive, malleable, and cheap. Such characteristics allow the two materials to be used in several applications in the industry (Pan et al., 2015). In the 1940s, a versatile form involving only one sheet of carbon atoms in a "beehive" arrangement (Fig. 11.1c) with sp^2 hybridization was theoretically proposed (Wallace, 1947). However, experimental success in its synthesis was not obtained in the following decades. The proposed structure would be a sheet of carbon atoms with covalent bonds with sp^2 hybridization, the graphite being generated by the stacking of these sheets, weakly bonded together by van der Waals interactions (Wallace, 1947). In 1986, the term "graphene" was used for the first time to describe this material of atomic thickness (Boehm et al., 1986).

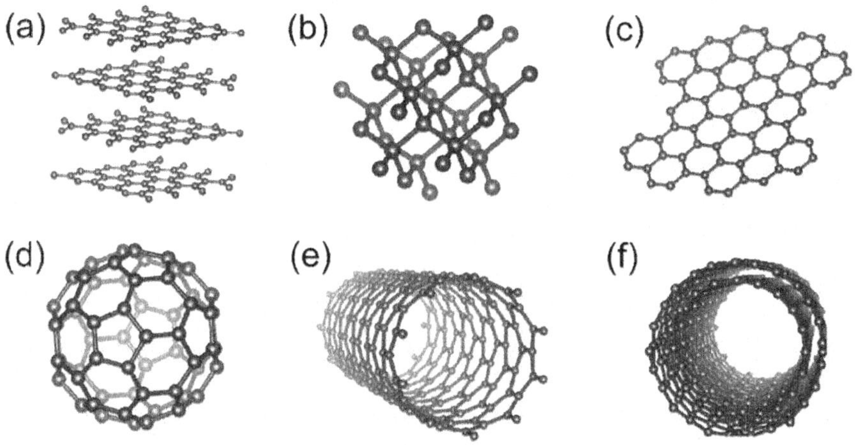

FIGURE 11.1 Crystalline structure representation of carbon materials and nanomaterials. (a) Graphite, (b) Diamond, (c) Graphene, (d) Fullerene, (e) Single-walled carbon nanotube, and (f) Multiwalled carbon nanotube.

In 1985, Kroto, Smalley, and Curl synthesized the first artificial structure formed by carbon atoms, fullerenes (Dresselhaus et al., 1996). Fullerenes are hollow spheres formed by sp^2 hybridized carbon atoms (Fig. 11.1d), being considered a zero-dimensional structure because the electrons confined within them have zero degrees of freedom (Birkett, 1998; Neto et al., 2009; Smalley, 1997). They can be understood as a sheet of graphene carefully shaped to form a spherical shell.

In 1991, Sumio Iijima called the world's attention to the result of his effective synthesis of carbon nanotubes (CNTs). CNTs are one-dimensional cylindrical nanostructures formed by the covalent bonds between C atoms (sp^2) with an average diameter of 1 nm and a length of the order of mm

(Iijima, 1991) and they can be understood as rolled graphene sheets. CNTs are named single-walled carbon nanotubes (SWCNTs) if only one layer of carbon atoms forms their structure (Fig. 11.1e) or multiwalled carbon nanotubes (MWCNTs) if concentric layers of carbon atoms are present (Fig. 11.1f). It is worth mentioning that the characteristics of the CNTs depend on the chirality presented by the carbon atoms in these structures, as well as on the number of concentric nanotubes and the pattern established in their interior (Iijima, 1991). Applications in the industry involving CNTs have been increasingly studied due to their physical and chemical properties (Dresselhaus et al., 1998), such as the high electrical conductivity (Ebbesen et al., 1996), mechanical resistance, and hardness (Musso et al., 2009).

Although fullerenes and nanotubes are formed by graphene sheets and graphite is formed by stacking graphene sheets, it was only in 2004 that the experimental isolation of graphene from a graphite sheet was carried out by André Geim and Konstantin Novoselov, being awarded in 2010 the Nobel Prize in Physics due to their contributions to the study of the physical and chemical properties of graphene (Novoselov et al., 2004).

In the graphene two-dimensional crystal, composed of carbon atoms in the sp^2 hybridization state, the so-called sigma (s) orbitals are responsible for the planar hexagonal patterns of the carbon sheets (Guimaraes, 2010). Because planar binding between the carbon atoms are covalent (Neto et al., 2009), this material has a high elastic constant. Figure 11.1a illustrates the structure of graphite with its interatomic distances between layers governed by van der Waals interactions, and the reduction of graphite to graphene at the limit of a single atomic layer is represented in Figure 11.1b.

Since its experimental synthesis in 2004, graphene became a material of great interest and importance due to its optical, mechanical, thermal, and electronic properties (Novoselov et al., 2004). From a mechanical point of view, graphene corresponds to the known available material with the highest elastic constant (Novoselov et al., 2004), which allows the application in different systems, such as soils (Kabiri et al., 2017), mortars (Barbosa, 2015), and concrete (Chuah et al., 2014; Mohammed et al., 2015; Pan et al., 2015) through its oxide or even its reduced oxide (Guimaraes, 2010).

GO is a derivative of graphene containing functional groups (hydroxyl, epoxy, carbonyl, and carboxyl) whose characteristics allow a broad range of applications (Maraschin, 2016). GO is usually obtained from a chemical process of graphite oxidation, in which only one graphite sheet incorporates heteroatoms, especially O and H. The graphene oxidation generates defects (such as holes, for example) in its structure, which allows the incorporation of functional groups, enabling its interaction with materials of organic or inorganic nature (Alves, 2013). The insertion of some of those functional groups can alter the carbon atoms hybridization. Thus, GO has a structure that contains carbon atoms in both sp^2 and sp^3 hybridization states (Lerf et al., 1998), as shown in Figure 11.2. In this perspective, the GO presents distinct regions, some with unoxidized aromatic rings and others with functional groups containing O and H through its all structure, including the edges and the top and bottom of the layer plane. The amount of these regions is determined according to the level of oxidation presented by the material (Alves, 2013; Lerf et al., 1998).

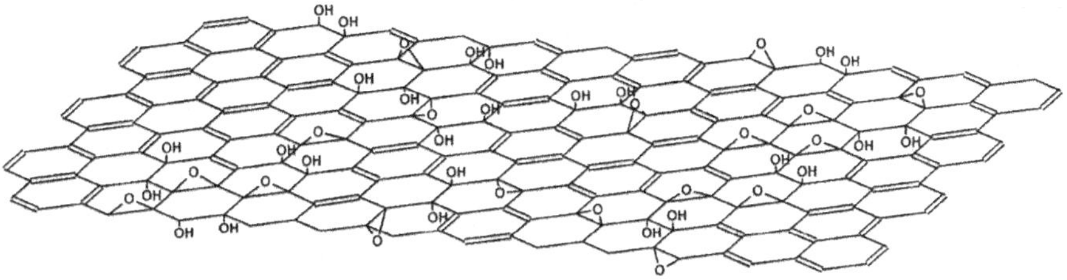

FIGURE 11.2 Representation of the basic structure of graphene oxide.

Source: Reprinted from He, H.; Klinowski, J.; Forster, M.; Lerf, A. A New Structural Model for Graphite Oxide. *Chem. Phys. Lett.* 1998, *287*, 53–56. Copyright (1998), with permission from Elsevier.

The level of oxidation present in the GO may vary according to the method used in the preparation of the material, as well as the purity of the precursor (Alves, 2013). The chemical properties of GO can be adjusted based on a possible variation in the amount of O through its chemical reduction (Dreyer et al., 2010). Such partial removal of O can enable improvements in semimetallic characteristics, making them close to those of graphene before oxidation, which increases its conductivity in comparison with the initial GO (Alves, 2013).

The GO has attracted the interest of researchers from different areas due to its multifunctional properties (Andelkovic et al., 2019; Andelkovic et al., 2018; Kabiri et al., 2017; Lerf et al., 1998; Zhang et al., 2014). Among the various possibilities of application of GO, it is worth highlighting its use in nanocomposites for performance in lithium batteries (Mussa et al., 2019), quantum dots (Li et al., 2019), and cement composites (Long et al., 2019). It has also been pointed as promising in the biomedical field as a drug transport agent in the body (Kazempour et al., 2019), especially in the fight against cancer (Ashjaran et al., 2019; Campbell et al., 2019).

11.1.3 ANTHROPOGENIC SOILS AND BIOCHAR STRUCTURAL MODEL: CONCEPT AND RELEVANCE FOR SUSTAINABLE SOIL REMEDIATION STRATEGIES

BC is the carbonaceous material produced by the heat treatment of different feedstocks, such as plant and animal residues, in the absence of oxygen, in a pyrolysis process (Lehmann, 2007). Their structures are irregularly shaped and contain several functional groups, having physical and chemical properties of great interest (Cho et al., 2015; Madari et al., 2010; Novotny et al., 2009). The study involving BC properties originates in the characterization of the "TPI" or Indian Dark Earths, which have properties desired by the agricultural sector (Glaser et al., 2000). Found further north of Brazil, especially in the Amazon region (tropical region), the emergence of TPI is linked to the activities of Pre-Columbian civilizations, such as the accumulation and burning of organic materials for many years (Glaser et al., 2000). According to Glaser et al. (2000), this region has tropical characteristics such as high temperature and frequent rainfall, which allows the rapid degradation of organic matter in the soil. Due to these conditions, nutrients are easily leached by the rains, characterizing poor soils.

Results obtained through radiocarbon dating found TPI with more than 7000 years, having a pre-Columbian origin (Glaser, 2007; Neves et al., 2004). These soils have about three times more organic matter, nearly 70 times more stable carbon, and higher nutrient retaining capacity than the surrounding soils (Glaser, 2007; Neves et al., 2004). This higher stable carbonaceous content favors the conditioning of their fertility. As it is a very fertile soil, TPI started to be the object of study in several research centers, especially due to its physical and chemical properties, such as thermal stability and high-nutrient content (Glaser, 2007). For the agricultural activities, the potential presented by the TPI is associated with their capacity to retain minerals and nutrients as well as their high cation exchange capacity, allowing their use as a direct natural fertilizer (Glaser, 2007). For Neves et al. (2004), the stable C content present in the TPI structure is similar to the carbonaceous content arrangement presented by the BC, whose structure varies according to the final temperature during the pyrolysis process.

BC properties have been studied for several applications, from amendment of soils to pollution control. Such applications involving BC depend on the parameters associated with their structure, which can be understood as graphene planes containing defects and nonlinear bonds stacked in a random manner. Besides, heteroatoms, such as O, P, N, and S are generally present in different functional groups or aromatic rings (Bourke et al., 2007). Covalent bonds between carbon atoms in the BC structure are responsible for their stability, making them materials with lower degradation rate in soil, unlike other forms of organic matter (Bourke et al., 2007).

Currently, studies have shown that BC can promote benefits to the soil, such as improvements in the use of nutrients by plants, greater productivity, and better quality of the harvested product, approaching the characteristics presented by the TPI (Speratti et al., 2018). In this context, they are being used in the recovery of depleted soils (Speratti et al., 2018) and as more stable slow-release fertilizers (Filho et al., 2020). The last is one of the main challenges encountered in relation to the availability of nutrients for longer periods in the soil. In addition, like other carbonaceous materials, BC has been proposed as adsorbent for removing or immobilizing pollutants in soils due to their adsorption properties. The fundamentals and use of those materials for soil remediation using adsorption are explored in the next section.

11.2 SOIL REMEDIATION USING ADSORPTION ON CARBONACEOUS NANOMATERIALS

Several contaminants are not efficiently retained by soil particles, so some technologies of soil remediation consist in modulating the availability and mobility of contaminants in soil. Many of these technologies are based on the stabilization of contaminants in the soil matrix, avoiding their leaching to groundwater. Among the different available options for soil remediation which are based on the stabilization of the contaminant in the soil matrix, technologies involving adsorption has gained visibility in view of several advantages, such as versatility, sustainability, and cost-effective character (Shen et al., 2018). In this scenario, carbonaceous nanomaterials have emerged as efficient adsorbents for the removal of different types of contaminants in soils because of their surface properties, including high specific surface area, reactivity, and versatility. Understanding some fundamental questions about adsorption phenomenon will be needed to evaluate how these carbonaceous nanomaterials surface properties determine their nanoremediation ability in soils. However, prior to discuss the adsorption process as a technology of soil remediation, the natural adsorption phenomenon which can occur when a contaminant enters in the soil matrix should be understood, as presented in Section 10.2.1.

11.2.1 ADSORPTION: GENERAL ASPECTS IN THE CONTEXT OF SOIL

Regarding the adsorption process, two concepts are of great importance: phase and interface. A phase denotes the macroscopic region of the system in which the intensive thermodynamic properties that do not depend on the extension of that region, are independent of the position in space. Thus, a phase is a homogeneous portion of the system. The contact region between two phases is called interface. From a microscopic point of view, the interface is characterized as the region of the system wherein the chemical composition continuously changes in space between the two phases forming the interface. In the environment, particles of the soil, aqueous solutions, and air are phases and the contact among them forms interfaces. For example, the contact between minerals in soils and aqueous solutions or between two different types of particles in the soil, as the mineral and organic fraction, forms interfaces with quite different properties.

An important property associated with the interface is the interfacial tension. Consider the interface formed in the contact between the soil particles and an aqueous solution, for instance. Due to aspects involving the different chemical nature of the species present in each phase, the interface has an excess of Gibbs free energy per unit of area compared with that of the phases in contact. Therefore, it must be recognized that increasing the interfacial area has an energetic cost. At constant temperature and pressure conditions, increasing the area of the interface by dA will lead to a variation of Gibbs free energy (dG) given by:

$$dG = \gamma dA \quad (11.1)$$

where γ, the interfacial tension, is a positive quantity. Equation 11.1 makes evident that the decrease in the interface area ($dA < 0$) promotes a thermodynamic stabilization (decrease in the Gibbs free-energy content, $dG < 0$) in a macroscopic system whose interfacial area is high. If this interface involves a phase comprising a porous solid material with a large specific surface area, as that involving the particles of soil, in contact with an aqueous solution, the surface area reduction is not possible to decrease the Gibbs free energy of the system due to the low mobility of chemical species in the solid structure. Then, the thermodynamic stabilization of the system will occur by a mechanism other than that of reduction of the interface area, that is, the interfacial tension will have to decrease, which can be achieved by the adsorption phenomenon.

When an aqueous phase (or air) containing a contaminant is in contact with particles of soil (solid), the contaminant can concentrate in the solid–fluid interface, altering the intensive thermodynamic properties of the interface. If the average equilibrium concentration of the contaminant in the interface region spontaneously become higher than the contaminant concentration in the phases forming the interface, the contaminant is adsorbed on the interface and the process is called adsorption. Figure 11.3 schematizes the adsorption of a contaminant on the interface formed between the soil particles and an aqueous solution in the soil.

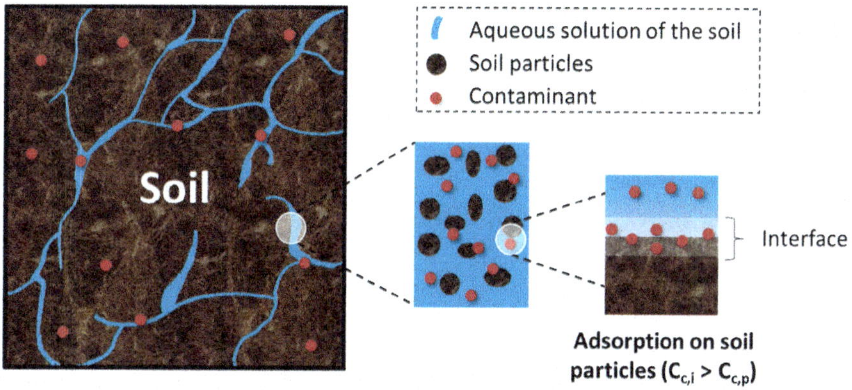

FIGURE 11.3 Adsorption of a contaminant on the interface formed between the soil particles and an aqueous solution in the soil. Adsorption is characterized by the higher contaminant concentration in the interface region ($C_{c,i}$) than in the phases forming the interface ($C_{c,p}$).

When a fluid phase is in contact with a solid phase, the solid phase is commonly named adsorbent and the adsorbed contaminant is called adsorbate. The functional groups or the group of atoms, located on the interface, with which the adsorbate interacts are the adsorption sites. Whether the immobilization process of the contaminant do not involve the interface, that is, the contaminant is transferred from one phase to another, then the process is called partition. Adsorption and partition are both referred as sorption processes, and depending on the adsorbent, both processes of immobilization can simultaneously occur.

The adsorption can be quantified by the amount of adsorbate present in the fluid-solid interface per mass unit of adsorbent, q_e, generally express in milligrams of adsorbate per gram of adsorbent. At constant temperature and pressure, q_e is a function of the adsorbate concentration that remains in the fluid phase at the thermodynamic equilibrium condition, C_e, and a plot of q_e versus C_e, at a fixed temperature, is named an isotherm of adsorption (Limousin et al., 2007). The isotherm profile is related to the intensity of the binding between the adsorbate and the adsorption sites, which can be modulated by changes in the conditions of the systems such as temperature, pH of the fluid phase, etc.

Usually, for a given C_e value, adsorbates that interact less intensely with the sites of one adsorbent are adsorbed with lower q_e values. Therefore, the less intense is the interaction between a contaminant and the chemical species forming the interface region, the greater will be the availability and mobility of this contaminant through the environmental compartments, determining the contaminant concentration in the dispersions of the soil. In other words, from an environmental perspective, weak adsorption of contaminants on soil particles will allow a contaminant to be easily transferred from the soil to air and water.

11.2.1.1 MECHANISMS DETERMINING CONTAMINANT ADSORPTION ON SOIL PARTICLES

As previously highlighted, the efficiency of the contaminant immobilization in soil will depend on the intensity with which the contaminant binds to the adsorption sites on the surface of the soil particles. The type of this binding (chemical binding or intermolecular interactions) determining the adsorption of the contaminant on the adsorbent is generally referred as the adsorption mechanism, although the term mechanism is more appropriate to refer to the set of steps by which a process occurs. Understanding the adsorption mechanism of contaminants on soil is a fundamental aspect in the design of technologies of soil remediation. Under this aspect, adsorption can be classified as specific or nonspecific. The dividing line between these two processes is not always sharp. In the specific adsorption, also called chemisorption, the adsorbate is bonded to the adsorbent through covalent or ionic bonds with the functional groups in the interface. The modules of the adsorption enthalpy changes associated with the chemisorption are generally higher than 40 kJ/mol. On the other hand, in the nonspecific adsorption, or physisorption, weaker interactions, as van der Waals or electrostatic ones, are involved in the transfer of the adsorbate from the fluid phase to the interface. The values of enthalpy changes for physisorption are usually smaller, not exceeding, in modulus, 40 kJ/mol (Levine, 2009). Both chemisorption and physisorption can be endothermic or exothermic.

Under certain conditions, chemisorption and physisorption of contaminants on the particles of the soil can occur simultaneously or alternatively, which will depend on the adsorbate molecular structure, the soil composition, and the nature of the functional groups on the surface of the soil particles. Moreover, changes in the nature of the adsorption process (physisorption or chemisorption) can also be caused by different geochemical condition, as the pH of the fluid phase, when an aqueous phase is involved, the presence of other contaminants in the soil or fluid phases, temperature, etc. For instance, artisanal and small-scale gold mining impacted soils contain particles with negatively charged surfaces of hydrous ferric oxides, kaolinite, and montmorillonite which could readily adsorb positively charged species, as metal cations, from aqueous solutions through nonspecific electrostatic interactions (Tabelin et al., 2020). If the soil, however, has a high content of natural organic matter (NOM), which comprises fulvic, humic, and humin materials with a great variety of functional groups (carbonyl, carboxyl, amino, thio, etc.), then metallic ions can bind forming specific interactions. Therefore, considering the nature of the adsorption processes of metal ions in different soil particles, it is expected that NOM-poor soils will retain toxic metal less intensely, increasing the contaminant mobility in the environment.

A combination of contaminants can also affect the adsorption process and it is needed to know the synergistic or nonsynergistic effect among those contaminants to enhance the effectiveness of remediation technologies in soils. For example, when metals and polycyclic aromatic hydrocarbons (PAH) are together in soils with high concentration of NOM, a synergistic effect potentializes the adsorption of the PAH (Saeedi et al., 2020). Particularly in remediation technologies based on adsorption, comprehending these aspects makes assertive the choice of adsorbents to be used properly in soil remediation. The next section presents the fundamentals on the use of adsorption for soil remediation.

11.2.2 ADSORPTION FOR NANOREMEDIATION OF SOILS

In the field of soil remediation technologies, there is an imperative need for the development of processes which can be employed for the treatment of soils contaminated with different classes of contaminants. Among these contaminants there are toxic metals (As, Cd, Ni, Pb, Cr, etc.) and organic compounds (chlorinated solvents, PAH, organophosphorus pesticides, polychlorinated biphenyls (PCB), poly-fluoroalkyl substances (PFAS), etc.) (Gong et al., 2018; Iavicoli et al., 2017; Mattina et al., 2003) as well as emergent contaminants (pharmaceuticals, personal care products, endocrine-disrupting compounds, etc.) are among the prime), which have caused great environmental concern in recent years (Gavrilescu et al., 2015). This wide range of chemical species, combined with the heterogeneity and complexity of the soil matrix, makes the development of remediation technologies for the treatment of contaminated soil a challenge. Several technologies have been proposed to solve soil clean up problem, including, for instance, chemical oxidation and reduction, soil flushing, leaching, thermal treatment, bioremediation, and adsorption, which can be applied *ex situ* and/or *in situ* (Koul and Taak, 2018a; Koul and Taak, 2018b).

The remediation technology based on adsorption has been widely evaluated and extensively proposed as alternative to clean up soils contaminated with toxic metals (Garcıa-Sanchez et al., 1999; Gong et al., 2016; Lei et al., 2020). However, its application in the treatment of soils contaminated with other classes of contaminants has been also highly explored (Gong et al., 2016; Pernyeszi et al., 2006). The interest in this technology relies on several advantages, such as (1) the contaminant can be immobilized in the soil for dozens or even hundreds of years, avoiding groundwater or atmosphere contamination; (2) the immobilization of organic contaminants and toxic metals in soils can reduce their bioavailability, impacting long-term changes on soil microorganisms; (3) low mass ratio between adsorbent and soil can be used, decreasing the costs of application; (4) versatility in terms of site size and contaminant type; (5) ease of implementation and simplicity of design and operation; and (6) less secondary contamination (Jones et al., 2012; Shen et al., 2018).

Fundamentally, adsorption for soil remediation can be classified as a stabilization technology in which the contaminant in the soil is immobilized on the surface of a material designedly mixed with the soil matrix. Thus, adsorbent materials added in contaminated soils actuate as a sorption barrier. Because of the nature of the adsorption process, which is determined by the intensity of the interaction between the adsorbate and the adsorbent, this treatment technology has as premise the fact that the contaminants in the soil to be treated must interact more strongly with the adsorbent used for the remediation than with the soil particles. However, only this is not enough, especially when nanomaterials are employed as adsorbents. Figure 11.4 illustrates other conditions that should be evaluated for the application of an adsorbent for soil nanoremediation, in which the term nanoremediation is used to design the remediation processes based on the use of nanomaterials.

Considering those aspects, three steps are recommended to propose the use of adsorption as viable solution for remediation of a soil: (1) investigation of the physicochemical and adsorptive properties of the adsorbent; (2) bench studies to evaluate the potential of the adsorbent to treat the contaminated soil of interest; and (3) field application followed by long-term monitoring. (Shen et al., 2018). The two first steps allow to determine if the adsorbent is stable and presents a high capacity to immobilize the contaminant, while the third one allows evaluating if environmental factors may affect the contaminant immobilization in the long-term. Figure 11.5 summarizes the main aspects regarding the use of adsorption in contaminated soil treatment, in which it is possible to understand how those steps contribute to develop a viable adsorption remediation process.

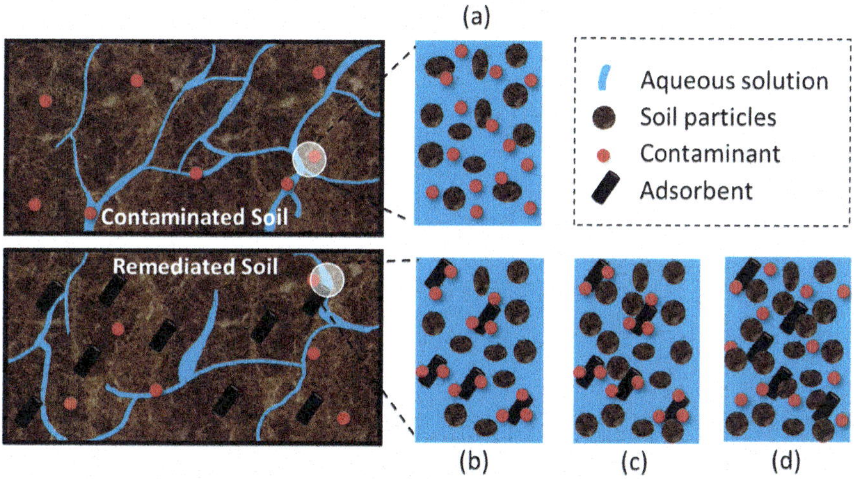

FIGURE 11.4 Different behaviors associated with the addition of an adsorbent in the soil matrix: if no adsorbent material is added, (a) the contaminant can adsorb in the soil–water interface or be leachate through the soil; when an adsorbent material with adsorption capacity higher than the soil particles is added, the adsorbent itself can (b) present high mobility through the soil matrix, being leached with the contaminant, or (c) remain immobilized in the soil particles, avoiding the leaching of both adsorbent and contaminant. Whether the interaction between the soil and the adsorbent is very favorable, (d) soil particles can disadvantage the contaminant adsorption on the adsorbent (Zhao et al., 2019b).

In a simplistic point of view, stable designed materials that can strongly interact with the contaminant, immobilizing a great amount of them in their surface even under condition of interaction with the soil particles, are needed. Under these conditions, the use of nanomaterials has been proved as feasible and efficient, highlighting the possibilities of the nanoremediation for soil cleanup. Among a variety of nanomaterials that can be proposed for soil cleanup, including metal oxides and nanocomposites (Qian et al., 2020), the carbonaceous nanomaterials, such as carbon nanotubes, graphene,

graphene oxides, and nanostructured biochar, have received attention due to their versatility and unique properties. Beyond their own surface chemical structure that can interact with different classes of molecules, carbonaceous nanomaterials may be manipulated at the atomic and molecular level by adding specific functionalities that can immobilize a particular contaminant within a complex soil mixture.

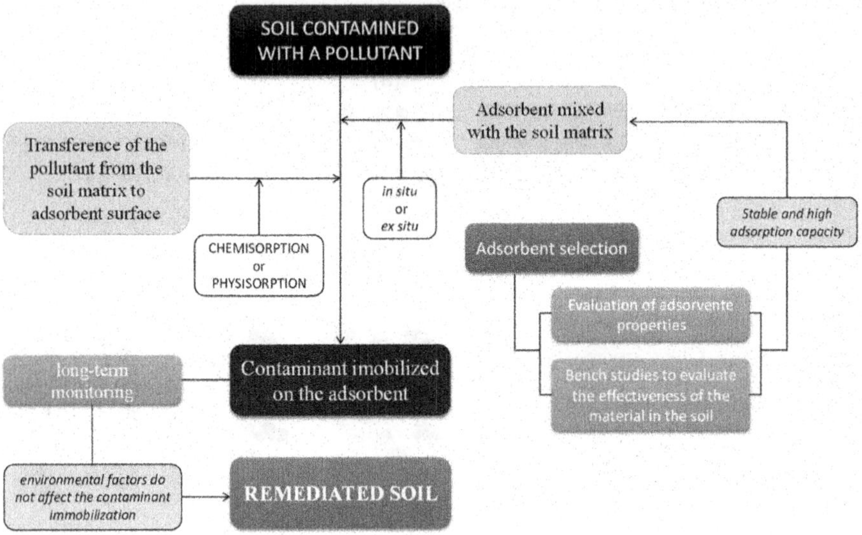

FIGURE 11.5 Flowchart illustrating the fundamental aspects associated with the use of adsorption as treatment technology for contaminated soils (Shen et al., 2018).

11.2.3 NANOREMEDIATION OF SOILS USING CARBONACEOUS NANOMATERIALS

Despite the common point in which carbon nanotubes, graphene, graphene oxides, and nanostructured biochar are essentially made up of carbon, their different structures and properties make them useful for distinct applications. However, some common properties of those carbonaceous nanomaterials, namely, their large specific superficial area, porous structure, and/or capacity to form composites with other materials to improve their adsorption capacity, make all of them widely used as adsorbents in the environmental application field for the removal of several contaminants. Thus, beyond their widespread application to remove contaminants from aqueous matrices, carbonaceous materials have been also used in soil remediation. In this section, the main advances in the field of soil remediation using different carbonaceous nanomaterials are presented.

11.2.3.1 CARBON NANOTUBES

After the discovery by Iijima (1991) and the elucidation of their properties, CNTs were extensively used as adsorbents for the removal of different contaminants from aqueous matrices. Although their use as adsorbent in soil remediation is still not fully explored, important advances have been made in this research field. In this section, firstly the fundamental aspects related with the use of CNTs for contaminant immobilization are discussed, and then the main applications of CNTs for soil remediation, including their application for sediment remediation, are presented.

11.2.3.1.1 Fundamental Aspects of the CNT Application for Soil Remediation

The potential of CNT to actuate for the treatment of contaminated soils can be attributed to two main aspects: (1) the effective retention of CNT by the soil matrix and (2) the great efficiency of CNT to adsorb large amounts of different classes of contaminants, especially hydrophobic organic compounds (HOC). The first aspect ensures that CNT will have a limited transport through the soil, avoiding its own leaching to groundwater; the second one guarantees the versatility of CNT to treat contaminated soils.

The high efficiency of soil to retain CNT has been attributed to morphological and surface properties of the CNT, such as irregular shape, large aspect ratio, and functionalization of the surface, as well as aspects of the soil, such as the high heterogeneity of particle size, high porosity and permeability, pore interconnectivity, surface potential, and hydrophobicity (Jaisi and Elimelech, 2009; Tian et al., 2012; Zhang et al., 2012). However, high CNT mobility can occur in some types of soil, favoring the co-transport of contaminants into deeper soil layers (Zhang et al., 2017b).

Regarding the great capacity of CNT to adsorb different contaminants, the capacity has been specially related with the chemical features of CNT surface that allow multiple mechanisms to control the adsorption process, including hydrophobic interactions as well as electrostatic interaction, hydrogen bonding, and π-π interaction.

Hydrophobic interaction is the main mechanism determining the adsorption of HOC on CNT, but it cannot completely explain the adsorption. Particularly for aromatic HOC, π-electron donor and acceptor interactions, formed between the aromatic structures of both CNT and adsorbate, can strongly contribute to the adsorption. In addition, if the functional groups are present in a large extension on the CNT surface and the HOC have polar or ionizable functional groups, the electrostatic interactions and hydrogen bonding can also influence the process of adsorption (Apul and Karanfil, 2015). Functional groups containing oxygen, for instance, can be present on the CNT surface due to the presence of defects, generated during the synthesis, in its nanostructure or they can be incorporated by functionalization. The presence of these oxygen-rich groups plays an important role in the adsorption of metal ions on CNT through electrostatic interactions, complexation, and/or ion exchange (Ihsanullah et al., 2016).

Summarily, adsorption on CNT is deeply dependent on the chemical nature of the functional groups presents in both CNT and adsorbate, making a challenge to the task of quantifying the individual contributions of the distinct mechanisms controlling the adsorption.

11.2.3.1.2 Applications of CNT for Soil Remediation

Despite the great potential of CNT application for remediation of soil contaminated with HOC, there are few studies evaluating the use of CNT to treat soils contaminated with these contaminants, such as that showing the effective potential of multiwalled carbon nanotubes (MWCNT) for immobilization of dichlorophenyl trichloroethane (DDT), an organochlorine pesticide (Taha and Mobasser, 2014). Majority of studies in this field are concerned to understand the effects of CNT on the soil properties, and consequently, on the mobility, toxicity, and bioavailability of HOC, such as naphthalene, fluorene, phenanthrene, and pyrene (Li et al., 2013; Towell et al., 2011; Zhang et al., 2019). In general, the results from these studies have shown that the presence of CNT increases the retention of HOC, changing their adsorption and desorption processes in soils. These processes are influenced by the type and concentration of CNT, as well as the adsorbate and soil properties. For example, dissolved organic matter in soil and coexistence of clay inhibit pyrene adsorption on CNT, either by pore blockage of CNT aggregates or by occupying adsorption sites on the CNT surface (Zhang et al., 2019).

For the purposes of remediation of soils contaminated with toxic metals, CNTs were only recently investigated as adsorbent for contaminant immobilization. Particularly, MWCNTs have been used to immobilize lead, copper, nickel, zinc, and antimony in contaminated soils (Matos et al., 2017; Vithanage et al., 2017). Despite the capacity that metal immobilization depends on the metal under investigation, in general, the presence of MWCNT increases the adsorption capacity of the soil, thus improving the immobilization of those toxic metals in the soil matrix. The potential of CNT to immobilize metals in soils has been also pointed in an interesting study from Gredilla et al. (2019) who, although have not applied CNT with soil remediation purpose, have evidenced the mercury sequestration by CNT present in agricultural soils from a coal-fired power plant exhaust.

Beyond the use of CNT as adsorbent, they have also been proposed as coadjutants in other processes of remediation in contaminated soils. For instance, the adsorption of Cr(VI) by functionalized MWCNT was investigated together with their catalytic effect on the metal reduction by citric acid, in which the addition of the MWCNT to an oxisol enhanced the yield of Cr(VI) reduction in the remediation process (Zhang et al., 2018c). Similarly, the use of a CNT barrier was shown to improve the electrokinetic remediation of 1,2-dichlorobenzene in clay (Yuan et al., 2009).

11.2.3.1.3 Applications of CNT for Sediment Remediation

The potential of CNT for remediation processes of solid matrices in the environment has been also demonstrated from their use for the remediation of sediments. Both MWCNT and SWCNT have been proposed to treat sediments containing residual contaminants, especially HOC, such as organochlorine pesticides, antibiotics, and drugs. However, as in studies involving soils, most of the studies in this field focus on understanding the effects of CNT on the physicochemical properties of sediments as well as the adsorption behavior of coexisting contaminants on the sediments themselves. In general, both SWCNT and MWCNT present a great potential to immobilize the contaminants in sediments. However, regarding the effect of CNT type on this process, SWCNT is more efficient than MWCNT to immobilize the contaminants, which is attributed to its larger surface area (Hua et al., 2017; Zhang et al., 2017a). Table 11.1 summarizes some studies, from the last decade, demonstrating the great potential of CNT to contaminant immobilization in sediments contaminated with organic or inorganic contaminants.

TABLE 11.1 Studies, from the Last Decade, Demonstrating the Great Potential of CNT to Contaminant Immobilization in Sediments.

CNT type	Contaminant	Sediment	Main result	Refs.
MWCNT	Chlorinated hydrocarbons	Sand	Sand textured with CNT was effective to reduce the migration of contaminants from the sediment to overlying water	Ma et al. (2010)
SWCNT	PCB[a]	New Bedford Harbor sediment	SWCNT reduced the PCB bioavailability, decreasing the bioaccumulation and toxicity of the contaminant to benthic organisms	Parks et al. (2014)
SWCNTs and MWCNTs	Sulfamethoxazole, sulfapyridine, and sulfadiazine antibiotics	Sediment from Xiangjiang river, China	CNTs strongly retained the antibiotics due to their high adsorption capacities and limited transport in the sediment	Su et al. (2016)
SWCNTs and MWCNTs	DDT and HCH[b] (Organochlorine pesticides)	Sediment from Dong-ting lake, China	Both SWCNTs and MWCNTs were effective to prevent the residual DDT and HCH being released from the sediment	Hua et al. (2017), Zhang et al. (2017a)
SWCNTs and MWCNTs	Paracetamol and diclofenac sodium	Sediment from Xiangjiang river, China	CNTs showed capacity to impede the transport of paracetamol and diclofenac sodium in the sediment	Yan et al. (2017)
MWCNT	Sodium dodecyl benzene sulfonate (surfactant)	Sediment from Xiangjiang river, China	MWCNT increased the adsorption capacity of the sediment for SDBS, affecting its transport in the sediment	Song et al. (2018)
MWCNT	Cadmium	Sediment from Xiangjiangriver, China	MWCNT increased the uptake of Cd, improving the phytoremediation efficiency in the restoration of the sediment	Gong et al. (2019)

[a]*PCB, Polychlorinated biphenyls;* [b]*DDT, Dichlorodiphenyltrichloroethane; HCH, hexachlorocyclohexane.*

The potential of CNT to be used in the development of technologies for the treatment of soils and sediments contaminated with different contaminants is an important issue that highlights the possibility of their use in different contexts of soils/sediments remediation, despite the challenges still associated with the use of these nanomaterials. For example, regarding the potential of CNT to immobilize toxic metals, specifically, a promising application could be associated with the development of technologies for the remediation of soils and sediments contaminated by toxic metals from mining tailings.

11.2.3.1.4 Challenges of the Use of CNT for Soil Remediation

Although CNTs have proved being effective to reduce the mobility of different contaminants in soils (or sediments), uncertainties about their environmental risks in natural ecosystems have limited their application. Most of the efforts in this field have been directed to understand the impacts of CNT on the natural ecosystems properties as well as the role of the aging process on the properties of CNT (Fan et al., 2017). Therefore, beyond simply evaluating the potential of CNT to immobilize a specific contaminant, a global comprehension about the influence of CNT on the physicochemical properties of the solid matrix is needed, which includes evaluating the effect of the presence of CNT on the potential of own soil particles to adsorb other contaminants. So that the presence of CNT cannot affect in a large extension the natural capacity of the soil to immobilize a contaminant despite their capacity to efficiently immobilize other one (Fang et al., 2008). In summary, to evaluate the ecological risks of CNT in the soil matrix is indispensable for the assessment of remediation potential of that nanomaterial, and consequently, the development of secure applications of CNT for soil cleanup (Fitz and Wenzel, 2002; Song et al., 2019; Towell et al., 2011).

11.2.3.2 NANOSTRUCTURED BIOCHARS

As other carbonaceous nanomaterials, biochars have amazing adsorption properties for a wide range of organic and inorganic contaminants (Zhang et al., 2013). Because of this, biochars produced from different feedstock and pyrolysis conditions have been proposed as material for soil remediation through adsorption technology. Beyond the own biochar ability to immobilize contaminants in soils, other processes associated with its use for soil remediation can be involved. For instance, biochars have shown to interact with microbes in soils, which would be probably associated with the degradation or transformation of contaminants (Ma et al., 2010). In addition, other aspects regarding mitigating sustainable practices on soil amendment are related to biochars usage, including carbon storage as well as fertility and quality increase of soil for agriculture. These aspects make biochars the most important carbonaceous material for soil remediation, although the benefits associate with the use of a specific biochar in soil, including their immobilization capacity of contaminants, are dependent on the biochar dosage in the soil matrix, remaining a challenge in this field to determine the optimized conditions for biochar applications (Gopinath et al., 2020).

The versatility of biochars to adsorb different classes of molecules is associated with their morphology which is characterized by the presence of different active functional groups spread on a large specific surface area rich in micropores. These aspects provide to biochar simultaneous distinct mechanisms to interact with contaminants in soils. In the next subsections, these mechanisms are highlighted for toxic metals and organic contaminants.

11.2.3.2.1 Soils Containing Toxic Metals

Interaction between toxic metals and biochars can occur through different types of mechanisms. The simplest involves the ion exchange between the metal cation and other cations associated with the biochar, such as Ca^{2+} and Mg^{2+}. Theses exchangeable cations in the biochars come from the complexed humic matter and mineral oxides generally incorporated to the biochar and can be quantified by the cation exchange capacity (CTC). In another mechanism, the mineral elements in the biochar, such as carbonates, sulfates, and phosphates, may combine with the toxic metals, forming poorly soluble salts.

Animal-derived biochars generally contain higher CTC than plant-derived ones due to their high Ca^{2+} contents, which makes ion exchange a dominant mechanism of animal-derived biochars in the immobilization of several metal cations. This can be observed, for instance, for immobilization of Cd^{2+} and Cu^{2+} on biochars derived from cattle and swine carcasses. However, the same biochars are effectives to immobilized Pb^{2+} via precipitation, showing that the mechanism for metal adsorption is not determined only by the biochar structure, but also depends on the metal adsorbed (Lei et al., 2019).

The adsorption of toxic metals on biochars can also occur by physisorption that comprises electrostatic, cation–π, and van der Waals interactions. Regarding the first one, when biochars are added in the soil, they can increase the surface negative charge density of the soil matrix, attracting more intensely the positively charged toxic metals (Liang et al., 2006). The relative contribution of the electrostatic interactions for metal adsorption is determined by the pH of the soil, which modulates the surface charge of both biochar and soil particles, beyond the charge of the own metal species. When the pH of the soil increases above the zero-charge point of the biochar (pH_{pzc}), the negative net charge on the surface of the biochar increases, favoring the electrostatic adsorption of positively charged metal ions. On the other hand, at high enough pH, some metals can form negative hydroxide complexes which interact through repulsive electrostatic interactions with the biochar surface, or insoluble neutral hydroxide complexes which precipitate may disadvantage the adsorption. In the first situation, leaching of the metals through the soil will increase, showing that the geochemical properties of the soil should be analyzed before biochar application for soil remediation.

Another mechanism for adsorption of toxic metals is associated with the capacity of the oxygen-containing functional groups of the biochar (carboxyl and hydroxyl, for instance) forming complexes with the metals, characterizing a chemisorption. Generally, biochars produced at lower temperatures of pyrolysis (200°C–400°C) are richer in oxygen-containing functional groups, presenting more efficiency to adsorb metal ions due to complexation mechanism. This is the main mechanism described for the adsorption of Pb^{2+} on variable charge soils amended with rice straw-derived biochar (300°C during pyrolysis) (Jiang et al., 2012) and Al^{3+} on biochar obtained from cattle manure (400°C during pyrolysis) for the alleviation of aluminum phytotoxicity in soils (Qian et al., 2013). Higher capacities of metal adsorption can also occur at high temperature biochars (500°C–900°C during pyrolysis), as observed for adsorption of nickel on biochars produced from wheat straw pellets and rice husk (Shen et al., 2017). In this case, for the same feedstock, a higher pyrolysis temperature results in the formation of more content of alkaline minerals, favoring the metal immobilization by precipitation and cation exchange mechanisms. In addition, more active aromatic structure in the biochar, generated at the higher temperatures, also favors the adsorption via cation–π interactions.

The mechanisms presented above generally act together. However, the relative importance of each one can depend on the properties of soil (pH, composition, etc.) and biochars, as well as all metal ions present in both soils and biochar. As a consequence, the success of the biochar usage for remediation of a contaminated soil should consider those mechanisms associated with the toxic metal immobilization on the biochar.

Another important issue beyond the own biochar adsorption capacity to immobilize toxic metals is the biochar ability to changing the adsorption capacity of the particles of the soil itself. Modification in the soil properties, specially pH and CEC, has been considered one of the most important contributions for the decline in the content of Al, Zn, Cu, Mn, and Cd in soils of different agricultural sites in the Czech Republic (Hailegnaw et al., 2020). Changes in the pH of soil induced by biochars can also reflect in undesirable process, as the increased As solubility with more alkaline pH already reported (Fitz and Wenzel, 2002).

11.2.3.2.2 Soil Containing Organic Compounds

Many works have shown that biochars have good adsorption capacities to several organic compounds, including PAH, benzene derivatives, phenols, agrochemicals, plasticizers, and pharmaceuticals, such as antibiotics and endocrine-disrupting chemicals (Dai et al., 2019; Gu et al., 2016; Regkouzas and Diamadopoulos, 2019; Zhang et al., 2016). Beyond other aspects referenced in previous sections, such as large specific surface area and high porosity, the great adsorption capacity of biochars for those organic contaminants has been mainly ascribed to their aromatic structure which can favorably interact with hydrophobic and aromatic organic compounds through hydrophobic, aromatic-π, and cation-π interactions (Jung et al., 2013). Generally, biochars obtained at pyrolysis temperatures higher than 500°C have a highly aromatic structure which favors those interactions. However, the mechanisms of adsorbing organic compounds also depend on the adsorbate structure and can involve other types of interactions, including electrostatic attraction, hydrogen bonding, and complexation.

Hydrogen bonding and π–π-electron donor–acceptor interactions are the dominant mechanisms for the adsorption of estrone, 17β-estradiol, estriol, 17α-ethynylestradiol, and bisphenol A on biochar from *Eucalyptus globulus* wood activated with phosphoric acid at pyrolysis temperature equal to 600°C and at nitrogen atmosphere. That adsorption is dependent on the change of the pH which modulates the presence of π-electron donor and π-electron acceptor domains on the biochar surface, modifying the adsorptive mechanism to guarantee an excellent removal efficiency of π-electron donors or π-electron acceptors hydrophobic organic contaminants (Ahmed et al., 2018).

Despite the fact that some charged organic contaminants can be adsorbed by electrostatic attraction (Hassan et al., 2020), electrostatic repulsion between functional groups charged with charges of the same sign in both biochar and adsorbate can also limit the adsorption capacity of the biochars. This can be observed, for example, for the adsorption of imazamox, metazachlor oxalic acid, and metazachlor sulfonic acid, polar pesticides in soil amendment with biochar produced from beech wood at temperatures up to 550°C (Dechene et al., 2014). Due to its low value of pH_{PCZ}, the biochar was negatively charged in the pH conditions investigated, promoting an electrostatic repulsion of the contaminants that were deprotonated and negatively charged at the same conditions. In this situation, the proposed biochar is not considered a viable strategy to reduce the leaching potential for those polar pesticides. However, if electrostatic interactions are not involved in the immobilization process, other interactions can take place in the adsorption of polar pesticides on biochars. For instance, adsorption of diuron, a nonionizable polar herbicide, was enhanced by the application of biochar from *Eucalyptus* in a Cerrado Haplic Plinthosol, in which hydrophobic interactions and hydrogen bonding formation were the main interactions governing the contaminant immobilization (Petter et al., 2019).

As observed, from a mechanistic point of view, biochars immobilize contaminants on their surface in a similar way of how CNTs do. However, higher porosity and more complex composition of the biochars determine extra mechanisms of sorption, such as partition and pore-filling. In the first, the contaminant is transferred from the fluid phase to the uncarbonized fraction of the carbonaceous material. In the last one, the contaminant fills entirely the pores of the biochar.

Partition as sorption mechanism is reported, for instance, in the remotion of PAH in soils using biochars made from pine needle in pyrolysis temperature from 100°C to 700°C. Particularly, partition is dominant at low pyrolysis temperatures and favored at high concentrations of the contaminant. However, sorption mechanism is changed to adsorption-dominant at higher pyrolysis temperatures and favored at low concentrations of contaminant (Chen and Yuan, 2010). Because PAH present low solubility in the aqueous solution of the soil, uncarbonized fraction of the biochar can solubilize those contaminants, favoring their partition to that fraction of the material. Pore-filling, on the other hand, is presented as one of the sorption mechanisms of thiacloprid, a neonicotinoid insecticide, on black soil amended by various biochars. Beyond pore-filling, the sorption mechanism on both biochars and biochar–soil mixtures are mainly through hydrophobic and π–π-electron donor–acceptor interactions (Zhang et al., 2018a).

In summary, the different types of structures of both contaminants and biochars make the mechanisms associated with immobilization of organic contaminants on biochar to be broad. In this context, evaluation of the surface properties of the biochars and knowledge on the molecular structure of the contaminant are indispensable to understand and predict the mechanisms of adsorption that will determine the efficiency of biochar to soil remediation.

11.2.3.2.3 Soils Containing Multicontaminants

As discussed in section 10.2.1.1, combination of different contaminants can alter the adsorption capacity of the soil regarding some of those contaminants. In this point of view, the evaluation of the potential of the adsorption for remediation of multicontaminated soil is a relevant issue, in which biochars have been shown to be promising choice. For instance, the simultaneous monitoring of toxic metals (Cu, As, Cd, and Zn) and PAH in a contaminated soil amended with biochar over 60 days field exposure shows that biochars effectively decrease the concentration of Cd and Zn as well as PAH in pore water. Although in the same condition the Cu and As concentrations in the soil pore water increases more than 30-fold after adding the biochar associated with significant increases in both dissolved organic carbon and pH (Beesley et al., 2010), those results show the potential of biochars to simultaneously immobilize different contaminants in the soil. However, it is also evident that the remediation of multicontaminated soils is more complicated and can require the combination of different technologies for remediation.

11.2.3.3 GRAPHENE AND GRAPHENE OXIDE

As CNT, graphene and graphene oxide in a short period of time since their discovery showed a prominent potential for water and air remediation, and also are acting as adsorbent for sequestration of toxic metal ions, rare-earth metal ions, organic compounds, etc. (Wang et al., 2013). However, the use of graphene-based materials for soil remediation is uncommon compared with those applications involving adsorptive remediation in water and air matrices.

The few works investigating the graphene-based materials usage for soil remediation are focused on graphene oxide (Baragano et al., 2020), which can be attributed to the surface properties of these materials. They have various oxygen-containing functional groups, such as hydroxyl, carbonyl, and carboxyl on their surface, which can directly interact with different contaminants or be covalently modified to interact with specific contaminants.

The mechanisms involving the adsorption of contaminants on these materials are similar to those on CNT, that is, electrostatic interaction, complexation, hydrophobic interaction, π–π interaction, and hydrogen bonding (Zhao et al., 2019b). In addition, because graphene oxides are nonporous adsorbents, adsorption occurs only on their outside surface, thus diminishing the control of internal diffusion mechanisms.

Regarding the application of graphene-based materials in soils, graphene oxide nanoparticles have been investigated for the remediation of soil contaminated with toxic metals and were compared with the performance of zero-valent iron nanoparticles (nZVI) (Baragano et al., 2020). While graphene nanoparticles have immobilized Cu, Pb, and Cd, and have mobilized As and P, the nZVI have promoted the immobilization of As and Pb, a less effective immobilization of Cd, and an increased availability of Cu in the soil. It is noteworthy that these opposite results for both nanoparticles indicate the possibility of applying hybrid approaches of phytoremediation and nanoremediation in a soil treated with both graphene oxide nanoparticles and nZVI.

The transport of contaminants through the soil matrix can also be affected in different extensions by graphene oxide nanoparticles, depending on the molecular properties of the contaminant and the level of carbonaceous material aggregation. Low concentrations of graphene oxide nanoparticles in a saturated soil, for instance, significantly enhance the transport of 1-naphthol, a polar aromatic compound, but affect the transport of phenanthrene, a hydrophobic aromatic one, to a much smaller extent. In addition, when the soil solution is subjected to chemistry conditions that favor the aggregation of the nanoparticles, the transport enhancement effects of the contaminants are increased (Qi et al., 2014).

One of the reasons that can explain the failure of graphene or graphene oxide for soil remediation is the blocking of adsorptive sites of graphene or graphene oxide due to intense interactions between these nanomaterials and the soil particles (Zhao et al., 2019b). To overcome this problem, 3D composites containing graphene oxide have been developed, in which their morphology allows the contaminant diffusion from the soil solution to the inner structure of the material, while the solid particles of soil would be retained at the exterior surface (Zhao et al., 2019a). This interesting approach has gained attention in the field of research about the application of carbonaceous materials for soil remediation, leading to the development of engineered carbonaceous nanomaterials, which are discussed in the next section.

11.2.4 ENGINEERED CARBONACEOUS NANOMATERIALS

Despite the great potential of conventional nanostructured carbonaceous materials (CNT, graphene, graphene oxide, and biochars) for contaminant immobilization in soils, further improvements in their remediation efficiency and environmental benefits are still needed. In this context, recent attention has been focused on the modification of carbonaceous materials or combination of them with other nanomaterials to obtain new ones with novel morphologies and surface properties as well as greater potential to be used as adsorbent in soil remediation. This class of new materials has been named engineered carbonaceous nanomaterials (ECN) and can be achieved using different approaches, such as acid/base treatment, chemical oxidation process, modification of functional groups (carboxylation and amination), coating, treatment with organic solvents, modification with surfactants, impregnation with mineral oxides, and magnetic modification (Madhavi et al., 2021; Rajapaksha et al., 2016).

ECN generally have increased specific surface area, reduced aggregation capacity, and greater density of active adsorption sites in their surface when compared with conventional carbonaceous nanomaterials, therefore, enhancing the adsorption capacity of those materials. These improved properties are compatible with a high performance of ECN application in environmental remediation.

Regarding the use of ECN for the soil remediation, recent works have shown the improvement of the sorption capacity of contaminants by those materials compared with the nonmodified ones. For instance, coating of pistachio-shell residues biochars with nZVI leads to higher efficiencies in cadmium stabilization compared with pristine biochars (Saffari et al., 2019). Other biochar-based engineered materials for the remediation of soil contaminated with toxic metal/metalloid include chitosan-modified biochars, which has been shown an effective, low-cost, and environment-friendly adsorbent for the removal of cadmium, lead, and cadmium (Zhou et al., 2013), as well as manganese oxide-modified biochar, which has been efficient to increase the removal of arsenic in soil (Yu et al., 2015).

Concerning the use of graphene-based materials, graphene oxide modified with sulfonate groups is a promising ECN for the nanoremediation of PAH-contaminated soils (Gan et al., 2017). Graphene and nZVI have been also integrated to a complex biochar-based engineered material for metal stabilization in soil using biochar from cornstalk. This new nanomaterial has a superior immobilization efficiency of Cu than pristine biochar, being a promising and effective amendment for immobilizing this metal in contaminated soils (Mandal et al., 2020).

Compared with biochars and graphene materials, the development of ECN based on CNT for soil remediation is less explored. However, effective CNT-based engineered materials have been proposed for the remediation of soil contaminated with organic compounds. For example, a composite formed by magnetic MWCNT and metal–organic framework, a porous crystalline material formed by metal units and organic linkers, have been shown as promising adsorbent for the removal of organophosphorus pesticides is soil (Liu et al., 2018).

The adsorption properties of the different carbonaceous materials explored in this section for contaminated soil remediation are also valuable for nutrient retention in poor soils. Also, their physical and chemical properties present fundamental information for the formulation of new strategies of their efficient use. In the next sections, characterization and performance aspects of these carbonaceous nanomaterials will be discussed in greater detail, broadening the understanding of their efficiency in nutrient retention.

11.3 NUTRIENT-DEPLETED SOILS REMEDIATION WITH CARBONACEOUS NANOFERTILIZERS

The characterization and understanding of physical–chemical phenomena involving carbon nanotubes, graphene, graphene oxide, and biochar are of fundamental importance to engineer their use. This process allows the verification of some properties of the material under analysis that corroborate the factors related to its applicability in the industry. The next sections of this chapter will be devoted to the exploitation of the main structural, chemical, and application properties of these materials and their relationship with soil remediation for both contaminated and nutrient-depleted soils.

11.3.1 GRAPHENE, GRAPHENE OXIDE, AND CARBON NANOTUBE-BASED FERTILIZERS

11.3.1.1 MORPHOLOGY AND STRUCTURE OF GRAPHENE, GRAPHENE OXIDE, AND CARBON NANOTUBE FERTILIZERS

There are several techniques, alone or combined with each other, that can be used to access the morphology and structure of CNM in smart fertilizers. Each one provides specific information in a microscopic level, contributing to the elucidation or improvement of the mechanism of slow release.

Atomic force microscopy (AFM) is a valuable technique to characterize carbonaceous nanomaterials down to the atomic thickness, and it was used to characterize the commercially available GO dispersion for the encapsulation of KNO_3 pellets (Zhang et al., 2014). Figure 11.6 displays the AFM image of GO sheets with micrometric lateral dimension on top of a mica substrate. The profile analysis indicates that single-layer graphene flakes with average 0.7 ± 0.2 nm in thickness were found, confirming the nanometric scale aimed to be used in the fertilizer formulation.

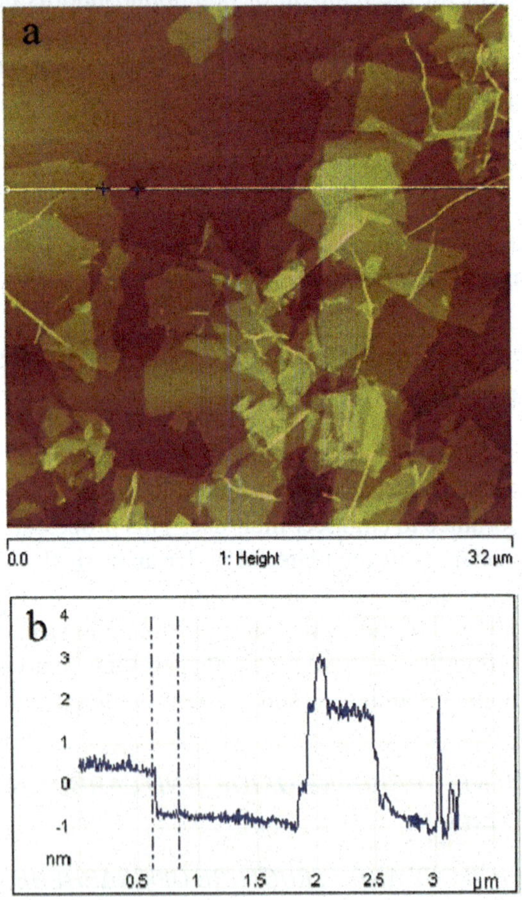

FIGURE 11.6 AFM of GO flakes on a mica substrate. (a) AFM image. (b) Profile of the image along the horizontal white line in (a). The two dots in (a) and the two dash vertical lines in (b) show the reference points for height measurement for the GO sheet.

Source: Reprinted from Zhang, M.; Gao, B.; Chen, J.; Li, Y.; Creamer, A. E.; Chen, H. Slow-Release Fertilizer Encapsulated by Graphene Oxide Films. *Chem. Eng. J.* 2020, 255, 107–113 Copyright (2020), with permission from Elsevier.

Scanning electron microscopy (SEM) is widely used to study the morphology of graphene-based fertilizers, as shown in Figure 11.7. Figure 11.7a is a representative cross-sectional image of a re-GO-coated KNO_3 pellet, with a KNO_3 core and re-GO shell (Zhang et al., 2014). It is worth noticing the smoothness of the exterior shell covering the fertilizer pellets fabricated. Figure 11.7b represents a higher resolution of the smooth shell, while (c) is a magnification showing the dense, smooth, and uniform shell with 20–30 μm thickness. Figure 11.7d shows the energy-dispersive X-ray (EDX) analysis of the re-GO-coated KNO_3 pellet, unveiling its chemical composition of mainly C, O, N, and K.

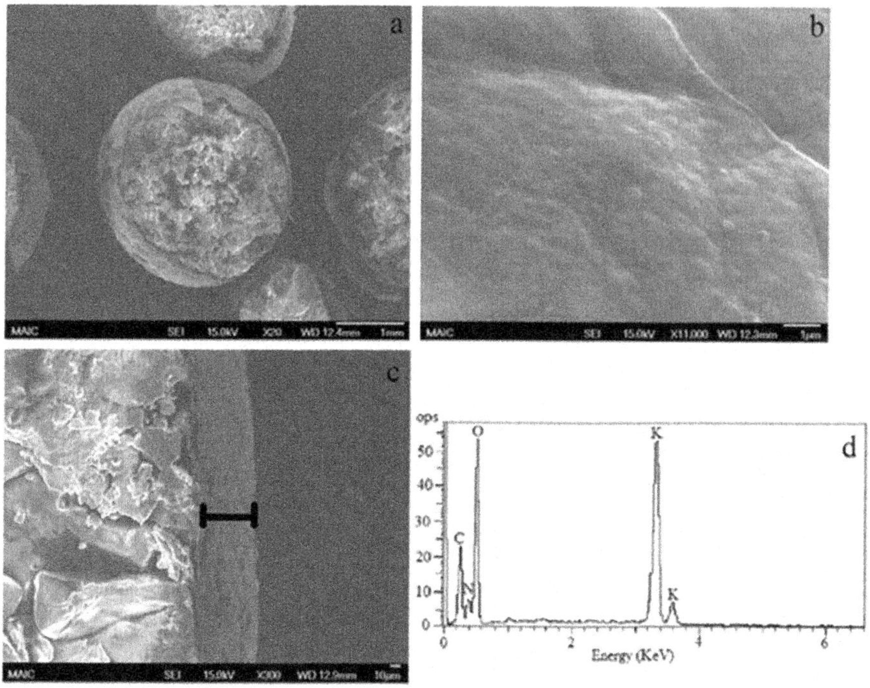

FIGURE 11.7 SEM-EDX analysis of (a) core and shell structure of a re-GO-coated KNO_3 pellet. (b) Higher resolution image of the surface exhibited in (a). (c) Higher resolution of the surface in (a), focusing on the shell structure. (d) EDX spectrum unveiling the chemical composition of the sample.

Source: Reprinted from Zhang, M.; Gao, B.; Chen, J.; Li, Y.; Creamer, A. E.; Chen, H. Slow-Release Fertilizer Encapsulated by Graphene Oxide Films. *Chem. Eng. J.* 2020, *255*, 107–113 Copyright (2020), with permission from Elsevier.

Figures 11.8a–i show the optical (a–c) and SEM (d–i) images of cogranulation of low rates of GN and GO with monoammonium phosphate (MAP) conventional fertilizer (Kabiri et al., 2018). While all samples granulated as round-shaped structures, the MAP control sample (Fig. 11.8d) presents a rough surface when compared with the MAP–graphene (MAP–GN, Figure 11.8e) and MAP–graphene oxide (MAP–GO, Figure 11.8f) cogranulations. Higher magnification SEM images (Figs. 11.8g–i) show homogeneous GN and GO embedment in the MAP–GN and MAP–GO composite. Stacked GN sheets were also detected in SEM cross-sections of MAP–GN granules when studying their crushing strength (Kabiri et al., 2020).

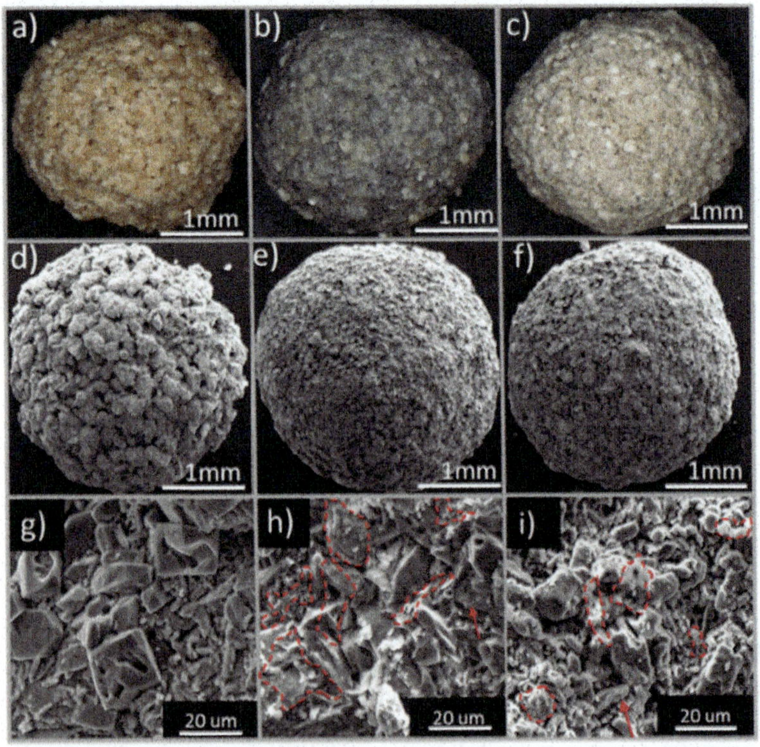

FIGURE 11.8 (a, b, c) Optical images of monoammonium phosphate (MAP) (a), MAP–graphene (MAP–GN) (b), and MAP–graphene oxide (MAP–GO) (c) cogranulations. (d–f) SEM images for these samples, in the same order. (g–i) are their respective magnifications, with red lines indicating the presence GN and GO in the MAP–GN, and MAP–GO cogranulations.

Source: Reprinted with permission from Kabiri, S.; Baird, R.; Tran, D. N.; Andelkovic, I.; McLaughlin, M. J.; Losic, D. Cogranulation of Low Rates of Graphene and Graphene Oxide with Macronutrient Fertilizers Remarkably Improves Their Physical Properties. *ACS Sustain. Chem. Eng.* 2018, *6* (1), 1299–1309. Copyright (2018) American Chemical Society.

SEM was used to study the composite obtained when loading Fe(III) and then P onto GO sheets ultrasonicated in a concentration of 1 g GO/L (Andelkovic et al., 2019). The observed planar and filamentous macroporous networks were attributed to the twisting and rolling of GO sheets in a stress stabilization process during the growth of ice crystals in the freeze-drying synthesis. The increase in GO concentration resulted also in stacked layers, which can be dispersed by ultrasonication, forming the nanoscrolls observed by the authors. Rolled and wrinkled GO sheets loaded with Cu and Zn micronutrients were also observed in the literature for other high-load slow-release fertilizer formulations (Kabiri et al., 2017).

Transmission electron microscopy (TEM) is routinely used to show the atomic structure of graphene and graphene oxide loaded with nutrients, as well as their electron diffraction representing their hexagonal arrangement at the nanoscale (for example, the reduced graphene oxide (re-GO) used to coat KNO_3 fertilizers (re-GO-coated KNO_3) (Zhang et al., 2014), or the pristine GO when exploring Cu- and Zn-loaded GO (Kabiri et al., 2017)). The morphology of urea-functionalized MWCNT for nitrogen delivery (Yatim et al., 2015) and its uptake into paddy roots was also examined with TEM (Yatim et al., 2019). The tubular structure of functionalized MWCNT was still clearly observed, but in a higher magnification TEM, the authors noticed the longitudinal rupture of these structures compared with pristine MWCNT. The functionalized MWCNTs were generally found separated as individual tubes and presenting ripped ends, with irregular appearance, and in some cases, their original lengths of the order of 0.5–1.0 μm were reduced to ~ 50 nm, and thinned in diameter to nearly 10 nm. Moreover, flat graphene-like structures with the width of ~ 80 nm were also found in the cells of these paddy root samples, probably from open tubular functionalized MWCNT structures. These observations confirmed the uptake of functionalized MWCNT resulted in increased defect density through the destruction of the graphitic structure, shedding light on how these materials modify when applied in fertilizer formulations.

Spectroscopic techniques have been increasingly used as methods of physical and chemical characterization of CNM. Among those techniques, Raman spectroscopy has proven to be a useful and effective characterization technique to obtain the structural information about CNM due to their simplicity (as it often does not require sample treatment), speed, high resolution, and because it is nondestructive, depending on the excitation energy and power used (Ferrari and Robertson, 2000). In the case of carbonaceous materials, the Raman spectrum presents broad peaks in two main regions (see Figure 11.9). The first region, known as the "D" band, near 1350 cm^{-1}, arises due to the breathing movement of the carbon atoms arranged in a hexagonal structure (predominantly sp^2) (Jorio et al., 2011; Tuinstra and Koenig, 1970), and is assigned as a k-point phonon of A_{1g} symmetry. This mode is activated by the presence of structural defects like vacancies or grain boundaries, providing information regarding the degree of disorder of the analyzed crystalline structure. The second region, called the "G" band, near 1600 cm^{-1}, corresponds to the stretch between planar C-C bonds (Tuinstra and Koenig, 1970), being assigned as a Γ-point E_{2g} phonon.

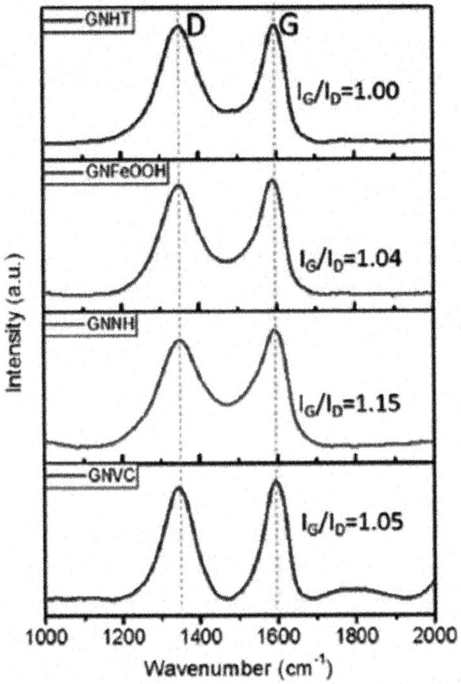

FIGURE 11.9 Raman spectra of hydrothermally reduced graphene oxide (GNHT), graphene reduced by hydrazine (GNNH), graphene reduced by vitamin C (GNVC), and graphene decorated with αFeOOH nanoparticles (GNFeOOH). Diverse I_G/I_D ratios represent different relative amounts of sp^2-hybridized carbon atoms.

Source: Reprinted (adapted) from Kabiri, S.; Tran, D. N.; Baird, R.; McLaughlin, M. J.; Losic, D. Revealing the Dependence of Graphene Concentration and Physicochemical Properties on the Crushing Strength of Co-Granulated Fertilizers by Wet Granulation Process. *Powder Technol.* 2020, *360*, 588–597. Copyright (2020), with permission from Elsevier.

Figure 11.9 represents the Raman spectra of chemically reduced graphene oxide samples formulated according to diverse methodologies aiming their use to improve the crushing strength of MAP fertilizer formulations (Kabiri et al., 2020). The intensity ratio between D and G integrated band intensities can be related to the different relative amounts of sp^2-hybridized carbon atoms. Although the I_D/I_G ratio is more widely used, the authors in this work analyze the I_G/I_D ratio to classify the differences in graphene oxide reduction processes and respective fertilizer crushing strength enhancements. They observed that the reduction by hydrazine (GNNH) and by vitamin C (GNVC) resulted in higher crushing strength when compared with hydrothermally reduced graphene oxide (GNHT), indicating that they are more effective in restoring the graphene original and less defective structure. Zhang et al. (2014) used the I_D/I_G ratio to show that the heat treatment of KNO_3 pellets encapsulated with graphene oxide is important to promote the metal cation reduction of graphene oxide to a much less defective structure. From another side, Kabiri et al. (2017) observed that the Raman spectra of both Zn- and Cu-functionalized graphene oxides (Zn–GO and Cu–GO) presented a blue shift of 9 cm^{-1} for the G band, while a red shift of 8 cm^{-1} for the D band was observed only for the Cu–GO composite, suggesting that these metal ions modified the structure of the original graphene oxide.

Raman spectroscopy is also valuable in evaluating MWCNT-based fertilizers. Kumar et al. (2018) proposed that the high I_D/I_G value of 1.15 is related to the abundance of defects in the commercially available MWCNT, confirmed by TEM images, showing distorted structures. The technique was also used to identify the incorporation of ZnO nanoparticles in these tubes, as well as MWCNT and ZnO/MWCNT presence inside onion seedlings.

11.3.1.2 GRAPHENE, GRAPHENE OXIDE, AND CARBON NANOTUBE-BASED FERTILIZERS: CHEMICAL PROPERTIES

In a chemical point of view, the atoms and the functional groups that they constitute on the surface of the CNM determine, in a large extension, the ability of the nanofertilizer to interact with the active compound, modulating its rate of release in the soil.

X-ray photoelectron spectra (XPS) is routinely used to inspect the chemical character of nanostructured fertilizers, being valuable to define the nature of the chemical binding in diverse fertilizer formulations. In the aforementioned work of Kabiri et al. (2017), the XPS spectra of GO samples were investigated before and after their loading with Cu and Zn metal ions in a micronutrient slow-release fertilizer, allowing to analyze the chemical binding between the metal ions and GO in the formed composites. In their supporting information, the authors highlighted the appearance of Zn (Zn 3s/3p) and Cu (Cu 3s) peaks in their survey spectra, corresponding to the presence of 4.34% Zn and 6.2% Cu. Figures 11.10a and b represent the data obtained for the C1s and O1s spectra of those composites, respectively. The GO sp^2-hybridized main carbon peak is present at 285 eV (Fig. 11.10a), with an asymmetric character due to the presence of carbon–oxygen terminations, such as C-OOH, C-O-C, C=O, and C-OH, and also sp^3-aliphatic hydrocarbons. The deconvolution of the C1s spectra for the Zn–GO and Cu–GO samples show peaks related to O=C-O, C=C or C-C, C-O, and C=O. The presence of the metal ions in the GO matrix is observed due to the shift of the C=O and O=C-O peaks toward higher binding energies after their loading. In addition, the O1s spectra presented in Figure 11.10b were fitted according to the presence of C=O, O=C-OH and C-O, differing considerably in intensity, shape, and position before and after Zn and Cu ion incorporation. From this analysis, it was concluded that the functional groups containing O represent the fundamental involvement on the sorption of the metal ions.

XPS was also used to explore the chemical changes induced by the heat treatment and consequent GO reduction of the KNO_3–GO encapsulated fertilizers (Zhang et al., 2014), in which the C–O peak near 286.5 eV decreased in intensity after reduction when compared with the C-C sp^2-hybridized peak, showing a more graphitic structure after this procedure. This result was also corroborated by the Raman analysis, as described in the previous section. Li et al. (2019) found that the XPS analysis clearly shows an increase in the abundance of carboxyl groups when GO is included in chitosan + GO fertilizer coatings (from 8% to 31%), indicating that it plays an important role in controlling the surface chemistry of these coatings.

Fourier transform infrared spectroscopy (FTIR) is another technique used in the characterization of organic or inorganic materials, valuable to obtain information about the present functional groups (chemical composition), as well as the possible types of bonds according to normal modes of vibration (stretching or angular deformations) (Pavia et al., 2010; Smith, 2011). In fertilizer formulation characterization, FTIR is commonly used to corroborate the XPS results. Figure 11.11 corresponds to the FTIR spectra of a GO and a Graphene oxide–Fe(III) (GO–Fe) composite containing phosphate for a slow-release fertilizer formulation (Andelkovic et al., 2018). Five spectral regions are of greater interest, specifically those close to 3350, 1730, 1618, 1220, and 1030 cm^{-1}. The broad band close to 3350 cm^{-1} corresponds to the vibrational stretching of hydroxyl groups (O-H) (Long et al., 2019). The peaks located at 1730 and 1618 cm^{-1} are related to the vibrational stretching of carbonyl groups (C = O) and aromatic ring (C = C), respectively (Zhang et al., 2014). The peaks at 1220 and 1030 represent C-O vibrations of epoxy and alkoxy groups (Mannan et al., 2018). These characteristic regions in the FTIR spectrum of the GO are also described in the works of Marcano et al. (2010), Zhang et al. (2014), and Mannan et al. (2018), although the single-layer isolation of the material was not carried out, and the measurements were made in its disordered bulk form. According to Zhang et al. (2014), when the domain of few layers approaches, the band corresponding to the vibrations of the epoxy groups tends to become more intense, which can be used as a strategy to detect the presence of GO of atomic thickness.

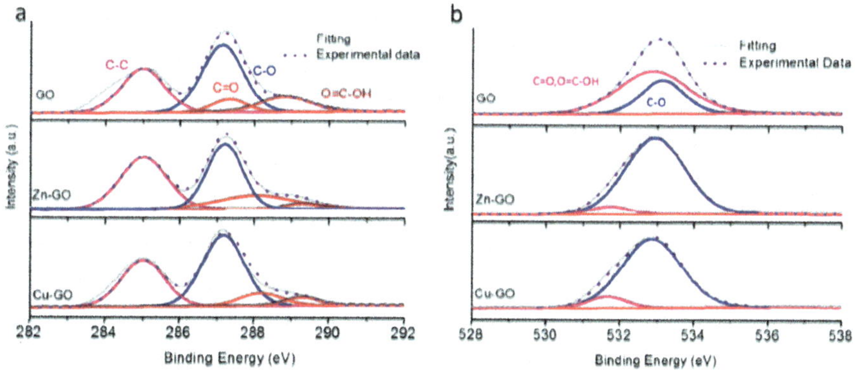

FIGURE 11.10 XPS spectra of GO and Zn- and Cu-loaded GO (Zn–GO and Cu–GO, respectively), for the C1s (a), and O1s (b) regions.

Source: Reprinted with permission from Kabiri, S.; Degryse, F.; Tran, D. N.; da Silva, R. C.; McLaughlin, M. J.; Losic, D. Graphene Oxide: A New Carrier for Slow Release of Plant Micronutrients. *ACS Appl. Mater. Interf.* 2017, *9* (49), 43325–43335. Copyright (2017) American Chemical Society.

The comparison of GO and GO–Fe spectra in Figure 11.11 unveils a decrease in the C=O intensity, and an increase in the O=C-O carboxyl stretch (1348 cm^{-1}) intensity. Shifts in these peaks are attributed to the coordination of Fe(III) ions to the GO carboxylic group (Andelkovic et al., 2018; Park et al., 2008), while the shift of the C=C peak from 1618 to 1606 cm^{-1} and its intensity increase are related to the transformation from the π electron interaction in the GO aromatic structure to cation-π interaction in GO–Fe (Andelkovic et al., 2018; Wang et al., 2013). Andelkovic et al. (2019) tested these hypotheses by optimizing the GO content in GO–Fe fertilizers for phosphorus loading. They found that for higher GO concentrations (1 gGO/L vs. 10 gGO/L), GO aggregation into stacked layers is higher, with Fe(III) ions intercalating between layers of GO and being weakly bonded to GO hydroxyl and alkoxy/alkoxide groups, which were much weaker in the 10 gGO/L formulation.

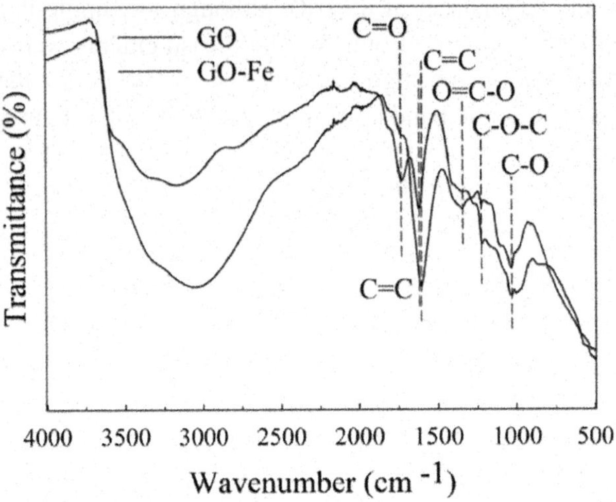

FIGURE 11.11 FTIR spectra of GO and GO–Fe used in the graphene oxide-Fe(III) composite containing phosphate for a slow-release fertilizer formulation.

Source: Reprinted from Andelkovic, I. B;, Kabiri, S.; Tavakkoli, E.; Kirby, J. K.; McLaughlin, M. J.; Losic, D. Graphene Oxide-Fe(III) Composite Containing Phosphate—A Novel Slow Release Fertilizer for Improved Agriculture Management. *J. Clean. Prod.* 2018, *185*, 97–104 Copyright (2018), with permission from Elsevier.

Chemical changes were described by using FTIR in the study of the use of nonfunctionalized and functionalized MWCNT interacting with urea on the growth of paddy (Yatim et al., 2019; Yatim et al., 2015; Yatim et al., 2018). Reactions between urea and functionalized MWCNT (fMWCNT) were expected through covalent bonding between NH_2 groups in urea and the carboxyl groups in fMWCNT in the urea-functionalized MWCNT (UF-MWCNT). Their FTIR spectra presented main features expressed by the O-H stretching vibration (3300–2500 cm^{-1}), C=O stretching (1700–1725 cm^{-1}), and C-O stretching vibration from carboxylic acid groups (1210–1320 cm^{-1}). For the case of UF-MWCNT, bands assigned to the C=O stretching vibration (1640–1690 cm^{-1}), N-H bending (1510–1600 cm^{-1}), CH_2 bending (1376–1388 cm^{-1}), and C-N stretching (1120–1290 cm^{-1}) were found. The disappearance of the band associated with the O-H stretching vibration at 3018 cm^{-1} in the UF-MWCNT, previously present in the fMWCNT sample, indicated that O-H groups reacted with amino groups in urea. In addition, peaks with altered shapes arose, being attributed to N-H bending (1519 cm^{-1}) and C-N stretching (1219 cm^{-1}) modes in the UF-MWCNT, with the consequent disappearance of the previously observed fMWCNT C=O (1739 cm^{-1}) and C-O (1219 cm^{-1}) modes. The production of amide groups through the reaction of NH_2 groups from urea and carboxyl groups (COOH) from the surface of fMWCNT were reported previously (Gao et al., 2005), and the FTIR results confirm the chemical reaction between them.

SEM morphological characterizations of carbon nanomaterials-based fertilizers usually occur with the simultaneous chemical characterization by using EDX analysis. Pellet (core and encapsulated shell) or nanomaterials chemical composition can be checked for different atomic constituents or contamination by means, for instance, of punctual analysis (Andelkovic et al., 2018; Zhang et al., 2014) and surface images in the case of GO (Kabiri et al., 2020) or punctual studies of urea-functionalized MWCNT (Yatim et al., 2018). The relative change in the proportion of chemical constituents after specific functionalization procedures in GN cogranulated fertilizers can also be studied (Kabiri et al., 2020). Due to the straightforward character of this technique, it will not be broadly discussed here.

11.3.1.3 PERFORMANCE OF GRAPHENE, GRAPHENE OXIDE, AND CARBON NANOTUBE FERTILIZERS

One of the most important improvement effects due to GN or GO incorporation in fertilizers is related to the crushing strength of these formulations. Kabiri et al. (2018) studied this improvement by cogranulating MAP fertilizer with GN and GO at different weight percentages (from 0.05% to 0.5%), observing that the carbonaceous materials improved the crushing strength of MAP (control) granules 3.0–18.2 times according to the different formulations. This improvement was explained by the high specific surface area and the pore-filling or interlocking properties of GN and GO, which are associated with their 2D planar and wrinkled structure, as indicated by SEM analysis, and their superior mechanical properties like high Young's modulus. Moreover, the GN cogranulated fertilizers presented superior crushing strength when compared with the GO formulations, with 10.8 kg force at 0.5 wt.%, 2.8 times higher than for a formulation with the same content of GO. The type and higher defect content of GO compared with GN can be related to this observation, resulting in inferior mechanical properties for GO formulation. It is worth mentioning that the improvement of MAP due to the addition of low percentages of those CNM is promising when compared with commercial strategies, like Norling A, in which 1% addition to MAP only doubles its crushing strength when compared with control samples. Other mechanical properties, such as abrasion resistance and impact resistance, important for fertilizer spread and storing, also presented substantial improvement. Kabiri et al. (2020) also expanded their study demonstrating the effects of GN concentration and physicochemical properties on the crushing strength of MAP and di-ammonium phosphate (DAP) fertilizers.

The slow-release behavior of fertilizer formulations based on carbonaceous materials is of paramount importance because the use of conventional products is usually damaged by lixiviation in soil, for example. To study this property, KNO_3 fertilizer pellets coated with re-GO were used in a soaking experiment monitoring the potassium release concentration over time (Zhang et al., 2014). It was shown that while the uncoated KNO_3 pellet reached equilibrium after only 1 h of potassium fast release, the re-GO-coated KNO_3 presented two different release stages. In the first one, from 0 to 7 h, only 34.5% of potassium ions were released, which was assigned to the establishment of "channels" through the shell and the diffusion of water into the core, in a slow process. In the second stage, from 7 to 8 h, corresponding to a burst release with the cracking of the re-GO shell, 93.8% of the initial potassium content is released. The visual inspection showed that the pellet was cracked after this stage. After this process, the release is again slow and stable, evidencing the promising slow-release character of this coating.

The cogranulated formulations of MAP with GN or GO described previously in this section (Kabiri et al., 2018) were also studied by mean of P diffusion experiments in moist soil, in which the movement of P from the granules of the cogranulated formulations to the soil was smaller than that in the case of reference MAP samples (see in Figure 11.12 the dark green zones (a) and their radius (b) for different GN and GO percentages and number of days). In other study, Kabiri et al. (2017) conducted experiments to determine the water solubility, dissolution, and release rate of Zn–GO and Cu–GO pellets formulated as slow-release micronutrient fertilizers, showing that their formulations released only ~ 55% of Zn and Cu after 72 h when compared with the almost 100% release within 20 h from the conventional commercially available $ZnSO_4$ and $CuSO_4$ micronutrient precursors. Andelkovic et al. (2018) and Andelkovic et al. (2019) also conducted slow-release experiments on GO–Fe composite loaded with P in three different soils, studying the kinetics by different models and engineering the material to a desired P release rate. A radius diffusion methodology similar to that of Kabiri et al. (2017) was performed and confirmed the slow-release properties for this formulation.

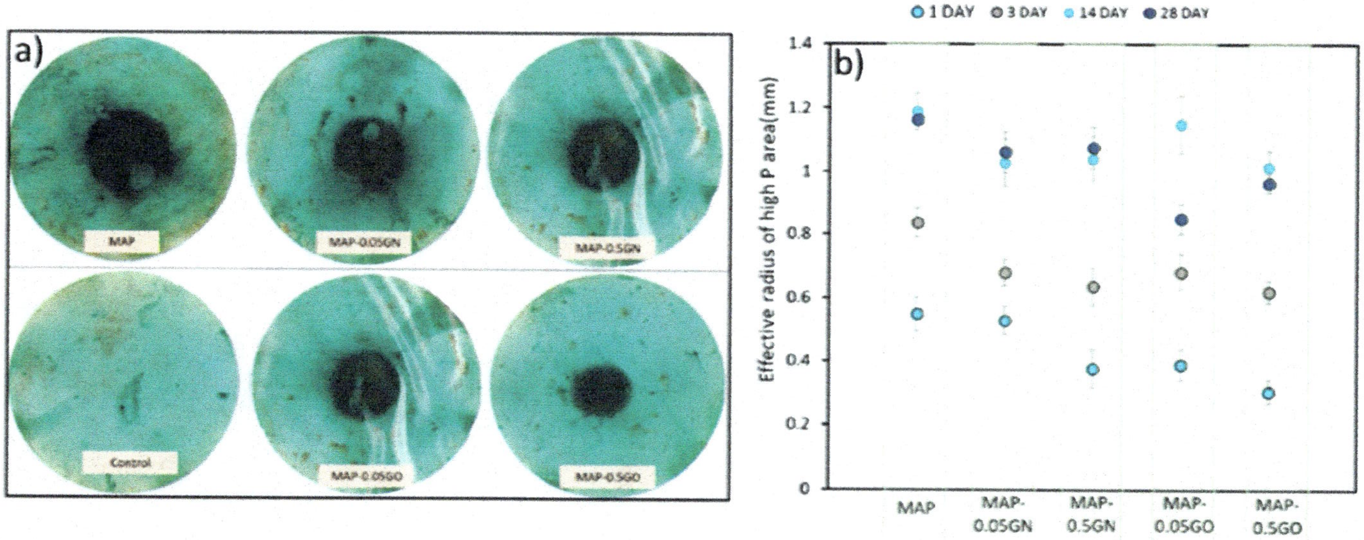

FIGURE 11.12 P diffusion zones visualization for MAP, MAP with 0.05% or 0.5% graphene oxide (MAP–0.05GO and MAP–0.5GO), and MAP with 0.05% or 0.5% graphene (MAP–0.05GN and MAP–0.5GN). (b) Radius of strong P release areas at 1, 3, 14, and 28 days for these formulations.

Source: Reprinted with permission from (Kabiri, S.; Baird, R.; Tran, D. N.; Andelkovic, I.; McLaughlin, M. J.; Losic, D. Cogranulation of Low Rates of Graphene and Graphene Oxide with Macronutrient Fertilizers Remarkably Improves Their Physical Properties. *ACS Sustain. Chem. Eng.* 2018, *6* (1), 1299–1309. Copyright (2018) American Chemical Society.

Plant yield studies are increasingly being used as direct measurements of carbon nanofertilizers efficiency. Andelkovic et al. (2019) choose three different soils (Port Wakefield—PW, Monarto, and Black Point—BP) to test the efficiency of their best formulation of GO-Fe composite loaded with P, proposed as a slow-release fertilizer, in wheat growth in pot experiments. Soil samples with MAP or 10GO–Fe-15P (a 10 mg/L initial GO concentration with the addition of 15 mg/kg of P, or 10GO–Fe-15P from now on) addition were compared with control soil samples (without P fertilizer addition). The soils were classified as highly responsive to P application due to the expressive yield improvement obtained for the different P fertilizer sources. Both Monarto and PW soils presented similar yield improvement of dry matter when MAP and 10GO–Fe-15P P sources were used, increasing from 0.2 to 0.8 g/pot and 0.3 to 0.8 g/pot, respectively. In the BP soil, MAP outperformed the 10GO–Fe-15P fertilizer in nearly 1 g/pot, with 3.5 g/pot for 10GO–Fe-15P, and nearly 4.5 g/pot for MAP. The wheat P uptake was also affected by the P source in some cases, being nearly 1.3 mg/pot for both MAP and 10GO–Fe-15P in PW soil, and with MAP outperforming 10GO–Fe-15P for both Monarto and BP soils (~ 1.6 mg/pot for 10GO–Fe-15P and 3.0 mg/pot for MAP in Monarto soil, and ~ 4.5 mg/pot for 10GO–Fe-15P and ~ 9.0 mg/pot for MAP in BP soil). The authors argue that lower dry yield is usually reported when slow-release P fertilizers are used in comparison to conventional soluble fertilizers in soils with high P-fixing capacity and deficiency in this nutrient, as in the BP soil. From this perspective, the use of these engineered fertilizers must be analyzed according to the specific crop and soil situation. In the best-case scenario, the performance of the slow-release 10GO–Fe-15P formulation was similar to that of the highly soluble MAP fertilizer. From this, it is speculated that their advantage can be in the case of applications in areas in which leaching of P is substantial, in which its slow-release character can maintain P available to the crops for a long period.

The agronomic performance of Zn–GO and Cu–GO micronutrient fertilizers has been also evaluated for cultivation of wheat in the study of Kabiri et al. (2017), comparing it with that of the conventional $CuSO_4$ and $ZnSO_4$ precursor salts. Dry mass and nutrient uptakes were checked. The Zn–GO formulation resulted in improved dry mass when compared with the $ZnSO_4$ salt, but for the Cu–GO and $CuSO_4$, there was not a significant improvement. The latter result is related to the fact that the soil used in this experiment was not Cu-responsive, once the yield of the control soil without Cu addition was similar. Moreover, higher Zn uptake was also detected for the Zn–GO formulation. When the conventional $ZnSO_4$ salt was used, the irreversible adsorption of Zn on carbonates and its precipitation as hydroxide or carbonate took place as Zn fixation mechanisms in the calcareous soil investigated by the authors. The Zn–GO formulation is expected to present smaller interaction with the soil components due to the strong bonding between the GO and the micronutrients, maintaining them in available forms for the plant uptake. Watts-Williams et al. (2020) also studied Zn-loaded GO fertilization in the growth of Barley and Medicago plants, with and without arbuscular mycorrhizal fungi inoculation. Greater amount of biomass and Zn uptake was detected for this formulation, substantially improved by the mycorrhizal inoculation, providing promising results for two distinct crops.

Urea-functionalized MWCNT (UF-MWCNT) fertilizers from the study of Yatim et al. (2018) were studied as a strategy to improve the yield of paddy. The UF-MWCNT formulation showed improvements of 28.6% and 36% for the number of panicles and grain yield, respectively. These results indicate the efficient delivery of N through the application of UF-MWCNT formulations, mainly attributed to the translocation of the MWCNT into the plant roots, increasing the number of pores and soil water containing nutrients.

11.3.2 BIOCHAR AND BIOCHAR-BASED FERTILIZERS

Two of the fundamental aspects widely discussed in the literature about the structure of BC are the influence of the pyrolysis temperature and feedstock kind used for their production, which will define the possible applications of the material (Lehmann and Joseph, 2015; Tomczyk et al., 2020). According to Song and Guo (2012), the rise in temperature gradually oxidizes aliphatic components contained in the original biomass, which in turn are converted into aromatic components. This increase in temperature also generates more organized structures in a graphitization process as well as promotes an increase in the ash content and pH. Different pyrolysis temperatures and atmosphere controls during the synthesis can guarantee the modulation of various characteristics of BC, such as the growth of pores in the material and increase in their specific surface area (Song and Guo, 2012). Shi et al. (2015) suggested a mechanism for this observation, stating that the rise in BC porosity is a consequence of the removal of volatile pores due to the temperature increase. Due to the large number of possibilities created by varying these parameters and the several mechanisms involved, the next sections will focus on the TPI characterization, as they represent structural models for BC and BBF. BC and BBF will also be discussed specially when synthesized at temperatures of 500°C–600°C, mimicking natural fires temperature range, and resulting in materials with better performance.

11.3.2.1 MORPHOLOGY AND STRUCTURE OF BIOCHAR

SEM is one of the first approaches in BC characterization, allowing the analysis of the morphology presented by the microscopic structure of these materials. Patterns of materials degradation and pore growth tendency according to synthesis conditions can be studied with small or without sample treatment (Ma et al., 2016). The SEM/EDX and TEM techniques allow mapping the concentration of chemical elements and their association with possible macro and micronutrients or minerals aggregated on the BC surface (Chia et al., 2012; Ma et al., 2016) or TPI samples (Jório et al., 2012). Cross-sectional studies can lead to the understanding of chemical composition distribution by using EDX, which is fundamentally influenced by structural aspects of these materials.

TPI samples envisioned as models for nature-inspired BC engineering were studied by Jório et al. (2012), and the SEM, TEM, and EDX results are shown in Figures 11.13a–d. In (a), (b), and (c), shell and core structures can be observed, with major shell structural disorder. The cross-section in (d) indicates different distribution for chemical elements, including the concentration of P, Fe, Al, and Si on the surface. The increase of the amorphous character on the surface of the structure was also shown and more dangling bonds as well as carboxyl and carbonyl terminations were present, suggesting that it can be a positive factor for nutrient aggregation. This is an important result indicating that improvements in BC efficiency can be engineered by mimicking this type of structure.

Raman spectroscopy has been routinely used in the analysis of the amount of structural defects in BC based on the simplification proposed in the in-plane crystallite size (L_a) theory (Ribeiro-Soares et al., 2013). This nanocrystalline structure is used to compare carbonaceous materials, allowing to classify them as possessing a more organized graphitic layout (sp^2) or an amorphous profile. A higher L_a indicates a predominance of sp^2 domains and well-structured graphite in "islands," while a lower L_a refers to a more amorphous pattern (Lucchese et al., 2010). The Raman spectra of TPI and BC are similar to those of GN and GO shown previously, being also dominated by the G and D bands. For TPI, the L_a size was modeled considering the full-width at half maximum of the G band of the Raman spectra (Ribeiro-Soares et al., 2013). The size of the crystallite L_a can be evaluated by the following equation:

$$L_a\,(\text{nm}) = 496/(\Gamma_G - 15) \quad (11.2)$$

where Γ_G corresponds to the full-width at half maximum of the G band, provided by a graphic adjustment of the G and D bands with two Lorentz peaks (Ribeiro-Soares et al., 2013). To obtain this equation, the authors used the analysis of Raman and X-ray crystallite size measurements of reference Diamond-like carbon samples with different structural organizations, being close to the results obtained by Cançado et al. (2007). L_a sizes in a broad range (3–8 nm) were found for TPI carbonaceous particles, while BC samples showed a more concentrated L_a distribution (8–12 nm). Additionally, a cross-section Raman experiment was performed in a TPI carbonaceous particle, revealing the decrease of L_a from the better organized graphitic core to the more amorphous shell. This is an important finding because smaller L_a is related with higher disorder and dangling bonds, increasing the bonding with nutrients, likely corroborating the EDX results from Figure 11.13.

The L_a approach has been used for other anthropogenic soils, BC, and BBF (Carneiro et al., 2018; Pagano et al., 2016; Pandey et al., 2020). For example, Carneiro et al. (2018) found that BBF presented a smaller L_a size when treated with MgO (~ 7 nm) when compared with the control samples (~ 10 nm), indicating that the O atoms can be responsible for the increase of nongraphitizing C. The increase in the CTC of these samples was also observed, probably due to the higher amount of defects for the smaller L_a, being a promising strategy for new fertilizer formulations.

11.3.2.2 BIOCHARS: CHEMICAL PROPERTIES

The pyrolysis temperature for BC production was shown to influence the surface concentration of different functional groups, increasing the number of basic and decreasing the acids ones with increasing temperature. These changes affect other properties, such as CTC and adsorption capacities of the BC (Shi et al., 2015). With the rise in pyrolysis temperature, elements, such as O and H volatize, promoting an increase in concentration of C. The other elements that do not volatilize in the same temperature range of O and H are retained with C until they reach their specific volatilization temperature, generating differences in CTC according to the increase in temperature (Novak et al., 2009). In this perspective, developing a soil conditioning material with chemically engineered properties resembling those of TPI would enable significant improvements in the agricultural sector.

Comparative XPS studies of charcoal from Amazonian native plant Embauba (*Cecropiahololeuca Miq.*), produced at 600°C, and TPI samples (Archanjo et al., 2014) show the presence of a broader range of elements on the last one. P, Al, Mg, Ca, and Na were detected in TPI carbonaceous particles, in addition to other elements, in quantities smaller than 1%

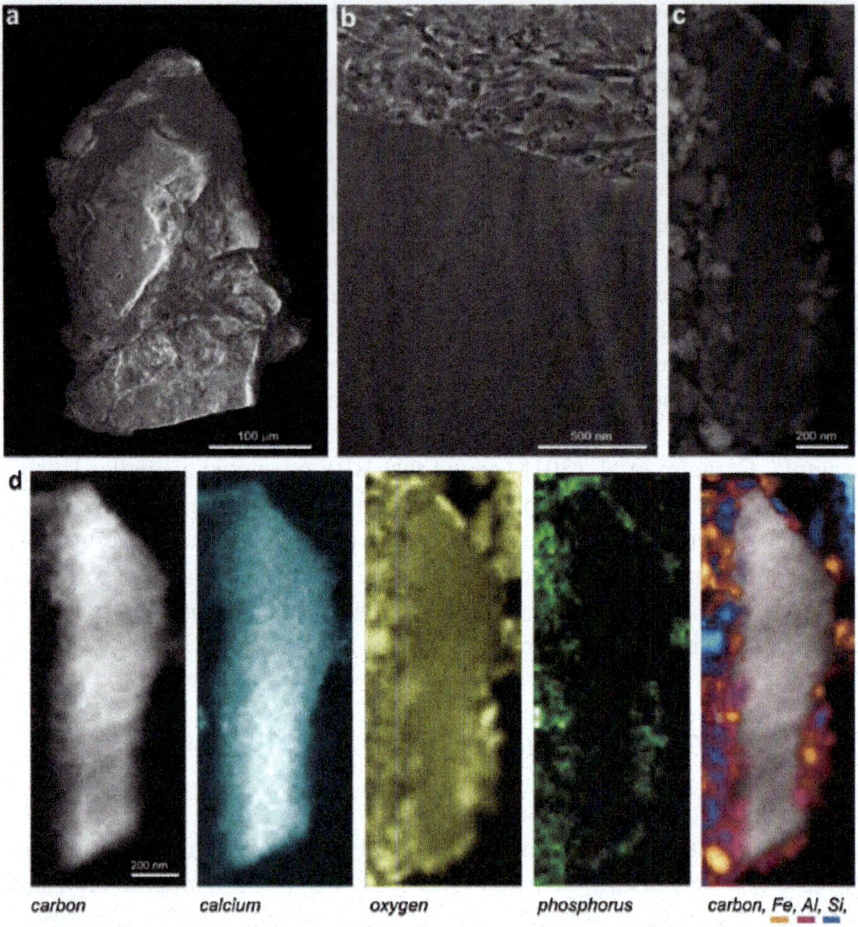

FIGURE 11.13 TPI carbonaceous grain studied by scanning electron microscopy (SEM) (a), thin cross-section transmission electron microscopy (TEM) study exploring the core shell (grafitic-porous) structure (b), magnification showing the scanning transmission electron microscopy (STEM) image of a nanostructured carbon particle (c), and different elements studied by energy-dispersive X-ray (EDX), labeled accordingly to each image (d).

Source: Reprinted from Jorio, A.; Ribeiro-Soares, J.; Cançado, L. G.; Falcão, N. P.; Dos Santos, H. F.; Baptista, D. L.; Martins Ferreira, E. H.; Archanjo, B. S.; Achete, C. A. Microscopy and Spectroscopy Analysis of Carbon Nanostructures in Highly Fertile Amazonian Anthrosoils. *Soil Tillage Res.* 2012, *122*, 61–66 Copyright (2012), with permission from Elsevier.

(small quantities of K, N, Ti, Cr, among others). The conventional charcoal presented mainly C and O in its composition. The analysis of C1s presented in chemical groups revealed that TPI carbonaceous particles are composed of C in sp^2 hybridization state, carbonyl, carboxylate, hydroxyl, and carbonate anions. These findings were also confirmed by using electron energy loss spectroscopy (EELS), with the different functional groups shown to be mainly located in the shell region (Archanjo et al., 2015). The XPS binding energies of O1s were less resolved than those of C1s, being associated with oxidized carbon and metal oxides. In the charcoal XPS spectra, the oxidized carbon species comprised only 27% in comparison with the 62% observed in TPI carbonaceous particles. From another side, the TPI carbons presented 38% of sp^2 carbon, while in charcoal, it composed the major contribution, with 73%. These results indicate that the conventional BC needs to be further engineered to present a higher chemical heterogeneity, being impregnated with nutrients, and a higher proportion of carbon oxidized forms with relation to the sp^2 carbon. This later issue is especially necessary to maintain the reactivity and resilience balance, important for its application as fertilizer.

The chemical composition of BC varies for different types of biomasses (Carneiro et al., 2018; Joseph, 2015; Kloss et al., 2012; Veiga et al., 2017). The synthesis of BC from agricultural (coffee husk) and forest (eucalyptus husk) production carried out by Veiga (2016) showed differences in BC properties, being directly related to the contents of cellulose, hemicelluloses, and lignin compounds present in the precursor biomasses. It was previously known that biomass from agricultural production presented higher levels of lignin, which may have contributed to an increase in fixed carbon and ash contents in the BC produced (Veiga et al., 2017), in which the difference among ash contents may be related to high concentrations of SiO_2 in these biomasses (Kloss et al., 2012). Kloss et al. (2012) also report the different properties of BC for varied biomasses with distinct chemical compositions (different C/N, H/C ratios, lignin contents, etc.), highlighting variations of C and N in BC, whose C/N ratio can influence the immobilization of N by the plant after the BC soil application. However, there are remaining bottlenecks regarding the biomass chemical composition factors that undergo changes during and after synthesis, directly affecting the properties of BC, and consequently their application (Carneiro et al., 2018; Kloss et al., 2012; Novak et al., 2009). This fact makes it clear the necessity of the use of various techniques providing relevant information that can be associated with the composition of these materials, such as FTIR, for example (Liang et al., 2018).

The FTIR biochar analysis is a powerful strategy to understand chemical variations according to precursor biomass, pyrolysis temperature, copyrolysis with other materials for biochar enrichment, or in the study of TPI composition (Nanda et al., 2013). The FTIR spectra of BC have bands associated with characteristic vibrations, which are already well described in the literature. They can present a broad band located in the 3100–3500

cm^{-1} region, corresponding to the stretching vibration of hydroxyl groups (-OH) linked to carboxylic groups and water molecules, especially for those cases presenting low thermal treatment level and original biomass content (Carneiro et al., 2018; Fu et al., 2011). The presence of this band can also correspond to the N-H elongation in amine compounds, which shows a weakening pattern after pyrolysis (Carneiro et al., 2018). Bands located around 2900 cm^{-1}, with a more explicit character in the biomass spectra correspond to the symmetrical and asymmetric vibrational stretching of the CH_2 groups, which tends to disappear as the temperature increases during pyrolysis, as well as the reduction of hydroxyl groups (Carneiro et al., 2018). Bands in the region around 2350 cm^{-1} can be related to stretching vibrations of CO_2, which is produced and released during the cellulose, hemicellulose, and lignin cross-link polymerization and dehydrogenation oxidation in the pyrolysis process (Liang et al., 2018).

In recent studies of BC from chicken litter, sewage sludge, and swine manure impregnated with $MgCl_2$ for P adsorption and release, Nardis et al. (2020) report the emergence of specific FTIR spectra bands (Mg-OH: 3743 cm^{-1}; $CaHPO_4$-$2H_2O$: 528 cm^{-1}, 871, 782, 522 and 576 cm^{-1}; OPO: 538 cm^{-1}; P=O: 985 cm^{-1}, among others) related to bonds containing P and Mg (combining these findings with Raman spectroscopy, XRD, SEM-EDX techniques, electrical conductivity, total nutrient content, among others). In this case, the appearance of bonds containing P occurred due to the increase of O during the impregnation and adsorption of P (Nardis et al., 2020), with the highest rates of P adsorption related to the swine manure BC (68 mg/g). Several research groups are focusing on the study of pre- and postprocessing of BC for nutrient adsorption (Cui et al., 2016; Nardis et al., 2020; Novais et al., 2018; Vikrant et al., 2018; Zhu et al., 2020), or for example, BC containing MgO for P removal from aqueous solutions (Zhu et al., 2020), demonstrating the great interest in physical and chemical properties of the promising BC subjected to pre-or posttreatment, which can perform equivalently to conventional fertilizers (Nardis et al., 2020; Vikrant et al., 2018; Zhu et al., 2020).

11.3.2.3 PERFORMANCE OF BIOCHAR FERTILIZERS

The BC addition to soils present changes in several of their chemical and physical properties, for example, nutrient and water retention, pH, and particle aggregation (Chew et al., 2020; Hagemann et al., 2017; Tomczyk et al., 2020). Different mechanisms take place and are viewed to have positive impacts on the reduction of N_2O emissions, improvement in soil N availability, reduction of N leaching, carbon sequestration, and promotion of the activity of soil microbes (Chew et al., 2020; Hagemann et al., 2017; Liao et al., 2020; Tomczyk et al., 2020). These observations are also related to studies of TPI due to their long resilience in tropical conditions and the fertility detected in their sites (Glaser, 2007; Glaser et al., 2000; Neves et al., 2004). In general, plant yield is improved because of these different factors when BC are applied observing the specificities of different soils (Guilhen, 2018; Guimaraes, 2010; Lehmann and Joseph, 2009; Nardis et al., 2020).

The structural properties of BC are generally related to their resilience in soil, which is also important for carbon sequestration, in addition to plant development (Archanjo et al., 2014; Archanjo et al., 2015; Jório et al., 2012; Tomczyk et al., 2020; Weng et al., 2017). The graphitized phase of these materials is considerably more inert and adapted to persist for long periods. As shown for the BC found in TPI, an efficient structure is composed by a graphitized core, guaranteeing the stability for millennia, and an amorphous shell with dangling bonds contributing to nutrient retention. The chemical properties of BC also confirmed the major C and O composition of pristine BC depending on the pyrolysis temperature, with a major chemical variety reported in the literature for TPI BC, especially in their surface, or BC generated from high-nutrient precursor biomasses (Carneiro et al., 2018; Kloss et al., 2012; Veiga et al., 2017). The presence of several functional groups, such as carboxyl, ketone, hydroxyl, and lactone results in the adsorption of several nutrient ions, for example, phosphate, nitrate, ammonium, and potassium (Tomczyk et al., 2020). BBF is invariably being inspired in these characteristics to design new strategies to improve plant yield and plant and soil health.

The influence of BC and BBF application in soil biology and microbiology is an essential point to be explored, although fundamental information as the role of nitrogen availability on soil bacterial communities is still scarce (Liao et al., 2020). It was previously pointed out that some researchers sought to relate the high fertility of TPI with a kind of special microorganism, however, no experimental evidence was obtained for its existence (Glaser, 2007). The pores of BC are routinely recognized as habitat for soil microorganisms, such as actinomycetes bacteria, or mycorrhizal fungi, promoting high levels of these microbiological populations, which are also food for other soil biota (Compant et al., 2010; Tomczyk et al., 2020). Soil enzymatic activity was also reported to be ameliorated using BC, producing soils with higher quality (Mierzwa-Hersztek et al., 2016; Ouyang et al., 2014). Improvement in the activity of dehydrogenase and urease in soil was shown previously (Ameloot et al., 2013; Mierzwa-Hersztek et al., 2016), with increases of 19.0% and 44.0%, respectively (Mierzwa-Hersztek et al., 2016). Weyers and Spokas (2011) show that BC application can result in short-term negative effects in earthworm soil activity, but that this effect is null in the long term. The increase of soil pH due to the application of BC was related to this negative effect (Haefele et al., 2011), again making explicit the necessity to combine BC and soil properties.

BC or BBF application can present different effects on plant yield, increasing, being neutral, or even decreasing it. Two fundamental aspects to be observed when crop yield improvement is aimed with BC and BBF addition are the pH and the cation exchange capacity (CEC) effects, which are reasonably known in literature (Rajkovich et al., 2012). The pH is usually negatively impacted by the application of typically acid BC, which are detrimental to crop development in soils which are already acidic. The CEC can be improved depending on the BC or BBF used (Liang et al., 2018), being valuable to nutrient retention by avoiding leaching (Lehmann et al., 2003). Other important factors under intense scrutiny recently present profound impacts in crop yield, such as feedstock, the pyrolysis temperature, and soil type (Purakayastha et al., 2019; Rajkovich et al., 2012; Tomczyk et al., 2020).

The pyrolysis temperature and feedstock type are important parameters to define the plant yield, and a fundamental work was performed by Rajkovich et al. (2012) for corn growth in on example of tempered soil (Alfisol). They found that in general, crop growth is reduced when low pyrolysis temperatures of the order of 300°C–400°C were used in food waste and paper mill waste during thermal treatment, with no difference being observed in yield when higher temperatures were used. In general, BC produced at 500°C was the most efficient to promote plant growth. Different feedstock was also pyrolyzed in this work, with an even higher influence on crop growth, reaching eight times more variation than pyrolysis temperature. Corn, hazelnut, oak, pine, diary manure, food waste, paper waste, and poultry BC were analyzed, with the higher corn crop yield improvement obtained by nutrient-rich BC such as those synthesized from animal manures or corn. Plant residues, such as hazelnut, pine, and oak produced little improvement in the studied situation, with relevant results only obtained in the corn stover case, which improved crop growth by 16%. The results were obtained for the relatively fertile Alfisol, so the result of its use in other situations can be distinct, going from beneficial to detrimental. It is important to stress that due to the large variation in corn crop yield improvement, it is also fundamental to test the advantages from BC obtained from different feedstock before field-scale application.

Several studies listing the crop yield improvement with the use of BC are being published recently, but the vast range of variables involved in each case prevented the conception of a definitive work. As an evolving field, the BC and BBF research provides several opportunities to explore the effects of temperature pyrolysis, soil type, feedstock, application rate, and test crop on crop yield, and also the engineering of physical and chemical properties of these BC to mimic the TPI natural case (Purakayastha et al., 2019; Rajkovich et al., 2012; Tomczyk et al., 2020).

11.4 SUMMARY AND PERSPECTIVES

Remediation of polluted or nutrient-depleted soils is a strategic issue with increasing importance today, especially due to the rising of worldwide population in humid tropical regions and the necessity of sustainable agricultural practices and food security. Different nanomaterials are envisioned as promising for this task due to their special properties that emerge due to their reduced dimensionality. Carbon nanotubes, graphene, graphene oxide, biochar, and biochar-based fertilizers are the principal carbonaceous materials with a plethora of relevant properties valuable to attack these issues. These properties include, for example, their high specific surface area, surface reactivity, mechanical resistance, structural stability, superior adsorption, and flexibility to be engineered. The use of materials science, chemistry, and engineering methodologies conventionally applied to comprehend these carbonaceous nanomaterials is becoming increasingly more common in soil science issues depending on several variables like the adsorption and immobilization of several contaminants and nutrient depletion.

The ability of all carbonaceous materials to adsorb a broad range of contaminants is a remarkable issue regarding the potentiality of these materials for use in soil remediation. While BC are generally rich in different functional groups, which can be modulated by the pyrolysis temperature involved in their production, CNT, GN, and OG can be functionalized to attain the desired surface properties. These possibilities of modulating the CNM properties open countless possibilities to increase the capacity of those materials to interact with a specific contaminant. Because of this, a variety of mechanisms can be involved in the capacity of CNM to retain some contaminants avoiding their leaching, such as hydrophobic, electrostatic, hydrogen bond, van der Waals, and π–π interaction. Despite all the amazing adsorption properties of CNM, current use of these materials in real ecosystems for in situ soil remediation, specialty pristine CNT and graphene-based materials, is still a promise because their toxic effects on different organisms is still an emerging area of study. In this sense, new engineered materials based on CNM are already seen as a possibility to overcome the toxicological problem. CNT or GO composites, for instance, can limit the leaching of own CNM in soil, avoiding their spread in the environment.

The stability of some carbonaceous materials is one of the valuable properties provided by their structural and chemical properties, and are especially useful as soil amendments. In addition to soil decontamination and nutrient retention, carbon sequestration is also expected for initiatives such as the use of biochar and biochar-based fertilizers with high stable C content. These artificial materials are bioinspired in the biochar particles extracted from the anthropogenic Terras Pretas de Índios, with notorious tuning possibilities. The in-plane crystallite size L_a estimated by Raman spectroscopy in biochars extracted from the Terras Pretas de Índios is a valuable and fast analysis to comparatively evaluate the structural organization level among different biochars. The 3–8 nm L_a found for TPI are lower and more broadly distributed than the typically 8–12 nm found for artificially synthesized biochars, indicating the necessity to further engineer their structure. Also, the core-shell structure studied by several techniques here indicates that new strategies aiming a more amorphous surface are needed in these materials for their efficient use in soils.

The structural properties studied by AFM, SEM, and Raman spectroscopy show that graphene and graphene oxide are valuable for fertilizer encapsulation and cogranulation, improving their mechanical resistance with the possibility to engineer their defect level by using different functionalization procedures. The chemical properties inspected by XPS, FTIR, and EDX show different compositions, including the core-shell structure with functional groups in the shell for TPI, for example. The functional groups are fundamental for carbonaceous nanomaterials interaction with conventional fertilizers, such as urea for carbon nanotubes, for MAP, and the loading of Zn and Cu micronutrients in graphene oxide composite slow-release fertilizers, and K for the encapsulation case. Although promising results in terms of crop yield, improvement of mechanical properties in fertilizers, and microbiology in soil were presented, it became obvious that they are dependent of soil properties like acidity or nutrient responsivity, for example, and a study prior to its use is necessary.

The interaction of carbonaceous nanomaterials with soil systems is complex and the analysis proposed in this chapter is limited due to the broad range of factors. On the other hand, the benefits of the use of carbonaceous materials for soil remediation are clear, with perspective to generate valuable strategies to solve one of the most relevant issues related to sustainable soil use. The several cases discussed here make it clear that new initiatives focused in the engineering of carbonaceous structures are needed, for both new adsorption and fertilizer conceptions arising for the low dimensionality of these systems, and to mimic the nanostructured biochars found in anthropogenic soils with notorious fertility. These certainly will be explored and concretized by multidisciplinary efforts form groups involving soil, materials science, and engineer professionals focusing on innovative approaches for this fundamental issue.

KEYWORDS

- **adsorption in carbon nanomaterials**
- **biochar**
- **graphene oxide fertilizer**
- **bioinspired carbon-based fertilizers**
- **Terras Pretas de Índios**

REFERENCES

Achari, G. A.; Kowshik, M. Recent Developments on Nanotechnology in Agriculture: Plant Mineral Nutrition, Health, and Interactions with Soil Microflora. *J. Agric. Food Chem.* **2018,** *66* (33), 8647–8661.

Ahmed, M. B.; Zhou, J. L.; Ngo, H. H.; Johir, M. A. H.; Sun, L.; Asadullah, M.; Belhaj, D. Sorption of Hydrophobic Organic Contaminants on Functionalized Biochar: Protagonist Role of Pi-Pi Electron-Donor-Acceptor Interactions and Hydrogen Bonds. *J. Hazard. Mater.* **2018,** *360*, 270–278.

Alves, D. C. B. *Estudo E Aplicações De Nanomateriais Multifuncionais: Propriedades De Transporte De Nanotubos De Titanato E Novos Materiais Baseados Em Óxido De Grafeno*; Universidade Federal de Minas Gerais: Belo Horizonte, 2013.

Ameloot, N.; Neve, S. D.; Jegajeevagan, K.; Yildiz, G.; Buchan, D.; Funkuin, Y. N.; Prins, W.; Bouckaert, L.; Sleutel, S. Short-Term Co2 and N2o Emissions and Microbial Properties of Biochar Amended Sandy Loam Soils. *Soil Biol. Biochem.* **2013,** *57*, 401–410.

Andelkovic, I. B.; Kabiri, S.; Silva, R. C.; Tavakkoli, E.; Kirby, J. K.; Losic, D.; McLaughlin, M. J. Optimisation of Phosphate Loading on Graphene Oxide–Fe (Iii) Composites–Possibilities for Engineering Slow Release Fertilisers. *New J. Chem.* **2019,** *43* (22), 8580–8589.

Andelkovic, I. B.; Kabiri, S.; Tavakkoli, E.; Kirby, J. K.; McLaughlin, M. J.; Losic, D. Graphene Oxide-Fe (Iii) Composite Containing Phosphate–a Novel Slow Release Fertilizer for Improved Agriculture Management. *J. Cleaner Prod.* **2018,** *185*, 97–104.

Apul, O. G.; Karanfil, T. Adsorption of Synthetic Organic Contaminants by Carbon Nanotubes: A Critical Review. *Water Res.* **2015,** *68*, 34–55.

Archanjo, B. S.; Araujo, J. R.; Silva, A. M.; Capaz, R. B.; Falcão, N. P.; Jorio, A.; Achete, C. A. Chemical Analysis and Molecular Models for Calcium–Oxygen–Carbon Interactions in Black Carbon Found in Fertile Amazonian Anthrosoils. *Environ. Sci. Technol.* **2014,** *48* (13), 7445–7452.

Archanjo, B. S.; Baptista, D. L.; Sena, L.; Cançado, L. G.; Falcão, N. P.; Jório, A.; Achete, C. A. Nanoscale Mapping of Carbon Oxidation in Pyrogenic Black Carbon from Ancient Amazonian Anthrosols. *Environ. Sci.: Processes Impacts* **2015,** *17* (4), 775–779.

Ashjaran, M.; Babazadeh, M.; Akbarzadeh, A.; Davaran, S.; Salehi, R. Stimuli-Responsive Polyvinylpyrrolidone-Nippam-Lysine Graphene Oxide Nano-Hybrid as an Anticancer Drug Delivery on Mcf7 Cell Line. *Artif. Cells, Nanomed., Biotechnol.* **2019,** *47* (1), 443–454.

Baragano, D.; Forjan, R.; Welte, L.; Gallego, J. L. R. Nanoremediation of as and Metals Polluted Soils by Means of Graphene Oxide Nanoparticles. *Sci. Rep.* **2020,** *10* (1), 1896.

Barbosa, D. F. *Influência Do Óxido De Grafeno Em Argamassas De Cal Hidráulica Natural*; Faculdade de Ciências e Tecnologia da Universidade Nova de Lisboa: Nova Lisboa, 2015.

Beesley, L.; Moreno-Jimenez, E.; Gomez-Eyles, J. L. Effects of Biochar and Greenwaste Compost Amendments on Mobility, Bioavailability and Toxicity of Inorganic and Organic Contaminants in a Multi-Element Polluted Soil. *Environ. Pollut.* **2010,** *158* (6), 2282–2287.

Birkett, P. R. Fullerene Chemistry. *Annu. Rep. Prog. Chem., Sect. A: Inorg. Chem.* **1998,** *94*, 55–84.

Boehm, H.; Setton, R.; Stumpp, E. *Nomenclature and Terminology of Graphite Intercalation Compounds*; Pergamon, 1986.

Bourke, J.; Manley-Harris, M.; Fushimi, C.; Dowaki, K.; Nunoura, T.; Antal, M. J. Do All Carbonized Charcoals Have the Same Chemical Structure? 2. A Model of the Chemical Structure of Carbonized Charcoal. *Ind. Eng. Chem. Res.* **2007,** *46* (18), 5954–5967.

Campbell, E.; Hasan, M. T.; Pho, C.; Callaghan, K.; Akkaraju, G.; Naumov, A. Graphene Oxide as a Multifunctional Platform for Intracellular Delivery, Imaging, and Cancer Sensing. *Sci. Rep.* **2019,** *9* (1), 1–9.

Cançado, L.; Jorio, A.; Pimenta, M. Measuring the Absolute Raman Cross Section of Nanographites as a Function of Laser Energy and Crystallite Size. *Phys. Rev. B* **2007,** *76* (6), 064304.

Carneiro, J. S. S.; Lustosa Filho, J. F.; Nardis, B. r. O.; Ribeiro-Soares, J.; Zinn, Y. L.; Melo, L. n. C. A. Carbon Stability of Engineered Biochar-Based Phosphate Fertilizers. *ACS Sustainable Chem. Eng.* **2018,** *6* (11), 14203–14212.

Chen, B.; Yuan, M. Enhanced Sorption of Polycyclic Aromatic Hydrocarbons by Soil Amended with Biochar. *J. Soils Sediments* **2010,** *11* (1), 62–71.

Chew, J.; Zhu, L.; Nielsen, S.; Graber, E.; Mitchell, D. R.; Horvat, J.; Mohammed, M.; Liu, M.; van Zwieten, L.; Donne, S. Biochar-Based Fertilizer: Supercharging Root Membrane Potential and Biomass Yield of Rice. *Sci. Total Environ.* **2020,** *713*, 136431.

Chia, C. H.; Gong, B.; Joseph, S. D.; Marjo, C. E.; Munroe, P.; Rich, A. M. Imaging of Mineral-Enriched Biochar by Ftir, Raman and Sem–Edx. *Vib. Spectrosc.* **2012,** *62*, 248–257.

Cho, D.-W.; Cho, S.-H.; Song, H.; Kwon, E. E. Carbon Dioxide Assisted Sustainability Enhancement of Pyrolysis of Waste Biomass: A Case Study with Spent Coffee Ground. *Bioresour. Technol.* **2015,** *189*, 1–6.

Chuah, S.; Pan, Z.; Sanjayan, J. G.; Wang, C. M.; Duan, W. H. Nano Reinforced Cement and Concrete Composites and New Perspective from Graphene Oxide. *Constr. Build. Mater.* **2014,** *73*, 113–124.

Compant, S.; Clément, C.; Sessitsch, A. Plant Growth-Promoting Bacteria in the Rhizo-and Endosphere of Plants: Their Role, Colonization, Mechanisms Involved and Prospects for Utilization. *Soil Biol. Biochem.* **2010,** *42* (5), 669–678.

Cui, X.; Dai, X.; Khan, K. Y.; Li, T.; Yang, X.; He, Z. Removal of Phosphate from Aqueous Solution Using Magnesium-Alginate/Chitosan Modified Biochar Microspheres Derived from Thalia Dealbata. *Bioresour. Technol.* **2016,** *218*, 1123–1132.

Dai, Y.; Zhang, N.; Xing, C.; Cui, Q.; Sun, Q. The Adsorption, Regeneration and Engineering Applications of Biochar for Removal Organic Pollutants: A Review. *Chemosphere* **2019,** *223*, 12–27.

Dechene, A.; Rosendahl, I.; Laabs, V.; Amelung, W. Sorption of Polar Herbicides and Herbicide Metabolites by Biochar-Amended Soil. *Chemosphere* **2014,** *109*, 180–186.

Dimkpa, C. O.; Bindraban, P. S. Nanofertilizers: New Products for the Industry? *J. Agric. Food Chem.* **2017,** *66* (26), 6462–6473.

Dresselhaus, G.; Dresselhaus, M. S.; Saito, R. *Physical Properties of Carbon Nanotubes*; World Scientific, 1998.

Dresselhaus, M. S.; Dresselhaus, G.; Eklund, P. C. *Science of Fullerenes and Carbon Nanotubes: Their Properties and Applications*; Elsevier, 1996.

Dreyer, D. R.; Jia, H. P.; Bielawski, C. W. Graphene Oxide: A Convenient Carbocatalyst for Facilitating Oxidation and Hydration Reactions. *Angew. Chem.* **2010,** *122* (38), 6965–6968.

Ebbesen, T.; Lezec, H.; Hiura, H.; Bennett, J.; Ghaemi, H.; Thio, T. Electrical Conductivity of Individual Carbon Nanotubes. *Nature* **1996,** *382* (6586), 54–56.

Fan, X.; Wang, C.; Wang, P.; Hou, J.; Qian, J. Effects of Carbon Nanotubes on Physicochemical Properties and Sulfamethoxazole Adsorption of Sediments with or without Aging Processes. *Chem. Eng. J.* **2017,** *310*, 317–327.

Fang, H.; Chen, M.; Chen, Z. Surface Pore Tension and Adsorption Characteristics of Polluted Sediment. *Sci. China, Ser. G: Phys., Mech. Astron.* **2008,** *51* (8), 1022–1028.

FAO, F. The Future of Food and Agriculture–Trends and Challenges. *Annu. Rep.* **2017**.

Ferrari, A. C.; Robertson, J. Interpretation of Raman Spectra of Disordered and Amorphous Carbon. *Phys. Rev. B* **2000,** *61* (20), 14095.

Filho, J. F. L.; Carneiro, J. S. S.; Barbosa, C. F.; Lima, K. P.; Leite, A. A.; Melo, L. C. A. Aging of Biochar-Based Fertilizers in Soil: Effects on Phosphorus Pools and Availability to Urochloa Brizantha Grass. *Sci. Total Environ.* **2020,** *709*, 136028.

Fitz, W. J.; Wenzel, W. W. Arsenic Transformations in the Soil-Rhizosphere-Plant System: Fundamentals and Potential Application to Phytoremediation. *J. Biotechnol.* **2002,** *99*, 259–278.

Fu, P.; Yi, W.; Bai, X.; Li, Z.; Hu, S.; Xiang, J. Effect of Temperature on Gas Composition and Char Structural Features of Pyrolyzed Agricultural Residues. *Bioresour. Technol.* **2011,** *102* (17), 8211–8219.

Gan, X.; Teng, Y.; Ren, W.; Ma, J.; Christie, P.; Luo, Y. Optimization of Ex-Situ Washing Removal of Polycyclic Aromatic Hydrocarbons from a Contaminated Soil Using Nano-Sulfonated Graphene. *Pedosphere* **2017,** *27* (3), 527–536.

Gao, C.; Jin, Y. Z.; Kong, H.; Whitby, R. L.; Acquah, S. F.; Chen, G.; Qian, H.; Hartschuh, A.; Silva, S.; Henley, S. Polyurea-Functionalized Multiwalled Carbon Nanotubes: Synthesis, Morphology, and Raman Spectroscopy. *J. Phys. Chem. B* **2005,** *109* (24), 11925–11932.

Garcıa-Sanchez, A.; Alastuey, A.; Querol, X. Heavy Metal Adsorption by Different Minerals: Application to the Remediation of Polluted Soils. *Sci. Total Environ.* **1999,** *242*, 179–188.

Gavrilescu, M.; Demnerova, K.; Aamand, J.; Agathos, S.; Fava, F. Emerging Pollutants in the Environment: Present and Future Challenges in Biomonitoring, Ecological Risks and Bioremediation. *N. Biotechnol.* **2015,** *32* (1), 147–156.

Glaser, B. Prehistorically Modified Soils of Central Amazonia: A Model for Sustainable Agriculture in the Twenty-First Century. *Philos. Trans. R. Soc., B* **2007,** *362* (1478), 187–196.

Glaser, B.; Balashov, E.; Haumaier, L.; Guggenberger, G.; Zech, W. Black Carbon in Density Fractions of Anthropogenic Soils of the Brazilian Amazon Region. *Org. Geochem.* **2000,** *31* (7–8), 669–678.

Gong, X.; Huang, D.; Liu, Y.; Peng, Z.; Zeng, G.; Xu, P.; Cheng, M.; Wang, R.; Wan, J. Remediation of Contaminated Soils by Biotechnology with Nanomaterials: Bio-Behavior, Applications, and Perspectives. *Crit. Rev. Biotechnol.* **2018,** *38* (3), 455–468.

Gong, X.; Huang, D.; Liu, Y.; Zeng, G.; Wang, R.; Xu, P.; Zhang, C.; Cheng, M.; Xue, W.; Chen, S. Roles of Multiwall Carbon Nanotubes in Phytoremediation: Cadmium Uptake and Oxidative Burst in Boehmeria Nivea (L.) Gaudich. *Environ. Sci.: Nano* **2019,** *6* (3), 851–862.

Gong, Y.; Tang, J.; Zhao, D. Application of Iron Sulfide Particles for Groundwater and Soil Remediation: A Review. *Water Res.* **2016,** *89*, 309–320.

Gopinath, K. P.; Vo, D.-V. N.; Gnana Prakash, D.; Adithya Joseph, A.; Viswanathan, S.; Arun, J. Environmental Applications of Carbon-Based Materials: A Review. *Environ. Chem. Lett.* **2020.**

Gredilla, A.; Fdez-Ortiz de Vallejuelo, S.; Rodriguez-Iruretagoiena, A.; Gomez, L.; Oliveira, M. L. S.; Arana, G.; de Diego, A.; Madariaga, J. M.; Silva, L. F. O. Evidence of Mercury Sequestration by Carbon Nanotubes and Nanominerals Present in Agricultural Soils from a Coal Fired Power Plant Exhaust. *J. Hazard. Mater.* **2019,** *378*, 120747.

Gu, J.; Zhou, W.; Jiang, B.; Wang, L.; Ma, Y.; Guo, H.; Schulin, R.; Ji, R.; Evangelou, M. W. Effects of Biochar on the Transformation and Earthworm Bioaccumulation of Organic Pollutants in Soil. *Chemosphere* **2016,** *145*, 431–437.

Guilhen, S. N. *Síntese E Caracterização De Biocarvão Obtido a Partir Do Resíduo De Coco De Macaúba Para Remoção De Urânio De Soluções Aquosas*; Universidade de São Paulo: São Paulo, 2018.

Guimaraes, M. H. D. *Aspectos Eletrônicos E Estruturais Do Grafeno E Derivados: Um Estudo Teórico-Experimental*, 2010.

Haefele, S.; Konboon, Y.; Wongboon, W.; Amarante, S.; Maarifat, A.; Pfeiffer, E.; Knoblauch, C. Effects and Fate of Biochar from Rice Residues in Rice-Based Systems. *Field Crops Res.* **2011,** *121* (3), 430–440.

Hagemann, N.; Joseph, S.; Schmidt, H.-P.; Kammann, C. I.; Harter, J.; Borch, T.; Young, R. B.; Varga, K.; Taherymoosavi, S.; Elliott, K. W. Organic Coating on Biochar Explains Its Nutrient Retention and Stimulation of Soil Fertility. *Nat. Commun.* **2017,** *8* (1), 1–11.

Hailegnaw, N. S.; Mercl, F.; Pračke, K.; Praus, L.; Száková, J.; Tlustoš, P. The Role of Biochar and Soil Properties in Determining the Available Content of Al, Cu, Zn, Mn, and Cd in Soil. *Agronomy* **2020,** *10* (6), 885.

Hassan, M.; Liu, Y.; Naidu, R.; Parikh, S. J.; Du, J.; Qi, F.; Willett, I. R. Influences of Feedstock Sources and Pyrolysis Temperature on the Properties of Biochar and Functionality as Adsorbents: A Meta-Analysis. *Sci. Total Environ.* **2020,** *744*, 140714.

Hua, S.; Gong, J. L.; Zeng, G. M.; Yao, F. B.; Guo, M.; Ou, X. M. Remediation of Organochlorine Pesticides Contaminated Lake Sediment Using Activated Carbon and Carbon Nanotubes. *Chemosphere* **2017,** *177*, 65–76.

Iavicoli, I.; Leso, V.; Beezhold, D. H.; Shvedova, A. A. Nanotechnology in Agriculture: Opportunities, Toxicological Implications, and Occupational Risks. *Toxicol. Appl. Pharmacol.* **2017,** *329,* 96–111.

Ihsanullah; Abbas, A.; Al-Amer, A. M.; Laoui, T.; Al-Marri, M. J.; Nasser, M. S.; Khraisheh, M.; Atieh, M. A. Heavy Metal Removal from Aqueous Solution by Advanced Carbon Nanotubes: Critical Review of Adsorption Applications. *Sep. Purif. Technol.* **2016,** *157,* 141–161.

Iijima, S. Helical Microtubules of Graphitic Carbon. *Nature* **1991,** *354* (56–58).

Jaisi, P. D.; Elimelech, M. Single-Walled Carbon Nanotubes Exhibit Limited Transport in Soil Columns. *Environ. Sci. Technol.* **2009,** *43,* 9161–9166.

Jiang, T. Y.; Jiang, J.; Xu, R. K.; Li, Z. Adsorption of Pb(Ii) on Variable Charge Soils Amended with Rice-Straw Derived Biochar. *Chemosphere* **2012,** *89* (3), 249–256.

Jones, D. L.; Rousk, J.; Edwards-Jones, G.; DeLuca, T. H.; Murphy, D. V. Biochar-Mediated Changes in Soil Quality and Plant Growth in a Three Year Field Trial. *Soil Biol. Biochem.* **2012,** *45,* 113–124.

Jorio, A.; Dresselhaus, M. S.; Saito, R.; Dresselhaus, G. *Raman Spectroscopy in Graphene Related Systems*; John Wiley & Sons, 2011.

Jório, A.; Ribeiro-Soares, J.; Cançado, L. G.; Falcão, N. P.; Dos Santos, H. F.; Baptista, D. L.; Ferreira, E. M.; Archanjo, B. S.; Achete, C. A. Microscopy and Spectroscopy Analysis of Carbon Nanostructures in Highly Fertile Amazonian Anthrosoils. *Soil Tillage Res.* **2012,** *122,* 61–66.

Joseph, S; Lehman, J. *Biochar for Environmental Management*; 2015.

Jung, C.; Park, J.; Lim, K. H.; Park, S.; Heo, J.; Her, N.; Oh, J.; Yun, S.; Yoon, Y. Adsorption of Selected Endocrine Disrupting Compounds and Pharmaceuticals on Activated Biochars. *J. Hazard. Mater.* **2013,** *263* (Pt 2), 702–710.

Kabiri, S.; Baird, R.; Tran, D. N.; Andelkovic, I.; McLaughlin, M. J.; Losic, D. Cogranulation of Low Rates of Graphene and Graphene Oxide with Macronutrient Fertilizers Remarkably Improves Their Physical Properties. *ACS Sustainable Chem. Eng.* **2018,** *6* (1), 1299–1309.

Kabiri, S.; Degryse, F.; Tran, D. N.; da Silva, R. C.; McLaughlin, M. J.; Losic, D. Graphene Oxide: A New Carrier for Slow Release of Plant Micronutrients. *ACS Appl. Mater. Interfaces* **2017,** *9* (49), 43325–43335.

Kabiri, S.; Tran, D. N.; Baird, R.; McLaughlin, M. J.; Losic, D. Revealing the Dependence of Graphene Concentration and Physicochemical Properties on the Crushing Strength of Co-Granulated Fertilizers by Wet Granulation Process. *Powder Technol.* **2020,** *360,* 588–597.

Kazempour, M.; Namazi, H.; Akbarzadeh, A.; Kabiri, R. Synthesis and Characterization of Peg-Functionalized Graphene Oxide as an Effective Ph-Sensitive Drug Carrier. *Artif. Cells Nanomed. Biotechnol.* **2019,** *47* (1), 90–94.

Kloss, S.; Zehetner, F.; Dellantonio, A.; Hamid, R.; Ottner, F.; Liedtke, V.; Schwanninger, M.; Gerzabek, M. H.; Soja, G. Characterization of Slow Pyrolysis Biochars: Effects of Feedstocks and Pyrolysis Temperature on Biochar Properties. *J. Environ. Qual.* **2012,** *41* (4), 990–1000.

Koul, B.; Taak, P. Ex Situ Soil Remediation Strategies. In *Biotechnological Strategies for Effective Remediation of Polluted Soils*; Springer: Singapore, 2018a; pp 39–57.

Koul, B.; Taak, P. In Situ Soil Remediation Strategies. In *Biotechnological Strategies for Effective Remediation of Polluted Soils*; Springer: Singapore, 2018b; pp 59–75.

Lehmann, J. Bio-Energy in the Black. *Front. Ecol. Environ.* **2007,** *5* (7), 381–387.

Lehmann, J.; da Silva, J. P.; Steiner, C.; Nehls, T.; Zech, W.; Glaser, B. Nutrient Availability and Leaching in an Archaeological Anthrosol and a Ferralsol of the Central Amazon Basin: Fertilizer, Manure and Charcoal Amendments. *Plant Soil* **2003,** *249* (2), 343–357.
Lehmann, J.; Joseph, S. *Biochar for Environmental Management: An Introduction Biochar for Environmental Management: Science and Technology*, 2nd ed.; Routledge, 2009.
Lehmann, J.; Joseph, S. *Biochar for Environmental Management: Science, Technology and Implementation*; Routledge, 2015.
Lei, C.; Chen, T.; Zhang, Q. Y.; Long, L. S.; Chen, Z.; Fu, Z. P. Remediation of Lead Polluted Soil by Active Silicate Material Prepared from Coal Fly Ash. *Ecotoxicol Environ Saf* **2020,** *206*, 111409.
Lei, S.; Shi, Y.; Qiu, Y.; Che, L.; Xue, C. Performance and Mechanisms of Emerging Animal-Derived Biochars for Immobilization of Heavy Metals. *Sci. Total Environ.* **2019,** *646*, 1281–1289.
Lerf, A.; He, H.; Forster, M.; Klinowski, J. Structure of Graphite Oxide Revisited. *J. Phys. Chem. B* **1998,** *102* (23), 4477–4482.
Levine, I. *Physical Chemistry*; McGraw-Hill Education, 2009.
Li, S.; Turaga, U.; Shrestha, B.; Anderson, T. A.; Ramkumar, S. S.; Green, M. J.; Das, S.; Canas-Carrell, J. E. Mobility of Polyaromatic Hydrocarbons (Pahs) in Soil in the Presence of Carbon Nanotubes. *Ecotoxicol. Environ. Saf.* **2013,** *96*, 168–174.
Li, T.; Gao, B.; Tong, Z.; Yang, Y.; Li, Y. Chitosan and Graphene Oxide Nanocomposites as Coatings for Controlled-Release Fertilizer. *Water, Air, & Soil Pollution* **2019,** *230* (7), 146.
Liang, B.; Lehmann, J.; Solomon, D.; Kinyangi, J.; Grossman, J.; O'Neill, B.; Skjemstad, J. O.; Thies, J.; Luizão, F. J.; Petersen, J.; Neves, E. G. Black Carbon Increases Cation Exchange Capacity in Soils. *Soil Sci. Soc. Am. J.* **2006,** *70* (5), 1719–1730.
Liang, F.; Wang, R.; Hongzhong, X.; Yang, X.; Zhang, T.; Hu, W.; Mi, B.; Liu, Z. Investigating Pyrolysis Characteristics of Moso Bamboo through Tg-Ftir and Py-Gc/Ms. *Bioresour. Technol.* **2018,** *256*, 53–60.
Liao, J.; Liu, X.; Hu, A.; Song, H.; Chen, X.; Zhang, Z. Effects of Biochar-Based Controlled Release Nitrogen Fertilizer on Nitrogen-Use Efficiency of Oilseed Rape (Brassica Napus L.). *Sci. Rep.* **2020,** *10* (1), 1–14.
Limousin, G.; Gaudet, J. P.; Charlet, L.; Szenknect, S.; Barthès, V.; Krimissa, M. Sorption Isotherms: A Review on Physical Bases, Modeling and Measurement. *Appl. Geochem.* **2007,** *22* (2), 249–275.
Liu, G.; Li, L.; Huang, X.; Zheng, S.; Xu, X.; Liu, Z.; Zhang, Y.; Wang, J.; Lin, H.; Xu, D. Adsorption and Removal of Organophosphorus Pesticides from Environmental Water and Soil Samples by Using Magnetic Multi-Walled Carbon Nanotubes @ Organic Framework Zif-8. *J. Mater. Sci.* **2018,** *53* (15), 10772–10783.
Liu, T.; Gao, B.; Fang, J.; Wang, B.; Cao, X. Biochar-Supported Carbon Nanotube and Graphene Oxide Nanocomposites for Pb (II) and Cd (II) Removal. *RSC Adv.* **2016,** *6* (29), 24314–24319.
Long, W.-J.; Gu, Y.-c.; Ma, H.; Li, H.-d.; Xing, F. Mitigating the Electromagnetic Radiation by Coupling Use of Waste Cathode-Ray Tube Glass and Graphene Oxide on Cement Composites. *Composites, Part B* **2019,** *168*, 25–33.
Lucchese, M. M.; Stavale, F.; Ferreira, E. M.; Vilani, C.; Moutinho, M. V. d. O.; Capaz, R. B.; Achete, C. A.; Jorio, A. Quantifying Ion-Induced Defects and Raman Relaxation Length in Graphene. *Carbon* **2010,** *48* (5), 1592–1597.
Ma, X.; Anand, D.; Zhang, X.; Tsige, M.; Talapatra, S. Carbon Nanotube-Textured Sand for Controlling Bioavailability of Contaminated Sediments. *Nano Res.* **2010,** *3* (6), 412–422.

Ma, X.; Zhou, B.; Budai, A.; Jeng, A.; Hao, X.; Wei, D.; Zhang, Y.; Rasse, D. Study of Biochar Properties by Scanning Electron Microscope–Energy Dispersive X-Ray Spectroscopy (Sem-Edx). *Commun. Soil Sci. Plant Anal.* **2016,** *47* (5), 593–601.

Madari, B. E.; Petter, F. A.; Carvalho, M. d. M.; Machado, D. M.; Silva, O. M.; Freitas, F. C.; Otoni, R. d. F. Biomassa Carbonizada Como Condicionante De Solo Para a Cultura Do Arroz De Terras Altas, Em Solo Arenoso, No Cerrado: Efeito Imediato Para a Fertilidade Do Solo E Produtividade Das Plantas *Embrapa Arroz e Feijão-Comunicado Técnico (INFOTECA-E)*, 2010.

Madhavi, V.; Reddy, A. V. B.; Madhavi, G. Synthesis, Characterization, and Properties of Carbon Nanocomposites and Their Application in Wastewater Treatment. In *Environmental Remediation through Carbon Based Nano Composites*; Jawaid, M., Ahmad, A., Ismail, N., Rafatullah, M., Eds.; Singapore: Springer, 2021.

Mandal, S.; Pu, S.; He, L.; Ma, H.; Hou, D. Biochar Induced Modification of Graphene Oxide & Nzvi and Its Impact on Immobilization of Toxic Copper in Soil. *Environ. Pollut.* **2020,** *259*, 113851.

Mannan, M.; Hirano, Y.; Quitain, A.; Koinuma, M.; Kida, T. Boron Doped Graphene Oxide: Synthesis and Application to Glucose Responsive Reactivity. *J. Mater. Sci. Eng.* **2018,** *7*, 1–6.

Maraschin, T. G. *Preparação De Óxido De Grafeno E Óxido De Grafeno Reduzido E Dispersão Em Matriz Polimérica Biodegradável*; Pontifícia Universidade Católica do Rio Grande do Sul, Rio Grande do Sul, 2016.

Marcano, D. C.; Kosynkin, D. V.; Berlin, J. M.; Sinitskii, A.; Sun, Z.; Slesarev, A.; Alemany, L. B.; Lu, W.; Tour, J. M. Improved Synthesis of Graphene Oxide. *ACS Nano* **2010,** *4* (8), 4806–4814.

Matos, M. P. S. R.; Correia, A. A. S.; Rasteiro, M. G. Application of Carbon Nanotubes to Immobilize Heavy Metals in Contaminated Soils. *J. Nanopart. Res.* **2017,** *19* (4).

Mattina, M. I.; Lannucci-Berger, W.; Musante, C.; White, J. C. Concurrent Plant Uptake of Heavy Metals and Persistent Organic Pollutants from Soil. *Environ. Pollut.* **2003,** *124* (3), 375–378.

Mierzwa-Hersztek, M.; Gondek, K.; Baran, A. Effect of Poultry Litter Biochar on Soil Enzymatic Activity, Ecotoxicity and Plant Growth. *Appl. Soil Ecol.* **2016,** *105*, 144–150.

Mohammed, A.; Sanjayan, J. G.; Duan, W.; Nazari, A. Incorporating Graphene Oxide in Cement Composites: A Study of Transport Properties. *Constr. Build. Mater.* **2015,** *84*, 341–347.

Mukherjee, A.; Majumdar, S.; Servin, A. D.; Pagano, L.; Dhankher, O. P.; White, J. C. Carbon Nanomaterials in Agriculture: A Critical Review. *Front. Plant Sci.* **2016,** *7*, 172.

Mussa, Y.; Ahmed, F.; Abuhimd, H.; Arsalan, M.; Alsharaeh, E. Enhanced Electrochemical Performance at High Temperature of Cobalt Oxide/Reduced Graphene Oxide Nanocomposites and Its Application in Lithium-Ion Batteries. *Sci. Rep.* **2019,** *9* (1), 1–10.

Musso, S.; Tulliani, J.-M.; Ferro, G.; Tagliaferro, A. Influence of Carbon Nanotubes Structure on the Mechanical Behavior of Cement Composites. *Compos. Sci. Technol.* **2009,** *69* (11–12), 1985–1990.

Nanda, S.; Mohanty, P.; Pant, K. K.; Naik, S.; Kozinski, J. A.; Dalai, A. K. Characterization of North American Lignocellulosic Biomass and Biochars in Terms of Their Candidacy for Alternate Renewable Fuels. *BioEnergy Res.* **2013,** *6* (2), 663–677.

Nardis, B. O.; Carneiro, J. S. d. S.; Souza, I. M. G.; Barros, R. G.; Melo, L. C. A. Phosphorus Recovery Using Magnesium-Enriched Biochar and Its Potential Use as Fertilizer. *Arch. Agron. Soil Sci.* **2020.**

Neto, A. C.; Guinea, F.; Peres, N. M.; Novoselov, K. S.; Geim, A. K. The Electronic Properties of Graphene. *Rev. Mod. Phys.* **2009,** *81* (1), 109.

Neves, E. G.; Petersen, J. B.; Bartone, R. N.; Heckenberger, M. J. The Timing of Terra Preta Formation in the Central Amazon: Archaeological Data from Three Sites. In *Amazonian Dark Earths: Explorations in Space and Time*; Springer: Singapore, 2004; pp 125–134.

Novais, S. V.; Zenero, M. D. O.; Tronto, J.; Conz, R. F.; Cerri, C. E. P. Poultry Manure and Sugarcane Straw Biochars Modified with Mgcl2 for Phosphorus Adsorption. *J. Environ. Manage.* **2018,** *214*, 36–44.

Novak, J. M.; Lima, I.; Xing, B.; Gaskin, J. W.; Steiner, C.; Das, K.; Ahmedna, M.; Rehrah, D.; Watts, D. W.; Busscher, W. J. Characterization of Designer Biochar Produced at Different Temperatures and Their Effects on a Loamy Sand. *Ann. Environ. Sci.* **2009.**

Novoselov, K. S.; Geim, A. K.; Morozov, S. V.; Jiang, D.; Zhang, Y.; Dubonos, S. V.; Grigorieva, I. V.; Firsov, A. A. Electric Field Effect in Atomically Thin Carbon Films. *Science* **2004,** *306* (5696), 666–669.

Novotny, E. H.; Hayes, M. H.; Madari, B. E.; Bonagamba, T. J.; Azevedo, E. R. d.; Souza, A. A. d.; Song, G.; Nogueira, C. M.; Mangrich, A. S. Lessons from the Terra Preta De Índios of the Amazon Region for the Utilisation of Charcoal for Soil Amendment. *J. Braz. Chem. Soc.* **2009,** *20* (6), 1003–1010.

Ouyang, L.; Tang, Q.; Yu, L.; Zhang, R. Effects of Amendment of Different Biochars on Soil Enzyme Activities Related to Carbon Mineralisation. *Soil Res.* **2014,** *52* (7), 706–716.

Pagano, M. C.; Ribeiro-Soares, J.; Cançado, L. G.; Falcão, N. P.; Gonçalves, V. N.; Rosa, L. H.; Takahashi, J. A.; Achete, C. A.; Jorio, A. Depth Dependence of Black Carbon Structure, Elemental and Microbiological Composition in Anthropic Amazonian Dark Soil. *Soil Tillage Res.* **2016,** *155*, 298–307.

Pan, Z.; He, L.; Qiu, L.; Korayem, A. H.; Li, G.; Zhu, J. W.; Collins, F.; Li, D.; Duan, W. H.; Wang, M. C. Mechanical Properties and Microstructure of a Graphene Oxide–Cement Composite. *Cem. Concr. Compos.* **2015,** *58*, 140–147.

Pandey, S. D.; Rocha, L. C.; Pereira, G.; Deschamps, C.; Campos, J. L. E.; Falcão, N.; Prous, A.; Jorio, A. Properties of Carbon Particles in Archeological and Natural Amazon Rainforest Soils. *Catena* **2020,** *194*, 104687.

Park, S.; Lee, K.-S.; Bozoklu, G.; Cai, W.; Nguyen, S. T.; Ruoff, R. S. Graphene Oxide Papers Modified by Divalent Ions—Enhancing Mechanical Properties Via Chemical Cross-Linking. *ACS Nano* **2008,** *2* (3), 572–578.

Parks, A. N.; Chandler, G. T.; Portis, L. M.; Sullivan, J. C.; Perron, M. M.; Cantwell, M. G.; Burgess, R. M.; Ho, K. T.; Ferguson, P. L. Effects of Single-Walled Carbon Nanotubes on the Bioavailability of Pcbs in Field-Contaminated Sediments. *Nanotoxicology* **2014,** *8* (Suppl 1), 111–117.

Pavia, D. L.; Lampman, G. M.; Kriz, G. S.; Vyvyan, J. R. *Introdução À Espectroscopia*; Cengage Learning, 2010.

Pernyeszi, T.; Kasteel, R.; Witthuhn, B.; Klahre, P.; Vereecken, H.; Klumpp, E. Organoclays for Soil Remediation: Adsorption of 2,4-Dichlorophenol on Organoclay/Aquifer Material Mixtures Studied under Static and Flow Conditions. *Applied Clay Science* **2006,** *32* (3–4), 179–189.

Petter, F. A.; Ferreira, T. S.; Sinhorin, A. P.; Lima, L. B.; Almeida, F. A.; Pacheco, L. P.; Silva, A. F. Biochar Increases Diuron Sorption and Reduces the Potential Contamination of Subsurface Water with Diuron in a Sandy Soil. *Pedosphere* **2019,** *29* (6), 801–809.

Purakayastha, T.; Bera, T.; Bhaduri, D.; Sarkar, B.; Mandal, S.; Wade, P.; Kumari, S.; Biswas, S.; Menon, M.; Pathak, H. A Review on Biochar Modulated Soil Condition Improvements and Nutrient Dynamics Concerning Crop Yields: Pathways to Climate Change Mitigation and Global Food Security. *Chemosphere* **2019,** *227*, 345–365.

Qi, Z.; Hou, L.; Zhu, D.; Ji, R.; Chen, W. Enhanced Transport of Phenanthrene and 1-Naphthol by Colloidal Graphene Oxide Nanoparticles in Saturated Soil. *Environ. Sci. Technol.* **2014,** *48* (17), 10136–10144.

Qian, L.; Chen, B.; Hu, D. Effective Alleviation of Aluminum Phytotoxicity by Manure-Derived Biochar. *Environ. Sci. Technol.* **2013,** *47* (6), 2737–2745.

Qian, Y.; Qin, C.; Chen, M.; Lin, S. Nanotechnology in Soil Remediation-Applications Vs. Implications. *Ecotoxicol. Environ. Saf.* **2020,** *201*, 110815.

Rajapaksha, A. U.; Chen, S. S.; Tsang, D. C.; Zhang, M.; Vithanage, M.; Mandal, S.; Gao, B.; Bolan, N. S.; Ok, Y. S. Engineered/Designer Biochar for Contaminant Removal/Immobilization from Soil and Water: Potential and Implication of Biochar Modification. *Chemosphere* **2016,** *148*, 276–291.

Rajkovich, S.; Enders, A.; Hanley, K.; Hyland, C.; Zimmerman, A. R.; Lehmann, J. Corn Growth and Nitrogen Nutrition after Additions of Biochars with Varying Properties to a Temperate Soil. *Biol. Fertil. Soils* **2012,** *48* (3), 271–284.

Raliya, R.; Saharan, V.; Dimkpa, C.; Biswas, P. Nanofertilizer for Precision and Sustainable Agriculture: Current State and Future Perspectives. *J. Agric. Food Chem.* **2017,** *66* (26), 6487–6503.

Regkouzas, P.; Diamadopoulos, E. Adsorption of Selected Organic Micro-Pollutants on Sewage Sludge Biochar. *Chemosphere* **2019,** *224*, 840–851.

Ribeiro-Soares, J.; Cançado, L. G.; Falcão, N. P.; Martins Ferreira, E.; Achete, C. A.; Jório, A. The Use of Raman Spectroscopy to Characterize the Carbon Materials Found in Amazonian Anthrosoils. *J. Raman Spectrosc.* **2013,** *44* (2), 283–289.

Saeedi, M.; Li, L. Y.; Grace, J. R. Effect of Co-Existing Heavy Metals and Natural Organic Matter on Sorption/Desorption of Polycyclic Aromatic Hydrocarbons in Soil: A Review. *Pollution* **2020,** *6* (1), 1–24.

Saffari, M.; Vahidi, H.; Moosavirad, S. M. Effects of Pristine and Engineered Biochars of Pistachio-Shell Residues on Cadmium Behavior in a Cadmium-Spiked Calcareous Soil. *Arch. Agron. Soil Sci.* **2019,** *66* (7), 942–956.

Shen, Z.; Li, Z.; Alessi, D. S. Stabilization-Based Soil Remediation Should Consider Long-Term Challenges. *Front. Environ. Sci. Eng.* **2018,** *12* (2).

Shen, Z.; Zhang, Y.; McMillan, O.; Jin, F.; Al-Tabbaa, A. Characteristics and Mechanisms of Nickel Adsorption on Biochars Produced from Wheat Straw Pellets and Rice Husk. *Environ. Sci. Pollut. Res. Int.* **2017,** *24* (14), 12809–12819.

Shi, K.; Xie, Y.; Qiu, Y. Natural Oxidation of a Temperature Series of Biochars: Opposite Effect on the Sorption of Aromatic Cationic Herbicides. *Ecotoxicol. Environ. Saf.* **2015,** *114*, 102–108.

Smalley, R. E. Discovering the Fullerenes. *Rev. Mod. Phys.* **1997,** *69* (3), 723.

Smith, B. C. *Fundamentals of Fourier Transform Infrared Spectroscopy*; CRC Press, 2011.

Song, B.; Chen, M.; Ye, S.; Xu, P.; Zeng, G.; Gong, J.; Li, J.; Zhang, P.; Cao, W. Effects of Multi-Walled Carbon Nanotubes on Metabolic Function of the Microbial Community in Riverine Sediment Contaminated with Phenanthrene. *Carbon* **2019,** *144*, 1–7.

Song, B.; Xu, P.; Zeng, G.; Gong, J.; Wang, X.; Yan, J.; Wang, S.; Zhang, P.; Cao, W.; Ye, S. Modeling the Transport of Sodium Dodecyl Benzene Sulfonate in Riverine Sediment in the Presence of Multi-Walled Carbon Nanotubes. *Water Res.* **2018,** *129*, 20–28.

Song, W.; Guo, M. Quality Variations of Poultry Litter Biochar Generated at Different Pyrolysis Temperatures. *J. Anal. Appl. Pyrolysis* **2012,** *94*, 138–145.

Speratti, A. B.; Johnson, M. S.; Sousa, H. M.; Dalmagro, H. J.; Couto, E. G. Biochars from Local Agricultural Waste Residues Contribute to Soil Quality and Plant Growth in a Cerrado Region (Brazil) Arenosol. *GCB Bioenergy* **2018,** *10* (4), 272–286.

Su, C.; Zeng, G.-M.; Gong, J.-L.; Yang, C.-P.; Wan, J.; Hu, L.; Hua, S.-S.; Guo, Y.-Y. Impact of Carbon Nanotubes on the Mobility of Sulfonamide Antibiotics in Sediments in the Xiangjiang River. *RSC Adv.* **2016,** *6* (21), 16941–16951.

Tabelin, C. B.; Silwamba, M.; Paglinawan, F. C.; Mondejar, A. J. S.; Duc, H. G.; Resabal, V. J.; Opiso, E. M.; Igarashi, T.; Tomiyama, S.; Ito, M.; Hiroyoshi, N.; Villacorte-Tabelin, M. Solid-Phase Partitioning and Release-Retention Mechanisms of Copper, Lead, Zinc and Arsenic in Soils Impacted by Artisanal and Small-Scale Gold Mining (Asgm) Activities. *Chemosphere* **2020,** *260,* 127574.

Taha, M. R.; Mobasser, S. Adsorption of Ddt from Contaminated Soil Using Carbon Nanotubes. *Soil Sediment Contam.* **2014,** *23* (7), 703–714.

Tian, Y.; Gao, B.; Wang, Y.; Morales, V. L.; Carpena, R. M.; Huang, Q.; Yang, L. Deposition and Transport of Functionalized Carbon Nanotubes in Water-Saturated Sand Columns. *J. Hazard. Mater.* **2012,** *213–214,* 265–272.

Tomczyk, A.; Sokołowska, Z.; Boguta, P. Biochar Physicochemical Properties: Pyrolysis Temperature and Feedstock Kind Effects. *Rev. Environ. Sci. Bio/Technol.* **2020,** *19,* 191–215.

Towell, M. G.; Browne, L. A.; Paton, G. I.; Semple, K. T. Impact of Carbon Nanomaterials on the Behaviour of 14c-Phenanthrene and 14c-Benzo-[a] Pyrene in Soil. *Environ. Pollut.* **2011,** *159* (3), 706–715.

Tuinstra, F.; Koenig, J. L. Raman Spectrum of Graphite. *J. Chem. Phys.* **1970,** *53* (3), 1126–1130.

Veiga, T. R. L. A.; Lima, J. T.; Dessimoni, A. L. d. A.; Pego, M. F. F.; Soares, J. R.; Trugilho, P. F. Different Plant Biomass Characterizations for Biochar Production. *Cerne* **2017,** *23* (4), 529–536.

Vikrant, K.; Kim, K.-H.; Ok, Y. S.; Tsang, D. C.; Tsang, Y. F.; Giri, B. S.; Singh, R. S. Engineered/Designer Biochar for the Removal of Phosphate in Water and Wastewater. *Sci. Total Environ.* **2018,** *616,* 1242–1260.

Vithanage, M.; Herath, I.; Almaroai, Y. A.; Rajapaksha, A. U.; Huang, L.; Sung, J. K.; Lee, S. S.; Ok, Y. S. Effects of Carbon Nanotube and Biochar on Bioavailability of Pb, Cu and Sb in Multi-Metal Contaminated Soil. *Environ. Geochem. Health* **2017,** *39* (6), 1409–1420.

Wallace, P. R. The Band Theory of Graphite. *Phys. Rev.* **1947,** *71* (9), 622.

Wang, S.; Sun, H.; Ang, H. M.; Tadé, M. O. Adsorptive Remediation of Environmental Pollutants Using Novel Graphene-Based Nanomaterials. *Chem. Eng. J.* **2013,** *226,* 336–347.

Watts-Williams, S. J.; Nguyen, T. D.; Kabiri, S.; Losic, D.; McLaughlin, M. J. Potential of Zinc-Loaded Graphene Oxide and Arbuscular Mycorrhizal Fungi to Improve the Growth and Zinc Nutrition of Hordeum Vulgare and Medicago Truncatula. *Appl. Soil Ecol.* **2020,** *150,* 103464.

Weng, Z. H.; Van Zwieten, L.; Singh, B. P.; Tavakkoli, E.; Joseph, S.; Macdonald, L. M.; Rose, T. J.; Rose, M. T.; Kimber, S. W.; Morris, S. Biochar Built Soil Carbon over a Decade by Stabilizing Rhizodeposits. *Nat. Clim. Change* **2017,** *7* (5), 371–376.

Weyers, S. L.; Spokas, K. A. Impact of Biochar on Earthworm Populations: A Review. *Appl. Environ. Soil Sci.* **2011,** *2011.*

Yan, J.; Gong, J. L.; Zeng, G. M.; Song, B.; Zhang, P.; Liu, H. Y.; Huan, S. Y.; Li, X. D. Carbon Nanotube-Impeded Transport of Non-Steroidal Anti-Inflammatory Drugs in Xiangjiang Sediments. *J. Colloid Interface Sci.* **2017,** *498,* 229–238.

Yatim, N. M.; Shaaban, A.; Dimin, M. F.; Mohamad, N.; Yusof, F. Urea Functionalized Multiwalled Carbon Nanotubes as Efficient Nitrogen Delivery System for Rice. *Adv. Nat. Sci.: Nanosci. Nanotechnol.* **2019,** *10* (1), 015011.

Yatim, N. M.; Shaaban, A.; Dimin, M. F.; Yusof, F. Statistical Evaluation of the Production of Urea Fertilizer-Multiwalled Carbon Nanotubes Using Plackett Burman Experimental Design. *Procedia Soc Behav Sci* **2015,** *195*, 315–323.

Yatim, N. M.; Shaaban, A.; Dimin, M. F.; Yusof, F.; Abd Razak, J. Effect of Functionalised and Non-Functionalised Carbon Nanotubes-Urea Fertilizer on the Growth of Paddy. *Trop. Life Sci. Res.* **2018,** *29* (1), 17.

Ye, L.; Camps-Arbestain, M.; Shen, Q.; Lehmann, J.; Singh, B.; Sabir, M. Biochar Effects on Crop Yields with and without Fertilizer: A Meta-Analysis of Field Studies Using Separate Controls. *Soil Use Manage.* **2020,** *36* (1), 2–18.

Yu, Z.; Zhou, L.; Huang, Y.; Song, Z.; Qiu, W. Effects of a Manganese Oxide-Modified Biochar Composite on Adsorption of Arsenic in Red Soil. *J. Environ. Manage.* **2015,** *163*, 155–162.

Yuan, C.; Hung, C.-H.; Huang, W.-L. Enhancement with Carbon Nanotube Barrier on 1,2-Dichlorobenzene Removal from Soil by Surfactant-Assisted Electrokinetic (Saek) Process–the Effect of Processing Fluid. *Sep. Sci. Technol.* **2009,** *44* (10), 2284–2303.

Zarbin, A. J.; Oliveira, M. M. Carbon Nanostructures (Nanotubes and Graphene): Quo Vadis? *Quím. Nova* **2013,** *36* (10), 1533–1539.

Zaytseva, O.; Neumann, G. Carbon Nanomaterials: Production, Impact on Plant Development, Agricultural and Environmental Applications. *Chem. Biol. Technol. Agric.* **2016,** *3* (1), 17.

Zhang, J.; Gong, J. L.; Zeng, G. M.; Yang, H. C.; Zhang, P. Carbon Nanotube Amendment for Treating Dichlorodiphenyltrichloroethane and Hexachlorocyclohexane Remaining in Dong-Ting Lake Sediment-an Implication for in-Situ Remediation. *Sci. Total Environ.* **2017a,** *579*, 283–291.

Zhang, L.; Petersen, E. J.; Zhang, W.; Chen, Y.; Cabrera, M.; Huang, Q. Interactions of 14c-Labeled Multi-Walled Carbon Nanotubes with Soil Minerals in Water. *Environ. Pollut.* **2012,** *166*, 75–81.

Zhang, M.; Engelhardt, I.; Simunek, J.; Bradford, S. A.; Kasel, D.; Berns, A. E.; Vereecken, H.; Klumpp, E. Co-Transport of Chlordecone and Sulfadiazine in the Presence of Functionalized Multi-Walled Carbon Nanotubes in Soils. *Environ. Pollut.* **2017b,** *221*, 470–479.

Zhang, M.; Gao, B.; Chen, J.; Li, Y.; Creamer, A. E.; Chen, H. Slow-Release Fertilizer Encapsulated by Graphene Oxide Films. *Chem. Eng. J.* **2014,** *255*, 107–113.

Zhang, P.; Sun, H.; Min, L.; Ren, C. Biochars Change the Sorption and Degradation of Thiacloprid in Soil: Insights into Chemical and Biological Mechanisms. *Environ. Pollut.* **2018a,** *236*, 158–167.

Zhang, W.; Lu, Y.; Sun, H.; Zhang, Y.; Zhou, M.; Song, Q.; Gao, Y. Effects of Multi-Walled Carbon Nanotubes on Pyrene Adsorption and Desorption in Soils: The Role of Soil Constituents. *Chemosphere* **2019,** *221*, 203–211.

Zhang, X.; Sarmah, A. K.; Bolan, N. S.; He, L.; Lin, X.; Che, L.; Tang, C.; Wang, H. Effect of Aging Process on Adsorption of Diethyl Phthalate in Soils Amended with Bamboo Biochar. *Chemosphere* **2016,** *142*, 28–34.

Zhang, X.; Wang, H.; He, L.; Lu, K.; Sarmah, A.; Li, J.; Bolan, N. S.; Pei, J.; Huang, H. Using Biochar for Remediation of Soils Contaminated with Heavy Metals and Organic Pollutants. *Environ. Sci. Pollut. Res. Int.* **2013,** *20* (12), 8472–8483.

Zhang, Y.; Cao, B.; Zhao, L.; Sun, L.; Gao, Y.; Li, J.; Yang, F. Biochar-Supported Reduced Graphene Oxide Composite for Adsorption and Coadsorption of Atrazine and Lead Ions. *Appl. Surf. Sci.* **2018b,** *427*, 147–155.

Zhang, Y.; Yang, J.; Zhong, L.; Liu, L. Effect of Multi-Wall Carbon Nanotubes on Cr(Vi) Reduction by Citric Acid: Implications for Their Use in Soil Remediation. *Environ. Sci. Pollut. Res. Int.* **2018c,** *25* (24), 23791–23798.

Zhao, L.; Guan, X.; Yu, B.; Ding, N.; Liu, X.; Ma, Q.; Yang, S.; Yilihamu, A.; Yang, S. T. Carboxylated Graphene Oxide-Chitosan Spheres Immobilize Cu(2+) in Soil and Reduce Its Bioaccumulation in Wheat Plants. *Environ. Int.* **2019a,** *133* (Pt B), 105208.

Zhao, L.; Yang, S.-T.; Yilihamu, A.; Wu, D. Advances in the Applications of Graphene Adsorbents: From Water Treatment to Soil Remediation. *Rev. Inorg. Chem.* **2019b,** *39* (1), 47–76.

Zhou, Y.; Gao, B.; Zimmerman, A. R.; Fang, J.; Sun, Y.; Cao, X. Sorption of Heavy Metals on Chitosan-Modified Biochars and Its Biological Effects. *Chem. Eng. J.* **2013,** *231*, 512–518.

Zhu, D.; Chen, Y.; Yang, H.; Wang, S.; Wang, X.; Zhang, S.; Chen, H. Synthesis and Characterization of Magnesium Oxide Nanoparticle-Containing Biochar Composites for Efficient Phosphorus Removal from Aqueous Solution. *Chemosphere* **2020,** *247*, 125847.

CHAPTER 12

Air Pollution Management by Nanomaterials

YASSINE SLIMANI[1*], ESSIA HANNACHI[2], and GHULAM YASIN[3]

[1]*Department of Biophysics, Institute for Research and Medical Consultations (IRMC), Imam Abdulrahman Bin Faisal University, Dammam 31441, Saudi Arabia*

[2]*Laboratory of Physics of Materials–Structures and Properties, Department of Physics, Faculty of Sciences of Bizerte, University of Carthage, Zarzouna 7021, Tunisia*

[3]*State Key Laboratory of Chemical Resource Engineering, College of Materials Science and Engineering, Beijing University of Chemical Technology, Beijing 100029, China*

Corresponding author. E-mail: yaslimani@iau.edu.sa; slimaniyassine18@gmail.com

ABSTRACT

Currently, air pollution is being one of the main concerns for the undesirable and harmful impacts on human health and environment. Nanomaterials and nanotechnology are broadly utilized for the management of air pollution. Reduced low-price nanostructured materials-based sensors could be hosted to monitor the levels of pollution of indoor and outdoor air. In this chapter, we present diverse nanostructured materials utilized as receptors in air/gas sensors. The preparation and the important characteristics of the nanostructured materials to be involved in sensing air pollution applications are briefly described. Then, different nanostructured materials that could contribute to

Nanotechnology for Environmental Pollution Decontamination: Tools, Methods, and Approaches for Detection and Remediation. Fernanda Maria Policarpo Tonelli, Rouf Ahmad Bhat, & Gowhar Hamid Dar (Eds.)

the adsorption of toxic gases are highlighted. Finally, some commercially available examples of sensors for air pollution are given.

12.1 INTRODUCTION

Nowadays, because of the development and modernization of industries in the world, our atmosphere became glutted with different kinds of pollutants released from manufacturing practices or activities of humans. Some of these kinds of contaminants are sulfur dioxide, organic compounds (dioxins and volatile organic compounds), nitrogen oxides, hydrocarbons, heavy metals (As, Cr, Pb, Cd, Hg, and Zn), chlorofluorocarbons (CFCs), and carbon monoxide (CO), etc. The activities made by humans, like the combustion of gas, coal and oil, displayed a great potential in altering the emissions from natural resources. Of course, besides the pollution of air, there exists also the pollution of water provoked by numerous causes, involving the extraction of fossil fuels, by-products of manufacturing processes and combustion, pesticides, herbicides, leakage of fertilizers, oil spills, and waste disposal.

Pollutants are essentially found mixed in the soil, water, and air. Therefore, it is essential to establish a technology which can inspect, detect, and preferably it can clean the pollutants from the soil, water, and air. In this regard, nanotechnology provides a broad variety of skills and tools to enhance the quality and condition of the current environment.

Nanotechnology provides the capability to manage the material at the nanoscale and develop compounds displaying particular characteristics with a specific role. A great enthusiasm concerning the chances and risk ratio related to the nanotechnology has been established, wherein the majority of the benefits have been attached to the outlook of improving the quality of life and health. Nanomaterials are very tiny, and display elevated surface-to-volume ratio, leading them to be utilized in the detection of very sensitive pollutants (Yunus et al., 2012). The nanotechnology is additionally utilized to avoid and stop the development of contaminants or pollutants by employing the nanomaterials, advanced nanotechnologies, manufacturing processes, and others. Therefore, the applications of nanotechnology in the areas of environment in general and of air in particular can be categorized into three main applications: (1) remediation and purification of polluted materials, (2) sensing and detection of pollution, and (3) prevention of pollution. With the quick raise of pollutants varieties and their amount, developing of advanced instruments and innovative technologies that are capable of treating and preventing pollution became mandatory.

12.2 AIR QUALITY

The quality of air could be identified as the measurements of the air condition respecting the requirements of humans (Santos et al., 2020). Frequently, the air quality is linked to the health, nevertheless, the smells could also affect the quality of air from a sensorial viewpoint, although both are frequently connected (SRF Consulting Group Inc., 2004; World Health Organization, 2019). Pollution is frequently utilized to designate the substances that reduce the quality of air.

Before industrialization, pollution has been a simple nuisance, then after the starting of the industrial development and industrial revolution, it is being a real trouble (Brimblecombe, 1976; Flick, 1980). The initial problems were principally sulfur and shoot arising from the initial manufactories positioned within cities. Rapidly, authorities constructed rules and protocols for protecting people. As industry processes transformed and developed, several new problems arose related to health and damages to ecosystems. As pollution costs do not affect the industries that produces them, new regulations are needed by the government. Lead (Pb) contamination is one of the earliest examples. Among several utilizations, Pb has been inserted in vehicle fuels in order to enhance their performances in combustion engines (Stroud, 2015). Lead (even for extremely low amounts) has shown to cause a long-term impact on kids which affects their cognitive progress (Grosse et al., 2002). This has been exposed to have an enormous effect on economy. Consequently, there have been rules prohibiting the utilization of Pb initially in the United States and in Europe and later all over the world (Grosse et al., 2002; Von Storch et al., 2003).

As stated by the World Health Organization (WHO), "the ambient pollution of air contributes to more than 7.5% of whole deaths and household pollution of air to more than 7.8% of all deaths" (World Health Organization, 2019). It impacts more than 90% of the worldwide people, with a specific impact on low- and medium-income areas in which indoor pollution provoked by biomass and carbon cooking and heating display a particular effect. The mortalities by indoor and ambient pollution, in 2016, are assessed as 3.8cmillion and 4.2 million, respectively. Pollution health impacts are not found to affect only the lungs, but also impact cardiovascular health (World Health Organization, 2019).

The consciousness of the harmful impact of pollution on health has driven policymakers and other organizations to minimize the chemical and gas emissions (Wettestad, 2018; Barrett and Therivel, 2020) as well as to promote the production and commercialization of electric vehicles (Bjerkan et al., 2016). Knowing about air quality pollution could change individuals' exposure. However, as the industries develop many novel chemical materials and/or release particulates into the environment, these pollutants and contaminants are enormous, and for several of them, their effects on health have not been completely assessed (Konduracka, 2019). Therefore, the difficulties and complications of air pollution will remain important.

Pollution has many resources, some of them are natural, nevertheless, the largest common source of pollution at the present time is fossil fuels burned in industries, housing or transportation. Certain of the natural sources are geological, such as dust storms, lightning storms, or volcanoes, which could create enormous amounts of particulates, nitrogen oxides, or sulfur oxides, etc. Biomasses like oceans or forests could engender particles of pollen, volatile organic compounds, or liberate carbon in fires. Although these sources could display great effect, in most of times, sources generated by humans are invasive. The incineration of fossil fuels generates volatile organic materials, particulate matters, carbon monoxide (CO), nitrogen oxides (NO_x), and sulfur dioxide (SO_2) in general energy production, industry, cooking, housing heating, or transportation. The burning of wastes is also a significant source. Some of the contaminants are liberated directly from the pollutant activities, whereas others are produced during the reaction of these products. This is for example, the event of ground level ozone, which is generated by the reaction of volatile organic compounds with nitrogen dioxide with sunlight intervention or particles creation by isoprene forest emissions (Yu et al., 2014).

To monitor the quality of air, numerous materials are typically designated based on numerous crucial points, such as the capability to measure them with precision, the pervasiveness of the gases in the indoor environments and cities, and also how they influence on the health. Some standard limits of pollutants utilized by diverse organizations are shown in Table 12.1. The majority of protocols involve permanent and nonstop monitors that average the amounts by the hour. The position of the fixed measuring points is monitored by signs to ensure that the measurements are significant and covered. The quality of air is summed in many reports by an index, which involves the diverse monitored pollutants, but no standardized equation is available for computing it. EPA utilized a color-coded index for informing about the quality of air on the basis of the concentrations of SO_2, CO, NO_2, and particles matter. As stated, the developments of industries and the investigations of the effect of air pollution and released substances on the health push the national and multinational organizations to bring up to date their protocols.

Although the majority of pollutants consist of gases, the particulates depress significantly the quality of air. Particulates are extremely hazardous pollutants which could be constituted by broad varieties of matters, such as nitrates, sulfates, black carbon, ClNa, mineral dust, etc. (Santos et al., 2020). Because of this broad diversity, their measurements are frequently performed only based on the density of particles and/or the quantity of mass. Pb could be measured, for example, using atomic absorption spectrometry, X-ray fluorescence spectrometry. Particles with dimensions in the range of 2.5–10 μm are more stacked in the upper part of the respiratory system, whereas particles with dimensions lower than 2.5 μm are probably stacked in the profounder parts of lungs (Williams et al., 2011). The tinier particles, if they are insoluble, could pass into the blood system or translocate in the body of human. To make measurements of PM, numerous instrumentations are utilized, such as laser particle counters, neutron activation or X-ray spectroscopy, microbalances, beta-attenuation by mass accumulation, and mass weight of filtered particles, etc. (Santos et al., 2020).

The other released gases like volatile organic compounds, CO, NO_x, and ozone cause diverse undesirable effects on the human health involving cardiopathologies, infections, bronchitis, inflammation in respiratory system, asthma increase, etc. They could be assessed using standard methods. UV fluorescence and absorbance can be used to measure SO_2 and O_3, respectively, with a standard lower detection limit of about 0.5 ppbv (Santos et al., 2020). The amounts of CO and NO_2 could be measured by means of infrared photometry technique with a standard lower detection limit of about 50 and 0.2 ppbv, respectively (Santos et al., 2020).

There exist several alternative approaches that are not yet validated. For instance, differential absorption LIDAR has been employed for detecting the CO_2 plume of the Etna with an assessed precision of tens of ppm or NO_2 with precisions of 0.9 ppbv (Santoro et al., 2017; Mei et al., 2017). Other techniques involve tiny microsensors principally founded on electrochemical reactions that produce currents or sensors based on metal oxides that vary their resistance (Aleixandre and Gerboles, 2012). These sensors showed elevated uncertainty mostly owing to the nonrelevance of their responses to gases, low sensitivity, and their drift resulted from the variations in their compositions. Although these techniques display some drawbacks, they showed some benefits, such as online measurement, ease of utilization, power consumption, size, and cost effective, which make them possible alternatives techniques to be further developed (Karagulian et al., 2019).

TABLE 12.1 Typical Lower Limits of Pollutants Provided by Diverse Organizations.

Pollutants	Typical lower limits			
	Japan (Ministry of the Environment Government of Japan, 2009)	**EU (Official Journal of the European Communities, 1999)**	**USA (EPA United States, 2020)**	**WHO (World Health Organization, 2006)**
Benzene	3 μg/m^3	2 μg/m^3	–	–
SO_2	0.04 ppm	125 μg/m^3	75 ppb	20 μg/m^3
PM10	100 μg/m^3	–	150 μg/m^3	20
PM2.5	15 μg/m^3	25 μg/m^3	12 μg/m^3	10
O_3	60 ppb	120 μg/m^3	0.07 ppm	100 μg/m^3
NO_2	40 ppb	40 μg/m^3	53 ppb	40 μg/m^3
Lead	–	0.5 μg/m^3	0.15 μg/m^3	–
CO	10 ppm	–	9 ppm	–

12.3 SENSOR DEVICES FOR AIR POLLUTION

There exist diverse applications of nanomaterials and nanotechnology in the environment field. One of the environmental applications for nanotechnology is the water remediation. With similar importance, nanotechnology can be used for air remediation to detect the toxic gases existing in the ambient air. Sensors are devices that provide apparent signals via entering in interaction with the analyte, for example, biospecies, chemical, and physical. Therefore, the well-known types of sensors are:

- Biosensors adopted for the analyses of the bioactivities or biomolecules.
- Chemical sensors utilized for the analyses of chemicals.
- Physical sensors adopted for the measurement of physical parameters such as pressure, distance, mass, and temperature.

The chemical sensors can be classified into three main classes depending on their principle work: (1) mass sensors, (2) optical sensors, and (3) electrochemical sensors. The sensors provide the benefits to measure the physical quantity and transform it to signals that could be subsequently seen by human or by instruments. Sensors collect the radiations and convert them to some other forms adequate to get data/info into a control signal, a warning, such pattern (a profile, an image, etc.), or some other signals. Therefore, by means of reduced low-price sensor devices, a warning bell or visible signal can be produced to monitor the level of indoor and outdoor air pollution. The operation of sensor networks is highly complicated. Generally, most sensors consist of a detection element (called receptor), a transducer, and a signal process. Figure 12.1 shows the main constituents of sensor utilized for air pollution.

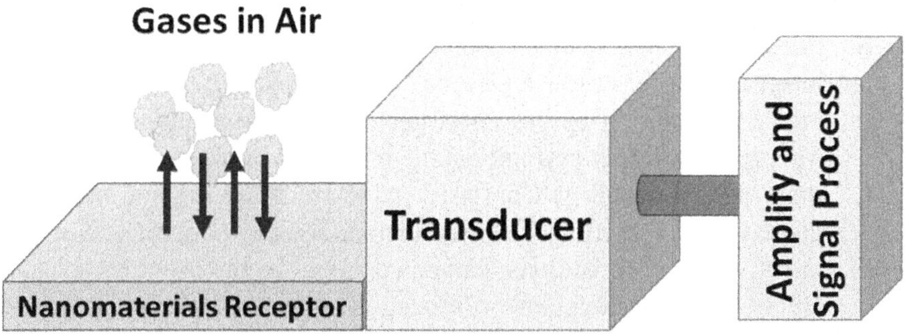

FIGURE 12.1 Main elements of sensor utilized for air pollution.

The gaseous contaminants existing in air enter in interaction with the receptor to engender a response. This response is received by the transducing element via various principles, which will be then amplified and converted into a signal, leading to interpret and quantify them in the microelectronic processor. In chemical sensors, diverse chemical detectors have been used, such as polymers, catalytic materials, insulators, solid electrolytes, metals, metal oxide semiconductors, and composites. The exploitation of different types and structures of nanomaterials employed in receptors is one of the main goals to develop innovative sensors for detecting air pollutants. Nowadays, several researchers seek to develop "nano"-sensors with high performances by optimizing different parameters:

i. **Processability:** The manufacture of sensor layers (playing the role of analyte receptors) using nanostructured materials is a very delicate and important procedure because the sensitivity of nanosensors is dependent on transducer features, morphology, crystallinity, chemical compositions, thickness, etc. Several studies have been made to fabricate sensing layers. The largest used techniques are self-assembly techniques, layer-by-layer deposition, printing, film casting, spin-coating, dip-coating, pellet formation, electrochemical deposition. Generally, there exist two key approaches for the fabrication of sensor layer based on nanostructured materials including thermal processability and solution processability.

ii. **Mechanical Strength:** The nanostructured materials to be employed in sensors should display satisfactory mechanical strength to resist the stresses and should be durable (Kar, 2020).
iii. **Operating Temperature:** This is another vital parameter that should be controlled for air pollutant sensor. In the best case, operating room temperature is preferable to be achieved.
iv. **Stability:** Thermal and environmental stability of the nanostructured materials in the sensor devices is another topic to be carefully taken into consideration before choosing them as air pollutant receptor. Mostly, nanostructured inorganic materials display an extremely high thermal stability. Contrarily, nanostructured organic materials display lower thermal stability than inorganic nanomaterials, but most of them can be considered as enough for the usage as sensing layers. This could explain the utilization of sensors based on polymers at ambient temperature and on ceramic semiconductors at elevated operating temperatures (Kar, 2020).
v. **Transducer Properties**

12.4 NANOMATERIALS USED FOR THE SENSORS OF AIR POLLUTION

Nanostructured materials are the greatest satisfactory approach to reduce the real difficulties of gas sensors: lack of stability, sensibility, and sensitivity. Currently, nanostructured materials and nanotechnology are the leading scientific research fields. Although nanoscience could be perceived as a trend to miniaturization with higher levels, it also involves new biological, chemical, and physical characteristics, all of them are still under investigations and developments. Through nanotechnology, the nanostructured materials could be designed and fabricated with characteristics that correspond to their corresponding applications (Brock, 2004). In this manner, owing to nanotechnologies, the chemical sensors have been revived during last years. The opportunity to create inexpensive miniscule tools capable of detecting lowest concentrations (ppm or ppb intervals) establishes the beginning of a novel sensors' generation, generally known as nanosensors, which are still under progress.

The characteristics of nanostructured materials are considerably different than those of present materials, mainly for the two following significant features: the quantum effects and the rise of the relative surface (Barth et al., 2010). These factors could alter and/or enhance certain characteristics of the nanostructured materials like their reactivity and electrical traits.

Gas sensors were initially put into commercialization in the 70s and 80s by enterprises that utilized the diverse detection principles, gravimetric, resistive, electrochemical, and optical. Electrochemical and optical sensors are perhaps the highly precise ones, but both gravimetric and resistive sensors provide a better sensitivity and a lower cost. Among these diverse kinds of gas sensors (Bhattacharyya et al., 2019), the resistive sensors based on metal oxide semiconductor (MOX) are the utmost fascinating ones because of their low construction and maintenance fee, ease portability and use, and simple electronic interface. Metal oxide semiconductor nanostructures have provided eminent prospects as sensitive layers because of their unique chemical, electronic, and optical features besides their feasible miniaturization. These whole characteristics appear to be the main keys to develop innovative and outstanding gas nanosensors.

Since the discovery of carbon nanotubes (CNTs), nanomaterials have fascinated broad interest because of their promising applications in a large diversity of fields as energy storage and conversion, optoelectronics, electronics, etc. (Slimani et al., 2019c; Nanda et al., 2020; Slimani and Hannachi, 2020; Hannachi et al., 2020; Yasin et al., 2020b; Yasin et al., 2020a; Ullah et al., 2020; Kumar et al., 2020; Gunasekaran et al., 2020; Slimani and Hannachi, 2021; Nadeem et al., 2021). Nanomaterials could be described as those that display at least one dimension in the range of 1–100 nm (Kreyling et al., 2010). Depending on their dimensional shapes, nanomaterials could be categorized into four classes: zero-dimensional (0D), one-dimensional (1D), two-dimensional (2D), and three-dimensional (3D) nanostructures. They could be prepared via top-down and bottom-up approaches (Gates et al., 2005; Biswas et al., 2012). Top-down method permits the creation of nanomaterials from primary larger dimensions down to the wanted nanoscale level. Bottom-up method consists of the growth from atom/ion to atom/ion until achieving the anticipated dimensions and shape. The benefits of this method are the evolution of nanomaterials with a superior order because of the synthesis conditions near to the thermodynamic balance, a high homogeneity, and a minimized density of defects. The chief disadvantages of top-down process are associated with the pollution on their surface, defects, and the internal tension of the products. Though these benefits and disadvantages conditions, the choice of the synthesis technique and the quality of the nanomaterials are also controlled by the availability of the technology and the implementation costs. Top-down processes mostly involve lithography, laser, etching, mechanical milling, etc. However, the bottom-up processes largely consist of molecular self-assembly, electrospinning, solution phase (like solvothermal, hydrothermal, sol–gel, precipitation, etc.), vapor phase (like PVD and CVD), etc. (Seevakan et al., 2019; Almessiere et al., 2019a; Slimani et al., 2019d; Almessiere et al., 2019c, 2020a, 2020b; Manikandan et al., 2020; Ajeesha et al., 2020).

During recent years, an immense progress has been done in the preparation techniques of nanostructured materials (Hamrita et al., 2014; Hannachi et al., 2018; Slimani et al., 2019a, 2019b; Almessiere et al., 2019b). To get nanostructures materials, advanced chemical and physical approaches have been established. To meet the requirements of commercialization, the method of preparation should be consistent, practicable, simple, and well-matched with the incorporation within the sensors. Nanostructured materials could be developed indirectly or directly on sensor devices. Irrespective to the procedure, a particular interest should be given to how to integrate them on the transducer. For indirect approaches, nanostructured materials are dissolved in a solvent and then the solvent is removed by drying, allowing the sensors coating. Some of these approaches are blade-coating, spin-coating, dip-coating, and drop-casting, etc. These techniques are time and cost-effective, and because they are versatile, they are greatly exploited in the development of nanosensors. Table 12.2 listed some examples of the nanosensors exploited for detecting gases pollution.

12.4.1 0D NANOMATERIALS

This class of nanomaterials involves nanocluster materials and nanoparticles (NPs). The most significant and common methods to synthesize them are the bottom-up processes. Nanoparticles are prepared by the heterogeneous nucleation on substrates, or by homogeneous nucleation from liquid or vapor. They could be also synthesized by segregating a phase via annealing properly at higher temperatures conceived solid materials (Cao and Wang, 2011). The agglomeration and the size of nanoparticles could be controlled to enhance the sensitivity (Gao et al., 2017). Generally, the performances of sensors can be improved by a uniform distribution of NPs. The size of nanoparticles controls the response of nanosensor. Indeed, if the size of nanoparticles is analogous to or lesser than the Debye length, the response of nanosensor enhances (Xu et al., 1991). Wang and Chen investigated and evaluated the better performances of semiconducting gas sensors based on V-SnO_2 NPs, synthesized through coprecipitation process, for the detection of carbon monoxide at low temperatures (Wang and Chen, 2010).

TABLE 12.2 Some Examples of the Nanosensors Exploited for Detecting Gases Pollution.

Sensors	Materials	Morphology	Gases	Refs.
Gravimetrics	ZnO	Nanorods/nanotubes	VOMs	Kilinc et al. (2014)
	Multiwalled CNTs	Nanotubes	C_6H_6, CO, NO_2	Clément et al. (2016)
	Graphene	Nanosheets	H_2, CO	Arsat et al. (2009)
	ZnO–CuO/CNTs	Nanocomposite	VOMs	Abraham et al. (2019)
	ZnO	Nanocrystalline	NO_2	Rana et al. (2017)
	SnO_2	Nanocrystalline	H_2S	Luo et al. (2013)
	Reduced graphene oxide/ZnO	Nanosheets/nanofibers	CO, VOMs	Abideen et al. (2018)
	SnO_2/reduced graphene oxide	Quantum wires/nanosheets	H_2S	Song et al. (2016)
	MoS_2	Nanosheets	NO_2	Donarelli et al. (2015)
	CuO/graphene	Nanoflowers/nanosheets	CO	Zhang et al. (2017)
	Reduced graphene oxide/ZnO	2D/nanoparticles	NO_2	Liu et al. (2014)
Resistive	SnO_2	3D hierarchical	VOMs	Li et al. (2017b)
	SnO_2@ZnO	Hierarchical	NO_2	Zhang et al. (2018)
	WO_3/porous silicon	Nanoparticles	NO_2	Yan et al. (2014)
	ZnO	Polygonal nanoflakes	NO_2	Chen et al. (2011a)
	WO_3	Flower-like	NO_2	Wang et al. (2015a)
	In_2O_3	Nanorod-flowers	NO_2	Xu et al. (2016)
	Graphene/Cu_2O	Nanosheet	H_2S	Zhou et al. (2013)
	SnO_2	Nanofibers	NO_2	Cho et al. (2011)
	SnO_2	Nanofibers	NO_2	Santos et al. (2014)
	Au-Carbon nanotubes	Nanotubes	CO, NO_2	Zanolli et al. (2011)
	Reduced graphene oxide/MoS_2	2D-nanosheets	NO_2	Zhou et al. (2017)
	TiO_2/SnO_2 core shell	Nanofibers	VOMs	Li et al. (2017a)
	SnO_2/ZnO core shell	Nanowires	NO_2	Park et al. (2013)
	SnO_2	Quantum dots	CO, CH_4	Mosadegh Sedghi et al. (2010)

12.4.2 1D NANOMATERIALS

One-dimensional nanostructured materials display various morphologies like nanotubes (NTs), nanoribbons (NRBs) or nanobelts (NBs), nanorods (NRs), nanowhiskers, nanofibers (NFs), and nanowires (NWs). One-dimensional nanomaterials exhibit a very high ratio of length-to-width and provide a greater density of integration (Comini et al., 2009). Consequently, these nanostructured materials are appropriate applicants for gas sensors. The diameter of one-dimensional nanomaterials shows a crucial impact on the processes of detection, since a lesser diameter includes extra surface atoms that participate in the reactions of gas detection (Lu et al., 2006). One of the largest utilized approaches to synthesize one-dimensional nanomaterials (NBs, NWs, NFs, etc.) is the chemical vapor deposition (CVD) technique, which is classified as a bottom-up approach.

Nowadays, the exploitation and construction of a single one-dimensional nanostructured material (NT, NF, NW, etc.) in a sensor is hard and costly. Thus, the majority of the investigated and examined nanosensors are a conglomeration of one-dimensional nanomaterials in which the different joints between nanostructured materials offer electrical paths that improve the conductivity. In these arrangements, the nanostructured materials are interlinked and create porous networks, favoring the gas penetration and diffusion.

Santos et al. (2020) reported in their chapter a study on nanowires of SnO_2 performed using low pressure-CVD in Ar atmosphere with oxygen traces at 800°C. The nanowires were developed on substrates of $Si–SiO_2$ by gold as catalysts. The nanowires were grown vertically to the substrate, display tiny diameters in the range of 50–200 nm and lengths around 30 μm. It was observed that the nanowires are branched and formed interlinked networks. Moreover, other morphology shapes, such as NRs or NBs could be noticed. The anisotropic growth is caused by the metal catalysts and the largest acceptable mechanism of growth is the vapor-liquid-solid one (Kolasinski, 2006). High-temperature superconductor of Bi-2212 nanowires, having a 250 nm in diameter and a 100 μm in length, have been fabricated by means of electrospinning technique (Koblischka et al., 2020).

Among the different one-dimensional nanostructures, nanotubes and nanofibers are being the most interested topics for scientists because of their distinctive morphologies. Nanofibers display large specific surface area and higher porosity, which are required for ultrasensitive nanosensors. The lower cross-sectional areas and higher ratio of surface-to-volume provide more efficient executions of electrochemical detectors like the signal-to-noise current ratio, transport of electric charge, and mass transport (Ding et al., 2010).

Electrospinning is one of the utmost cheap, versatile and simple techniques used to produce nanofibers of inorganic or organic compounds (Huang et al., 2003; Kim and Rothschild, 2011; Alahmari et al., 2020c). Semiconducting nanofibers could be produced using metal precursors and polymer solutions and a following heat treatment. By utilizing the electrospinning technique and adjusting the procedure parameters or the precursor solutions, one can obtain ultrafine nanostructures of core sheath, hollow, and porous nanofibers. Therefore, porous nanofibers could be prepared in diverse approaches, viz., from polymer blends, or two-component (polymer/chloride or acetate) and subsequent heat treatment (Cho et al., 2011; Cheng et al., 2014; Santos et al., 2014). Santos et al. reported in their chapter a study on nanofibers of SnO_2 having 40–60 nm in diameter, which were performed by heat treatment at 500 °C (in air) of electrospun fibers of tin chloride pentahydrate and polyvinyl alcohol (Santos et al., 2020). It is obvious that the nanofibers are constructed of linked NPs that comprise smaller grains. The nanofibers were formed from a precursor solution flow. By adjusting the flow, distinctive morphologies could be achieved. Hence, nanorods having 500–700 in diameters were gotten at very small flows. NFs of TiO_2 have been prepared through sol–gel and electrospinning processes under air–argon environments (Ansari et al., 2020). Er substituted $NiFe_2O_4$ NFs were successfully produced through electrospinning procedure using metal nitrates and polyvinyl pyrrolidone polymer, followed a calcination stage (Albetran et al., 2020). Nanofibers Cd–Nd co-substituted CoNi spinel ferrites were similarly fabricated through electrospinning technique (Alahmari et al., 2020a, 2020b, 2020c). The diverse NFs showed a grain size dependence of the magnetic, optical, electrical, dielectric properties as well as the antibacterial and anticancer activities.

Nanotubes and nanofibers are porous nanostructures, but they are also hollow ones. Thus, their morphologies have large surface area, leading them appropriate for the adsorption of gases. Different approaches could be used to produce nanotubes, such as electrospinning, sol–gel template processes, hydrothermal process, etc. Wang et al. prepared In_2O_3 nanotubes by a facile coaxial electrospinning process and a posterior calcination (Wang et al., 2016). The produced nanotube products comprise of indium oxide nanocrystals with primary 10–25 nm grain sizes of show rough surfaces. Furthermore, it was shown that the control of the size of grains and the calcination temperature could be controlled to enhance the performance of HCHO gas nanosensor.

Kolmakov et al. (2003) reported the high performances of sensing CO and O_2 by means of SnO_2 NWs. The mechanism of sensing was described based on the depletion of electrons by O_2; however, the reduction of CO gas is resulted from the removal of electrons (Fig.12.2). Multiwalled carbon nanotubes-doped SnO_2 materials revealed extremely high specific surface area in comparison to pristine SnO_2 produced by simple in situ technique (Zhao et al., 2007). These showed excellent sensing performances of CO gas at ambient temperature. Wanna et al. explored the CO gas sensing properties of CNTs–polyaniline composites and they showed promising reversible responses to CO at concentrations interval of 100–500 ppm (Wanna et al., 2006). Star et al. constructed sensor arrays of nanocomposites of single-walled CNTs and Pt or Rh metals (Star et al., 2006). These sensors revealed their ability for detecting lower concentrations of CO gas. Chen et al. studied the sensing responses of ZnO nanomaterials with different shapes regarding the sensing of NO_2 gas (Chen et al., 2011b). It was demonstrated that gas sensor of ZnO NTs present shorter response time and greater sensitivity compared with that of ZnO nanoflowers.

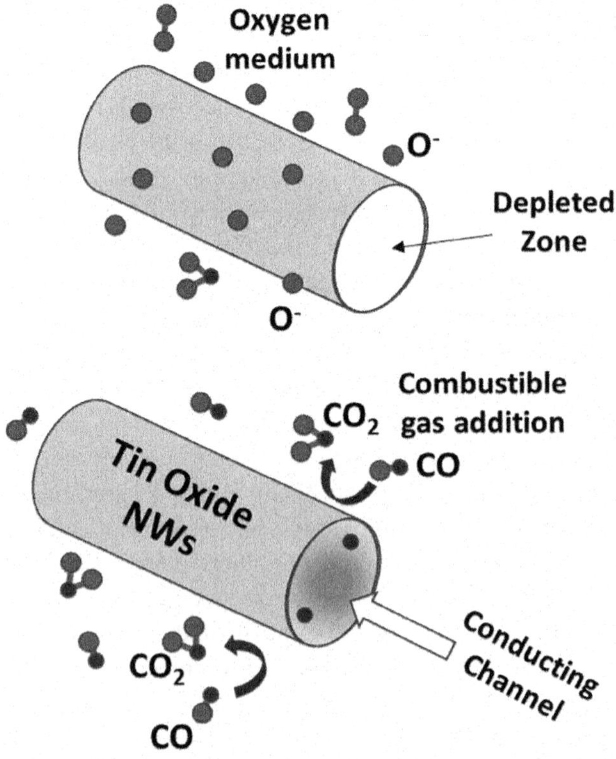

FIGURE 12.2 Mechanism of sensing of SnO_2 NWs for O_2 and CO detection.

12.4.3 2D NANOSTRUCTURED MATERIALS

Two-dimensional nanostructured materials display an extremely thin thickness and a wide lateral size, which offer them extremely high specific surface area, providing a very high amount of surface atoms. Particularly, few layers or single-layered two-dimensional nanostructured materials, in which the interactions between the layers are minimized or nonexistent, attain outstanding and promising performances for the applications in gases detection (Zhang, 2015; Neri, 2017). Graphene presents tremendously high chemical stability, optical transparency, mechanical flexibility, and carrier mobility, which delivers an excellent occasion to develop novel electronic materials, and hence innovative sensor technologies (Sun et al., 2014). Graphene and their derivatives (functionalized graphene, graphene oxide, and reduced graphene oxide) are among the largest investigated two-dimensional nanomaterials for the applications in chemical gas sensors. Nowadays, other ultrathin two-dimensional nanostructured materials displaying similar nanostructures, such as the case of transition metal dichalcogenides (WSe_2, $MoSe_2$, WS_2, TiS_2, MoS_2, etc.) have been also of great importance (Duan et al., 2015; Joshi et al., 2018). Principally, there exist three methods to develop two-dimensional nanosheets: wet chemical self-assembly, CVD growth, and liquid/chemical exfoliation of layered host materials (Santos et al., 2020). Tai et al. studied the NH_3 gas responses of thin films of PANI/TiO_2 nanocomposite coated on a Si substrate with Au electrodes (Tai et al., 2008). The findings revealed that the prepared thin films fabricated at 10 °C display better gas sensing responses in terms of stability, selectivity, and reproducibility. Bittencourt et al. prepared tungsten oxide films modified with multiwalled CNTs (Bittencourt et al., 2006). The nanocomposite films of tungsten films impregnated with CNT showed good sensing responses toward NH_3, NO_2, and CO gases. For instance, the sensitivity is about 500 ppb for NO_2 at room circumstances.

12.4.4 3D NANOSTRUCTURES MATERIALS

Three-dimensional nanostructured materials involve powders, polycrystalline, multilayer, and fibrous materials wherein the zero-dimensional, one-dimensional, and two-dimensional structural elements are connected with each other and create interfaces. Commonly, nanoparticles have the tendency to create strong aggregations since the attraction of Van der Waals between the nanoparticles and the particles size are inversely proportional

(Gao et al., 2017). For dense and large aggregations, only the nanoparticles close to the superficial area participate in the detection of gases. Furthermore, the slow diffusion of gas via the aggregated nanostructured materials minimizes the velocity of nanosensor response (Korotcenkov, 2005). Hence, the aggregation of nanoparticles is not a suitable morphology for functional sensor layers.

Nanomaterials with hierarchical structure are higher dimensional nanostructures, which are gathered from low dimensional nanostructure blocks, such as zero-dimensional nanoparticles, one-dimensional (NFs, NTs, NRs, and NWs), and two-dimensional nanosheets. Hollow and porous hierarchical nanostructured materials are very interesting to attain a great surface area and hence a quick and good responses for the detection of gases (Lee, 2009). Up-to-date, numerous hierarchical nanostructured materials (tower-like nanocolumns, flower-like nanocolumns, urchin-like spheres, mesoporous, nanohelixes, and NWs arrays) have been synthesized using diverse approaches (electrospinning, template-assisted growth, CVD, thermal evaporation, ...) and they appear to be encouraging for innovative gas sensors displaying excellent performances (Wang et al., 2015b; Liu et al., 2016).

12.5 NANOTECHNOLOGY FOR THE ADSORPTION OF TOXIC GASES

12.5.1 ADSORPTION OF DIOXINS

Dioxin and related materials are a set of extremely poisonous chemical compounds that are dangerous to health. Dibenzo-p-dioxins are a class of materials containing two rings of benzene linked by two atoms of oxygen. It displays zero to eight atoms of chlorine linked to the ring. Dibenzofuran is an analogous but different material, in that only one of the bonds between two rings of benzene is bound by oxygen. The poisonousness of dioxins differs with the number of chlorine atoms. For example, dioxins with no or only one chlorine atom are not poisonous, while the dioxins displaying more than one chlorine atom are poisonous. 2,3,7,8-Tetrachlorodibenzodioxin (TCDD) is a carcinogen to humans. Dioxins also alter the endocrine system, immune system and embryo development. Dioxin materials are mostly engendered from the combustion of organic substances in burning waste. The concentration of dioxins engendered from the combustion is ranging from 10 to 500 ng/m^3. Regulations on the emissions of dioxin are complicated and differ from country to atoner. Yet, it is commonly essential to minimize the concentration of dioxin to less than 1 ng/m^3. The minimization and prevention of dioxins have been deeply discussed by previous reports (Kulkarni et al., 2008; Wielgosiński, 2010). To remove dioxins from burning waste, adsorption by means of activated carbons (ACs) has been broadly adopted in Europe and Japan. The removal of dioxins using AC adsorbents is more efficient than other adsorbents, such as zeolites and γ-Al_2O_3 due to the high bond energy between ACs and dioxins compared with the other adsorbents (Cudahy and Helsel, 2000). Many attempts have been made in order to find a more efficient adsorbent instead of ACs, so that the poisonousness of dioxin will further be reduced to the minimum level. In 2001, Long and Yang have used CNTs as adsorbent for the elimination of dioxins (Long and Yang, 2001). The authors have demonstrated that the interaction of dioxins with CNTs is approximately three times stronger than the interaction of dioxins with ACs. This enhancement is possibly due to the crooked surface of the nanotubes, which offer more robust interaction forces between the dioxins and the carbon nanotubes compared with the flat sheets (Bhushan, 2010).

12.5.2 NO_X ADSORPTION

NO_X is a general term for the oxides of nitrogen that are most related to the air pollution, namely, nitric oxide (NO) and nitrogen dioxide (NO_2). These gases participate the development of exhaust fumes and acid rains and altering the tropospheric ozone. Many attempts have been made to remove NO_X emissions from the combustion of fossil fuels. Ion exchange zeolites, ACs, and FeOOH dispersed on ACs fibers are the known adsorbents utilized to eliminate NO_X at lower temperatures. The nitrogen oxide can be effectually absorbed on ACs owing to the interaction of the surface functional groups, even though the quantity of the absorbed kinds is still not important. Long and Yang showed that CNTs can be utilized as an adsorbent for the elimination of NO at room temperature (Long and Yang, 2001). The authors demonstrated that an adsorption amount of NO_X of about 75 mg/g was obtained when the CNTs were subjected to 1000 ppm NO + 5% O_2/He for 2 h. The adsorption of nitrogen oxides may be correlated to the exceptional structural, electronic properties, and surface functional groups of CNTs. When O_2 and NO pass across CNTs, NO is oxidized to NO_2 and then adsorbed onto the surface of nitrate kinds. This thought was proved by Mochida and coworkers (Mochida et al., 1997). The authors showed that NO is oxidized to NO2 at room temperature on ACs fiber.

12.5.3 CARBON DIOXIDE CAPTURE

Since the Kyoto Protocol was taken into consideration in 2005, the capture and the storage of CO_2 generated from fossil fuels power plants have gained more interest. Diverse technologies based on the capture of CO_2 such as adsorption, cryogenic, absorption, etc. have been studied (White et al., 2003; Aaron and Tsouris, 2005). The adsorption technology is one of the most adopted processes. It consists of amine or ammonia-based adsorption process. Yet, other processes are presently being explored worldwide owing to the high energy requirement. The Intergovernmental Panel on Climate Change (IPCC) assumed that the development of the adsorption process could be achievable and the introduction of an innovative generation of materials capable of effectively absorbing carbon dioxide would definitely improve the effectiveness of adsorption separation in the flue gases (Metz et al., 2005). The adsorbents involve ACs, silica, zeolite, SWNTs, etc. The chemical change of CNTs would have a good capacity for the capture of the greenhouse of CO_2 gas. The efficiency of CO_2 adsorption can be enhanced after the CNTs were modified with other solutions, including polyethyleneimine (PEI), 3-aminopropyltriethoxysilane (APTS), and ethylene diamine (EDA) (Long and Yang, 2001). The solution comprises groups of amines that can enter in reaction with CO_2 to produce carbamate in the absence of H_2O, consequently increasing the efficiency value of CO_2. CNTs modified with APTS present the value of efficiency of CO_2 of about 41 mg/g. This value is the highest compared with those obtained in the CNTs modified with EDA (38.8 mg/g) and CNTs modified with PEI 39.0 mg/g). Generally, some factors affect the efficiency of CO_2 adsorption based on modified CNTs, such as humidity and temperature. For example, the performances rise with the rise in humidity, nevertheless, it shows an opposite tendency with the rise in temperature.

12.5.4 ELIMINATION OF VOLATILE ORGANIC MATERIALS FROM AIR

Numerous chemicals can be produced by atmospheric reactions, including nitrous acid (Indarto, 2012), polyaromatic compounds (Natalia and Indarto, 2008), soot (Indarto, 2009), and VOMs. The regulations of clean air are being more and more severe since these materials are possibly harmful to human health. Most current systems of air purification are founded on adsorbents or photocatalysts, like ACs or Ozonolysis. Nevertheless, traditional systems are not excellent for the removal of organic pollutant at ambient temperature.

Sinha et al. proposed a novel material, which is extremely efficient for the elimination VOMs from air at ambient temperature (Sinha and Suzuki, 2007). The material covers chromium classes with combined oxidation. While the three-dimensional mesoporous Cr_2O_3 presented around 0.1 times the surface area than that of mesoporous Si, the Cr_2O_3 displayed better ability to eliminate toluene. At ambient circumstances, the materials work as an eclectic sorbent for VOMs, whereas at high temperature, it can be an efficient and eclectic catalyst for the intense oxidation of VOMs.

12.5.5 ADSORPTION OF ISOPROPYL ALCOHOL

In addition to its usage as a solvent, adsorption of isopropyl alcohol (IPA) is frequently utilized in the production of optoelectronic apparatuses. Due to the deficiency of air pollution monitoring, the vapor of IPA is liberated into the atmosphere without any remediation. The liberation of IPA is risky for the health of human and cause cancer and is irritating. For instance, Hsu and Lu performed an investigation on SWCNTs modified by HNO_3 and NaClO solutions (Hsu and Lu, 2007). This material was then utilized to adsorb the vapor IPA. The oxidation of SWCNTs by NaClO, HNO_3, and HCl solutions lead to the enhancement of the physical and chemical features of SWCNTs resulting in the minimization of pores size. Therefore, SWCNTs can act effectively as IPA vapor adsorbent from the air flow. Among the different proposed adsorbents, SWNTs oxidized by NaClO solution present the better performance for the adsorption of IPA vapor.

Throughout the adsorption process, the chemical and physical interactions attract IPA. The chemical adsorption happens as a result of the chemical interaction among adsorbent molecules and adsorbent surface functional groups. The physical adsorption is arising from forces type van der Waals among adsorbates and adsorbent. The dissimilarity between these processes is very beneficial for comprehending the influences that modify the adsorption rate.

By oxidizing SWCNTs by HNO_3 and NaClO solutions, the physical adsorption capability value augmented from 29.5 mg/g (in raw adsorbent) to 42.7 (NaClO adsorbent) and 39.5 mg/g (in HNO_3 adsorbent), while the chemical adsorption capability value increased from 10.8 mg/g (in raw adsorbent) to 26.8 and 43.5 mg/g, for HNO_3 and NaClO solutions, respectively. The enhancement in physical adsorption capacity was possibly ascribed to the size of pores of SWCNTs that reduced to closely the size of IPA, thus strengthening the physical bonds among SWCNTs and vapor of IPA. Further, the enlarged surface area of micrometric pores may rise the bonds force.

The enhancement in chemical adsorption capacity might be attributable to a rise in the sites of basical surface. For a quite low inlet content of IPA, the adsorption mechanism of IPA vapor on SWCNTs and SWNTs oxidized with NaClO is principally engendered by physical strength, while for a quite elevated inlet content, adsorption mechanism of the IPA vapor on SWNTs/NaClO can be engendered either by chemical or physical strengths.

12.6 SOME COMMERCIALIZED NANOSENORS FOR AIR POLLUTION

Resistive sensors based on nanomaterials have been evidenced for the detection of atmospheric contaminants (like CO and NO_2) in laboratory environments with a great specificity and sensitivity (Panda et al., 2016; Rickerby and Skouloudis, 2017). Nevertheless, the concentrations of gases are frequently in the range of ppm and above, thus, they are not appropriate for controlling the quality of air (sub-ppm). This task is still a challenge to be further investigated and validated. Yet, there are numerous commercialized gas nanosensors, principally are founded on thin films, which have been exploited for controlling pollutants. Furthermore, numerous studies are focusing on nanomaterials-based sensors to monitor the quality air are being rising.

Commercially, the number of companies working on gas nanosensors are starting to boost. Several of these establishments are university spin-offs and remain under progress. Although there exist some nanomaterials-based sensors in the market, their selling has scarcely commenced. Companies are in continuous transformation, some vanish practically as quickly as they established, and others combined or changed their pathways. For instance, one of the companies that changed their directions is SGX SENSORTECH S.A. SGX is a recently established corporation that acquired over the activities of MicroChemical Systems SA (MiCS) in the construction and advancement of gas sensors. MiCS was initiated in 98s by purchasing the business of Motorola's gas sensors. Since 2001, the company based in Switzerland has founded its reputation in the market of automotive as supplier of semiconducting gas nanosensors. SGX possesses the MOX patented technology, which uses the latter MEMS technology generation that merges specialized nanoparticles sensing layers with a patented polysilicon heater. These nanosensors are able to detect and measure easily explosive and pollutant gases in parts per billions (ppb) and are adequate for numerous industrial, automotive, and environmental applications (Sensortech,).

Dentoni and co-workers investigated the testing performances of a commercialized electronic nose with regard to three criteria (1) the limit of detection of the instrument, (2) the invariability response to changeable atmospheric conditions, and (3) the accuracy of classifying the odor (Dentoni et al., 2012). The measurements were performed in a zone of various industrial facilities. The electronic nose is constructed on six metal oxide sensors. Fan and co-workers carried out a classification of recorded information from numerous commercialized gas microsensors (TGS, MICS) in outside conditions (Fan et al., 2018). Their objective has been to assess the unverified algorithms of gas discrimination.

Sensors based on MOX thin films made by SGX Sensotech (MiCS 2610 (MiCS-2610,) and FIS SP-61 (A1320301-SP61 series,)) were exploited to measure the concentrations of low ozone (0–110 ppb) in existence of other interferents, such as humidity, ammonia, nitrogen oxides, carbon dioxide, and carbon monoxide (Sironi et al., 2016). MiCS-2610 sensor showed a lower response time compared with SP61 sensor. The detection limit is found to be of some ppb. Though short-term drifts were decided as satisfactory, long-term drifts make the recalibration mandatory.

Spinelle et al. have made a calibration of commercialized MOX sensors in country zones (Spinelle et al., 2015; Spinelle et al., 2017). SGX-Sensotech sensors, such as MiCS-4514 (MiCS-4514,) and MiCS-2710 (MiCS-2710,) were utilized in combination with diverse electrochemical sensors. By utilizing reference gas analyzers, these sensors were calibrated with respect to CO_2, CO NO_2, NO, and O_3. The researchers adopted numerous calibration models, such as artificial neural networks (ANN), multivariate linear regression (MLR), and simple linear regression (LR) in order to estimate the uncertainty of measurements. ANN has shown the excellent performances. The problems of cross-sensitivity have been resolved and the humidity as well as the temperature impacts have been adjusted by means of electrochemical and MOX sensors. The researchers concluded that the usage of cluster of sensors could meet the data quality purposes made by the European Instruction for O_3 at quasi-rural places.

The evaluation of AQ microsensors for air quality with respect to reference approaches, known as the EuNetAir joint exercise, was performed for the first time in an urban zone (Aveiro, Portugal) (Borrego et al., 2016). Fifteen groups from diverse centers of research and companies, from 12 diverse nations have joined the movement. More than 100 sensors assembled with numerous platforms were installed to control diverse parameters (meteorological variables and atmospheric pollutants) by means of diverse principles of measurements (PID, NDIR, optical, electrochemical, MOX, etc.). It was concluded that the most appropriate microsensos, if they are endorsed by the appropriate post-treating and tools of data modeling, could be employed to provide temporally and spatially valuable information related to the levels of air quality. The program made by the Air Quality Sensor Performance Evaluation Center (AQ-SPEC) has been established in order to perform a systematic investigation and evaluation of presently constructed low-cost sensors in ambient and controlled (laboratory) conditions (AQ-SPEC,). This center is focusing largely in complete systems like AQmesh, Air quality egg, and Aeroqual S500 among others that involve electrochemical and MOX sensors. CAIRSENSE (community air sensor network) project has been implemented to evaluate the performances of commercial and low-cost sensors in a suburban zone of about 2 km in the southeastern United States (Jiao et al., 2016). Rather than particle sensors, the scientists employed MOX and electrochemical-based sensors like the above-mentioned AQmesh, Air quality egg and Aeroqual S500. The association of diverse sensors that measure particles, sulfur dioxide, carbon monoxide, ozone, and nitrogen oxides demonstrated extremely high variable performances under real environments. The European COST Action TD1105 are developing a novel prototype for detecting gas pollutants founded on low-cost sensing technologies for monitoring the quality of air and are establishing interdisciplinary coordinated networks with highest level to perform novel processes in wireless sensor systems, sensing devices, gas sensors, sensor nanomaterials, distributed computing, models, approaches, standards, and protocols for the sustainability of environment in the European research region (Penza, 2015). Lately, EURAMET MACPoll research project has been initiated in order to construct gas sensors with low price and evaluate their performances for the monitoring of air quality (Publications Office of the EU). Nanosen-AQM project proposed to develop gas nanosensors-based electronic devices with low price for measuring the quality of air in the southwest Europe areas and providing data in a real-time. This project will permit, for example, costumers to have an idea about the quality of the air if it is good before exercising outside (NanoSen-AQM,). Additional examples of different types of commercialized low-cost nanosensors for monitoring air pollution can be found in a recent review made by Karagulian et al. (Karagulian et al., 2019).

12.7 CONCLUSION

Nanostructured materials and nanotechnology are widely used for air pollution management through enormous efforts of researchers worldwide. Nanosensors of air pollutants should meet particular properties to be efficient for sensing, such as noncomplexity and reduced arrangement of sensor devices, reliability, reproducibility, reversibility, durability, low-cost, stability, the nanomaterials must interact with the analyte at the room conditions, the sensing circumstances including moisture, pressure, electrical current supply, temperature, etc. must be satisfactory. The availability of commercialized nanostructured material-based sensors is quite scarce. Furthermore, certain serious problems in utilizing nanostructured materials as sensing materials for air contaminants should be resolved. Scientists around the world are working hard to develop nanostructured materials as sensing materials with much better performances. Additionally, the obstacles for the utilization of nanostructured materials are mainly related with the difficulties faced in process, morphologies, size distribution, thermal and chemical stability, mechanical strength, processability problems, reproducibility problems, etc. Accordingly, researchers tend to employ more efficient and simple techniques. All these requirements remain a challenge to be achieved.

KEYWORDS

- **air pollution**
- **sensors**
- **nanomaterials nanotechnology**
- **adsorption**

REFERENCES

A1320301-SP61 series. New Ozone Sensor Modules. http://www.fisinc.co.jp/en/common/pdf/A1320301-SP61 seriesE_P.pdf (accessed Jan 16, 2021).

Aaron, D.; Tsouris, C. Separation of CO_2 from Flue Gas: A Review. *Sep. Sci. Technol.* **2005**, *40*, 321–348. https://doi.org/10.1081/SS-200042244

Abideen, Z. U.; Kim, J. H.; Mirzaei, A.; Kim, H. W.; Kim, S. S. Sensing Behavior to ppm-Level Gases and Synergistic Sensing Mechanism in Metal-Functionalized rGO-Loaded ZnO Nanofibers. *Sens. Actuat. B Chem.* **2018**, *255*, 1884–1896. https://doi.org/10.1016/j.snb.2017.08.210

Abraham, N.; Reshma Krishnakumar, R.; Unni, C.; Philip, D. Simulation Studies on the Responses of ZnO-CuO/CNT Nanocomposite Based SAW Sensor to Various Volatile Organic Chemicals. *J. Sci. Adv. Mater. Devices* **2019**, *4*, 125–131. https://doi.org/10.1016/j.jsamd.2018.12.006

Ajeesha, T.; A, A.; George, M.; Manikandan, A.; Mary, J. A.; Slimani, Y.; Almessiere, M. A.; Baykal, A. Nickel Substituted $MgFe_2O_4$ Nanoparticles via Co-precipitation Method for Photocatalytic Applications. *Phys. B Condens. Matter.* **2020**, 412660. https://doi.org/10.1016/j.physb.2020.412660

Alahmari, F.; Almessiere, M. A.; Slimani, Y.; Güngüneş, H.; Shirsath, S. E.; Akhtar, S.; Jaremko, M.; Baykal, A. Synthesis and Characterization of Electrospun Ni0.5Co0.5−xCdxNd0.02Fe1.78O4 Nanofibers. *Nano-Struct. Nano-Objects* **2020a**, *24*, 100542. https://doi.org/10.1016/j.nanoso.2020.100542

Alahmari, F.; Almessiere, M. A.; Ünal, B.; Slimani, Y.; Baykal, A. Electrical and Optical Properties of Ni0·5Co0.5-xCdxNd0.02Fe1·78O4 ($x \leq 0.25$) Spinel Ferrite Nanofibers. *Ceram. Int.* **2020b**, *46*, 24605–24614. https://doi.org/10.1016/j.ceramint.2020.06.249

Alahmari, F.; Rehman, S.; Almessiere, M.; Khan, F. A.; Slimani, Y.; Baykal, A. Synthesis of Ni0.5Co0.5-xCdxFe1.78Nd0.02O4 ($x \leq 0.25$) Nanofibers by Using Electrospinning Technique Induce Anti-Cancer and Anti-Bacterial Activities. *J. Biomol. Struct. Dyn.* **2020c**, 1–8. https://doi.org/10.1080/07391102.2020.1761880

Albetran, H.; Slimani, Y.; Almessiere, M. A.; Alahmari, F.; Shirsath, S. E.; Akhtar, S.; Low, I. M.; Baykal, A.; Ercan, I. Synthesis, Characterization and Magnetic Investigation of Er-Substituted Electrospun $NiFe_2O_4$ Nanofibers. *Phys. Scr.* **2020**, *95*, 075801. https://doi.org/10.1088/1402-4896/ab8b7d

Aleixandre, M.; Gerboles, M. Review of Small Commercial Sensors for Indicative Monitoring of Ambient Gas. *Chem. Eng. Trans.* **2012**, *30*, 169–174. https://doi.org/10.3303/CET1230029

Almessiere, M. A.; Slimani, Y.; Güner, S.; Nawaz, M.; Baykal, A.; Aldakheel, F.; Akhtar, S.; Ercan, I.; Belenli.; Ozçelik, B. Magnetic and Structural Characterization of Nb 3+ -Substituted $CoFe_2O_4$ Nanoparticles. *Ceram. Int.* **2019a**, *45*, 8222–8232. https://doi.org/10.1016/j.ceramint.2019.01.125

Almessiere, M. A.; Slimani, Y.; Guner, S.; Sertkol, M.; Demir Korkmaz, A.; Shirsath, S. E.; Baykal, A. Sonochemical Synthesis and Physical Properties of Co0.3Ni0.5Mn0.2EuxFe2−xO4 Nano-Spinel Ferrites. *Ultrason. Sonochem.* **2019b**, *58*, 104654. https://doi.org/10.1016/j.ultsonch.2019.104654

Almessiere, M. A.; Slimani, Y.; Güngüneş, H.; Kostishyn, V. G.; Trukhanov, S. V.; Trukhanov, A. V.; Baykal, A. Impact of Eu3+ Ion Substitution on Structural, Magnetic and Microwave Traits of Ni–Cu–Zn Spinel Ferrites. *Ceram. Int.* **2020a**, *46*, 11124–11131. https://doi.org/10.1016/j.ceramint.2020.01.132

Almessiere, M. A.; Trukhanov, A. V.; Khan, F. A.; Slimani, Y.; Tashkandi, N.; Turchenko, V. A.; Zubar, T. I.; Tishkevich, D. I.; Trukhanov, S. V.; Panina, L. V.; Baykal, A. Correlation between Microstructure Parameters and Anti-Cancer Activity of the [Mn0.5Zn0.5](EuxNdxFe2-2x)O4 Nanoferrites Produced by Modified Sol-Gel and Ultrasonic Methods. *Ceram. Int.* **2020b**, *46*, 7346–7354. https://doi.org/10.1016/j.ceramint.2019.11.230

Almessiere, M. A. A.; Slimani, Y.; Korkmaz, A. D. D.; Taskhandi, N.; Sertkol, M.; Baykal, A.; Shirsath, S. E. S. E.; Ercan, İ.; Ozçelik, B.; Ercan.; Ozçelik, B. Sonochemical Synthesis of Eu3+ Substituted $CoFe_2O_4$ Nanoparticles and Their Structural, Optical and Magnetic Properties. *Ultrason. Sonochem.* **2019c**, *58*, 104621. https://doi.org/10.1016/j.ultsonch.2019.104621

Ansari, M. A.; Albetran, H. M.; Alheshibri, M. H.; Timoumi, A.; Algarou, N. A.; Akhtar, S.; Slimani, Y.; Almessiere, M. A.; Alahmari, F. S.; Baykal, A.; Low, I.-M. Synthesis of Electrospun TiO_2 Nanofibers and Characterization of Their Antibacterial and Antibiofilm Potential against Gram-Positive and Gram-Negative Bacteria. *Antibiotics* **2020**, *9*, 572. https://doi.org/10.3390/antibiotics9090572

AQ-SPEC. Air Quality Sensor Performance Evaluation Center. http://www.aqmd.gov/aq-spec (accessed Jan 16, 2021).

Arsat, R.; Breedon, M.; Shafiei, M.; Spizziri, P. G.; Gilje, S.; Kaner, R. B.; Kalantar-zadeh, K.; Wlodarski, W. Graphene-Like Nano-Sheets for Surface Acoustic Wave Gas Sensor Applications. *Chem. Phys. Lett.* **2009**, *467*, 344–347. https://doi.org/10.1016/j.cplett.2008.11.039

Barrett, B. F. D.; Therivel, R. *Environmental Policy and Impact Assessment in Japan*; Routledge, 2020.

Barth, S.; Hernandez-Ramirez, F.; Holmes, J. D.; Romano-Rodriguez, A. Synthesis and Applications of One-Dimensional Semiconductors. *Prog. Mater. Sci.* **2010**, *55*, 563–627. https://doi.org/10.1016/j.pmatsci.2010.02.001

Bhattacharyya, P.; Acharyya, D.; Dutta, K. Resistive and Capacitive Measurement of Nano-Structured Gas Sensors. In *Environmental Nanotechnology*; Dasgupta, N., Ranjan, S., Lichtfouse, E., Eds.; Springer: Cham, 2019; pp 25–62.

Bhushan, B. *Springer Handbook of Nanotechnology*, 3rd ed.; Springer: New York, 2010.

Biswas, A.; Bayer, I. S.; Biris, A. S.; Wang, T.; Dervishi, E.; Faupel, F. Advances in Top-Down and Bottom-Up Surface Nanofabrication: Techniques, Applications & Future Prospects. *Adv. Colloid Interf. Sci.* **2012**, *170*, 2–27. https://doi.org/10.1016/j.cis.2011.11.001

Bittencourt, C.; Felten, A.; Espinosa, E. H.; Ionescu, R.; Llobet, E.; Correig, X.; Pireaux, J. J. WO_3 Films Modified with Functionalised Multi-Wall Carbon Nanotubes: Morphological, Compositional and Gas Response Studies. *Sens. Actuat. B Chem.* **2006**, *115*, 33–41. https://doi.org/10.1016/j.snb.2005.07.067

Bjerkan, K. Y.; Nørbech, T. E.; Nordtømme, M. E. Incentives for Promoting Battery Electric Vehicle (BEV) Adoption in Norway. *Transp. Res. Part D Transp. Environ.* **2016**, *43*, 169–180. https://doi.org/10.1016/j.trd.2015.12.002

Borrego, C.; Costa, A. M.; Ginja, J.; Amorim, M.; Coutinho, M.; Karatzas, K.; Sioumis, T.; Katsifarakis, N.; Konstantinidis, K.; De Vito, S.; Esposito, E.; Smith, P.; André, N.; Gérard, P.; Francis, L. A.; Castell, N.; Schneider, P.; Viana, M.; Minguillón, M. C.; Reimringer, W.; Otjes, R. P.; von Sicard, O.; Pohle, R.; Elen, B.; Suriano, D.; Pfister, V.; Prato, M.; Dipinto, S.; Penza, M. Assessment of Air Quality Microsensors Versus Reference Methods: The EuNetAir Joint Exercise. *Atmos. Environ.* **2016**, *147*, 246–263. https://doi.org/10.1016/j.atmosenv.2016.09.050

Brimblecombe, P. Attitudes and Responses Towards Air Pollution in Medieval England. *J. Air Pollut. Control Assoc.* **1976**, *26*, 941–945. https://doi.org/10.1080/00022470.1976.10470341

Brock, S. L. Nanostructures and Nanomaterials: Synthesis, Properties and Applications. *J. Am. Chem. Soc.* **2004**, *126*, 14679–14679. https://doi.org/10.1021/ja0409457

Cao, G.; Wang, Y. *Nanostructures and Nanomaterials*; World Scientific, 2011.

Chen, M.; Wang, Z.; Han, D.; Gu, F.; Guo, G. High-Sensitivity NO_2 Gas Sensors Based on Flower-Like and Tube-Like ZnO Nanomaterials. *Sens. Actuat. B Chem.* **2011b**, *157*, 565–574. https://doi.org/10.1016/j.snb.2011.05.023

Chen, M.; Wang, Z.; Han, D.; Gu, F.; Guo, G. Porous ZnO Polygonal Nanoflakes: Synthesis, Use in High-Sensitivity NO_2 Gas Sensor, and Proposed Mechanism of Gas Sensing. *J. Phys. Chem. C* **2011a**, *115*, 12763–12773. https://doi.org/10.1021/jp201816d

Cheng, L.; Ma, S. Y.; Li, X. B.; Luo, J.; Li, W. Q.; Li, F. M.; Mao, Y. Z.; Wang, T. T.; Li, Y. F. Highly Sensitive Acetone Sensors Based on Y-Doped SnO_2 Prismatic Hollow Nanofibers Synthesized by Electrospinning. *Sens. Actuat. B Chem.* **2014**, *200*, 181–190. https://doi.org/10.1016/j.snb.2014.04.063

Cho, N. G.; Yang, D. J.; Jin, M. J.; Kim, H. G.; Tuller, H. L.; Kim, I. D. Highly Sensitive SnO_2 Hollow Nanofiber-Based NO_2 Gas Sensors. *Sens. Actuat. B Chem.* **2011**, *160*, 1468–1472. https://doi.org/10.1016/j.snb.2011.07.035

Clément, P.; Del Castillo Perez, E.; Gonzalez, O.; Calavia, R.; Lucat, C.; Llobet, E.; Debéda, H. Gas Discrimination Using Screen-Printed Piezoelectric Cantilevers Coated with Carbon Nanotubes. *Sens. Actuat. B Chem.* **2016**, *237*, 1056–1065. https://doi.org/10.1016/j.snb.2016.07.163

Comini, E.; Baratto, C.; Faglia, G.; Ferroni, M.; Vomiero, A.; Sberveglieri, G. Quasi-One Dimensional Metal Oxide Semiconductors: Preparation, Characterization and Application as Chemical Sensors. *Prog. Mater. Sci.* **2009**, *54*, 1–67. https://doi.org/10.1016/j.pmatsci.2008.06.003

Cudahy, J. J.; Helsel, R. W. Removal of Products of Incomplete Combustion with Carbon. In *Waste Management*; Elsevier Science Ltd, 2000; pp 339–345.

Dentoni, L.; Capelli, L.; Sironi, S.; Rosso, R.; Zanetti, S.; Torre, M. Development of an Electronic Nose for Environmental Odour Monitoring. *Sensors* **2012**, *12*, 14363–14381. https://doi.org/10.3390/s121114363

Ding, B.; Wang, M.; Wang, X.; Yu, J.; Sun, G. Electrospun Nanomaterials for Ultrasensitive Sensors. *Mater. Today* **2010**, *13*, 16–27. https://doi.org/10.1016/S1369-7021(10)70200-5

Donarelli, M.; Prezioso, S.; Perrozzi, F.; Bisti, F.; Nardone, M.; Giancaterini, L.; Cantalini, C.; Ottaviano, L. Response to NO_2 and Other Gases of Resistive Chemically Exfoliated MoS_2-Based Gas Sensors. *Sens. Actuat. B Chem.* **2015**, *207*, 602–613. https://doi.org/10.1016/j.snb.2014.10.099

Duan, X.; Wang, C.; Pan, A.; Yu, R.; Duan, X. Two-Dimensional Transition Metal Dichalcogenides as Atomically Thin Semiconductors: Opportunities and Challenges. *Chem. Soc. Rev.* **2015,** *44*, 8859–8876.

EPA United States. NAAQS Table | Criteria Air Pollutants | US EPA. In Epa, 2020. https://www.epa.gov/criteria-air-pollutants/naaqs-table (accessed Jan 16, 2021).

Fan, H.; Bennetts, V. H.; Schaffernicht, E.; Lilienthal, A. J. A Cluster Analysis Approach Based on Exploiting Density Peaks for Gas Discrimination with Electronic Noses in Open Environments. *Sens. Actuat. B Chem.* **2018**, *259*, 183–203. https://doi.org/10.1016/j.snb.2017.10.063

Flick, C. The Movement for Smoke Abatement in 19th-Century Britain. *Technol. Cult.* **1980**, *21*, 29–50. https://doi.org/10.2307/3103986

Gao, H.; Jia, H.; Bierer, B.; Wöllenstein, J.; Lu, Y.; Palzer, S. Scalable Gas Sensors Fabrication to Integrate Metal Oxide Nanoparticles with Well-Defined Shape and Size. *Sens. Actuat. B Chem.* **2017**, *249*, 639–646. https://doi.org/10.1016/j.snb.2017.04.031

Gates, B. D.; Xu, Q.; Stewart, M.; Ryan, D.; Willson, C. G.; Whitesides, G. M. New Approaches to Nanofabrication: Molding, Printing, and Other Techniques. *Chem. Rev.* **2005**, *105*, 1171–1196. https://doi.org/10.1021/cr030076o

Grosse, S. D.; Matte, T. D.; Schwartz, J.; Jackson, R. J. Economic Gains Resulting from the Reduction in Children's Exposure to Lead in the United States. *Environ. Health Perspect.* **2002**, *110*, 563–569. https://doi.org/10.1289/ehp.02110563

Gunasekaran, S.; Thanrasu, K.; Manikandan, A.; Durka, M.; Dinesh, A.; Anand, S.; Shankar, S.; Slimani, Y.; Almessiere, M. A.; Baykal, A. Structural, fabrication and Enhanced Electromagnetic Wave Absorption Properties of Reduced Graphene Oxide (rGO)/Zirconium Substituted Cobalt Ferrite (Co0·5Zr0·5Fe2O4) Nanocomposites. *Phys. B Condens. Matter* **2020**, 412784. https://doi.org/10.1016/j.physb.2020.412784

Hamrita, A.; Slimani, Y.; Ben Salem, M. K.; Hannachi, E.; Bessais, L.; Ben Azzouz, F.; Ben Salem, M. Superconducting Properties of Polycrystalline YBa2Cu 3O7–D Prepared by Sintering of Ball-Milled Precursor Powder. *Ceram. Int.* **2014**, *40*, 1461–1470. https://doi.org/10.1016/j.ceramint.2013.07.030

Hannachi, E.; Almessiere, M. A. A.; Slimani, Y.; Baykal, A.; Ben Azzouz, F. AC Susceptibility Investigation of YBCO Superconductor Added by Carbon Nanotubes. *J. Alloys Compd.* **2020**, *812*, 152150. https://doi.org/10.1016/j.jallcom.2019.152150

Hannachi, E.; Slimani, Y.; Ben Azzouz, F.; Ekicibil, A.; E Hannachi, Y. S. F. B. A. A. E. Higher Intra-Granular and Inter-Granular Performances of YBCO Superconductor with TiO_2 Nano-Sized Particles Addition. *Ceram. Int.* **2018**, *44*, 18836–18843. https://doi.org/10.1016/j.ceramint.2018.07.118

Hsu, S.; Lu, C. Modification of Single-Walled Carbon Nanotubes for Enhancing Isopropyl Alcohol Vapor Adsorption from Air Streams. *Sep. Sci. Technol.* **2007**, *42*, 2751–2766. https://doi.org/10.1080/01496390701515060

Huang, Z. M.; Zhang, Y. Z.; Kotaki, M.; Ramakrishna, S. A Review on Polymer Nanofibers by Electrospinning and Their Applications in Nanocomposites. *Compos. Sci. Technol.* **2003**, *63*, 2223–2253. https://doi.org/10.1016/S0266-3538(03)00178-7

Indarto, A. Heterogeneous Reactions of HONO Formation from NO_2 and HNO_3: A Review. *Res. Chem. Intermed.* **2012**, *38*, 1029–1041. https://doi.org/10.1007/s11164-011-0439-z

Indarto, A. Soot Growth Mechanisms from Polyynes. *Environ. Eng. Sci.* **2009**, *26*, 1685–1691. https://doi.org/10.1089/ees.2007.0325

Jiao, W.; Hagler, G.; Williams, R.; Sharpe, R.; Brown, R.; Garver, D.; Judge, R.; Caudill, M.; Rickard, J.; Davis, M.; Weinstock, L.; Zimmer-Dauphinee, S.; Buckley, K. Community Air Sensor Network (CAIRSENSE) Project: Evaluation of Low-Cost Sensor Performance in a Suburban Environment in the Southeastern United States. *Atmos. Meas. Tech.* **2016**, *9*, 5281–5292. https://doi.org/10.5194/amt-9-5281-2016

Joshi, N.; Hayasaka, T.; Liu, Y.; Liu, H.; Oliveira, O. N.; Lin, L. A Review on Chemiresistive Room Temperature Gas Sensors Based on Metal Oxide Nanostructures, Graphene and 2D Transition Metal Dichalcogenides. *Microchim. Acta* **2018**, *185*, 1–16. https://doi.org/10.1007/s00604-018-2750-5

Kar, P. Nanomaterials Based Sensors for Air Pollution Control. In *Environmental Nanotechnology*; Dasgupta, N., Ranjan, S., Lichtfouse, E., Eds., Vol. 4; Springer: Cham, 2020, pp 349–403.

Karagulian, F.; Barbiere, M.; Kotsev, A.; Spinelle, L.; Gerboles, M.; Lagler, F.; Redon, N.; Crunaire, S.; Borowiak, A. Review of the Performance of Low-Cost Sensors for Air Quality Monitoring. *Atmosphere (Basel).* **2019**, *10*, 506. https://doi.org/10.3390/atmos10090506

Kilinc, N.; Cakmak, O.; Kosemen, A.; Ermek, E.; Ozturk, S.; Yerli, Y.; Ozturk, Z. Z.; Urey, H. Fabrication of 1D ZnO Nanostructures on MEMS Cantilever for VOC Sensor Application. *Sens. Actuat. B Chem.* **2014**, *202*, 357–364. https://doi.org/10.1016/j.snb.2014.05.078

Kim, I. D.; Rothschild, A. Nanostructured Metal Oxide Gas Sensors Prepared by Electrospinning. *Polym. Adv. Technol.* **2011**, *22*, 318–325. https://doi.org/10.1002/pat.1797

Koblischka, M. R.; Koblischka-Veneva, A.; Zeng, X.; Hannachi, E.; Slimani, Y. Microstructure and Fluctuation-Induced Conductivity Analysis of Bi2Sr2CaCu2O8+δ (Bi-2212) Nanowire Fabrics. *Crystals* **2020**, *10*, 986. https://doi.org/10.3390/cryst10110986

Kolasinski, K. W. Catalytic Growth of Nanowires: Vapor-Liquid-Solid, Vapor-Solid-Solid, Solution-Liquid-Solid and Solid-Liquid-Solid Growth. *Curr. Opin. Solid State Mater. Sci.* **2006**, *10*, 182–191. https://doi.org/10.1016/j.cossms.2007.03.002

Kolmakov, A.; Zhang, Y.; Cheng, G.; Moskovits, M. Detection of CO and O_2 Using Tin Oxide Nanowire Sensors. *Adv. Mater.* **2003**, *15*, 997–1000. https://doi.org/10.1002/adma.200304889

Konduracka, E. A Link between Environmental Pollution and Civilization Disorders: A Mini Review. *Rev. Environ. Health* **2019**, *34*, 227–233. https://doi.org/10.1515/reveh-2018-0083

Korotcenkov, G. Gas Response Control through Structural and Chemical Modification of Metal Oxide Films: State of the Art and Approaches. *Sens. Actuat. B Chem.* **2005**, *107*, 209–232. https://doi.org/10.1016/j.snb.2004.10.006

Kreyling, W. G.; Semmler-Behnke, M.; Chaudhry, Q. A Complementary Definition of Nanomaterial. *Nano Today* **2010**, *5*, 165–168. https://doi.org/10.1016/j.nantod.2010.03.004

Kulkarni, P. S.; Crespo, J. G.; Afonso, C. A. M. Dioxins Sources and Current Remediation Technologies—A Review. *Environ. Int.* **2008**, *34*, 139–153. https://doi.org/10.1016/j.envint.2007.07.009

Kumar, A.; Yasin, G.; Korai, R. M.; Slimani, Y.; Ali, M. F.; Tabish, M.; Tariq Nazir, M.; Nguyen, T. A. Boosting Oxygen Reduction Reaction Activity by Incorporating the Iron Phthalocyanine Nanoparticles on Carbon Nanotubes Network. *Inorg. Chem. Commun.* **2020**, *120*, 108160. https://doi.org/10.1016/j.inoche.2020.108160

Lee, J. H. Gas Sensors Using Hierarchical and Hollow Oxide Nanostructures: Overview. *Sens. Actuat. B Chem.* **2009**, *140*, 319–336. https://doi.org/10.1016/j.snb.2009.04.026

Li, F.; Gao, X.; Wang, R.; Zhang, T.; Lu, G. Study on TiO_2-SnO_2 Core-Shell Heterostructure Nanofibers with Different Work Function and Its Application in Gas Sensor. *Sens. Actuat. B Chem.* **2017a**, *248*, 812–819. https://doi.org/10.1016/j.snb.2016.12.009

Li, Y. X.; Guo, Z.; Su, Y.; Jin, X. B.; Tang, X. H.; Huang, J. R.; Huang, X. J.; Li, M. Q.; Liu, J. H. Hierarchical Morphology-Dependent Gas-Sensing Performances of Three-Dimensional SnO_2 Nanostructures. *ACS Sens.* **2017b**, *2*, 102–110. https://doi.org/10.1021/acssensors.6b00597

Liu, J.; Huang, H.; Zhao, H.; Yan, X.; Wu, S.; Li, Y.; Wu, M.; Chen, L.; Yang, X.; Su, B. L. Enhanced Gas Sensitivity and Selectivity on Aperture-Controllable 3D Interconnected Macro-Mesoporous ZnO Nanostructures. *ACS Appl. Mater. Interf.* **2016**, *8*, 8583–8590. https://doi.org/10.1021/acsami.5b12315

Liu, S.; Yu, B.; Zhang, H.; Fei, T.; Zhang, T. Enhancing NO_2 Gas Sensing Performances at Room Temperature Based on Reduced Graphene Oxide-ZnO Nanoparticles Hybrids. *Sens. Actuat. B Chem.* **2014**, *202*, 272–278. https://doi.org/10.1016/j.snb.2014.05.086

Long, R. Q.; Yang, R. T. Carbon Nanotubes as a Superior Sorbent for Nitrogen Oxides. *Ind. Eng. Chem. Res.* **2001**, *40*, 4288–4291. https://doi.org/10.1021/ie000976k

Lu, J. G.; Chang, P.; Fan, Z. Quasi-One-Dimensional Metal Oxide Materials-Synthesis, Properties and Applications. *Mater. Sci. Eng. R Reports* **2006**, *52*, 49–91. https://doi.org/10.1016/j.mser.2006.04.002

Luo, W.; Fu, Q.; Zhou, D.; Deng, J.; Liu, H.; Yan, G. A Surface Acoustic Wave H_2S Gas Sensor Employing Nanocrystalline SnO_2 Thin Film. *Sens. Actuat. B Chem.* **2013**, *176*, 746–752. https://doi.org/10.1016/j.snb.2012.10.086

Manikandan, A.; Yogasundari, M.; Thanrasu, K.; Dinesh, A.; Raja, K. K.; Slimani, Y.; Jaganathan, S. K.; Srinivasan, R.; Baykal, A. Structural, Morphological and Optical Properties of Multifunctional Magnetic-Luminescent ZnO@Fe_3O_4 Nanocomposite. *Phys. E Low-Dimensional Syst. Nanostruct.* **2020**, *124*, 114291. https://doi.org/10.1016/j.physe.2020.114291

Mei, L.; Guan, P.; Kong, Z. Remote Sensing of Atmospheric NO_2 by Employing the Continuous-Wave Differential Absorption Lidar Technique. *Opt. Express* **2017**, *25*, A953. https://doi.org/10.1364/oe.25.00a953

Metz, B.; Davidson, O.; de Coninck, H.; Loos, M.; Meyer, L. *Carbon Dioxide Capture and Storage: Intergovernmental Panel on Climate Change special Report*; Cambridge University Press: Cambridge, **2005**.

MiCS-2610. MiCS-2610 O3 Sensor. https://www.cdiweb.com/datasheets/e2v/mics-2610.pdf (accessed Jan 16, 2021)

MiCS-2710. MiCS-2710 NO2 Sensor. https://www.cdiweb.com/datasheets/e2v/mics-2710.pdf (accessed Jan 16, 2021)

MiCS-4514. The MiCS-4514 Is a Compact MOS Sensor with Two Fully Independent Sensing Elements on One Package. Features Detectable Gases. https://www.sgxsensortech.com/content/uploads/2014/08/0278_Datasheet-MiCS-4514-rev-16.pdf (accessed Jan 16,2021)

Ministry of the Environment Government of Japan. Environmental Quality Standards in Japan—Air Quality [MOE], 2009. https://www.env.go.jp/en/air/aq/aq.html (accessed Jan 16, 2021)

Mochida, I.; Kawabuchi, Y.; Kawano, S.; Matsumura, Y.; Yoshikawa, M. High Catalytic Activity of Pitch-Based Activated Carbon Fibres of Moderate Surface Area for Oxidation of NO to NO_2 at Room Temperature. *Fuel* **1997**, *76*, 543–548. https://doi.org/10.1016/S0016-2361(96)00223-2

Mosadegh Sedghi, S.; Mortazavi, Y.; Khodadadi, A. Low temperature CO and CH4 dual selective gas sensor using SnO2 quantum dots prepared by sonochemical method. *Sens. Actuat. B Chem.* **2010**, *145*, 7–12. https://doi.org/10.1016/j.snb.2009.11.002

Nadeem, M.; Yasin, G.; Arif, M.; Tabassum, H.; Bhatti, M. H.; Mehmood, M.; Yunus, U.; Iqbal, R.; Nguyen, T. A.; Slimani, Y.; Song, H.; Zhao, W. Highly Active Sites of Pt/Er Dispersed N-Doped Hierarchical Porous Carbon for Trifunctional Electrocatalyst. *Chem. Eng. J.* **2021**, *409*, 128205. https://doi.org/10.1016/j.cej.2020.128205

Nanda, A.; Nanda, S.; Nguyen, T. A.; Rajendran, S.; Slimani, Y. *Nanocosmetics: Fundamentals, Applications and Toxicity*; Elsevier, 2020.

NanoSen-AQM. nanosenaqm.eu | Desarrollo y validación en campo de un sistema de nanosensores de bajo consumo y bajo coste para la monitorización en tiempo real de la calidad del aire ambiente. https://www.nanosenaqm.eu/ (accessed Jan 16, 2021).

Natalia, D.; Indarto, A. Aromatic Formation from Vinyl Radical and Acetylene. A Mechanistic Study. *Bull. Korean Chem. Soc.* **2008**, *29*, 319–322. https://doi.org/10.5012/bkcs.2008.29.2.319

Neri, G. Thin 2D: The New Dimensionality in Gas Sensing. *Chemosensors* **2017**, *5*, 21. https://doi.org/10.3390/chemosensors5030021

Official Journal of the European Communities. Council Directive 1999/30/EC, 1999. https://eur-lex.europa.eu/legal-content/EN/TXT/PDF/?uri=CELEX:31999L0030&from=EN (accessed Jan 16, 2021).

Panda, D.; Nandi, A.; Datta, S. K.; Saha, H.; Majumdar, S. Selective Detection of Carbon Monoxide (CO) Gas by Reduced Graphene Oxide (rGO) at Room Temperature. *RSC Adv.* **2016**, *6*, 47337–47348. https://doi.org/10.1039/c6ra06058g

Park, S.; An, S.; Mun, Y.; Lee, C. UV-Enhanced NO_2 Gas Sensing Properties of SnO_2-Core/ZnO-Shell Nanowires at Room Temperature. *ACS Appl. Mater. Interfaces* **2013**, *5*, 4285–4292. https://doi.org/10.1021/am400500a

Penza, M. COST Action TD1105—European Network on New Sensing Technologies for Air Pollution Control and Environmental Sustainability. Overview and Plans. In *Procedia Engineering*; Elsevier Ltd, 2015; pp 476–479.

Publications Office of the EU. Protocol of Evaluation and Calibration of Low-Cost Gas Sensors for the Monitoring of Air Pollution—Publications Office of the EU. https://op.europa.eu/en/publication-detail/-/publication/bf86f3fe-c036-42cd-a347-f777230238a6/language-en (accessed Jan 16, 2021).

Rana, L.; Gupta, R.; Tomar, M.; Gupta, V. ZnO/ST-Quartz SAW Resonator: An Efficient NO_2 Gas Sensor. *Sens. Actuat. B Chem.* **2017**, *252*, 840–845. https://doi.org/10.1016/j.snb.2017.06.075

Rickerby, D. G.; Skouloudis, A. N. Nanostructured Metal Oxides for Sensing Toxic Air Pollutants. In *RSC Detect. Sci*; Royal Society of Chemistry, 2017; pp 48–90.

Santoro, S.; Parracino, S.; Fiorani, L.; D'Aleo, R.; Di Ferdinando, E.; Giudice, G.; Maio, G.; Nuvoli, M.; Aiuppa, A. Volcanic Plume CO_2 Flux Measurements at Mount Etna by Mobile Differential Absorption Lidar. *Geosciences* **2017**, *7*, 9. https://doi.org/10.3390/geosciences7010009

Santos, J.; Fernández, M.; Fontecha, J.; Matatagui, D.; Sayago, I.; Horrillo, M.; Gracia, I. Nanocrystalline Tin Oxide Nanofibers Deposited by a Novel Focused Electrospinning Method. Application to the Detection of TATP Precursors. *Sensors* **2014**, *14*, 24231–24243. https://doi.org/10.3390/s141224231

Santos, J. P.; Sayago, I.; Aleixandre, M. Air Quality Monitoring Using Nanosensors. In *Nanomaterials for Air Remediation*; Elsevier, 2020; pp 9–31.

Seevakan, K.; Manikandan, A.; Devendran, P.; Slimani, Y.; Baykal, A.; Alagesan, T. Structural, Magnetic and Electrochemical Characterizations of Bi<inf>2</inf>Mo<inf>2</inf>O<inf>9</inf> Nanoparticle for Supercapacitor Application. *J. Magn. Magn. Mater.* **2019**, *486*. https://doi.org/10.1016/j.jmmm.2019.165254

Sensortech, S. Metal Oxide Sensors for Combustible Gases. https://www.sgxsensortech.com/productsservices/industrial-safety/metal-oxide-sensors/ (accessed Jan 16, 2021).

Sinha, A. K.; Suzuki, K. Novel Mesoporous Chromium Oxide for VOCs Elimination. *Appl. Catal. B Environ.* **2007**, *70*, 417–422. https://doi.org/10.1016/j.apcatb.2005.10.035

Sironi, S.; Capelli, L.; Spinelle, L.; Gerboles, M.; Aleixandre, M.; Bonavitacola, F. Evaluation of Metal Oxides Sensors for the Monitoring of O3 in Ambient Air at Ppb Level. *Chem. Eng. Trans.* **2016**, *54*, 319–324. https://doi.org/10.3303/CET1654054

Slimani, Y.; Almessiere, M. A.; Güner, S.; Tashkandi, N. A.; Baykal, A.; Sarac, M. F.; Nawaz, M.; Ercan, I. Calcination Effect on the Magneto-Optical Properties of Vanadium Substituted $NiFe_2O_4$ Nanoferrites. *J. Mater. Sci. Mater. Electron.* **2019a**, *30*, 9143–9154. https://doi.org/10.1007/s10854-019-01243-x

Slimani, Y.; Almessiere, M. A.; Korkmaz, A. D.; Guner, S.; Güngüneş, H.; Sertkol, M.; Manikandan, A.; Yildiz, A.; Akhtar, S.; Shirsath, S. E.; Baykal, A. Ni0.4Cu0.2Zn0.4TbxFe2-xO4 Nanospinel Ferrites: Ultrasonic Synthesis and Physical Properties. *Ultrason. Sonochem.* **2019b**, *59*, 104757. https://doi.org/10.1016/j.ultsonch.2019.104757

Slimani, Y.; Hannachi, E. Magnetic Nanosensors and Their Potential Applications. In *Nanosensors for Smart Cities*; Elsevier, 2020, pp 143–155.

Slimani, Y.; Hannachi, E. Ru-Based Perovskites/RGO Composites for Applications in High Performance Supercapacitors. In *Hybrid Perovskite Composite Materials*; Elsevier, 2021, pp 335–354.

Slimani, Y.; Hannachi, E.; Tombuloglu, H.; Güner, S.; Almessiere, M. A. A.; Baykal, A.; Aljafary, M. A. A.; Al-Suhaimi, E. A. A.; Nawaz, M.; Ercan, I. Magnetic Nanoparticles Based Nanocontainers for Biomedical Application; Elsevier, 2019c.

Slimani, Y.; Unal, B.; Hannachi, E.; Selmi, A.; Almessiere, M. A.; Nawaz, M.; Baykal, A.; Ercan, I.; Yildiz, M. Frequency and dc Bias Voltage Dependent Dielectric Properties and Electrical Conductivity of $BaTiO_3$-$SrTiO_3$/$(SiO_2)_x$ Nanocomposites. *Ceram. Int.* **2019d**, *45*, 11989–12000. https://doi.org/10.1016/j.ceramint.2019.03.092

Song, Z.; Wei, Z.; Wang, B.; Luo, Z.; Xu, S.; Zhang, W.; Yu, H.; Li, M.; Huang, Z.; Zang, J.; Yi, F.; Liu, H. Sensitive Room-Temperature H2S Gas Sensors Employing SnO2 Quantum Wire/Reduced Graphene Oxide Nanocomposites. *Chem. Mater.* **2016**, *28*, 1205–1212. https://doi.org/10.1021/acs.chemmater.5b04850

Spinelle, L.; Gerboles, M.; Villani, M. G.; Aleixandre, M.; Bonavitacola, F. Field Calibration of a Cluster of Low-Cost Available Sensors for Air Quality Monitoring. Part A: Ozone and Nitrogen Dioxide. *Sens. Actuat. B Chem.* **2015**, *215*, 249–257. https://doi.org/10.1016/j.snb.2015.03.031

Spinelle, L.; Gerboles, M.; Villani, M. G.; Aleixandre, M.; Bonavitacola, F. Field Calibration of a Cluster of Low-Cost Commercially Available Sensors for Air Quality Monitoring. Part B: NO, CO and CO_2. *Sens. Actuat. B Chem.* **2017**, *238*, 706–715. https://doi.org/10.1016/j.snb.2016.07.036

SRF Consulting Group Inc. A Review of National and International Odor Policy, Odor Measurement Technology and Public Administration, 2004. https://www.pca.state.mn.us/sites/default/files/p-gen2-02.pdf (accessed Jan 16, 2021)

Star, A.; Joshi, V.; Skarupo, S.; Thomas, D.; Gabriel, J. C. P. Gas Sensor Array Based on Metal-Decorated Carbon Nanotubes. *J. Phys. Chem. B* **2006**, *110*, 21014–21020. https://doi.org/10.1021/jp064371z

Stroud, D. A. Regulation of Some Sources of Lead Poisoning: A Brief Review. In *Proceedings of the Oxford Lead Symposium*; Edward Grey Institute: University of Oxford Oxford, UK, pp 8–26.

Sun, Z.; Liao, T.; Dou, Y.; Hwang, S. M.; Park, M. S.; Jiang, L.; Kim, J. H.; Dou, S. X. Generalized Self-Assembly of Scalable Two-Dimensional Transition Metal Oxide Nanosheets. *Nat. Commun.* **2014**, *5*, 1–9. https://doi.org/10.1038/ncomms4813

Tai, H.; Jiang, Y.; Xie, G.; Yu, J.; Chen, X.; Ying, Z. Influence of Polymerization Temperature on NH3 Response of PANI/TiO_2 Thin Film Gas Sensor. *Sens. Actuat. B Chem.* **2008**, *129*, 319–326. https://doi.org/10.1016/j.snb.2007.08.013

Ullah, S.; Yasin, G.; Ahmad, A.; Qin, L.; Yuan, Q.; Khan, A. U.; Khan, U. A.; Rahman, A. U.; Slimani, Y. Construction of Well-Designed 1D Selenium-Tellurium Nanorods Anchored on Graphene Sheets as a High Storage Capacity Anode Material for Lithium-Ion Batteries. *Inorg. Chem. Front.* **2020**, *7*, 1750–1761. https://doi.org/10.1039/c9qi01701a

Von Storch, H.; Costa-Cabral, M.; Hagner, C.; Feser, F.; Pacyna, J.; Pacyna, E.; Kolb, S. Four Decades of Gasoline Lead Emissions and Control Policies in Europe: A Retrospective Assessment. *Sci. Total Environ.* **2003**, *311*, 151–176. https://doi.org/10.1016/S0048-9697(03)00051-2

Wang, C.; Li, X.; Feng, C.; Sun, Y.; Lu, G. Nanosheets Assembled Hierarchical Flower-Like WO_3 Nanostructures: Synthesis, Characterization, and Their Gas Sensing Properties. *Sens. Actuat. B Chem.* **2015a**, *210*, 75–81. https://doi.org/10.1016/j.snb.2014.12.020

Wang, C. T.; Chen, M. T. Vanadium-Promoted Tin Oxide Semiconductor Carbon Monoxide Gas Sensors. *Sens. Actuat. B Chem.* **2010**, *150*, 360–366. https://doi.org/10.1016/j.snb.2010.06.060

Wang, L.; Cao, J.; Qian, X.; Zhang, H. Facile Synthesis and Enhanced Gas Sensing Properties of Grain Size-Adjustable In_2O_3 Micro/Nanotubes. *Mater. Lett.* **2016**, *171*, 30–33. https://doi.org/10.1016/j.matlet.2016.02.053

Wang, S.; Yang, J.; Zhang, H.; Wang, Y.; Gao, X.; Wang, L.; Zhu, Z. One-Pot Synthesis of 3D Hierarchical SnO_2 Nanostructures and Their Application for Gas Sensor. *Sens. Actuat. B Chem.* **2015b**, *207*, 83–89. https://doi.org/10.1016/j.snb.2014.10.032

Wanna, Y.; Srisukhumbowornchai, N.; Tuantranont, A.; Wisitsoraat, A.; Thavarungkul, N.; Singjai, P. The Effect of Carbon Nanotube Dispersion on CO Gas Sensing Characteristics of Polyaniline Gas Sensor. *J. Nanosci. Nanotechnol.* **2006**, *6*, 3893–3896. https://doi.org/10.1166/jnn.2006.675

Wettestad, J. *Clearing the Air: European Advances in Tackling Acid Rain and Atmospheric Pollution*; Taylor and Francis, 2018.

White, C. M.; Strazisar, B. R.; Granite, E. J.; Hoffman, J. S.; Pennline, H. W. Separation and Capture of CO_2 from Large Stationary Sources and Sequestration in Geological Formations—Coalbeds and Deep Saline Aquifers. *J. Air Waste Manag. Assoc.* **2003**, *53*, 645–715. https://doi.org/10.1080/10473289.2003.10466206

Wielgosiński, G. The Possibilities of Reduction of Polychlorinated Dibenzo-p-Dioxins and Polychlorinated Dibenzofurans Emission. *Int. J. Chem. Eng.* **2010**, Article ID 392175. https://doi.org/10.1155/2010/392175

Williams, R. O.; Carvalho, T. C.; Peters, J. I. Influence of Particle Size on Regional Lung Deposition—What Evidence Is There? *Int. J. Pharm.* **2011**, *406*, 1–10. https://doi.org/10.1016/j.ijpharm.2010.12.040

World Health Organisation. Air Quality Guidelines Global Update 2005—World Health Organisation, 2006. https://www.who.int/phe/health_topics/outdoorair/outdoorair_aqg/en/ (accessed Jan 16, 2021).

World Health Organization. Air Pollution, 2019. https://www.who.int/airpollution/en/ (accessed Jan 16, 2021).

Xu, C.; Tamaki, J.; Miura, N.; Yamazoe, N. Grain Size Effects on Gas Sensitivity of Porous SnO_2-Based Elements. *Sens. Actuators B. Chem.* **1991**, *3*, 147–155. https://doi.org/10.1016/0925-4005(91)80207-Z

Xu, X.; Zhang, H.; Hu, X.; Sun, P.; Zhu, Y.; He, C.; Hou, S.; Sun, Y.; Lu, G. Hierarchical Nanorod-Flowers Indium Oxide Microspheres and Their Gas Sensing Properties. *Sens. Actuat. B Chem.* **2016**, *227*, 547–553. https://doi.org/10.1016/j.snb.2015.12.085

Yan, W.; Hu, M.; Zeng, P.; Ma, S.; Li, M. Room Temperature NO_2 -Sensing Properties of WO_3 Nanoparticles/Porous Silicon. *Appl. Surf. Sci.* **2014**, *292*, 551–555. https://doi.org/10.1016/j.apsusc.2013.11.169

Yasin, G.; Anjum, M. J.; Malik, M. U.; Khan, M. A.; Khan, W. Q.; Arif, M.; Mehtab, T.; Nguyen, T. A.; Slimani, Y.; Tabish, M.; Ali, D.; Zuo, Y. Revealing the Erosion-Corrosion Performance of Sphere-Shaped Morphology of Nickel Matrix Nanocomposite Strengthened with Reduced Graphene Oxide Nanoplatelets. *Diam. Relat. Mater.* **2020a**, *104*, 107763. https://doi.org/10.1016/j.diamond.2020.107763

Yasin, G.; Arif, M.; Mehtab, T.; Shakeel, M.; Mushtaq, M. A.; Kumar, A.; Nguyen, T. A.; Slimani, Y.; Nazir, M. T.; Song, H. A Novel Strategy for the Synthesis of Hard Carbon Spheres Encapsulated with Graphene Networks as a Low-Cost and Large-Scalable Anode Material for Fast Sodium Storage with an Ultralong Cycle Life. *Inorg. Chem. Front.* **2020b**, *7*, 402–410. https://doi.org/10.1039/c9qi01105f

Yu, H.; Ortega, J.; Smith, J. N.; Guenther, A. B.; Kanawade, V. P.; You, Y.; Liu, Y.; Hosman, K.; Karl, T.; Seco, R.; Geron, C.; Pallardy, S. G.; Gu, L.; Mikkilä, J.; Lee, S.-H. New Particle Formation and Growth in an Isoprene-Dominated Ozark Forest: From Sub-5 nm to CCN-Active Sizes. *Aerosol Sci. Technol.* **2014**, *48*, 1285–1298. https://doi.org/10.1080/02786826.2014.984801

Yunus, I. S.; Harwin.; Kurniawan, A.; Adityawarman, D.; Indarto, A. Nanotechnologies in Water and Air Pollution Treatment. *Environ. Technol. Rev.* **2012**, *1*, 136–148. https://doi.org/10.1080/21622515.2012.733966

Zanolli, Z.; Leghrib, R.; Felten, A.; Pireaux, J. J.; Llobet, E.; Charlier, J. C. Gas Sensing with Au-Decorated Carbon Nanotubes. *ACS Nano* **2011**, *5*, 4592–4599. https://doi.org/10.1021/nn200294h

Zhang, D.; Jiang, C.; Liu, J.; Cao, Y. Carbon Monoxide Gas Sensing at Room Temperature Using Copper Oxide-Decorated Graphene Hybrid Nanocomposite Prepared by Layer-by-Layer Self-Assembly. *Sens. Actuat. B Chem.* **2017**, *247*, 875–882. https://doi.org/10.1016/j.snb.2017.03.108

Zhang, H. Ultrathin Two-Dimensional Nanomaterials. *ACS Nano* **2015**, *9*, 9451–9469. https://doi.org/10.1021/acsnano.5b05040

Zhang, Z.; Xu, M.; Liu, L.; Ruan, X.; Yan, J.; Zhao, W.; Yun, J.; Wang, Y.; Qin, S.; Zhang, T. Novel SnO2@ZnO Hierarchical Nanostructures for Highly Sensitive and Selective NO_2 Gas Sensing. *Sens. Actuat. B Chem.* **2018**, *257*, 714–727. https://doi.org/10.1016/j.snb.2017.10.190

Zhao, L.; Choi, M.; Kim, H. S.; Hong, S. H. The Effect of Multiwalled Carbon Nanotube Doping on the CO Gas Sensitivity of SnO_2-Based Nanomaterials. *Nanotechnology* **2007**, *18*, 445501. https://doi.org/10.1088/0957-4484/18/44/445501

Zhou, L.; Shen, F.; Tian, X.; Wang, D.; Zhang, T.; Chen, W. Stable Cu_2O Nanocrystals Grown on Functionalized Graphene Sheets and Room Temperature H_2S Gas Sensing with Ultrahigh Sensitivity. *Nanoscale* **2013**, *5*, 1564–1569. https://doi.org/10.1039/c2nr33164k

Zhou, Y.; Liu, G.; Zhu, X.; Guo, Y. Ultrasensitive NO_2 Gas Sensing Based on rGO/MoS2 Nanocomposite Film at Low Temperature. *Sens. Actuat. B Chem.* **2017**, *251*, 280–290. https://doi.org/10.1016/j.snb.2017.05.060

PART V
Nanobioremediation

CHAPTER 13

Nano-Phytoremediation: Using Plants and Nanomaterials to Environmental Pollution Remediation

AHMED ALI ROMEH*

Plant Production Department, Faculty of Technology and Development, Zagazig University, Zagazig, Egypt

**E-mail: ahmedromeh2006@yahoo.com*

ABSTRACT

Among the most serious threats to soil and water supply are persistent organic compounds and heavy metals that pollute the soil, which accumulate in the soil of agricultural fields and reach the food chain, causing serious health effects. Recently, scientists have tried to search for advanced methods and continuity between them, such as treatment with nano-phytoremediation, in order to demonstrate real efficiency in removing persistent organic pollutants and heavy metals in soil and water. Phytoremediation is a mechanism that uses the green plants to decontaminate, uptake, stabilize, metabolize, or detoxify contaminants from soil, water, and waste. Plants with unusual metal-accumulating ability of metals in their above-ground parts are known as hyperaccumulator plants, such as *Alyssum bertolonii, Thlaspi caerulescens, Calendula officinalis,* and *Tagetes erecta.* Significant surface areas, a high number of active surface sites and high adsorption capabilities make nanomaterials a favorable alternative for polluted soil remediation. The different physical and chemical properties of heavy metal ions and interactions with nano zerovalent iron (nZVI)-based materials play an important role in different adsorption processes, including

Nanotechnology for Environmental Pollution Decontamination: Tools, Methods, and Approaches for Detection and Remediation. Fernanda Maria Policarpo Tonelli, Rouf Ahmad Bhat, & Gowhar Hamid Dar (Eds.)

adsorption, redox, aggregation, ion exchange, hydroxylation, and subsequent precipitation. Unfortunately, the direct use of nZVI is restricted by powder state as a result of the small particle size and, and the intrinsic characteristics of nZVI to react with surrounding media or agglomerate during the preparation process decreases the reactivity of nanoparticles and also results in poor mobility and efficient transfer of nZVI to the contaminated area for continuous in situ remediation. To address this problem, integrating phytoremediation with nanotechnology provides a better solution for heavy metals and hydrophobic organic pollutants remediation, through various mechanisms. Nanomaterials can enhance phytoremediation by direct pollutant removal by nanomaterials through redox reactions, surface processes, adsorption, ion exchange, surface complexation, and electrostatic interaction, or by improving plant growth and development, such as plant-growth promoting rhizobacteria (PGPR), applying plant growth regulators and using transgenic plants, and increasing the phytoavailability. Several authors studied the combination between nanotechnology and phytotechnology for detoxification of organic pollutants and remediation of heavy metals in contaminated environments. Integration of organic acids (i.e., citric, malic, tartaric, and oxalic acids), organic amendments, and nZVI with phytoremediation are feasible practices for the repair of metal polluted soils, which can be because of the resulting increase in reactive surface locations, higher nZVI volume to surface area, which might bind to more metal ions and enhancing availability of heavy metals for phytoremediation. Moreover, recent studies have shown beneficial effects of microorganisms in the rhizosphere with nanotechnology on the performance of phytoremediation in removing heavy metals. The combination of bioamendments with nZVI remediation techniques maximizes the favorable effect of nZVI and minimizes its toxicity and improves growth parameters and gas exchange, leading to new insight into nZVI heavy metal remediation. Green nanotechnology can provide safe and environmentally friendly possibilities to clean and manage the ecosystem without harming the nature. The appropriate types of plants and nanomaterials must be selected to uptake the pollutants, along with improving the agricultural management of the high-precision cleaning process. For example, the combination of Ficus ZVI (F–Fe^0), ipomoea–silver (Ip–Ag^0), and brassica–silver nanoparticles (Br–Ag^0) as green nanotechnology, and *Plantago major* as phytoremediation has played a principal role in the clean-up of water and soil polluted with chlorfenapyr (Romeh and Saber, 2020). In addition, the contribution of adsorbents assisted by F–Fe^0, in particular, wheat bran (WB) and *P. major* would play an effective role in the complete removal of chlorpyrifos from the water with a significant reduction in the dangerous degradation product TCP (Rady et al., 2019). It can be concluded that the impact of Fe^0 on phytoremediation may be an important issue, and alternatives are required to mitigate its potential negative effects on phytoremediation or accelerate its positive effects.

13.1 INTRODUCTION

The main source of pollution of organic compounds is accidental spills of petroleum-based products used in transportation (usually diesel fuel), while other activities such as mining, agriculture, and fossil fuel use lead to a large amount of metal pollutants being discharged into the soil (Barrutia et al., 2011; Khan and Kathi, 2014). These pollutants in the soils need to search for new technologies for their decontamination (Agnello et al., 2016).

Among the most serious threats to soil and water supply are persistent organic compounds and heavy metals that pollute the soil, which accumulate in the soil of agricultural fields and reach the food chain, causing serious health effects (Verma et al., 2021). The movement of these contaminants through the soil into noncontaminated areas as dust or leachates leads to ecosystem pollution (Tangahu et al., 2011).

Heavy metal ions, such as Cr, Pb, Fe, Cu, Zn, Hg, and Ni, cannot be broken down into cleaning materials compared to conventional organic contaminants (Hu et al., 2012), which have accumulated in living organisms and are considered to be highly toxic or carcinogenic to most of them (Liu et al., 2013). In addition, reduced crop productivity and productivity can occur in heavily polluted soils due to changes in the morphological, physiological, and chemical processes vital to plant species (Fabbricino et al., 2018; Rehman et al., 2019; Feller et al., 2019). Also, with the rapid advancement in industrialization, the disposal of industrial effluents poses serious threats to the environment and becomes the greatest concern of the sustainable development of human society. Several conventional techniques are already being used to remediate the environment from these contaminants including soil washing, soil flushing, thermal desorption, incineration, chemical precipitation, electrochemical treatment, electrodialysis, evaporative recovery, ultrafiltration, ion-exchange, oxidation/reduction, reverse osmosis, filtration, adsorption, and membrane technologies (Román-Ross et al., 2006; Bouhamed et al., 2012; Azarudeen et al., 2013; Ashraf et al., 2017, 2019). Most of these methods, however, have several drawbacks, such as high costs, long response times, poor degradation efficiency, far from ideal performance, and are not an effective method of removing dangerous heavy metals and other organic pollutants from agricultural land (Muszynska and Hanus-Fajerska, 2015; Kołaci ́nski et al., 2017; Bao et al., 2019). Therefore, it is important to research new cost-effective and safe treatment technologies to remove organic pollutants or detoxify water and soil. More recently, scientists have tried to look for many methods and continuity between them in order to show the excellent removal capacity of heavy metals and organic compounds in soil and water. Of these attempts, the use of nanotechnology, phytoremediation, and nano-phytoremediation with the assistance of other processes that enhance the performance of phytoremediation such as green nanotechnology helped by adsorbent materials to improve the process of adsorption and reduction of both organic and metals, oxidation process with strong oxidants such as H_2O_2, persulfate, and permanganate to improve the dissipation of organic pollutants.

13.2 PHYTOREMEDIATION TECHNOLOGIES OF POLLUTANTS

Phytoremediation is a technique that uses plants to remove, uptake, stabilize, metabolize, and detoxify contaminants from soil, water, and waste (Kuo et al. 2014; Ali et al., 2020). Phytoremediation is an environmentally sustainable pollution reduction, control, and remediation technology. Phytoremediation for heavy metals is part of the new green technologies that uptake metals from the soil to store them in the upper parts of plants, which can then be harvested (Chaturvedi et al., 2016). The efficiency of this technique depends on the plants' growth rates and their ability to absorb and uptake metals in their shoots (Romeh et al., 2016). This emerging technology is a promising remediation tool that is effective, inexpensive, fast, safe, economical, compatible, environmental friendly, and low cost. It is one of the main components of green technology, and it allows cheap decontamination of hazardous waste sites. It can be applied directly to polluted sites where other methods of treatment are too costly (Ashraf et al., 2019; Saleem et al., 2020; Ali et al., 2020). For agricultural land and water bodies, there are different forms of phytoremediation, such as phytotransformation, bioremediation of the rhizosphere, phytostabilization, phytoextraction (phytoaccumulation), rhizofiltration, phytovolatilization, phytodegradation, and hydraulic regulation. Plants with extraordinary metal-accumulating capacity in their above- ground portions are known as hyperaccumulator plants (Usman et al., 2018; Verma et al., 2021). These include *Alyssum bertolonii, Thlaspi caerulescens, Calendula officinalis,* and *Tagetes erecta* (Assunçao et al., 2003). Hyperaccumulating plants are recognized as containing more than 1000 mg/kg dry weight of Ni, Cu, and Pb, and 100 mg/kg dry weight of Cd (Hasan et al., 2019). Because plants take time to grow and thrive, it is regarded as a slow process. Remediation and heavy metal extraction by plants are typically limited by the availability of heavy metals in soils (Jiang et al., 2019). It is, therefore, necessary to increase the adequacy of phytoremediation through the technique of heavy metal phytoextraction. One common approach is the use of nanotechnology, which allows the uptake and translocation of heavy metals in the above-ground parts of plants. However, phytoremediation takes characteristics of the specific and selective accumulation capabilities of plant root systems, with regard to organic contaminants, along with the translocation and contaminant dissipation capacities of the plant body (Tangahu et al., 2011). The efficiency of the remediation of organic contaminated soil is affected by the solubility and bioavailability of the contaminants. Phytoremediation is not an effective remedy for soils polluted with highly hydrophobic compounds, which are not easily transported within the plant. These chemicals include polychlorinated biphenyls (PCBs), polycyclic aromatic hydrocarbons (PAHs), petroleum hydrocarbons, radio nucleosides, and explosives with log Kow higher than 3.0. They are bound (and are, therefore, unable to dissipate in the rhizosphere) so strongly to the surface of roots that they are not easily translocated to aerial tissues (Germida et al., 2002). In addition, its strong hydrophobic nature, which results from low solubility, is correlated with soil organic matter and minerals, making it less bioavailable and more recalcitrant (Megharaj et al., 2011). The low water solubility of highly hydrophobic organic contaminants is calculated by the rate of uptake and metabolism, restricts mobility, and bioavailability (Shekhar et al., 2015). For moderately hydrophobic compounds with a log Kow range of 0.5–3.0, plant roots tend to be picked up and join the xylem stream for eventual accumulation or degradation (Chakraborty and Das, 2016). Water soluble chemicals with log Kow lower than 0.5 are not sufficiently absorbed to roots or actively transported through plant membranes (Chirakkara et al., 2016). In order to resolve these issues, plants and other technologies such as the presence of nanotechnologies that enhance soil bioavailability are combined in phytoremediation to promote the removal of such organic contaminants in soils by enhancing their bioavailability, plant uptake, or microbial degradation, thus improving the efficacy of *in situ* phytoremediation.

13.3 NANOTECHNOLOGY IN THE REMEDIATION OF ENVIRONMENTAL SITES

The nano zerovalent iron (nZVI) application will influence metal accumulator plants as an essential part of ecosystems and the phytoremediation process in heavy metal-contaminated sites. Due to its large surface areas, increased number of active surface sites and excellent absorption capabilities, nanomaterials are the promising alternative for polluted soil remediation. nZVI has shown tremendous potential for the reduction of heavy metal ion transformation, which is due to adequate mobility and excellent efficiency (Zhang, 2003; Kirschling et al., 2010). In addition, because of their smaller size and large surface area, nZVI has a good affinity with metals, so they easily penetrate into the contamination zone of metal-challenged ecosystems. Nanoscale nZVI, through its controllable particle size, high reactivity and abundant reactive surface sites, has been successfully used to treat various metal ions in aqueous solutions and stabilize biosolids (Yan et al., 2010; Huang et al., 2013; Zou et al., 2016; Li et al., 2017). Unfortunately, the direct use of nZVI is restricted by the small particle size and powder state, and the intrinsic characteristics of nZVI to react with surrounding media or agglomerate during the preparation process decreases the reactivity of nanoparticles and also results in poor mobility and efficient transfer of nZVI to the polluted area for continuous *in situ* remediation (Grieger et al., 2010; O'Carroll et al., 2013). Furthermore, the use of nZVI to actual polluted soil is too costly to be likely to be a beneficial remediation practice. There are so many questions regarding its use, including possible conflict with other phytotechnologies of remediation and potential risk to both human and environmental ecosystem (Patil et al., 2016). To address this problem, nZVI-based materials, including surface-modified nZVI (SM-nZVI), nZVI-supported porous material, activated carbon, and multi-walled carbon nanotubes, CMC-nZVI, and nZVI-supported inorganic clay minerals (e.g., kaolinite, zeolite, clay, montmorillonite, rectorite, palygorskite, and bentonite) were successfully synthesized as effective adsorbents to isolate the contaminants from the environment (Shu et al., 2010; Kim et al., 2013; Wang et al., 2014; Zou et al., 2016).

13.4 REMEDIATION OF HEAVY METAL CONTAMINATED SITES BY NANO-PHYTOREMEDIATION

Integrating phytoremediation with nanotechnology provides a better solution for heavy metal remediation. The integration strategy of phytoremediation and nZVI (nano-phytoremediation) is a new emerging technology for remediation of pollutants (Gil-Diaz et al., 2016; Gong et al., 2017), where nZVI is able to reduce the bioavailability of heavy metals in the soil and improve the growth of plants (Tafazoli et al., 2017). The usage of nanoparticles can also enhance the stress tolerance of plants in different conditions due to improving phytoremediation potential (Pillai and Kottekottil, 2016; Souri et al., 2017). The amendment of phytoremediation with nanoparticles include an important step in the progress of soil decontamination. Several authors studied the combination between nanotechnology and phytotechnology for detoxification or remediation the organic, inorganic, and heavy metal pollutants in contaminated environments. Harikumar et al., (2019) showed that both phytoremediation and nano-phytoremediation (nZVI) resulted in phytostabilization of Pb in contaminated soil. Gil-Dıaz et al., (2016) found that the use of 10% nZVI significantly lowered the accumulate and uptake of As in roots and shoots plant, respectively. Also, the height and dry weight of barley plants under metal stress has increased, while Singh and Lee (2016) reported that a usage of nano-TiO_2 can improve Cd uptake from 128.5 to 507.6 μg/plant due to addition of 100–300 mg/kg TiO_2 NPs to soil and minimize Cd stress in soybean plants. Since nano-TiO_2 particles are small (<5 nm) in size and are able to form a covalent bond with most nonconjugated forms of normal living matter and translocations into the roots and shoot tissues of plants upon a specific distribution (Aslani et al., 2014). In addition, nano-TiO_2 application restricts Cd toxicity as a result of an increase in the rate of photosynthetic rate and growth rates of plants (Singh and Lee, 2016). The decrease concentrations of nZVI (100–500 mg/kg) could effectively provide high accumulation capacity for Pb with the increase of biomass, and also promoted the plant growth indicated by the lower oxidative stress in plants. Moreover, the application of nZVI-treated sediment reduced the toxicity of heavy metal, and this was due to the low acidity lower acid soluble fraction and higher residual fraction of Pb (Huang et al., 2018). Also, Huang et al. (2018) showed a stimulatory response of the nZVI in low doses in integration with phytoremediation in Pb-contaminated soil by ryegrass (*Lolium perenne*) plant. Using nano-hydroxyapatite (0.2% w/w) has been reported effective in promoting Pb phytoextraction efficiency by ryegrass. With the assistance of nano-hydroxyapatite, ryegrass removed over 30% of Pb in the soil after 1 month and 44.39% after 3 months (Liang et al., 2017). Some nanomaterials have been demonstrated to enhance the phytoextraction of Cd in soil. Singh and Lee (2016) reported the favorable response of TiO_2 nanoparticles on Cd accumulation in soybean plants. Studies showed that the phytoremediation of soil contaminated with cadmium, chromium, lead, nickel, and zinc could be enhanced by applying nanomaterials (Tripathi et al., 2015; Singh and Lee, 2016; Liang., et al., 2017; Vítková et al., 2018), for example, application of low to moderate concentrations of nZVI (100–500 mg/kg) enhanced Pb accumulation capacity in roots and shoots of *Kochia scoparia* (Daryabeigi., et al., 2020). The efficiency of *Trifolium repens* substantially improved using nZVI for Sb with the greatest accumulation capacity of 3896.4 μg per pot gained in the "PGPR+500 mg/kg nZVI" treatment (Zand et al., 2020).

13.5 INTERACTION MECHANISM BETWEEN NZVI-BASED MATERIALS, POLLUTANTS, AND PLANTS IN PHYTOREMEDIATION

The effect of nZVI on phytoremediation may be an important issue, and alternatives are required to either mitigate its potential negative effects on phytoremediation or accelerate its positive effects (Mokarram-Kashtiban et al., 2019). Nanoparticles suitable for phytoremediation should be nontoxic for plants; increase germination, growth of seedlings, root–shoot elongation, plant height, and biomass; increase plant development of phytoenzymes; improve plant growth hormones and contaminant binding capabilities; and increase bioavailability for plant and enhanced phytoremediation technology. In plants (Srivastav et al., 2018), the uptake of nanoparticles is principally due to their size, form, and chemical composition. Size is the nanoparticle's key factor in entering the plants and moving from the roots to other areas of the plants (Ma et al., 2010; Cornelis et al., 2014). Plants can accumulate nanomaterials by mycorrhizal fungi (Whiteside et al., 2009) or by osmotic pressure (e.g., through pores in cell walls), by intercellular plasmodesmata, or by symplastic uptake, linked to their particular uptake (Lin et al., 2009). One hypothesis is that, by making holes and permeating into the cells through these holes, ZnO nanoparticles might be able to increase the permeability of plant cell walls and move them through plasmodesmata between cells (Lin and Xing, 2008). In addition, nanoparticles will be strongly adsorbed in the roots, thereby exerting effects—even without actually being taken up as particles (Navarro et al., 2012). The nano-phytoremediation technique for the remediation of polluted soil is based on the application of plant-based nanoparticles (NP type, dosage, and speciation) and methods of phytoremediation (e.g., phytostabilization, phytoextraction, and phytodegradation). The different plant species and varieties of nanoparticles are used in contaminant nano-phytoremediation experiments (Srivastav et al., 2018). The mechanism of interaction for pollutants with materials based on nZVI is debatable. Heavy metal ions express different physicochemical properties and interactions with nZVI-based materials in different adsorption processes by physical and chemical reactions, including adsorption, redox, aggregation, ion exchange, hydroxylation, and subsequent precipitation. The interaction mechanism could be considered the phase of adsorption, reduction, and oxidation (Zou et al., 2016). The smaller size of nanomaterials allows them to penetrate various barriers in the plant systems, helping them enter particular locations, while their additional surface area helps to increase their adsorption and deliver the desired substances in a targeted manner (Kashyap et al., 2015). The positive effect of a low nZVI dose may be due to iron entry into the soil, which is an important nutrient for the growth of plants and soil microorganisms (Souza et al., 2015). Nanomaterials can enhance phytoremediation by direct pollutant removal through adsorption or redox reactions, or by promoting plant growth—such as inoculating plant growth promoting rhizobacteria (PGPR), applying plant growth regulators and using transgenic plants, and increasing the phyto-availability (Mueller and Nowack, 2010; Yadu et al., 2018; Song et al., 2019) (Fig. 13.1). In addition, many studies apply nZVI for reductive dechlorination of chlorinated organic pollutants and nZVI can also act by adsorbing and co-precipitating inorganic ions (Li et al., 2018). By adsorption through carbon nanotubes, contaminants can be immobilized. This is similar to phytostabilization, as carbon nanotubes can stabilize organic contaminants through electrostatic attraction, hydrophobic interaction, and p–p bonding interactions, while complexation, electrostatic attraction, physical adsorption, and surface precipitation are involved in the interactions between carbon nanotubes and heavy metals (Song et al., 2018). Nanomaterials—such as graphene quantum dots, carbon nanotubes, Ag nanoparticles, ZnO nanoparticles, nZVI particles, and nanoparticle upconversion, could improve plant growth. The mechanisms of these nanomaterials in enhancing plant growth are different. Carbon nanotubes, for example, may enable the reproductive system of plants, contributing to improved tomato growth (Khodakovskaya et al., 2013), while graphene quantum dots might serve as nanofertilizers and pesticides to improve the growth rates of *Coriandrum sativum* and *Allium sativum* (Chakravarty et al., 2015). Furthermore, ZnO nanoparticles can increase the plant tolerance by regulating the gene expression of enzymes (Praveen et al., 2018). Applied nano-hydroxyapatite by Jin et al. (2016) led to an increase in phosphorus concentration in soil resulting in improved efficiency of lead phytoremediation by ryegrass. As Cd uptake increased with TIO_2 nanoparticles, the authors suggested a potential mechanism for

TiO_2 nanoparticles to join chloroplasts and speed up light adaptation and electron transfer (Singh and Lee, 2016). In addition, Timmusk et al. (2018) concluded that TiO_2 nanoparticles could enhance the performance of PGPR phytoavailability of pollutants as a key factor affecting the_efficiency of phytoremediation, especially for phytoextraction. For example, lead usually exists in insoluble forms in soil due to adsorption, complexation, and precipitation, which makes it difficult for phytoextraction (Zaier et al., 2014). Several techniques, including agronomic management (e.g., fertilization), treatment with chemical additives (e.g., a chelating agent), inoculation of rhizospheric microorganisms, and the use of genetic engineering—have also been suggested to improve the phytoavailability of contaminants (Glick, 2010; Habiba et al., 2015; Jacobs, et al., 2018). Nanomaterials may serve as a carrier of pollutants when they penetrate the cell, thereby increasing the bioavailability (Su et al., 2013). On the one hand, adsorption of pollutants into nanomaterials outside organisms may reduce free pollutants, thereby decreasing the bioavailability (Glomstad et al., 2016).

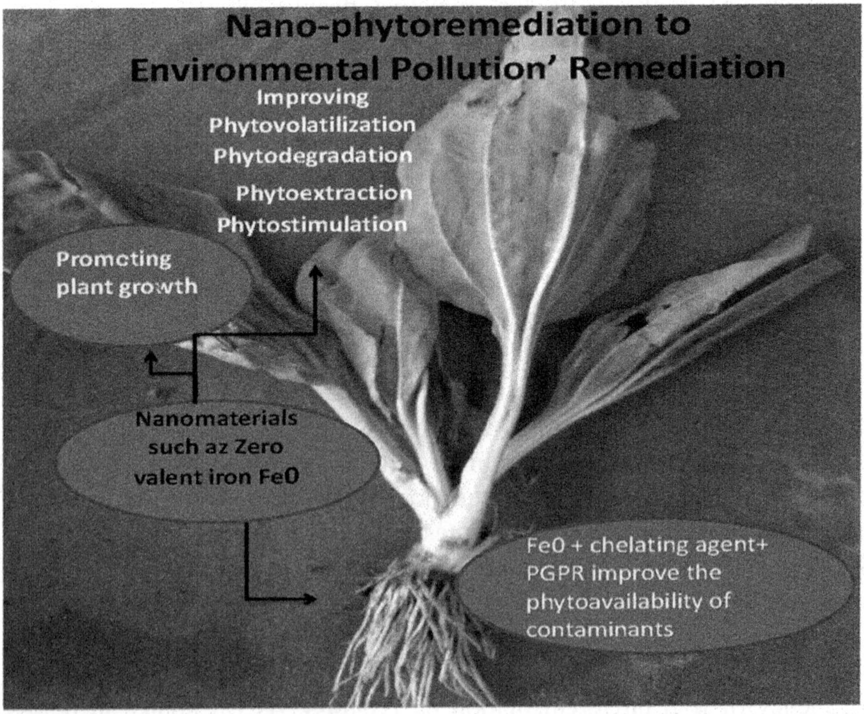

FIGURE 13.1 Nano-phytoremediation to environmental remediation of contaminants.

13.6 INTEGRATION OF NANO ZEROVALENT IRON (NZVI OR FE^0) AND OTHER TECHNOLOGIES FOR ENHANCING PHYTOREMEDIATION OF HEAVY METALS

The addition of organic acids (i.e., citric, malic, tartaric, and oxalic acids), organic amendments, and nZVI with phytoremediation are suitable practices for the uptake of heavy metal in polluted soils, because of increasing reactive surface locations and higher nZVI volume to the ratio of surface area (Singh et al., 2012), which are associated with more metal ions and increasing availability of heavy metals for phytoremediation process (Lacalle et al., 2018). The efficiency of *Brassica napus* L. enhanced with an organic amendment and nZVI for the uptake of soils polluted with both organic (diesel) and inorganic pollutants (Zn, Cu, and Cd) were tested by Cao et al., 2018. Data showed that organic amendment contributed into diesel degradation and improved the health and biomass of *B. napus*. This nZVI was ineffectual in soil remediation without inducing any toxicity. The combination of nZVI nanoparticles with phytoremediation in improving the performance of sunflower plants reduced Cr uptake in the polluted soil, followed by increased activity of cell detoxification enzymes, SOD, CAT, POX, and APX (Mohammadi et al., 2020). Recent studies demonstrated the beneficial effects of nZVI and rhizosphere microorganisms on the efficiency of phytoremediation on heavy metal removal. The combination of bio-amendments with nZVI remediation techniques increased growth parameters, gas exchange, and useful effects of nZVI and minimized its toxicity. This led to new insight into Fe^0 heavy metal remediation (He and Yang 2007; Srivastav et al., 2018; Mokarram-Kashtiban et al., 2019). The inoculation by PGPR, especially *Pseudomonas fluorescens* and arbuscular mycorrhizal fungus (AMF), especially *Rhizophagus irregularis* in the combination with nZVI enhanced plant growth, physiological and biochemical parameters of white willow (*Salix alba* L.). This also increased the bioconcentration factor (BCF) of Pb, Cu, and Cd. To gain a better understanding of the possible mechanism of this phenomenon, further studies are needed (Mokarram-Kashtiban et al., 2019). The rhizosphere microorganisms decreased nZVI stress in plants under high dose conditions of nZVI, (Mokarram-Kashtiban et al., 2019). The phytoremediation potential of *Sorghum bicolor's* for antimony (Sb) removal from the soils was decreased by co-application of TiO_2 NPs and biochar (BC). Using TiO_2, NPs significantly improved the accumulation ability of *S. bicolor* for Sb with the highest accumulation efficiency of 1624.1 μg per pot achieved in "250 mg/kg TiO_2 NP + 2.5% BC" treatment (Song et al., 2019). The use of multi-walled carbon mnanotubes (MWCNTs) at acceptable levels could increase the potential of phytoremediation in the restoration of heavy metal contaminated river sediments, and the inevitable release of MWCNTs at high contents would exacerbate metal-induced toxicity to plants (Gong et al., 2019).

13.7 IMPROVING PHYTOREMEDIATION OF ORGANIC POLLUTANTS IN SOIL AND WATER BY NANOTECHNOLOGY

One of the biggest solutions for hydrophobic organic pollutant remediation is provided by the integration of phytoremediation with nanotechnology (phytonanotechnology) (Table 13.1). Nanomaterials are involved in photocatalysis or oxidation, reduction, hydrolysis, or organic contaminant elimination reactions (Tiwari et al., 2008; Jassal et al., 2016). Through various mechanisms such as redox reactions, surface processes, adsorption, ion exchange, surface complexation, and electrostatic interaction, the nanoparticles are able to adsorb and facilitate pollutant degradation (Trujillo-Reyes et al., 2014; Medina-Pérez et al., 2019). Organic pollutants such as chlorpyrifos, molinate, atrazine, lindane, pentachlorophenol, trichloroethylene (TCE), pyrene, PCBs, 2,4-dinitrotoluene, and ibuprofen in polluted soil environment were quickly degraded by nZVI, magnetite nanoparticles (nFe_3O_4), and bimetallic nanoparticles (Pd/Fe) (Zhang et al., 2011; Singh et al., 2012; Gomes et al., 2014; Srivastav et al., 2018; Verma et al., 2021). Excellent photocatalysts with advanced features (size, morphology, and high adsorption capacity) are found to be nanomaterials of TiO_2, ZnO as well as several metal oxides and sulfides and appear as the most emerging destructive technology (Rani and Shanker, 2018). Nanomaterials make electron-hole pairs through the exposure of light sources, which move to the semiconductor surface and degrade organic pollutants via photo-oxidation into nontoxic products (Lavand and Malghe, 2015). However, these methods are usually expensive and require special, high-energy equipment and contain hazardous, corrosive and flammable chemical substances such as sodium borohydride and hydrazine hydrate or organic solvents as reducing agents that cause unwanted adverse environmental effects. In addition, nanoparticles appear to aggregate if not properly coated, leading to decreased reactivity and stability of these nanoparticles (Shahwan et al., 2011; O'Carroll et al., 2013). Environmental applications of nZVI (Fe^0) have been widely adopted by several users and government agencies, mainly because of the low costs and nontoxicity of iron (Rani and Shanker, 2018). Several consumers and government agencies have widely embraced environmental applications of nZVI (Fe^0), primarily due to the low cost and nontoxicity of iron (Rani and Shanker, 2018). Another catalyst most studied is nZVI, which reduced

TABLE 13.1 Using Nano-Phytoremediation for Remediating Organic Pollutants in Water and Soil.

Pollutants	Control (without any treatment)	Pytoremediation	Nano material	Nano-phytoremediation	References
Organic pollutants (mg/kg)					
Trinitrotoluene (soil) 120 days		*Panicum maximum* Jacq.	nZVI	Zerovalent iron nanoparticles enhanced the removal efficiency of TNT from 85.7 to 100% after 120 days	Jiamjitrpanich et al. (2012)
Endosulfan removal (%)	7.61–20.13 (7–28 days)	*Alpinia calcarata* 51.74–81.20 (7–28 days) *Ocimum sanctum* 8.25–20.76 (7–28 days) *Cymbopogon citratus* 4.81–65.08 (7–28 days)	17.29–55.09	*Alpinia calcarata* 82.20–94.92(7–28 days) *Ocimum sanctum* 10.91–76.28 (7–28 days) *Cymbopogon citratus* 62.53–86.16 (7–28 days)	Pillai, and Kottekottil (2016)
Fipronil (water) removal (%) through 144 h	39.71		Brassica-AgNps (98.21) Ipomoea-AgNps (95.93) Camellia-AgNps (82.44) Plantago-AgNps (89.20)		Romeh (2018)
Fipronil (flooded soil) removal (%) through 144 h	3.6	13.76		*Plantago major* + Brassica-AgNps (82.56) Ipomoea-AgNps (57.51) Camellia-AgNps (44.75) Plantago-AgNps (68.41)	Romeh (2018)
Chlorpyrifos (water) removal (%) 24 h	17.88	43.76	F–Fe^0 (81.69)	F–Fe^0 supported on (wheat bran, cement kiln dust, rice straw ash) +*Plantago major* nearly 100%	Romeh (2020)

TABLE 13.1 *(Continued)*

Pollutants	Control (without any treatment)	Pytoremediation	Nano material	Nano-phytoremediation	References
Chlorfenapyr (water) removal (%) 24 h	6.17	69.27	F–Fe^0 Ach (86.0) Ip–Ag^0 Ach (79.70) Br–Ag^0 Ach (79.70)	*P. major* plus F–Fe^0 Ach(93.7) *P. major* plus Ip–Ag^0 Ach(91.30) *P. major* plus Br–Ag^0 Ach (92.92)	Romeh and Saber (2020)
Chlorfenapyr (soil) removal (%) 24 h	12.40	25.83		*P. major* plus F–Fe^0 Ach (71.22) *P. major* plus Ip–Ag^0 Ach (57.32) *P. major* plus Br–Ag^0 Ach (73.10)	Romeh and Saber (2020)
Heavy metals (mg/Kg) dw					
Cr		*Pisum sativum* root (1472.6) *P sativum* shoot (62.5)	Silicon nanoparticles (SiNp)	SiNp + *Pisum sativum* root (516.6) SiNp + *P sativum* shoot (35.2)	Tripathi et al. (2015)
Cd		*Glycine max* (root and shoot) at 100 cd (131.9) *Glycine max* shoot (28.0) *Glycine max* root (448.1)	nano-TiO_2	At 100 cd + 300 TiO_2 (1534.7) *Glycine max* shoot + 300 TiO_2 (73.2) *Glycine max* root + 300 TiO_2 (1461.5)	Singh and Lee (2016)
As		*Hordeum vulgare*	nZVI	*Hordeum vulgare* shoots + 1% nZVI (as reduced by 66%) *Hordeum vulgare* roots + 10% nZVI (as reduced by 97%)	Gil-Díaz et al. (2016)
As		*Lolium perenne*	nZVI	*Lolium perenne* shoots + nZVI (as reduced by 62%) *Lolium perenne* roots + nZVI (as reduced by 75%)	Vítková et al. (2018)

TABLE 13.1 *(Continued)*

Pollutants	Control (without any treatment)	Pytoremediation	Nano material	Nano-phytoremediation	References
Pb		*Lolium perenne*	nZVI	*Lolium perenne* (total) + nZVI 100 (assist the phytoremediation of Pb-polluted sediment)	Huang et al. (2018)
Pb		*Alternanthera dentata* 35.00 *Wedelia cheninsis* 27.41	nZVI	*Alternanthera dentata* + nZVI (38.80) *Wedelia cheninsis* (Total) + nZVI (39.13)	Harikumar et al. (2019)
Pb		*Kochia scoparia* (more than 350)	nZVI	*Kochia scoparia* (total) + nZVI 500 (857.18)	Daryabeigi et al. (2020)
Sb		*Trifolium repens*	nZVI +PGPR	*Trifolium repens*+ PGPR + 500 mg/kg nZVI (greatest accumulation capacity of 3896.4 µg per pot)	Zand et al. (2020)

almost all halogenated hydrocarbons to benign hydrocarbons within 24 h by rapid and full dechlorination (Shih et al., 2011). The degree of dechlorination of lindane was greater (95%) for the nZVI (60 nm) (Elliott et al., 2009). A mixture of nanoparticles and plant species (nanophytoremediation) has removed essential contaminants from contaminated soil. Plant species have increased the removal of toxins from contaminated soil, which takes less time than normal. Different plant species and varieties of nanoparticles are used in nano-phytoremediation studies for water and soil remediation technologies (Srivastav et al., 2018). Nano-phytoremediation was more successful than either phytoremediation or nano-remediation for TNT-contaminated soil degradation (Jiamjitrpanich et al., 2012; Verma et al., 2021). In the presence of nZVI, the phytoremediation of endosulfan-contaminated soil with three plant species, *Alpinia calcarata, Ocimum sanctum, and Cymbopogon citratus* increased from 81.2 to 100%, from 20.76 to 76.28%, and from 65.08 to 86.16%, respectively. Small amounts of endosulfan have accumulated in the plants due to decrease endosulfan dichlorination (Pillai and Kottekottil, 2016). Silver nanoparticles (AgNps) and *Plantago major* also play an substantial role in the remediation of fipronil-contaminated water and flooded soil, while *P. major* played an significant role in the uptake of polar break product, fipronil-amide (Romeh, 2018). Silver nanoparticle significantly (>90%) increased the development of ABA and GA phytohormones in plants, whih allows plants to withstand stresses and increase the uptake of nutrients and water to enhance growth (Khan and Bano, 2016). Although nZVI has many benefits, but its efficiency decreases due to the formation of oxide layers that block its surface active sites, its efficiency decreases over time, expensive and requires special equipment, high energy and includes toxic, corrosive, and flammable chemical substances such as sodium borohydride and hydrazine hydrate or organic solvents as reducing agents that cause unnecessary adverse environmental impacts (Bardos et al., 2011; O'Carroll et al., 2013). Researchers are, therefore, continuing their efforts to develop simple, effective, and reliable green chemistry processes to solve this problem by developing nanomaterials. In order to make the synthesis process green, researchers worldwide are currently working on the synthesis of nanoparticles using sunlight, or plant-based surfactants or microorganisms, and water or a mixture of them. Nanomaterials generated through the green route were found to be cheap and efficient catalysts for environmental remediation (Rani and Shanker, 2018). In particular, little research focuses on water and soil remediation technologies with the combined application of green nano-phytoremediation. Green nanotechnology can provide green and eco-friendly alternatives for environmental cleanup and management. Appropriate plant species and nanomaterials need to be selected for the absorption of pollutants, along with agronomic management optimization for the high-resolution cleanup process. The integration of green nanotechnology such as Ficus iron nanoparticles (F–Fe^0), ipomoea–silver (Ip–Ag^0) and brassica–silver nanoparticles (Br–Ag^0) and phytoremediation technology using *P. major* has played a major role in the remediation of soil and water polluted with chlorfenapyr (Romeh and Saber, 2020). Nanomaterials can improve phytoremediation by directly acting on the contaminants and plants, promoting plant growth by promoting enzymatic activity in plants, increasing chlorophyll content, regulating nutrient release in the soil system, and increasing the phytoavailability of pollutants (Varma and Khanuja, 2017; Subramanian et al., 2015; Song et al., 2019) or by indirectly influencing the final efficiency of the remediation (Zand et al., 2020). Adsorbents assisted b-y F–Fe^0 in particular wheat bran (WB) as a green nanotechnology and *P. major* as phytoremediation, would play a significant role in the complete removal of chlorpyrifos from water with a significant reduction in the toxic degradation product TCP (Romeh, 2020). Also, *P. major* plus marjoram-prepared nZVI (Mar-nZVI) and moringa-prepared nZVI (Mor-nZVI) aided by activated charcoal (Ach), and bentonite (Bent) play a principal role in the remediation of flonicamid-contaminated water (Rady et al., 2020). For nZVI-reduced graphene oxide (rGO), the removal rate and adsorption capability were found to be the maximum, which shows that this adsorbent could be used possible futuristic adsorbent for explosive removal (Khurana et al., 2018).

13.8 CONCLUSION

Compared to conventional organic contaminants, heavy metal ions are difficult to degrade into cleaning materials. Several traditional and modern techniques are using to decontaminate the environment from these contaminants. Most of these methods, however, have several drawbacks, such as high costs, long response times, poor degradation efficiency, poor performance, and ineffective removal of toxic heavy metals and organic persistent from agricultural land. As discussed earlier, the technology of nano-phytoremediation is comparatively a new field for the remediation of ecosystems polluted with heavy metals and hydrophobic organic pollutants. The effect of nZVI on phytoremediation may be an important issue, and alternatives are required to mitigate its potential negative effects on phytoremediation and to accelerate its positive effects. Integration of organic acids (i.e., citric, malic, tartaric, and oxalic acids), organic amendments, and nZVI with phytoremediation are feasible practices for the repair of metal-polluted soils, which can be because of the resulting increase in reactive surface locations, higher nZVI volume to the surface area, which might bind to more metal ions and enhancing the availability of heavy metals for phytoremediation. It can be concluded that the selective interaction of plants and nZVI has great application prospects in the context of soil remediation.

KEYWORDS

- **phytoremediation**
- **nanomaterials**
- **nano-phytoremediation**
- **heavy metals**
- **organic pollutants**

REFERENCES

Agnello, A. C.; Bagard, M.; van Hullebusch, E. D.; Esposito, G.; Huguenot, D. Comparative Bioremediation of Heavy Metals and Petroleum Hydrocarbons Co-Contaminated Soil by Natural Attenuation, Phytoremediation, Bioaugmentation and Bioaugmentation- Assisted Phytoremediation. *Sci. Total Environ.* **2016,** *563*, 693–703.

Ali, S., Abbas, Z., Rizwan, M., Zaheer, I. E., Yavaş, İ., Ünay, A., Kalderis, D. Application of Floating Aquatic Plants in Phytoremediation of Heavy Metals Polluted Water: A Review. *Sustainability* **2020,** *12* (5), 1927.

Ashraf, M. A.; Hussain, I.; Rasheed, R.; Iqbal, M.; Riaz, M.; Arif, M. S. Advances in Microbe-Assisted Reclamation of Heavy Metal Contaminated Soils over the Last Decade: A Review. *J. Environ. Manag.* **2017,** *198*, 132–143.

Ashraf, S.; Ali, Q.; Zahir, Z. A.; Ashraf, S.; Asghar, H. N. Phytoremediation: Environmentally Sustainable Way for Reclamation of Heavy Metal Polluted Soils. *Ecotoxicol. Environ. Saf.* **2019,** *174*, 714–727.

Aslani, F.; Bagheri, S.; Muhd Julkapli, N.; Juraimi, A. S.; Hashemi, F. S. G.; Baghdadi, A. Effects of Engineered Nanomaterials on Plants Growth: An Overview. *Sci. World J.* **2014,** *2014*.

Assunçao, A.; Schat, H.; Aarts, M. *Thlaspi caerulescens*, an Attractive Model Species to Study Heavy Metal Hyperaccumulation in Plants. *New Phytol.* **2003,** *159*, 351–360.

Azarudeen, R. S.; Subha, R.; Jeyakumar, D.; Burkanudeen, A. R. Batch Separation Studies for the Removal of Heavy Metal Ions Using a Chelating Terpolymer: Synthesis, Characterization and Isotherm Models. *Sep. Purif. Technol.* **2013,** *116*, 366–377.

Bao, T.; Jin, J.; Damtie, M. M.; Wu, K.; Yu, Z. M.; Wang, L.; Frost, R. L. Green Synthesis and Application of Nanoscale Zero-Valent Iron/Rectorite Composite Material for P-Chlorophenol Degradation via Heterogeneous Fenton Reaction. *J. Saudi Chem. Soc.* **2019,** *23* (7), 864–878.

Bardos, P.; Bone, B.; Elliott, D.; Hartog, N.; Henstock, J.; Nathanail, P. A Risk/Benefit Approach to the Application of Iron Nanoparticles for the Remediation of Contaminated Sites in the Environment. *Dep. Environ. Food Rural Aff.* **2011**.

Barrutia, O.; Garbisu, C.; Epelde, L.; Sampedro, M. C.; Goicolea, M. A.; Becerril, J. M. Plant Tolerance to Diesel Minimizes Its Impact on Soil Microbial Characteristics during Rhizoremediation of Diesel-Contaminated Soils. *Sci. Total Environ.* **2011,** *409*, 4087–4093.

Bouhamed, F.; Elouear, Z.; Bouzid, J. Adsorptive Removal of Copper(II) from Aqueous Solutions on Activated Carbon Prepared from Tunisian Date Stones: Equilibrium, Kinetics and Thermodynamics. *J. Taiwan Inst. Chem. Eng.* **2012,** *43*, 741−749.

Cao, Y.; Zhang, S.; Zhong, Q.; Wang, G.; Xu, X.; Li, T.; Li, Y. Feasibility of Nanoscale Zero-Valent Iron to Enhance The Removal Efficiencies of Heavy Metals from Polluted Soils by Organic Acids. *Ecotoxicol. Environ. Safety* **2018,** *162*, 464–473.

Chakraborty, J.; Das, S. Molecular Perspectives and Recent Advances in Microbial Remediation of Persistent Organic Pollutants. *Environ. Sci. Pollut. Res.* **2016,** *23* (17), 16883–16903.

Chaturvedi, R.; Varun, M.; Paul, M. S. Phytoremediation: Uptake and Role of Metal Transporters in Some Members of Brassicaceae. In *Phytoremediation*; Springer International Publishing: Cham, Switzerland, 2016; pp 453–468.

Chirakkara, R. A.; Cameselle, C.; Reddy, K. R. Assessing the Applicability of Phytoremediation of Soils with Mixed Organic and Heavy Metal Contaminants. *Rev. Environ. Sci. Bio/Technol.* **2016,** *15* (2), 299–326.

Cornelis, G.; Hund-Rinke, K.; Kuhlbusch, T.; Van den Brink, N.; Nickel, C. Fate and Bioavailability of Engineered Nanoparticles in Soils: A Review. *Crit. Rev. Environm. Sci. Technol.* **2014,** *44* (24), 2720–2764.

Daryabeigi Zand, A.; Mikaeili Tabrizi, A.; Zand, A. D.; Tabrizi, A. M. Effect of Zero-Valent Iron Nanoparticles on the Phytoextraction Ability of Kochia Scoparia and Its Response in Pb Contaminated Soil. *Environ. Eng. Res.* **2020,** *26* (4).

Elliott D. W.; Lien, H. L.; Zhang, W. X. Degradation of Lindane by Zero-Valent Iron Nanoparticles. *J. Environ. Eng.* **2009,** *135*, 317–324.

Fabbricino, M.; Ferraro, A.; Luongo, V.; Pontoni, L.; Race, M. Soil Washing Optimization, Recycling of the Solution, and Ecotoxicity Assessment for the Remediation of Pb-Contaminated Sites Using EDDS. *Sustainability* **2018,** *10* (3), 636.

Feller, U.; Anders, I.; Wei, S. Distribution and Redistribution of 109Cd and 65Zn in the Heavy Metal Hyperaccumulator *Solanum nigrum* L.: Influence of Cadmium and Zinc Concentrations in the Root Medium. *Plants* **2019,** *8* (9), 340.

Germida, J. J.; Frick, C. M.; Farrell, R. E. Phytoremediation of Oil-Contaminated Soils. *Dev. Soil Sci.* **2002,** *28*, 169–186.

Gil-Díaz, M.; Diez-Pascual, S.; González, A., Alonso, J.; Rodríguez-Valdés, E.; Gallego, J. R.; Lobo, M. C. A Nanoremediation Strategy for the Recovery of an As-Polluted Soil. *Chemosphere* **2016,** *149*, 137–145.

Glick, B. R. Using Soil Bacteria to Facilitate Phytoremediation. *Biotechnol. Adv* **2010,** *28* (3), 367–374.

Glomstad, B.; Altin, D.; Sørensen, L.; Liu, J.; Jenssen, B. M.; Booth, A. M. Carbon Nanotube Properties Influence Adsorption of Phenanthrene and Subsequent Bioavailability and Toxicity to *Pseudokirchneriella subcapitata. Environ. Sci. Technol.* **2016,** *50* (5), 2660–2668.

Gomes, H. I.; Fan, G.; Mateus, E. P.; Dias-Ferreira, C. Ribeiro, A. B. Assessment of Combined Electro–Nanoremediation of Molinate Contaminated Soil. *Sci. Total Environ.* **2014,** *493*, 178–184.

Gong, X.; Huang, D.; Liu, Y.; Zeng, G.; Wang, R.; Wan, J.; Xue, W. Stabilized Nanoscale Zerovalent Iron Mediated Cadmium Accumulation and Oxidative Damage of *Boehmeria nivea* (L.) Gaudich Cultivated in Cadmium Contaminated Sediments. *Environ. Sci. Technol.* **2017,** *51* (19), 11308–11316.

Gong, X.; Huang, D.; Liu, Y.; Zeng, G.; Wang, R.; Xu, P.; Chen, S. Roles of Multiwall Carbon Nanotubes in Phytoremediation: Cadmium Uptake and Oxidative Burst in *Boehmeria nivea* (L.) Gaudich. *Environ. Sci. Nano* **2019,** *6* (3), 851–862.

Grieger, K. D.; Fjordboge, A.; Hartmann, N. B.; Eriksson, E.; Bjerg, P. L.; Baun, A. Environmental Benefits and Risks of Zero-Valent Iron Nanoparticles (nZVI) for in Situ Remediation: Risk Mitigation or Trade-Off? *J. Contam. Hydrol.* **2010,** *118*, 165−183.

Habiba, U.; Ali, S.; Farid, M.; Shakoor, M. B.; Rizwan, M.; Ibrahim, M.; Ali, B. EDTA Enhanced Plant Growth, Antioxidant Defense System, and Phytoextraction of Copper by *Brassica napus* L. *Environ. Sci. Pollut. Res.* **2015,** *22* (2), 1534–1544.

Harikumar, P. S.; Lamya, T. V.; Shalna,T. Enhanced Phytoremediation Efficiency of Lead Contaminated Soil by Zero Valent Nano Iron. *Int. J. Innov. Eng. Technol.***2019,** *12* (2), 44–50.

Hasan, M.; Uddin, M.; Ara-Sharmeen, I. F.; Alharby, H.; Alzahrani, Y.; Hakeem, K. R.; Zhang, L. Assisting Phytoremediation of Heavy Metals Using Chemical Amendments. *Plants* **2019,** *8* (9), 295.

He, Z. L.; Yang, X. E. Role of Soil Rhizobacteria in Phytoremediation of Heavy Metal Contaminated Soils. *J. Zhejiang Univ. Sci. B* **2007,** *8* (3), 192–207.

Hu, J.; Yang, S. T.; Wang, X. K. Adsorption of Cu(II) on β- Cyclodextrin Modified Multiwall Carbon Nanotube/Iron Oxides in the Absence/Presence of Fulvic Acid. *J. Chem. Technol. Biotechnol.* **2012,** *87*, 673−681.

Huang, D.; Qin, X., Peng, Z.; Liu, Y.; Gong, X.; Zeng, G.; Hu, Z. Nanoscale Zero-Valent Iron Assisted Phytoremediation of Pb in Sediment: Impacts on Metal Accumulation and Antioxidative System of Lolium Perenne. *Ecotoxicol. Environ. Saf.* **2018,** *153*, 229–237.

Huang, P. P.; Ye, Z. F.; Xie, W. M.; Chen, Q.; Li, J.; Xu, Z. C.; Yao, M. S. Rapid Magnetic Removal of Aqueous Heavy Metals and Their Relevant Mechanisms Using Nanoscale Zero Valent Iron (nZVI) Particles. *Water Res.* **2013,** *47*, 4050−4058.

Jacobs, A.; De Brabandere, L.; Drouet, T.; Sterckeman, T.; Noret, N. Phytoextraction of Cd and Zn with *Noccaea caerulescens* for Urban Soil Remediation: Influence of Nitrogen Fertilization and Planting Density. *Ecol. Eng.* **2018,** *116*, 178–187.

Jassal, V.; Shanker, U.; Kaith, B. S. Aegle Marmelos Mediated Green Synthesis of Different Nanostructured Metal Hexacyanoferrates: Activity Against Photodegradation of Harmful Organic Dyes. *Scientifica* **2016,** 1–13.

Jiamjitrpanich, W.; Parkpian, P.; Polprasert, C.; Kosanlavit, R. Enhanced Phytoremediation Efficiency of TNT-Contaminated Soil by Nanoscale Zero Valent Iron. *2nd Int. Conf. Environ. Industr. Innov. IPCBEE*, **2012,** *35*, 82-86.

Jiang, M.; Liu, S.; Li, Y.; Li, X.; Luo, Z.; Song, H.; Chen, Q. EDTA-Facilitated Toxic Tolerance, Absorption and Translocation and Phytoremediation of Lead by Dwarf Bamboos. *Ecotoxicol. Environ. Saf.* **2019,** *170*, 502–512.

Jin, Y.; Liu, W.; Li, X. L.; Shen, S. G.; Liang, S. X.; Liu, C.; Shan, L. Nano-Hydroxyapatite Immobilized Lead and Enhanced Plant Growth of Ryegrass in a Contaminated Soil. *Ecol. Eng.* **2016,** *95*, 25–29.

Kashyap, P. L.; Xiang, X.; Heiden, P. Chitosan Nanoparticle Based Delivery Systems for Sustainable Agriculture. *Int. J. Biol. Macromol.* **2015,** *77*, 36–51.

Khan, A. B.; Kathi, S. Evaluation of Heavy Metal and Total Petroleum Hydrocarbon Contamination of Roadside Surface Soil. *Int. J. Environ. Sci. Technol.* **2014,** *11*, 2259–2270.

Khan, N., Bano, A. Modulation of Phytoremediation and Plant Growth by the Treatment with PGPR, Ag Nanoparticle and Untreated Municipal Wastewater. *Int. J. Phytoremed.* **2016,** *18* (12), 1258–1269.

Khodakovskaya, M. V.; Kim, B. S.; Kim, J. N.; Alimohammadi, M.; Dervishi, E.; Mustafa, T., Cernigla, C. E. Carbon Nanotubes as Plant Growth Regulators: Effects on Tomato Growth, Reproductive System, and Soil Microbial Community. *Small* **2013,** *9* (1), 115–123.

Khurana, I.; Shaw, A. K.; Saxena, A.; Khurana, J. M.; Rai, P. K. Removal of Trinitrotoluene with Nano Zerovalent Iron Impregnated Graphene Oxide. *Water Air Soil Pollut.* **2018,** *229* (1), 17.

Kim, S. A.; Kannan, S. K.; Lee, K. J.; Park, Y. J.; Shea, P. J.; Lee, W. H.; Kim, H. M.; Oh, B. T. Removal of Pb(II) from Aqueous Solution by a Zeolite-Nanoscale Zero-Valent Iron Composite. *Chem. Eng. J.* **2013,** *217*, 54–60.

Kirschling, T. L.; Gregory, K. B.; Minkley, J. E. G.; Lowry, G. V.; Tilton, R. D. Impact of Nanoscale Zero Valent Iron on Geochemistry and Microbial Populations in Trichloroethylene Contaminated Aquifer Materials. *Environ. Sci. Technol.* **2010,** *44*, 3474–3480.

Kołacínski, Z.; Rincón, J.; Szymánski, T.; Sobiecka, E. Thermal Plasma Vitrification Process as the effective Technology for Hospital Incineration Fly Ash Immobilization. In *Vitrification and Geopolimerization of Wastes for Immobilization or Recycling*, Vol. 51; Universidad Miguel Hernandez: Alicante, Spain, 2017.

Kuo, H. C.; Juang, D. F.; Yang, L.; Kuo, W. C.; Wu, Y. M. Phytoremediation of Soil Contaminated by Heavy Oil With Plants Colonized by Mycorrhizal Fungi. *Int. J. Environ. Sci. Technol.* **2014,** *11* (6), 1661–1668.

Lacalle, R. G.; Gómez-Sagasti, M. T.; Artetxe, U.; Garbisu, C.; Becerril, J. M. Brassica Napus Has a Key Role in the Recovery of the Health of Soils Contaminated with Metals and Diesel by Rhizoremediation. *Sci. Total Environ.* **2018,** *618*, 347–356.

Li, S.; Wang, W.; Liang, F.; Zhang, W. X. Heavy Metal Removal Using Nanoscale Zero-Valent Iron (nZVI): Theory and Application. *J. Hazard. Mater.* **2017,** *322*, 163–171.

Li, Z.; Wang, L.; Meng, J.; Liu, X.; Xu, J.; Wang, F.; Brookes, P. Zeolite-Supported Nanoscale Zero-Valent Iron: New Findings on Simultaneous Adsorption of Cd (II), Pb (II), and As (III) in Aqueous Solution and Soil. *J. Hazard. Mater.* **2018,** *344*, 1–11.

Liang, C.; Xiao, H.; Hu, Z.; Zhang, X.; Hu, J. Uptake, Transportation, and Accumulation of C60 Fullerene and Heavy Metal Ions (Cd, Cu, and Pb) in Rice Plants Grown in an Agricultural Soil. *Environ. Pollut.* **2018,** *235*, 330–338.

Liang, J.; Yang, Z.; Tang, L.; Zeng, G.; Yu, M.; Li, X.; Luo, Y. Changes in Heavy Metal Mobility and Availability from Contaminated Wetland Soil Remediated with Combined Biochar-Compost. *Chemosphere* **2017,** *181*, 281–288.

Lin, S.; Hu, Q.; Hudson, J. S.; Reid, M. L.;, Ratnikova, T. A.; Rao, A. M., Luo, H.; Ke, P. C. Uptake, Translocation, and Transmission of Carbon Nanomaterials in Rice Plants. *Small* **2009,** *5*, 1128–1132.

Liu, Z. J.; Chen, L.; Zhang, Z. C.; Li, Y. Y.; Dong, Y. H.; Sun, Y. B. Synthesis of Multi-Walled Carbon Nanotube-Hydroxyapatite Composites and Its Application in the Sorption of Co(II) from Aqueous Solutions. *J. Mol. Liq.* **2013,** *179*, 46–53.

Ma, X. M.; Geiser-Lee, J.; Deng, Y.; Kolmakov, A. Interactions between Engineered Nanoparticles (ENPs) and Plants: Phytotoxicity, Uptake and Accumulation. *Sci. Total Environ.* **2010,** *408*, 3053–3061.

Medina-Pérez, G.; Fernández-Luqueño, F.; Vazquez-Nuñez, E.; López-Valdez, F.; Prieto Mendez, J.; Madariaga-Navarrete, A.; Miranda-Arámbula, M. Remediating Polluted Soils Using Nanotechnologies: Environmental Benefits and Risks. *Polish J. Environ. Stud.* **2019,** *28* (3).

Megharaj, M.; Ramakrishnan, B.; Venkateswarlu, K.; Sethunathan, N.; Naidu, R. Bioremediation Approaches for Organic Pollutants: A Critical Perspective. *Environ. Int.* **2011,** *37*, 1362–1375.

Mohammadi, H.; Amani-Ghadim, A. R.; Matin, A. A.; Ghorbanpour, M. Fe 0 Nanoparticles Improve Physiological and Antioxidative Attributes of Sunflower (*Helianthus annuus*) Plants Grown in Soil Spiked with Hexavalent Chromium. *3 Biotech.* **2020,** *10*, (1), 19.

Mokarram-Kashtiban, S.; Hosseini, S. M.; Kouchaksaraei, M. T.; Younesi, H. The Impact of Nanoparticles Zero-Valent Iron (nZVI) and Rhizosphere Microorganisms on the Phytoremediation Ability of White Willow and Its Response. *Environ. Sci. Pollut. Res.* **2019,** *26* (11), 10776–10789.

Mueller, N. C.; Nowack, B. Nanoparticles for Remediation: Solving Big Problems with Little Particles. *Elements* **2010,** *6* (6), 395–400.

Muszynska, E.; Hanus-Fajerska, E. Why Are Heavy Metal Hyperaccumulating Plants So Amazing? *Biotechnol. J. Biotechnol. Comput. Biol. Bionanotechnol.* **2015,** *96*, 265–271.

Navarro, D. A.; Bisson, M. A.; Aga, D. S. Investigating Uptake of Waterdispersible CdSe/ZnS Quantum Dot Nanoparticles by *Arabidopsis thaliana* Plants. *J. Hazard. Mater.* **2012,** *211*, 427–435.

O'Carroll, D.; Sleep, B.; Krol, M.; Boparai, H.; Kocur, C. Nanoscale Zero Valent Iron and Bimetallic Particles for Contaminated Site Remediation. *Adv. Water Resour.* **2013,** *51*, 104–122.

Patil, S. S.; Shedbalkar, U. U.; Truskewycz, A.; Chopade, B. A.; Ball, A. S. Nanoparticles for Environmental Clean-Up: A Review of Potential Risks and Emerging Solutions. *Environ. Technol. Innov.* **2016,** *5*, 10–21.

Pillai, H. P.; Kottekottil, J. Nano-Phytotechnological Remediation of Endosulfan Using Zero Valent Iron Nanoparticles. *J. Environ. Protect* **2016,** *7*, (05), 734.

Praveen, A.; Khan, E.; Perwez, M.; Sardar, M.; Gupta, M. Iron Oxide Nanoparticles as Nano-Adsorbents: A Possible Way to Reduce Arsenic Phytotoxicity in Indian Mustard Plant (Brassica juncea L.). *J. Plant Growth Regulation* **2018,** *37* (2), 612–624.

Rady, M.; Romeh, A.; Muhanna, A. E. H. Using Green Nano-Phytoremediation of Water Polluted with Flonicamid. *J. Prod. Dev.* **2019,** *24* (3), 571–593.

Rani, M.; Shanker, U. Degradation of Traditional and New Emerging Pesticides in Water by Nanomaterials: Recent Trends and Future Recommendations. *Int. J. Environ. Sci. Technol.* **2018,** *15* (6), 1347–1380.

Rehman, M.; Liu, L.; Wang, Q.; Saleem, M. H.; Bashir, S.; Ullah, S.; Peng, D. Copper Environmental Toxicology, Recent Advances, and Future Outlook: A Review. *Environ. Sci. Pollut. Res.* **2019,** 1–14.

Román-Ross, G.; Cuello, G. J.; Turrillas, X.; Fernández-Martínez, A.; Charlet, L. Arsenite Sorption and Co-Precipitation with Calcite. *Chem. Geol.* **2006,** *233*, 328–336.

Romeh, A. A. A. Green Silver Nanoparticles for Enhancing the Phytoremediation of Soil and Water Contaminated by Fipronil and Degradation Products. *Water Air Soil Pollut.* **2018,** *229* (5), 147.

Romeh, A. A. Synergistic Effect of Ficus-Zero Valent Iron Supported on Adsorbents and *Plantago Major* for Chlorpyrifos Phytoremediation from Water. *Int. J. Phytoremed.* **2020,** 1–11.

Romeh, A. A.; Khamis, M. A.; Metwally, S. M. Potential of Plantago major L. for Phytoremediation of Lead-Contaminated Soil and Water. *Water Air Soil Pollut.* **2016,** *227* (1), 9.

Romeh, A. A.; Saber, R. A. Green Nano-Phytoremediation and Solubility Improving Agents for the Remediation of Chlorfenapyr Contaminated Soil and Water. *J. Environ. Manage* **2020,** *260*, 110104.

Saleem, M.; Ali, S.; Rehman, M.; Rana, M.; Rizwan, M.; Kamran, M.; Imran, M.; Riaz, M.; Hussein, M.; Elkelish, A. et al. Influence of Phosphorus on Copper Phytoextraction via Modulating Cellular Organelles in Two Jute (*Corchorus capsularis* L.) Varieties Grown in a Copper Mining Soil of Hubei Province, China. *Chemosphere* **2020,** *248*, 126032.

Shahwan, T., Sirriah, S. A., Nairat, M., Boyacı, E., Eroğlu, A. E., Scott, T. B., Hallam, K. R. Green Synthesis of Iron Nanoparticles and Their Application as a Fenton-Like Catalyst for the Degradation of Aqueous Cationic and Anionic Dyes. *Chem. Eng. J.* **2011,** *172*, (1), 258–266.

Shekhar, S.; Sundaramanickam, A.; Balasubramanian, T. Biosurfactant Producing Microbes and Their Potential Applications: A Review. *Crit. Rev. Environ. Sci. Technol.* **2015,** *45*, 1522–1554.

Shu, H. Y.; Chang, M. C.; Chen, C. C.; Chen, P. E. Using Resin Supported Nano Zero-Valent Iron Particles for Decoloration of Acid Blue 113 Azo Dye Solution. *J. Hazard. Mater*. **2010,** *184*, 499–505.

Singh, J.; Lee, B. K. Influence of Nano-TiO_2 Particles on the Bioaccumulation of Cd in Soybean Plants (*Glycine max*): A Possible Mechanism for the Removal of Cd from the Contaminated Soil. *J. Environ. Manage.* **2016,** *170*, 88–96.

Singh, R.; Misra, V.; Mudiam, M. K. R.; Chauhan, L. K. S.; Singh, R. P. Degradation of γ-HCH Spiked Soil Using Stabilized Pd/Fe0 Bimetallic Nanoparticles: Pathways, Kinetics and Effect of Reaction Conditions. *J. Hazard. Mater.* **2012,** *237*, 355–364.

Song, B.; Xu, P.; Chen, M.; Tang, W.; Zeng, G.; Gong, J.; Ye, S. Using Nanomaterials to Facilitate the Phytoremediation of Contaminated Soil. *Crit. Rev. Environ. Sci. Technol* **2019,** *49* (9), 791–824.

Song, B.; Xu, P.; Zeng, G.; Gong, J.; Zhang, P.; Feng, H.; Ren, X. Carbon Nanotube-Based Environmental Technologies: The Adopted Properties, Primary Mechanisms, and Challenges. *Rev. Environ. Sci. Bio/Technol.* **2018,** *17* (3), 571–590.

Souri, Z.; Karimi, N.; Sarmadi, M.; Rostami, E. Salicylic Acid Nanoparticles (SANPs) Improve Growth and Phytoremediation Efficiency of Isatis cappadocica Desv., under as stress. *IET Nanobiotechnolo.* **2017,** *1* (6), 650–655.

Souza, R. D.; Ambrosini, A.; Passaglia, L. M. Plant Growth-Promoting Bacteria as Inoculants in Agricultural Soils. *Genet. Mol. Biol.* **2015,** *38* (4), 401–419

Srivastav, A.; Yadav, K. K.; Yadav, S.; Gupta, N.; Singh, J. K.; Katiyar, R.; Kumar, V.; Nano-Phytoremediation of Pollutants from Contaminated Soil Environment: Current Scenario and Future Prospects. In *Phytoremediation*; Springer: Cham, 2018; pp 383–401.

Su, Y.; Yan, X.; Pu, Y.; Xiao, F.; Wang, D.; Yang, M. Risks of Single-Walled Carbon Nanotubes Acting as Contaminants-Carriers: Potential Release of Phenanthrene in Japanese medaka (Oryzias latipes). *Environ. Sci. Technol* **2013,** *47* (9), 4704–4710.

Subramanian, K. S.; Manikandan, A.; Thirunavukkarasu, M.; Rahale, C. S. Nano-Fertilizers for Balanced Crop Nutrition. In *Nanotechnology Food Agriculture*; Rai, M., Ribeiro, C., Mattoso, L., Duran, N., Eds.; Springer: Cham, **2015**; pp 69–80.

Tafazoli, M.; Hojjati, S. M.; Biparva, P.; Kooch, Y.; Lamersdorf, N. Reduction of Soil Heavy Metal Bioavailability by Nanoparticles and Cellulosic Wastes Improved the Biomass of Tree Seedlings. *J. Plant Nutr. Soil Sci.* **2017,** *180*, 683–693.

Tangahu, B. V.; Sheikh Abdullah, S. R.; Basri, H.; Idris, M.; Anuar, N.; Mukhlisin, M. A Review on Heavy Metals (As, Pb, and Hg) Uptake by Plants through Phytoremediation. *Int. J. Chem. Eng.* **2011**.

Timmusk, S.; Seisenbaeva, G.; Behers, L. Titania (TiO_2) Nanoparticles Enhance the Performance of Growth-Promoting Rhizobacteria. *Sci. Rep.* **2018,** *8* (1), 1–13.

Tiwari, D. K.; Behari, J.; Sen, P. Application of Nanoparticles in Waste Water Treatment. *World Appl. Sci. J.* **2008,** *3*, 417–433.

Tripathi, D. K.; Singh, V. P.,; Prasad, S. M.; Chauhan, D. K.; Dubey, N. K. Silicon Nanoparticles (SiNp) Alleviate Chromium (VI) Phytotoxicity in *Pisum sativum* (L.) Seedlings. *Plant Physiol. Biochem.* **2015,** *96*, 189–198.

Trujillo-Reyes, J.; Peralta-Videa, J. R.; Gardea-Torresdey, J. L. Supported and Unsupported Nanomaterials for Water and Soil Remediation: Are They a Useful Solution for Worldwide Pollution? *J. Hazard. Mater.* **2014,** *280*, 487–503.

Usman, K.; Al-Ghouti, M. A.; Abu-Dieyeh, M. H. Phytoremediation: Halophytes as Promising Heavy Metal Hyperaccumulators. *Heavy Metal* **2018.**

Varma, A.; Khanuja, M. Role of Nanoparticles on Plant Growth with Special Emphasis on *Piriformospora indica*: A Review. In *Nanoscience and Plant–Soil Systems*; Springer: Cham., 2017; pp 387–403.

Verma, A.; Roy, A.; Bharadvaja, N. Remediation of Heavy Metals Using Nanophytoremediation. In *Advanced Oxidation Processes for Effluent Treatment Plants*; Elsevier, 2021; pp 273–296.

Vítková, M.; Puschenreiter, M.; Komárek, M. Effect of Nano Zero-Valent Iron Application on As, Cd, Pb, and Zn Availability in the Rhizosphere of Metal (Loid) Contaminated Soils. *Chemosphere* **2018,** *200*, 217–226.

Wang, Y.; Fang, Z.; Kang, Y.; Tsang, E. P. Immobilization and Phytotoxicity of Chromium in Contaminated Soil Remediated by CMC-Stabilized nZVI. *J. Hazard. Mater.* **2014,** *275*, 230–237.

Whiteside, M. D.; Treseder, K. K.; Atsatt, P. R. The Brighter Side of Soils: Quantum Dots Track Organic Nitrogen through Fungi and Plants. *Ecology* **2009,** *90*, 100–108.

Yadu, B.; Chandrakar, V.; Korram, J.; Satnami, M. L.; Kumar, M.; Keshavkant, S. Silver Nanoparticle Modulates Gene Expressions, Glyoxalase System and Oxidative Stress Markers in Fluoride Stressed *Cajanus cajan* L. *J. Hazard. Mater.* **2018,** *353*, 44–52.

Yan, W.; Herzing, A. A.; Kiely, C. J.; Zhang, W. X. Nanoscale Zerovalent Iron (nZVI): Aspects of the Core-Shell Structure and Reactions with Inorganic Species in Water. *J. Contam. Hydrol.* **2010,** *118*, 96–104.

Zaier, H.; Ghnaya, T.; Ghabriche, R.; Chmingui, W.; Lakhdar, A., Lutts, S.; Abdelly, C. EDTA-Enhanced Phytoremediation of Lead-Contaminated Soil by the Halophyte *Sesuvium portulacastrum*. *Environ. Sci. Pollut. Res.* **2014,** *21*, 7607–7615.

Zand, A. D.; Tabrizi, A. M.; Heir, A. V. The Influence of Association of Plant Growth-Promoting Rhizobacteria and Zero-Valent Iron Nanoparticles on Removal of Antimony from Soil by *Trifolium repens*. *Environ. Sci. Pollut. Res.* **2020,** 1–15.

Zhang, W. X. Nanoscale Iron Particles for Environmental Remediation: An Overview. *J. Nanopart. Res.* **2003,** *5*, 323–332.

Zhang, Y.; Li, Y. M.; Zheng, X. M. Removal of Atrazine by Nanoscale Zero Valent Iron Supported on Organobentonite. *Sci. Total Environ.* **2011,** *409*, 625–630.

Zou, Y.; Wang, X.; Khan, A.; Wang, P.; Liu, Y.; Alsaedi, A.; Wang, X. Environmental Remediation and Application of Nanoscale Zero-Valent Iron and Its Composites for the Removal of Heavy Metal Ions: A Review. *Environ. Sci. Technol.***2016,** *50* (14), 7290–7304.

CHAPTER 14

Mass Production of Arbuscular Mycorrhizal Fungus Inoculum and Its Use for Enhancing Biomass Yield of Crops for Food, for In Situ Nano-Phyto-Mycorrhizo Remediation of Contaminated Soils and Water, and for Sustainable Bioenergy Production

A. G. KHAN[1*] and A. MOHAMMAD[2]

[1]*Grad Life Member Western Sydney University, Sydney, Australia*

[2]*W. Booth School of Engineering Practice and Technology, Engineering Technology Building, McMaster University, Hamilton, Ontario, Canada*

**Corresponding author. E-mail: lasara@gmail.com*

ABSTRACT

Food and energy resources are gradually becoming scarce due to ever-increasing consumption and demand and the world faces crises. Resources like land, air, and water are increasingly becoming contaminated with heavy metal contaminants produced by wastages due to municipal, industrial, and mining activities by humans. There is an urgent need to address these problems by reclaiming the HM-contaminated soil and aquatic ecosystems for food and energy production. Various physicochemical remediation strategies are being developed and tested, but they are all expensive and applicable

Nanotechnology for Environmental Pollution Decontamination: Tools, Methods, and Approaches for Detection and Remediation. Fernanda Maria Policarpo Tonelli, Rouf Ahmad Bhat, & Gowhar Hamid Dar (Eds.)

to small areas. During the past couple of decades, eco-friendly plant-based strategy, phytoremediation, is gaining popularity due to its economic and aesthetic values. But, it is very time-consuming and often hampered by the plant involved being slow-growing, producing contaminant-laded biomass with the added cost of safely disposing, and requiring the use of costly fertilizers. To overcome all these drawbacks, the process can be hastened by using plant growth-promoting soil microbiota, especially universal and ubiquitous arbuscular mycorrhizal fungi (AMF) and plant growth-promoting rhizobacteria (PGPR). However, AMF being obligate symbionts, and production of its mass-scale inoculum for field application is an obstacle. Various commercial AMF inoculants, mixed with PGPR, developed by in vitro and in vivo production methods using substrate-based and substrate-free (hydroponics and aeroponics) base systems, are available in the international market, but they were reported to produce mixed, variable, and inconsistent results. We developed in our laboratory using a substrate-free system, Ultrasonic Nebuliser Aeroponic System. Pre-inoculated with AMF propagule (isolated from the rhizospheres of plants grown in the HM-contaminated sites to be remediated) was grown in the aeroponic system to produce adequate adventitious roots with extensive extraradical mycelium network outside the root surface of the test plant used. The aeroponically grown mycorrhizal roots were then sheared, air-dried, and used as inoculum to test its infectivity and efficiency in the roots of the plants in the field, in competition with the indigenous and not efficient AMF propagules in the field soil, causing intensive mycorrhization of root cortices, uptaking contaminants into their aerial parts (phytoremediation) and producing high biomass for bioenergy production, that is, Nano-Mycorrhizo-PhytoRemediation.

14.1 INTRODUCTION

Soil and its plant-root-associated microbiota population perform various functions such as nutrient (including heavy metals) sequestration and their biogeocycling, climate and water regulation, and are critical tools in ecosystem restoration and function, especially restoration of degraded ecosystems (Munoz-Rojas et al., 2018; Balser et al., 2002; Khan, 2002, 2008). These microorganisms, such as plant-growth-promoting rhizobia (PGPR), ubiquitous and universal root-associated symbiont, that is, arbuscular mycorrhizal fungi (AMF), constitute one of the major biodiversity reservoirs on earth (Kushwaha et al., 2020). These endophytes also play an integral and a unique and significant role in the functioning of ago-ecosystems, as noted by

Sansanwal et al., (2017), by promoting plant growth by direct mechanisms (e.g., N-fixation, K- and P-solubilization, siderophore production, production of indolic-compounds, ACC utilization, ammonia excretion, etc.) as well as indirect mechanisms by acting as biocontrol agents (e.g., production of metabolites and antibiotic substances).

The significance of the worldwide distributed soil fungi that form a symbiosis with roots of most land plants, that is, AMF, is now accepted that they improve plant growth and productivity, including bioenergy crops (Khan, 2020a), as well as increase plant resistance against abiotic and biotic stresses (Khan, 2005a, 2020b). Mycorrhiza-associated plants have been reported growing on contaminated soils (Chaudhary et al., 1999; Khan et al., 2000; Hayes et al., 2003). To improve bioenergy-crop-plant health and growth on contaminated sites, and to increase biomass production of non-food energy crops for bioenergy purposes, we need to consider the potential of nanomaterials secreted by bioenergy crop plants and these AMF, PGPR, and Mycorrhiza-Helping Bacteria (MHB), in our efforts toward achieving our goal of using them for the dual purpose of nano-mycorrhizo-phytoremediation (NMPR) of contaminated soils and producing greater biomass for bioenergy production. Many different techniques such as substrate (soil and sand)-based, substrate-free (hydroponic/aeroponic), and in vitro (excised or root-organ cultures) products, with advantages and disadvantages, have been developed during the last few years. Ijdo et al. (2011) reviewed these methods, probable sectors of their application and discussed their positive and negative aspects. Compared to the availability of P and trace metal fertilizers, AM fungal inoculants do not seem to encourage growers to them as they fail to appeal significant grower needs and broad-scale efficacy in cost-effective manners. Several researchers, including private companies, government industries, and technologists, have attempted to develop commercial AM inoculants. Various soil-less media, such as expanded clay aggregates (Dehne and Backhaus, 1986), peat, perlite, or vermiculite (Sreeramulu and Bagyaraj, 1986), sand (Millner and Kitt, 1992).

Being obligate symbionts, AMF cannot be grown in pure cultures and the techniques used for producing AMF inoculum are different from those usually employed for other biotechnologically interesting fungi. AMF cannot grow in the absence of a living root partner. However, culturing the fungi with living roots is hampered by contamination with other soil-borne pathogens that affect their viability and efficacy. This hampers the mass production of AMF inoculum (Hepper, 1984). Therefore, cost-effective and large-scale production methods of high-quality AMF inoculum production for their commercial exploitation are still not commercially utilized due to their non-consistent and unstable performances in field situations. There are several reports indicating that AMF are adaptive to different edaphic conditions and seem to possess unique properties based on their cell biology (Bethlenfalvay et al., 1990). The literature also reveals that fungal isolates from fertilized or heavy metal contaminated sites are better adapted to elevated levels of P (Jasper et al., 1979) plant growth by mitigating the negative effects of HMs and also to enhance the uptake and translocation of various minerals from HM-enriched soils (for reference see Khan, 2020a, 2020b).

Due to inherent efficacy, fungal aggressiveness, the number of infections, or a combination of the three (Borges and Chaney, 1988), the introduction of AM fungal isolates alien to particular soil conditions may not be effective in causing mycorrhization of roots, thereby, invalidating the beneficial effects of inoculation on plant growth. Therefore, it is important to the culture and exploits the use of AM isolates which may be well suited and efficient in the soils where they will be introduced, that is, AMF are site-specific and not plant-specific. Mohammad and Mittra (2013) assessed the efficacy of the AMF ecotype of stress-adapted AMF (*Glomus deserticola*) isolated from the rhizospheres of grasses growing along with the industrial waste in India on the growth of eggplant (*Solanum melongena* L.) and sorghum (*Sorghum Sudanese* Staph. seedlings in soils amended with various stress levels of NaCl, Zn, and Cd. The authors reported that the AMF-ecotype responds best to these soils, as evident from the significantly greater growth response and its aggressiveness in colonizing roots in all soil types tested. These results suggest that this AMF-ecotype can be used as an effective tool to alleviate the adverse effects of excessive salinity and HM toxicity on plant growth, that is, it has a potential in the NMPR of HM-polluted soils.

Faye et al. (2020) assessed the potential of commercial mycorrhizal inoculants and a rhizobial inoculant to improve soybean yield in Kenya both in greenhouse and field conditions and reported variable results, that is, significant or insignificant effect on plant P uptake, biomass production, leaf–chlorophyll index, and grain yield. The authors concluded that not only the inoculant type, soil type, and P source in soil are critical factors to evaluate mycorrhizal inoculants, additional criteria such as inoculant type, soil type, and P source are needed to be predictive of the response, without which the economic value of commercial inoculants will remain elusive. Soil microbes that improve and enhance soil fertility (N-fixing symbiotic and free-living rhizobia) and nutrient uptake efficiency by universal symbiotic mycorrhizal fungi – AMF) offer valuable alternative technologies. These microbes play an important role in nutrient acquisition and cycling, in situ nanoremediation of contaminated soils and waters, increasing plant biomass, promoting plant growth including those of high biomass producing nonfood bioenergy crops, reducing the translocation of heavy metals from contaminated soils and water to aerial parts of plants growing in contaminated habitats, etc., by synthesizing nanomaterials for their NMPR (Khan, 2020a, 2022a). NMPR offers a new window for future research.

A wide variety of commercial AMF inoculants, mixed with PGPR, are available in the international market, but their qualities are found to be sensitive to subtle changes in processes and transportation (Faye et al., 2020). These authors recommend that commercial inoculants need to be pre-evaluated on selected crops and regional soils before undertaking large field-scale use. Ijdo et al. (2011) reviewed the methods for large-scale production of AM fungi–past, present, and future. The authors attempted to describe and compare the principal in vivo and in vitro production methods, critically discussed the parameters criteria for optimal production, advantages and disadvantages of the methods, and highlighted their most probable sectors of the application of both substrate-based and substrate-free systems (hydroponics and aeroponics) and in vitro cultivation methods for large-scale developed AMF inoculum. As pointed out by Ijdo et al. (2011), there are almost as many techniques as there are laboratories working with AMF, since it is a prerequisite to fundamental research as well as for application purposes. Furthermore, in the presence of universal and ubiquitous AMF propagules in the field soil, tracing and relating positive effects of inoculation to the introduced AMF strain are not possible, and interaction of the indigenous AMF population and the introduced exotic strain poses risks associated with the introduction of unwanted microbes associated with the inoculum (Pringle et al., 2009). Voets et al. (2009) studied the role of extraradical mycelium network (ERMN) of AMF extending from mycorrhizal plants often not considered in vitro to inoculate young plantlets, as an important source of AMF inoculum in soils allowing fast colonization of plants under in vitro conditions. The authors grew young seedlings of *Medicago truncatula* in vitro in the ERMN extending from mycorrhizal plants, and after a few days of contact with ERMN seedlings were transplanted in vitro on a suitable growth medium for 4 weeks. They found that the seedlings were readily colonized by AMF and reproduced 1000s spores within 4 weeks. This fast mycorrhization process developed by Voets et al. (2009) has an open door for the mass production of AMF inoculum required for field application.

14.2 PRODUCING LARGE-SCALE AMF-COLONIZED-ROOTS WITH EXTENSIVE EXTRA RADICAL MYCELIUM NETWORK INOCULUM

We, in our laboratory at Western Sydney University, in an attempt to large-scale production of AMF inoculum for field studies, employed two techniques, that is, pot culture and aeroponic culture techniques. The specific objectives were to (1) preinoculate test plant with native AMF spores collected by wet sieving and decanting technique, from the rhizospheres of mycorrhizal plants growing on the HM-contaminated sites in New South Wales, Australia, in pot culture using sand-based media (Fig. 14.1) transferring the 4-week-old healthy-looking infected test plants from the pot cultures randomly into the perforated lid of the Atomizing Disk Aeroponic System (Fig. 14.2) or the Ultrasonic Nebuliser Aeroponic System (Fig. 14.2B), to produce adequate adventitious root systems with good AMF colonization levels of the root cortices harboring intra- and inter-cellular fungal hyphae, vesicles and arbuscules (Fig. 14.3C and D), and extensive ERMN outside the root surface of the grassroots from the atomizing disc aeroponic system as well as atomizing disc aeroponic system (Fig. 14.3A, B) (Khan, 2007).

AMF-inoculum for field application was developed by harvesting and shearing the ERMN and the adventitious roots excised from the plants growing in aeroponic chambers and blending them in an electric blender to produce a slurry that is air-dried to be used as inoculum for field application (for details see Khan, 2007). Comparatively, the inoculum potentials of the inoculum developed by the aeroponic culture techniques were significantly greater than those developed by the pot culture in all the inoculation treatments (Asif, 1997).

14.3 FIELD TESTING INFECTIVITY AND EFFICIENCY OF SHEARED-ROOT AMF-INOCULUM PRODUCED BY AEROPONIC TECHNIQUES TO YIELD GREATER BIOMASS FOR

(a) Food Crops for Enhancing Yield

To date, very few field trials have been conducted to evaluate the ability of introduced AM isolates to enhance plant growth in competition with the indigenous AMF propagules. Sylvia and his associates (Jarstfer and Sylvia, 1992; Sylvia, 1998, Sylvia and Hubbell, 1986); Sylvia and Jarstfer, 1992, 1994, Sylvia et al., 1993) were the first who used aeroponically grown sheared root

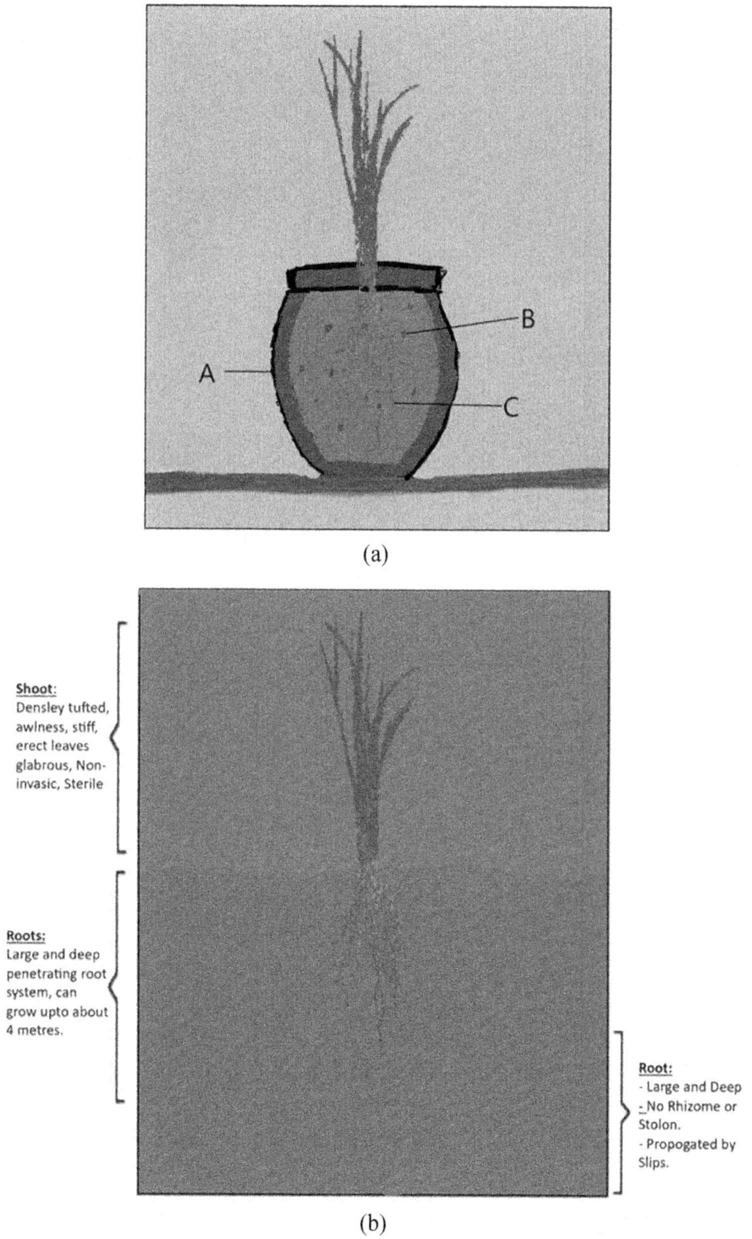

FIGURE 14.1 Establishing pot cultures (a) of surface sterilised vetiver grass 'slips' into sterilised sand inoculated with indigenous arbuscular mycorrhizal fungal spores (B), extracted by wet sieving and decanting technique from the indigenous contaminated soils, and producing mycorrhizal roots (C) used for producing large scale AMF inoculum in aeroponic culture systems.

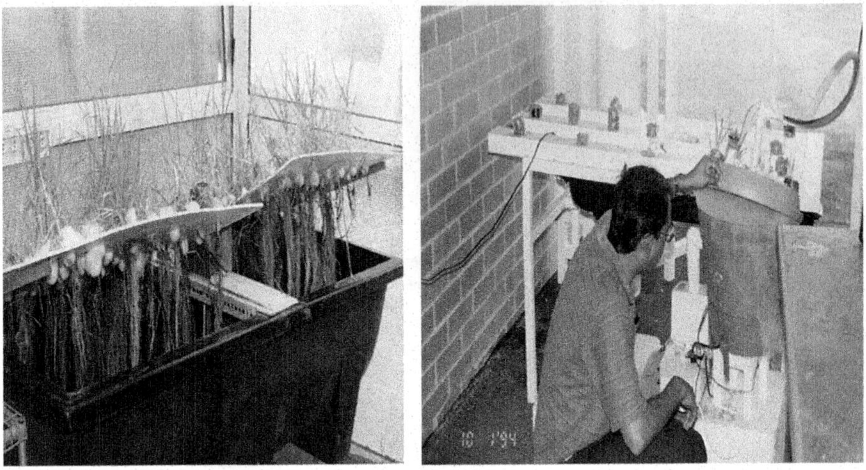

(a) (b)

FIGURE 14.2 Pre AMF-colonised rooted slipe from the pot culture are transferred to the lids of the Atomising-disk Aeroponic System (a) or Ultrasonic Aeroponic System (b), to mass produce AMF inoculum for Field application (Khan, 2007).

AMF-inoculum (consisting of root fragments and AMF propagules including in a commercial nursery and in a field and reported enhanced growth of transplanted sea oats at Miami Beach). We assessed, under field conditions, the ability of introduced AM isolate in competition with the indigenous AM fungal propagules produced by the ultrasonic aeroponic system under field conditions (Asif et al., 1995; Asif, 1997; Mohammad et al., 2004). These studies reported improved growth of plants in a field containing low levels of P and a low population of indigenous AMF propagules, when inoculated with a commercially produced sheared-root inoculum of *Glomus intraradices* consisting of root fragments and AMF propagules including ERMN, indicating that the introduced AMF can compete with its indigenous AMF and benefit plant growth. Khan (1975a, 1975b) may have been the first to demonstrate the potential of pre-inoculating plants with AMF propagules and transplanting them into nutrient-deficient field soil to compete with its indigenous AMF population, but it is not known how long such introduced strains persist. The composition of soil microbiota, including both indigenous and introduced AMF community, and their interactions clearly have relevance to mycorrhizoremediation of contaminated soils, but yet to be elucidated (Khan, 2005a, 2005b). Further research is needed to integrate in situ soil fertility enhancing technologies by using AMF and associated microbiota as plant biostimulants and the nano-phytoremediation technology, in order to

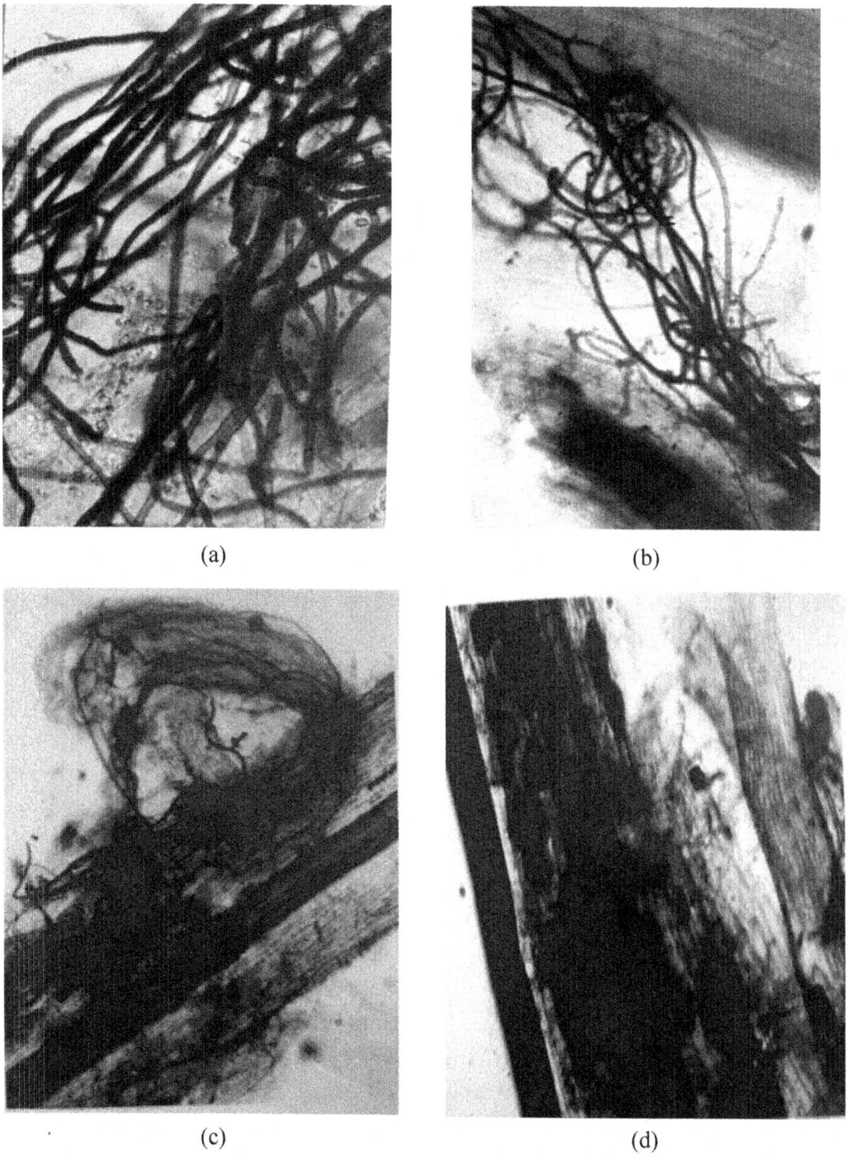

FIGURE 14.3 Large scale AMF-colonized adventitious roots, produced by Atomizing Disk Aeroponic system (A), and by Ultrasonic Aeroponic System (B), with extensive Extraradical Mycelium Network (ERMN) outside the root surfaces. Root cortices of the Adventitious Roots produced by both aeroponic systems had good AMF-colonization levels harboring inter- and intra-cellular fungal hyphae, vesicles and arbuscules, and Extraradical Mycelium Network (ERMN) on the root surfaces (C and D).

globally decontaminate and reclaim derelict land and water ecosystems and to increase crop yield, especially in the underdeveloped world economies. There is a need to investigate the dynamics and persistence or decomposition of the efficient introduced AMF inoculum in the mycorrhizosphere of crops growing in P-deficient soils with indigenous non-efficient AMF propagules. This knowledge may enable us to understand and develop these aspects and their input into developing future strategies to be used to optimize AMF-inoculum production for NPMR of contaminated soils and waters.

(b) Nonfood Crops for In Situ Nano-Mycorrhizo-Phytoremediation of HM-Contaminated Soils and Waters and Increasing Biomass for Bioenergy Production

Due to increasing energy demand for the rapidly growing human population and due to climate change and limited fossil fuel resources, accelerated efforts to develop alternative sustainable energy resources have prompted scientists and governments to explore other options such as the use of green plants for bioethanol and biodiesel production. However, there are concerns about overusing agricultural land and food crops for biofuel production. This dilemma developed interests in the use of (1) nonfood bioenergy crops capable of producing large biomass under stressed conditions in a short period of time such as Vetiver grass, (2) plant growth-enhancing microbes such as AMF, and (3) land such as derelict contaminated wastelands for growing such non-food crops. Kumar et al. (2018) suggested that a solution to carbon emissions and energy crisis is the production of microbial biofuels and that microbial-biotechnology might substitute crop-based biofuel production.

As stated above, most of the plants growing on disturbed, derelict, and HM-contaminated soils and water, have a strong dependency on AMF for optimal growth and biomass production. Both plant roots and their symbiont AMF synthesize nanoparticles (NPs) to protect themselves from the toxicity of contaminants by secreting HM-affinity transporter nanomolecules that immobilize or translocate HMs into root cells, that is, NMPR (Khan, 2020a). We propose that coupling sustainable NMPR with bioenergy by using non-food crops and derelict land is an integral approach to address the issue of reclamation of HM-contaminated and derelict land and simultaneously producing greater biomass for bio-energy production. Vetiver grass is one of such plants, that is, it is a fast-growing, mycorrhizal, high biomass producing, with dense-root systems, secreting high levels of HM-degrading NPs, capable of growing as halophyte, hydrophyte or xerophyte and with a potential for

NMPR as an ideal plant to be used in in situ phytoremediation purposes and ecological restoration of derelict land contaminated land (Khan, 2022, 2007, 2009). Phytoremediation of HM-contaminated wastewater ecosystems and wetlands and to produce large biomass for bioenergy production is also proposed by using NMPR strategy and constructed-wetlands planted with vetiver grass plants with roots pre-occupied with water-logging tolerant AM fungi extracted from the rhizosphere of HM-adapted aquatic macrophytes growing at the land–water interface of the contaminated water ecosystem (Fig.14.4) (Khan, 2021b).

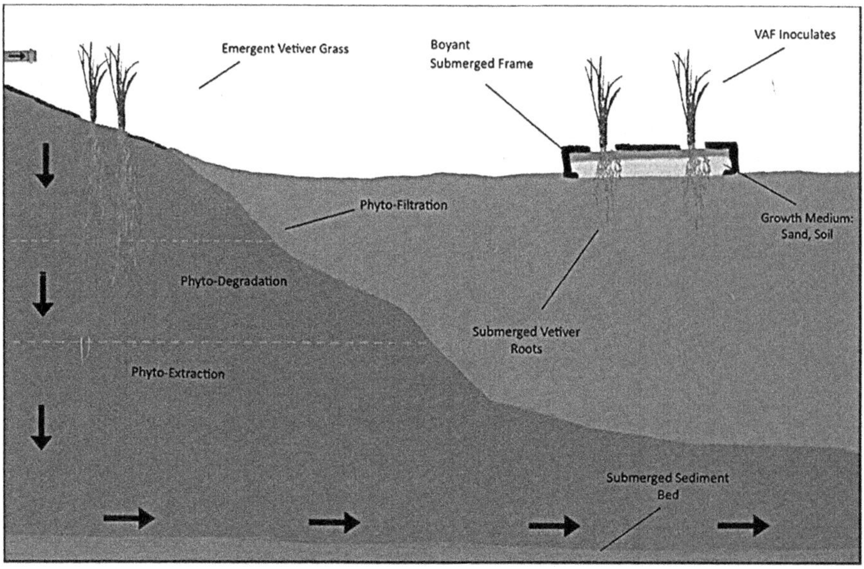

FIGURE 14.4 Schematic view of the Constructed Wetlands (CWs) planted with water-logging-tolerant and Pre-occupied with indigenous arbuscular mycorrhizal fungal (AMF) infected rooted slips of the Vetiver grass from the pot cultures (Fig. 14.1 a) transferred to the lids of the constructed wetlands (AM-CWs) for the Nano-Mycorrhizo-Phyto-Remediation (NMPR) of HM-Contaminated Wastewater Ecosystems and Wetlands. (Khan, 2021).

14.4 CONCLUDING REMARKS AND FUTURE DIRECTIONS

Production of large-scale AMF inoculum for applying in situ nanophytoremediation strategy to mycorrhizoremediate HM-contaminated soils and water depends upon the AMF isolate, the physicochemical characteristic of the contaminated ecosystem, and concentration of the contaminating

HM (Khan, 2020b). The mycorrhizal technology needs to be commercially exploited, but more information is required to understand the role played by AMF in ecosystem management, sustainability, and restoration of degraded and contaminated ecosystems before adopting this strategy. Using non-agricultural land and greater biomass producing mycorrhizal plants such as Vetiver grass seems to be the answer to achieve this goal.

As the use of a commercially produced AMF inoculum available in the International markets produces mixed results, using stress-tolerant and indigenous, site adaptive AMF/PGPR microbes from the rhizospheres of plants growing on the contaminated sites to be remediated is recommended. NMPR is an emerging strategy. Future of AMF/PGPR assisted nano-phytoremediation, using native site-adaptive microbes, and simultaneously producing greater biomass for bioenergy production, is the way to go. However, there are still many issues, such as the optimization of the processes and factors affecting NMPR, to be resolved before applying these plant–microbe interactions at field-scale using mass-produced AMF inoculum using various techniques discussed in this article. NPMR technology is a potential mechanism to improve survival.

ACKNOWLEDGMENTS

AGK thanks his grandson Rana Usman Khan for producing figures and for assisting with computer/printer-related troubleshooting.

MA greatly appreciates the constructive and invaluable advice provided by various faculty members of WSU for retechnical and linguistic contents of his PhD thesis.

CONFLICT OF INTEREST

The authors declare no conflict of interest.

AUTHORS' CONTRIBUTIONS

Both authors contributed equally to the manuscript and have read and agreed to the published version of the manuscript.

KEYWORDS

- **arbuscular mycorrhizal fungi**
- **inoculum production**
- **hydroponics and aeroponics**
- **field applications**
- **sustainable bioenergy production**
- **nano-phytoremediation**
- **plant growth-promoting rhizobia**
- **nano-mycorrhizo-phytoremediation**

REFERENCES

Asif, M. Comparative Study of Production, Infectivity, and Effectiveness of Arbuscular Mycorrhizal Fungi Produced by Soil-Based and Soil-Less Techniques. PhD Thesis, University of Western Sydney Macarthur, Faculty of Business and Technology, Department of Biological Sciences. Sydney, July 1997.

Asif, M.; Khan, A. G.; Khan, M. A.; Kueck, C. Growth Responses of Wheat to Sheared Root Vesicular Arbuscular Inoculum Under Field Condition. In *Mycorrhiza: Biofertilizer for the Future. Proceedings 3rd International Conference on Mycorrhiza*; Adholeya, A., Singh, S., Eds.; New Delhi, India, March 13–15, 1995. Tata Energy Research Institute, New Delhi, India; pp 432–437.

Balser, T. C.; Kinzig, A. P.; Firestone, M. K. Linking Soil Microbial Communities and Ecosystem Functioning. In *The Functioning Consequences of Biodiversity: Empirical Progress and Theoretical Extensions*; Kinzig, A. P. et al., Eds.; Princeton University Press, 2002.

Bethlenfalvay, G. J.; Brown, M. S.; Franson, R. L.; Mihara, K. L. The *Glycin-Glomus-Bradyrhizobium* Symbiosis. X. Relationship between Leag Gas Exchange and Plant and Soil Water Status in Nodulated, Mycorrhizal Soybean under Drought Stress. *Plant Physiol.* **1990,** *94*, 723–728.

Borges, R. G.; Chaney, W. R. The Response of *Acacia scleroxyla* Tuss. To Mycorrhizal Inoculation. *Int. Tree Crops J.* **1988,** *5*, 191–201.

Chaudhary, T. M.; Hill, K.; Khan, A. G.; Kuek, C. Colonization of Non- and Zn Contaminated Dumped Filter Cake Waste by Microbes, Plants and Associated Mycorrhizae. In *Remediation and Management of Degraded Lands*; Wong, M. H., Wong, J. W. C., Baker, A. J. M., Eds.; CRC Press LLC: Boca Raton, 1999; Chapter 27, pp 275–283.

Dehne, H. W.; Backhaus, G. F. The Use of Vesicular-Arbuscular Mycorrhizal Fungi in Plant Production. 1. Inoculum Production. *J. Plant Dis. Plant Prot.* **1986,** *93* (4), 415–424.

Faye, A.; Stewart, Z. I.; Ndung-u-Magioi, K.; Diouf, M.; Ndoye, I. Testing of Commercial Inoculants to Enhance P Uptake and Grain Yield of Promiscuous Soybean in Kenya. *Sustainability* **2020,** *12*, 3808. https://Doi:10.3390/su12093803

Giovannini, L.; Palla, M.; Agnolucci, M.; Agnolucci, M.; Avio, L.; Sbrana, C.; Turrini, A.; Giovannetti, M. Arbuscular Mycorrhizal Fungi and Associated Microbiota as Plant

Biostimulantss: Research Strategies for the Selection of the Best Performing Inocula. *Agronomy* **2020,** *10*, 106. https://doi.3390/agronong10010106

Hayes, W. J.; Chaudhry, T. M.; Buckney, R. T.; Khan, A. G. Phytoaccumulation of Trace Metalsat the Sunny Corner Mine, New South Wales, with Suggestion for Apossible Remediation Strategy. *Aust. J. Ecotoxicol.* **2003,** *9*, 69–82.

Hepper, C. M. Isolation and Culture of VA Mycorrhizal (VAM) Fungi. In *VA Mycorrhiza*; Li Powell, G., Bagyaraj, D. J., Eds.; CRC Press Inc.: Boca Raton, FL, 1984; pp 95–112.

Ijdo, M.; Cranenbrouck, S.; Declerck, S. Methods for Large-Scale Production of AM Fungi: Past, Present, and Future. *Mycorrhiza* 2011, *21*,1016. https://doi.10.1007/a00572.010,0337-z

Jarstfer, A. G.; Sylvia, D. M. Aeroponic Culture of VAM Fungi. In *Mycorrhiza–Structure, Function, Molecular Biology, and Biotechnology*; Varma, A., Hock, B., Eds.; Springer: Heidelberg, 1995; pp 427–441.

Jarstfer, A. G.; Sylvia, D. M. Inoculum Production and Inoculation Strategies for Vesicular-Arbuscular Mycorrhizal Fungi. In *Soil Microbial Techniques Applications in Agriculture, Forestry and Environmental Management*; Metting, B., Ed.; Marcel Dekker Inc.: New York, 1992; pp 349–377.

Jasper, D. A.; Robson, A. D.; Abbott, L. K. Phosphorus and the Formation of Vesicular Arbuscular Mycorrhizas. *Soil Biol. Biochem.* **1979,** *11*, 501–505.

Khan, A. G. The Significance of Microbes in Soil Rehabilitation. In *The Restoration and Management of Derilict Land: Modren Approaches*; World Scientific Publishing: Singapore, 2002; Chapter 8, pp 80–92.

Khan, A. G. Vetiver Grass as an Ideal Phycosymbiont for Glomalian Fungi for Ecological Restoration of Derelict Land. In *Proceedings 3rd International Conference on Vetiver and Exibition: Vetiver and Water*; Troung, P.; Hanping, X., Eds.; Guangzhou, China, October 2003. China Agricultural Press: Beijing, China; pp 466–474.

Khan, A. G. Role of Soil Microbiota in the Rhizosphere of Plants Growing on Heavy Metal Contaminated Soils in Phytoremediation of Trace Metals. *J. Trace Metal Elements Med. Biol.* **2005a,** *18*, 355–364.

Khan, A. G. Mycorrhizoremediation—An Enhanced Form of Phytoremediation. *Int. Symp. Phytoremediation Ecosyst. Health.* Hanghzhou, China, Sept 10–13, 2005b; p 42.

Khan, A. G. Producing Mycorrhizal Inoculum for Phytoremediation. In *Phytoremediation: Methods in Biotechnology*; Willey, N., Ed., Vol. 23; Humana Press Inc.: Totowa, NJ, 2007. https://doi.org/10.1007/978-1-59745-00980_7

Khan, A. G. Microbial Dynamics in the Mycorrhizosphere with Special Reference to Arbuscular Mycorrhizae. In *Plant Bacteria Interactions—Strategies and Techniques to Promote Plant Growth*; Ahmad, I., Pitchtel, J., Hayat, S., Eds.; Wiley-VCH/Verlag: Weinheim, 2008; Chapter 13, pp 245–256.

Khan, A. G. Role of Vetiver Grass and Arbuscular Mycorrhizal Fungi in Improving Crops against Abiotic Stresses. In *Salinity and Water Stress. Tasks for Vegetation Sciences*; Ashraf, M., Athar, H., Eds., Vol. 44; Springer: Dordrecht, 2009. https://doi.org/10.1007/978-1-4020-9065-3_12

Khan, A. G. In Situ Nano-Phytoremediation Strategy to Mycorrhizo-Remediate Heavy Metal Contaminated Soils Using Bioenergy Crops. In *Abstract 16th International Phytotechnology Conference: Phytotechnologies for Food Safety and Environmental Health*; Changsha, China, Sept 23–27, 2019; p 170.

Khan, A. G. In Situ Phytoremediation of Uranium Contaminated Soils. In *Phytoremediation: Concepts and Strategies in Plant Sciences*; Shmacfsky, B. R., Ed.; Springer Link, 2020a; Chapter 5, pp 123–151. https://doi.org/10.1007/978-3-030-00099-8_5

Khan, A. G. Promises and Potential of in Situ Nano-Phytoremediation Strategy to Mycorrhizoremediate Heavy Metal Contaminated Soils Using Non-Food Bioenergy Crops (*Vetiver zizinoides* & *Cannabis sativa*. *Int. J. Phytoremediation* **2020b**. https://doi.org/10.1080/15226514.2020.1774504

Khan, A. G. Potential Nano-Phytoremediation of Heavy-Metal Contaminated Aquatic Ecosystems via Planting of Vetiver Grass Pre-Colonized by Indigenous Water-Logging Resistant Arbuscular Mycorrhizal Fungi. *Water Air Soil Pollut.* **2022** (Submitted).

Khan, A. G. Nano-Phytoremediation of Heavy Metal Contaminated Wastewater Ecosystems and Wetlands by Constructed-Wetlands Planted with Water-Logging-Tolerant Mycorrhizal Fungi and Vetiver Grass. *2nd International Electronic Conference on Mineral Science Section Environmental Mineralogy and Biomineralization*, 2021b. http://sciforum.net/paper/view/conference/9385

Khan, A. G.; Chaudhary, T. M.; Khoo, C. S.; Hayes, W. J. The Role of Plants, Mycorrhizae, and Phytochelators in Heavy Metal Contaminated Land Remediation. *Chemosphere, Special Issue: Environmental Contamination, Toxicology, and Health* **2000,** *41*, 197–207.

Kumar, M.; Sundaram, S.; Gnansounou, E.; Larroche, C.; Thakur, I. S. Carbon Dioxide Capture, Storage and Production of Biofuel and Biomaterials by Bacteria. *Bioresour. Technol.* **2018.** https://dx.doi.org/10.1016/j.biortech.2017.09.050

Kushwaha M, Surabhi, Marwa N, Pandey V, Singh NA. 2020. Advanced tools to Assess Microbial Diversity and Their Functions in Restoration of Degraded Ecosystems. In *Microbial Services in Restoration Ecology*; Elesvier Inc., 2020; Chapter 6, pp 83–97. https://doi.org/10.1016/B978-0-819978.7.00006-3

Millner, P. D.; Kitt, D. G. The Beltsville Method for Soilless Production of Vesicular-Arbuscular Mycorrhizal Fungi. *Mycorrhiza* **1992,** *2*, 9–15.

Mohammad, A.; Khan, A. G. Monoxenic in Vitro Production and Colonization Potential of AM Fungus *Glomus intraradices*. *Indian J. Exp. Bot.* **2002,** *40*, 1087–1091.

Mohammad, A.; Khan, A. G.; Kuek, C. Improved Aeroponic Culture of Inocula of Arbuscular Mycorrhizal Fungi. *Mycorrhiza* **2000,** *9*, 337–339. https://doi: 10.1007/s0055720050278

Mohammad, A.; Mitra, B.; Khan, A. G. Effects of Sheared-Root Inoculum of *Glomus intraradices* on Wheat Growth at Different Phosphorus Levels in the Field. *Agric. Ecosyst. Environ.* **2004,** *103*, 245–249. https://doi.10.1016/j.agec.2003.09.017

Mohammad, A.; Mittra, B. Effects of Inoculation with Stress Adapted Arbuscular Mycorrhizal Fungus *Glomus deserticola* on Growth of *Solanum melogena* L. and Sorghum *sudanese* Staph. Seedlings under Salinity and Heavy Metal Stress Conditions. *Arch. Agron. Soil Sci.* **2013,** *59* (2), 173–183. https://doi.org/10.1080/03650340.2011.610029

Munoz-Rojas, M.; Chilton, A.; Ooi, M. K. J. Effects of Indigenous Soil Cyanobacteria on Seed Germination and Seedling Growth of Arid Species Used in Restoration. *Plant Soil* **2018,** *429*, 91–100. https://doi.org./10.1007/s11104-018-3607-8

Pringle, A.; Bever, J. D.; Gardes, M.; Parrent, J. L.; Rillig, M. C.; Klironomos, J. N. Mycorrhizal Symbioses and Plant Invasions. *Ann. Rev. Ecol. Evol. Syst.* **2009**. https://doi.10.1146/annurev.ecolsys.39.110707.173454

Sansanwal, R.; Ahlawat, U.; Wati, L. Role of Endophytes in Agriculture. *Rev. Lett.* **2017,** *6* (24), 2397–2407.

Sreeramulu, K. R.; Bagyaraj, D. J. Field Responses of Chilli to VA Mycorrhiza in Black Clayey Soil. *Plant Soil* **1986,** *92*, 299–304.

Sylvia, D. M. Mycorrhizal Symbiosis. In *Principles and Applications of Soil Microbiology*, Sylvia, D. M. et al., Eds.; Prentice Hall: Upper Saddle River, 1998; pp 408–426.

Sylvia, D. M. Nursery Inoculation of Sea Oats with Vesicular Arbuscular Mycorrhizal Fungi and Out Planting Performance of Florida Beaches. *J. Coastal Res.* **1989,** *5*, 747–754.

Sylvia, D. M.; Hubbell, D. H. Growth and Sporulation of Vesicular Arbuscular Mycorrhizal Fungi. *New Phytol.* **1986,** *95*, 655–661.

Sylvia, D. M.; Jarstfer, A. G. Sheared Root Inoculum of Vesicular Arbuscular Mycorrhizal Fungi in Aeroponic and Membrane Systems. *Symbiosis* **1992,** *1*, 259–267.

Sylvia, D. M.; Jarstfer, A. G.; Vosalka, M. Comparisons of Vesicular Arbuscular Mycorrhizal Species and Iocula Formulations in a Commercial Nursery and on Diverse Florida Beaches. *Biol. Fertil. Soil* 1993, *16*, 139–144.

Voets, L.; de la Providencia, T. E.; Fernandez, K.; Ijdo, M.; Cranenbrou, S.; Declerch, S. Extraradical Mycelium Network of Arbuscular Mycorrhizal Fungi Allows Fast Colonization of Seedlings under in Vitro Conditions. *Mycorrhiza* **2009,** *19*, 347–356. https://doi.10.1007/s00572-009-0233-6

PART VI
Nanomaterials Feasibility

CHAPTER 15

Hazardous and Safety and Management of Nanomaterials for the Personal Health and Environment

J. IMMANUEL SURESH* and A. JUDITH

PG Department of Microbiology, The American College, Madurai 625002, Tamil Nadu, India

**Corresponding author. E-mail: immanuelsuresh1978@gmail.com*

ABSTRACT

Nanoparticles have wide range of applications in various fields such as biomedical, environmental remediation, etc. High surface area and high reactivity properties of nanoparticles make it suitable for nanoremediation of environmental pollutants. Despite the magnificent properties of nanoparticles and their wide range of applications, the process of engineering and handling of nanoparticles pose considerable risk to health and environment safety. The toxicity of nanoparticles proves that some of these products become toxic at the cellular level in the tissues and organs after entering into human body. Inhalation of nanoparticles leads to various pulmonary illness in mammals including asthma. Nanoparticles introduced in water could pose high risk to living beings upon consumption. The highly potential field of nanotechnology can be exploited with safety by facilitating the development of high-performance filter media, respirators, dust-repellants, self-cleaning garments, and frequent work place risk assessment.

Nanotechnology for Environmental Pollution Decontamination: Tools, Methods, and Approaches for Detection and Remediation. Fernanda Maria Policarpo Tonelli, Rouf Ahmad Bhat, & Gowhar Hamid Dar (Eds.)

15.1 INTRODUCTION

Nanotechnology deals with the ultra small nanoparticles. The rapidly advancing field of nanotechnology has remarkable applications in various diverse fields. It has been widely studied for its potential to act as an efficient carrier for drug delivery, diagnostics, food pathogen detection, wearable sensors, and numerous other biomedical applications. Along with the above, nanomaterials also have shown positive results in environmental remediation. It can be used to clean up polluted media like soil, air, and water. Nanoremediation is possible because of the unique properties of nanoparticles such as high surface area, high reactivity, etc. This enables them to eliminate a wide range of toxic environmental pollutants. Despite the magnificent properties of nanoparticles and its wide range applications, the process of engineering and handling of nanoparticles poses considerable risk to health and environment safety. This review deals with the possible health safety, environmental risk associated with nanoparticles and risk management (Karn et al., 2009; Vishwakarma et al., 2010).

15.2 HEALTH HAZARDS

The excessive exposure of nanoparticles is considered to be unsafe for the biological system. Researches on the toxicity of nanoparticles indicate that some of these products become toxic at the cellular level in the tissues and organs after entering into human body (Rachel, 2006; Jin et al., 2005). Nanoparticles may enter into the body upon exposure via inhalation, dermal, oral, and parenteral in the case of biomedical applications (Stern and McNeil, 2008). Their smaller dimensions and higher specific surface area enable them to easily bind and transport toxic contaminants. They can cause numerous pulmonary diseases in mammals. If the nanoparticles enter the body, they will circulate freely across the body in the blood and reach organs such as the liver or brain. It may move deeper into the lungs and the bloodstream and also may cross the blood–brain barrier. During the handling of the nanoparticles, skin contact can easily occur. The possible toxicity of the materials of these particles could vary, but even inert nanoparticles are considered to be dangerous. They may get converted into toxic items on reaction with the body fluids. In order to produce highly reactive types of oxygen that can cause tissue damage, including inflammation and other toxic effects, certain nanoparticles can display improved catalytic properties. This damage will translate into asthma and atherosclerotic heart disorders in airborne particles.

In terrestrial organisms, uptake of nanoparticles by inhalation or ingestion is likely to be the major route (U.S. National Institute for Occupational Safety and Health, 2006). Nanoparticle deposition in the lung rises with exercise due to increased breathing rate and transition from nasal to mouth breathing and among people with known lung diseases or conditions (British Standard Institute [BSI]).

Threadlike nanotubes are structurally similar to asbestos fibers. According to a report by the Royal Society, the United Kingdom's National Science Academy, upon inhalation in large amounts for longer periods they are capable of causing lung fibrosis. Lower sized nanoparticles function more like a gas and are capable of moving through skin and lung tissue thereby penetrating into cell membranes. Once they enter into the cell, they may become toxic and interfere with the normal cell chemistry (Vishwakarma et al., 2010).

Particles with diameter of 1 nm or 0.001 μm do not penetrate the alveoli instead they get deposited in the nose and pharynx. In the tracheobronchial zone, the other 20% are deposited. The retention of inhaled nanoparticles at this size is nearly 100% (Claude, 2006a). Deposition is primarily in the alveolar zone of the lungs for particles larger than 5 nm. The deposition fraction of inhaled nanoparticles is higher in the alveolar and tracheo-bronchial regions of human lungs leading to the development of airways diseases, such as chronic obstructive pulmonary disease (COPD) or asthma. In addition, studies support the direct role of inhaled nanoparticles in systemic diseases such as cardiovascular diseases. Study on CNTs show that they tend induce platelet aggregation and enhance thrombosis (Stern and McNeil, 2008). The toxicity and health risk of nanoparticles may also be due to its various physical and chemical factors (Claude, 2006a; British Standard Institute [BSI]).

Dermal exposure to nanoparticles may lead to direct penetration of nanoparticles into the epidermis and lead it into the bloodstream; however, there is no remarkable evidence (U.S. National Institute for Occupational Safety and Health, 2006; Aitken et al., 2004). Only mild irritation was reported as an adverse reaction to topical nanomaterial application in limited in vivo studies conducted to resolve the issue of cutaneous toxicity. For example, nanoscale metal oxides are currently used in commercially available sunscreens and have been extensively tested by animals and clinicians to satisfy regulatory requirements. These studies found minimal irritancy potential, and no evidence of photoirritation, sensitization etc. (Stern and Mc Neil, 2008). In contrast, some recent studies have revealed some cutaneous penetration by ultrafine beryllium particles and the formation of cutaneous nodules (Claude, 2006a).

Ingestion is another route by which nanoparticles can penetrate the body. Ingested particles smaller than 20 μm (20,000 nm) can pass through the intestinal barrier and enter the bloodstream. Ingestion can occur from unintentional events and direct ingestion of contaminated drinking water or particles absorbed on vegetables, etc. (Claude, 2006a; U.S. National Institute for Occupational Safety and Health, 2006).

15.3 ENVIRONMENTAL HAZARDS

Nanoparticles in water sources that are either present or introduced can lead to secondary toxic effects and potentially endanger human health. This problem drags the awareness of the scientific community. Ensuring their protection, possible health and environmental effects, is a key challenge for emerging nanomaterials. Ecological researchers are facing a major challenge on determining the toxicity levels of nanomaterials; also, they have to investigate whether biomarkers of harmful effects currently used would also function in the study of environmental nanotoxicity. The practicality of using natural nanomaterials as sorbents is therefore investigated by many research groups. For instance, allophone is an excellent sorbents for copper and surface-modified smectite adsorbs naphthalene and 17β- estradiol (Yuan 2004). Both of these nano-sized minerals are soil-based and are of geological and pedological origin.

While there are insufficient data to predict the risk of fire and explosion associated with nanoparticles, the general trend is to increase the violence of the dust explosion and the ease of ignition as the particle size decreases or the specific area increases. Decreasing the particle size of fuel materials will reduce the minimum ignition energy and increase the potential for combustion and the rate of combustion, contributing to the likelihood of highly combustible materials being relatively inert (Claude, 2006a; Pritchard, 2004).

Nanoparticles exist in the atmosphere in large concentrations, still the release of manufactured nanoparticles into atmosphere and aquatic environment is yet unknown. Scientific community has raised concern that nanoparticles are potential of causing brain damage in aquatic environment (Rick, 2004).

In addition, due to their high surface areas, nanomaterials released to the soil can be heavily absorbed into the soil and become immobile. Nanosized particles may be taken up by bacteria and living cells, providing the basis for possible bioaccumulation in the food chain. Certain nanomaterials are engineered to be released into the atmosphere as reactants and are therefore required to undergo chemical transformation. One instance of this is iron

(FeO). Biodegradation of nanoparticles by some kinds of fungi leads to generation of toxic metabolites to microorganisms under various conditions (J et al, 2007).

15.4 EXPOSURE AND HANDLING

Occupational environment pose high risk of continuous exposure to nanoparticles. Followed by the removal of nanomaterials from a closed system, it can be dispersed in the gas phase. Nanomaterials can also be formulated and used as powders, suspensions, etc. Nanomaterials in gas phase and powdered state pose the greatest risk. Cleaning and handling of nanomaterials engineering system also paves way for the exposure to risk. The amount of release of nanoparticles and duration of exposure widely influences the level of risk. Handling nanomaterials in liquid media, in gas phase under non enclosed system, in powders and cleaning dust collection system without adequate protection are certain workplace factors that increase the exposure to risk (Vishwakarma et al., 2010).

Some organizations have started an active program to research the safe handling of nanomaterials in the workplace, such as the National Institute of Occupational Safety and Health (Schulte and Salamanca-Buentello, 2007). There is only little information available on the risks of handling these products, so employees should apply strict control procedures and safety engineering features to minimize exposure when working with them and not allow them to eat or drink in the laboratory. During manufacture and handling of these materials, there are high chances of release and exposure of nanoparticles to workers which can get inside their body through various routes (Kevin, 2005).

When workers handle the nanomaterials they should use laboratory safety practices such as Personnel Protective Equipment including gloves, lab coats, safety glasses, face shields, closed-toed shoes, etc. These safety precautions will help in avoiding the skin contact with nanoparticles or nanoparticles containing solutions. Workers should wear adequate respiratory protection, if it is required to treat nanoparticle powders with exhaust laminar flow hoods. It is very important to use fume exhaust hoods to evict fumes from tube furnaces or chemical reaction vessels. Laboratory workers should be educated on occupational danger risk, Material Safety Data Sheets (MSDS), marking, signage, etc., periodically. The disposal of nanoparticles also focuses on environmental protection. It should be in accordance with recommendations for hazardous chemical waste (Vishwakarma et al., 2010) (Table 15.1).

TABLE 15.1 Nanomaterial State, Potential Occupational Exposure, and Recommended Engineering Control.

Nanomaterial	Exposure	Control
Fixed nanostructures	Nanomaterials may be released during grinding, drilling, and sanding.	• Local exhaust ventilation • Laboratory chemical hood (with HEPA filtered exhaust) • HEPA-filtered exhausted enclosure (glovebox) • Biological safety cabinet class II type A1, A2, vented via thimble connection, or B1 or B2 • Appropriate equipment for monitoring toxic gas (e.g., CO)
Liquid suspension	Process like sonication or spraying, equipment cleaning and maintenance may result in aerosolization of nanoparticles.	
Dry dispersible nanoparticles	Handling dry powder formulations	
Nanoaerosols	Occurrence of leakage from the reactor, product recovery, processing, etc., results in exposure.	

15.5 MANAGEMENT OF RISK

An efficient dust collection system with HEPA filter installed in it must be kept in the workplace where there is high risk of generation of nanoparticles in the air (gas phase). HEPA filters are known to effectively filter nanoparticles, provided the HEPA filter should be fitted properly. Otherwise, the nanoparticles will bypass the filter and disperse in the air. Workers should be educated about the safe handling of nanoparticles. Workers should be encouraged to wash their hands thoroughly before eating or leaving the work site in order to avoid the ingestion of toxic nanoparticles. The unnecessary transport of nanoparticles outside the work area, through clothing and skin as a medium, can be avoided by providing showering and changing facility. Food should not be stored or consumed near the work place. Work place must be cleaned with HEPA filter based vacuum or damp cleaning method. Nanoparticles should not be handled as free particulate in open air. If the worker has the risk of being exposed to nanoparticles dispersed in air during handling or engineering, appropriate respirators can be used as safety precaution. Frequent risk assessment tests should be carried out for continuous monitoring and control of the risk (British Standard Institute [BSI]).

15.6 CONCLUSION

Recent research is showing that when harmless bulk materials are rendered into ultrafine particles, they appear to become harmful and toxic. These ultrafine particles could certainly enter into the human body via lungs, intestine, and less evidently through skin. The smaller the particles, the more reactive and harmful their effects are. Since ultrafine particles could pose a human health hazard, more research is required in this field (Hoet et al., 2004). Through the growing field of nanoscience if we get to understand the root cause of toxicity in these products, it will be possible to engineer better materials that will save human lives. Industry consortiums, actual environmental toxicity data for iterative groups and particular firms need to take strong steps before they enter the market to assess the quality and safety of goods and products. Nanomaterials established today also aid in dealing with emergencies. Sensors and communication systems based on nanotechnology may decrease their exposure to risk of injury and help in preventing diseases at the early stage. In conjunction with wireless technologies, nanosize will promote the production of wearable sensors and applications for workplace safety and health monitoring in real time. The highly potential field of nanotechnology can be exploited with safety by facilitating the development of high-performance filter media, respirators, dust-repellants, self-cleaning garments, etc. In order to minimize its effect on the environment, due consideration should be taken with respect to nanoparticles and nanotechnology safety concerns for the personal health and safety of staff engaged in nano-manufacturing processes and even customers.

KEYWORDS

- **nanoparticles**
- **nanoremediation**
- **environment safety**

REFERENCES

Aitken, R. J.; Creely, K. S.; Tran, C. L. *Nanoparticles: An Occupational Hygiene Review*; HSE, UK, 2004.

British Standard Institute (BSI). Nanotechnologies—Part 2: Guide to Safe Handling and Disposal of Manufactured Nanomaterials, PD 6699-2:2007, UK, 2007.

Claude, O. Nanoparticles—Actual Knowledge about Occupational Health and Safety Risks and Prevention Measures, 2006a.

Claude, O. Health Effects of Nanoparticles, 2006b.

E 2535. Standard Guide for Handling Unbound Engineered Nanoscale Particles in Occupational Settings. ASTM, USA, 2007.

Hoet, P. H.; Brüske-Hohlfeld, I.; Salata, O. V. Nanoparticles—Known and Unknown Health Risks. *J. Nanobiotechnol.* **2004,** *2* (1), 12.

J. M.; J. W. Nanotechnology White Paper. U S Environmental Protection Agency, 2007.

Jin, Y.; Wu, M.; Zhao, X. Toxicity of Nanomaterials to Living Cells, University of North Dakota, US, 2005, 274–277.

Karn, B.; Kuiken, T.; Otto, M. Nanotechnology and in Situ Remediation: A Review of the Benefits and Potential Risks. *Environ. Health Perspect.* **2009,***117* (12), 1813–1831.

Kevin, D. L. Health and Environmental Impact of Nanotechnology: Toxicological Assessment of Manufactured Nanoparticles. *Toxicol. Sci.* **2005,** *77*, 3–5.

Pritchard, D. K. *Literature Review—Explosion Hazards Associated with Nanopowders*; Health & Safety Laboratory (HSL), UK, 2004.

Rachel, C. Assessing Safety. Health Risks Nanomater. **2006,** *15* (05).

Rick, W. Nanoparticles Toxic in Aquatic Habitat. Study Finds, Washington Post Staff Writer, 2004.

Schulte, P. A.; Salamanca-Buentello, F. Ethical and Scientific Issues of Nanotechnology in the Workplace. Environ. Health Perspect. 2007, 115, 5–12.

Stern, S. T.; McNeil, S. E. Nanotechnology Safety Concerns Revisited. *Toxicol Sci.* **2008,** *101* (1), 4–21.

Texas A&M University. *Interim Guideline for Working Safely with Nanotechnology;* USA, 2005.

U S. National Institute for Occupational Safety and Health, Centers for Disease Control and Prevention. Approaches to Safe Nanotechnology—An Information Exchange with NIOSH, 2006.

Vishwakarma, V.; Samal, S. S.; Manoharan, N. Safety and Risk Associated with Nanoparticles—A Review. *J. Miner. Mater. Charact. Eng.* **2010,** *09* (05), 455–459.

Yuan, G. Natural and Modified Nanomaterials as Sorbents of Environmental Contaminants. *J. Environ. Sci. Health A Tox. Hazard. Subst. Environ. Eng.* **2004,** *39* (10), 2661–2670.

CHAPTER 16

Economic Impact of Applied Nanotechnology: An Overview

MIR ZAHOOR GUL* and BEEDU SASHIDHAR RAO

Department of Biochemistry, University College of Sciences, Osmania University, Hyderabad 500007, Telangana, India

**Corresponding author. E-mail: ziahgul@gmail.com*

ABSTRACT

As a result of the recent world industrial and technological revolution, scientists have reassessed the relationship between socioeconomic development and environmental ethics. It has long been acknowledged that adapting to the sustainable use of natural resources is critical for the survival of the future generations of human society. Nanotechnology is one of the most prominent emerging technologies and it has revolutionized all fields of medicine, agriculture, environmental science, etc., by offering abilities that have never been imagined. Nanotechnology has ushered the world materials and devices that are already used in a multitude of platforms. One of the most intriguing features of nanotechnology is its incredible scope of real and potential applications for fostering long-term economic growth. As a result, the propagation effect of nanotechnology is immense and the commercial opportunities it creates are both numerous and diverse. A broad alliance of industry leaders, scientists, and politicians has recognized nanotechnology as a pivotal technology for the 21st century that will lead to economic growth and sustainable development. In recent years, nanotechnology has significantly contributed to benefiting society and transforming the nature of modern life, and at the same time boosting economic growth as well as

Nanotechnology for Environmental Pollution Decontamination: Tools, Methods, and Approaches for Detection and Remediation. Fernanda Maria Policarpo Tonelli, Rouf Ahmad Bhat, & Gowhar Hamid Dar (Eds.)

improving the capacity and quality in industrial sectors. This chapter focuses on the economic impacts of nanotechnology by using market volumes as an accurate measure of its economic importance.

16.1 INTRODUCTION

Science has taken great strides, enabling what is perhaps the most incredible transformation in human productivity in the entire history. The field of economics has also made significant progress in understanding the application of such scientific innovations, that is, technology, which transforms society at large and creates a continuous chain of wealth and advancement. Technology is critical for the sustainable growth of mankind (Ottman, 2005). On the one hand, the technology strongly influences the demand for energy and raw materials, infrastructure and transport needs, the mass flow of materials, emission composition, and volume of waste materials. Technology, on the other hand, is also a central determinant in the innovation process and directly impacts prosperity, consumption habits, ways of living, interpersonal interactions, and cultural changes. Consequently, the advancement, fabrication, usage, and disposal of technical products and systems have an impact on environmental, social, and economic dimensions of sustainability. Nanotechnology is among the most powerful and influential evolving technologies. It has been heralded as a vital twenty-first-century technology that will contribute to economic growth and sustainability by a diverse coalition of politicians, scientists, and industry groups. Nanotechnology has always been at the frontier between scientific reality and optimistic ambitions, between first milestones and promising aspirations, between progressive advances and innovative discoveries. Most assessments and evaluations of current and potential nanotechnology developments, directly or indirectly, identify this spectrum of opportunities. It is a revolutionary technology and has transformed all fields, like medicine, agricultural, and environment, etc., by rendering skill and knowledge that would never have dreamt of before. It is a distinctive platform for multidisciplinary approaches in the field of physics, chemistry, biology, and engineering. New areas of nanomedicine, cancer nanotechnology, and environmental nanotechnology have surfaced due to the impact of nanotechnology approaches and are thriving with the advances in this expanding field. This technology is rapidly changing the ways in which devices and materials will be developed in the future.

Nanotechnology involves two major strategies for the synthesis of materials: (1) the "bottom up" approach that leads to the formation of nanostructured basic components and then assembles them into final material based on the molecular recognition principles, and (2) the "top-down" process includes the development of nanostructures from the bigger entities without atomic level control. The letter is similar to the methods used by the semiconductor industry to fabricate electronic devices using pattern formation, like pattern transfer processes (reactive ion etching) and electron beam lithography, thus creating nanoscale constructs (Ahmed et al., 2010; Iqbal et al., 2012). Analytical science provides nanotechnology a boost by combining it with new-generation analytical tools like atomic force microscopy (AFM) and tunnel scanning microscopy (STM) with processes like molecular beam epitaxy and electron beam lithography, which enables nanostructure manipulation with unique phenomena (Ahmed et al., 2010). The analytical chemistry is, therefore, crucial for the development of structures in the nano-systems and devices. The super interdisciplinary nature plays a vital role in the progression of nanotechnology. It helps to establish principles and methods for the application of nanotechnology with the unique characteristics of nanomaterials. The chemical composition, size, and morphology are characterized with the help of analytical tools (Ahmed et al., 2010). Moreover, the chemical synthesis leads to the fabrication of unique nanomaterials with new analytical possibilities. The engineering of molecular functional materials involves highly specialized concepts. It is an imagined capacity to generate materials using a bottom-up approach using existing technologies and techniques to synthesize complete high products. The ability to synthesize nanoscale blocks and then arrange them in a large structure with unique characteristics and functions, with precisely controlled size and structure, revolutionizes materials, their production, as well as their various applications.

16.1.1 COMMERCIALIZATION OF NANOTECHNOLOGY

Since its advent, nanotechnology has opened up new areas for future research, and researchers expect that it will have a profound impact on industry and technology, human health, social and economic growth, and the climate. Furthermore, since nanotechnology is expected to have a significant impact on the global economic system, market figures are good indicators of its commercial significance. Many nanotechnology industry predictions derive from the early 2000s, with a period of up to 2015. As over 2 million research articles related to nanotechnology have been published worldwide annually in recent years, and more than 3000 patents have been filed, both the volume of research and the interest it has attracted has shown surprising progress (Science and Scoreboard, 2016). Over the past few years, reports have usually been euphoric about nanotechnology and all the advantages it will offer in the future. Due to its immense potential, public and private investments in nanotechnology are substantial and increasing. This is an area with highly promising prospects for transforming basic research into effective inventions, not only to improve the competitiveness of the real industry but also to develop new products that will bring about positive impact in our daily lives, especially in the fields of healthcare, the environment, electronics, or any other sector. Plenty of nanomaterials, both in their pure state and as composite materials, are currently in use or under production, and the number of applications for nanomaterials is increasing swiftly. A continuing rise in the number of patents is the consequence of the successful technological advancements in nanotechnology. According to StatNano, in 2016, a total of 19,563 nanotechnology patents were registered by the U. S. Patent and Trademarks Office (USPTO). The European Patent Office (EPO) issued 3589 nanotechnology patents in the year 2016 (Science and Scoreboard, 2016). Around half of these patents belong to the United States, while the next ranks are held by South Korea, Japan, and Taiwan. Interestingly, the number of international patents for nanoproducts and nanotechnologies attained 189,000 in 2017, a rise of 31,000 compared to the year 2016. China, with more than 88,000 patents in nanotechnology, is the market leader, followed by the USA (86,000), Japan (25,000), and South Korea (22,000). Over the last 20 years, there have been about 620,000 patent applications submitted in the field of nanotechnology (Inshakova and Inshakova, 2020).

Nanotechnology is one of the most influential examples of new technologies and it increases high standards in a wide variety of areas impacting everyday life. Although the marketing of nanotechnology products has been relatively modest so far, recent and ongoing research efforts have made it possible to forecast excellent outcomes for the benefit of humanity in the years ahead. Large-scale production possibilities for the application of nanotechnology predict the possibilities of its growth in market share. Nanotechnology continues to have a pervasive effect on almost all aspects of the world economy, with market volumes being an acceptable measure of its economic value. The global nanotechnology market is valued at nearly USD 1165.90 million in 2019 and is projected to rise over the 2020–2027 forecast period with a healthy growth rate of more than 10.50% (Market Watch, 2021). Nanomaterials, with an 85% share, had the highest share of the global nanotechnology industry by component. Nanotools had the second-largest share of the global nanotechnology industry, while nanodevices had the least. Electronics, energy, and biomedical are the top three areas of nanotechnology and, together, they represent more than 70% of the global nanotechnology market. Scientific innovations, increased government and private sector R&D investment, increased demand for equipment miniaturization, and strategic partnerships between economies are expected to propel global nanotechnology market development.

16.1.2 GREEN ECONOMY AND NANOTECHNOLOGY

The word "green economy" was coined to discuss global concerns and challenges such as climate change, growing energy needs, and health issues (Gaurav et al., 2017; Organisation for Economic Co-operation and Development, 2015). It refers to a broad concept focused on sustainable development principles. It encompasses strategies built on the foundation of acknowledging social and ecological resources, continuity in environmental sustainability, preserving the socioecological harmony, which leadto improved human well-being through the development of green jobs, increased productivity, environmental accountability, eradicating poverty, and continued growth in important areas through sustainable approach (Loiseau et al., 2016; Dickel, and Petschow, 2013). This also involves socially comprehensive, source-competent, and carbon-conserving methods, actions, objectives, and values (Loiseau et al., 2016). Nanotechnology sets up modern and creative green methods for developing innovative devices and materials with unique physicochemical properties that are both economically and environmentally friendly (Barbier, 2011; Caprotti and Bailey, 2014; Hamdouch and Depret, 2010). Green-economy theory is characterized by a significant increase in investments and new projects in emerging markets, either contributing to an enhancement and reinforcement of the Earth's natural resources or contributions to reducing threats and shortcomings in environmental sectors. Sustainable agriculture and energy, clean technologies, minimal emission transportation methods, and waste management, are all being prioritized. Without a doubt, the energy sector is the one that plays a very important role as a source of energy. All economic trends hinge on consumption and management, and as a result, the idea of using renewable energy is strongly promoted. To provide a more comprehensive and integrated solution to the fit-in environment in economic theories, sustainable energy necessitates the development, adoption, and marketing of a green economy model. Nanotechnology aims to find new ways to improve economic growth while keeping environmental and ecological considerations in mind, as well as leading to the mitigation of natural resource challenges, primarily through energy-efficient technologies. Nanomaterials are anticipated to have a significant impact on industries and economic spheres.

16.2 NANOTECHNOLOGY AS AN ENABLING TECHNOLOGY

Nanotechnology is fuelling the development of several established nanotechnology-specific industrial sectors and new ones as it advances from the concept of nanotechnology to application in generic markets. According to government and industry, nanotechnology has enormous economic potential by improving the life cycle of materials and components, improving efficiency, and breaking the link between the ecological consequences and sustainable development. Apart from the financial implications, nanotechnology offers a plethora of great possibilities for improving healthcare and overall quality of life. The scientists' ability to envision and control the behavior of materials and nanoscales provides them with the tools for creating unique new products. Nanoscale materials with new and unexpected characteristics offer a unique opportunity to build "intelligent" materials that create products with completely different characteristics. The production of these products is less energy and resource intensive. Through nanotechnology, the perception is that production will become progressively cleaner and greener and that products will be inexpensive and will have more functionalities and many applications have been already proposed for nanotechnology. Nanotechnology will also facilitate technological advancements that minimize the environmental footprints of current applications in the industrialized nations and allow emerging economies to utilize nanotechnology to address some of their most urgent requirements (El Naschie, 2006; Schulte, 2005). Nanotechnology's market share growth prospects are determined by its huge production application possibilities. Nanotechnology is expected to have an impact in few areas. The following are some examples.

16.2.1 NANOTECHNOLOGY IN HEALTHCARE

In recent years, the healthcare industry has seen an increase in demand, and innovations such as nanoscience/nanotechnology have opened new doors in the fields of drug discovery and delivery, cancer treatment, gene therapy, and diagnostics (Patel and Nanda, 2015). Nanotechnology will facilitate the development of personalized medicines that are delivered precisely at the site of infection or disease; new and improved surgical procedures, frequently involving robotics; retinal and cochlear implants, and the ability to re-join damaged nerves will all become prevalent. Nanotechnology in medicine has emerged as a promising area of research. Nanotechnology has a major pharmacological benefit in that it helps researchers to ensure that drugs are administered more accurately to particular areas of the body and that drugs can be rendered such that the active ingredient reduces the necessary dose by optimizing cell membranes. The implementation of nanotechnology in healthcare will revolutionize the way we identify and treat damaged human tissues and organs in the future, therefore innovative technologies that evolved a few years ago have made great strides in becoming a reality. Healthcare professionals can diagnose and treat patients with a broad range of ailments using nanomedicine. By 2028, configurable nanorobotic systems and nano-hormatics will be able to reverse the effects of atherosclerosis and heart disease, assist the immune system in combating diseases, eradicate cancer, and correct congenital abnormalities in cells and the next 15–20 years will demonstrate an outstanding leap in health infrastructure attributable to nanotechnology (Heath, 2015). A team of researchers who have mapped out the application of nanotechnology and the demands of global health argue that nanomedicine is significant for third-world countries. Studies showing the production of methods are being undertaken by major research institutes in collaboration with some private corporations. Scientists and researchers are experimenting with and studying these outstanding features, as well as their implementation in modern medicine. Nanoparticles and treatment methods, diagnostic methods, antimicrobial methods, and their applications in cellular repair are all actively being investigated.

16.2.2 NANOTECHNOLOGY IN ENERGY

After more than a decade of global stability, several countries' political and scientific interests have resurfaced in recent years. Global energy consumption is growing, with emerging economies driving much of it. Between 2015 and 2040, global energy demand is projected to grow by 28%, according to several estimates [US Energy Information Administration (EIA)], 2017). On the other hand, over 940 million people (13% of the global population) lack access to stable energy sources (Ritchie and Roser, 2019). Since coal and petroleum products will continue to influence the global energy system, this has direct implications on the global greenhouse gas emissions balance. This is especially true for global greenhouse gas emissions, which are expected to increase rapidly than energy consumption since the highest growth rates are seen in the regions where fossil fuels are extensively used. A breakthrough in nanotechnology could result in new technologies that help to ensure global energy protection and distribution (Hussein, 2015; Serrano et al., 2009). According to a Rice University study (Texas, 2005), certain areas have been established in which nanotechnology can facilitate low-cost, effective, and environmentally friendly technologies (Shafiei and Salim, 2014). Nanotechnology's benefits are being recognized in several fields, including solar power, wind energy, clean coal, fusion and fission reactors, hydrogen production, storage, shipping and transportation, fuel cells, and batteries. One of the most promising and adaptable eco-friendly technologies is the direct conversion of solar energy into electrical energy (Abdin et al., 2013). Quantum dots-based-nanostructured photovoltaic devices have significant potential for cost reduction. Unlike traditional solar cells, they absorb visible light from a broader range of wavelengths in the sun's spectrum. Hetero-structured absorption layers can boost cell efficiency even more or allow for the use of inferior quality materials in the cells. Hydrogen is another significant forthcoming energy alternative; however, it still faces some major technical challenges. Among them are strategies for producing and storing hydrogen, as well as converting it to electricity. A successful hydrogen production strategy requires technologies that exploit renewable resources. One of them is the direct catalytic transformation of water into oxygen and hydrogen through the use of a nanostructured semiconductor catalyst or nanoscale additives (Candelaria et al., 2012). Nanotechnology may also aid in the development of new technologies for capturing biogenic hydrogen (Paniagua-Michel et al., 2015). To develop a hydrogen energy infrastructure, new portable, reliable, and secure hydrogen storage systems are required, especially for transportation applications. Carbon nanotubes, and complex metal hydrides, like alanates, are among the materials being researched for hydrogen storage. Nanoscale constructs can be added to improve their properties even further (Fichtner, 2005). Other energy technology opportunities include advanced catalysts for fuel cells, high-efficiency lighting, or appliance systems, modern technologies for low-loss power transmission, or high-strength lightweight materials for transportation, construction, or electric power applications (Seitz et al., 2013).

16.2.3 NANOTECHNOLOGY IN ENVIRONMENT AND AGRICULTURE

The current human economy is highly dependent on a constant flow of natural resources, involving large quantities of non-renewable material, which, after a relatively brief period of use, return to the ecosphere in some form or another. The chemical industry is a major contributor to this approach, as it is often followed by unplanned contamination from processing, material use as well as disposal. As a result, "conventional" chemistry has become a popular priority for environmentalists as well as the focus of many regulations over the years. As a result, the principle of green chemistry was established. Green chemistry, in particular, is an environmentally sustainable design of synthetic products and procedures that minimize or eliminate the utilisation of toxic substances and pollutants. This concept, which began as a grassroots movement led by a few chemists, is now being adopted by the chemical and other industries. Among the 12 green chemistry principles, there are many aspects where nanotechnology could play a critical role (Anastas and Warner, 2000). Catalysts are one of the first nanotechnologies to be used in the industry. Nanoparticles and nanostructured materials open up new possibilities for developing and regulating catalytic functions, such as increasing activity and selectivity for particular reactions. Since the behavior and selectivity of catalyst nanoparticles are strongly influenced by their surface structure, shape, size as well as their bulk and surface composition, the ability to synthesize particles with specified physical and chemical properties at the nanoscale is a critical step toward realizing the goal of catalysis by design. Nanotechnology is one of the many areas where green chemistry and green nanotechnology are plausible. Green nanotechnology is a prominent approach for the development of new, environmentally friendly products. This green synthesis transforms the fate of nanoparticles synthesizing and benefits future nanotechnology applications (Lu and Ozcan, 2015). There has been a slew of modern nanotechnology applications introduced to fix current sustainability issues. Nanoporous zeolites can allow for a slow release and efficient dosage of water and fertilizers for plants, allowing for improved agricultural production in countries where droughts last for extended periods. Food can be stored for longer periods in smart packaging using nanocomposites, particularly in areas where cooling is difficult to come by. Nanoparticles can increase the performance of catalytic converters in automobiles and reduce unique emissions. Nano-sensors can enhance the quality of ongoing environmental monitoring thereby lowering the cost. The use of iron nanoparticles in soil remediation at polluted sites should be encouraged as a technique that outperforms traditional methods in terms of productivity and pace.

16.2.4 NANOTECHNOLOGY IN WATER TREATMENT AND PURIFICATION

Water is vital to socioeconomic growth, energy and food production, safe ecosystems, and human survival. Water is also central to climate change adaptation, operating as a critical link between society and the environment. Water, on the other hand, is a global issue, mainly in developing nations. Bad water quality and unsustainable sources stymie overall economic development and can inflict damage on people's health and livelihoods. The WHO/UNICEF statistics for 2019 indicate that more than half of the global population, or 4.2 billion people, have no access to safe and hygienic sanitation facilities, and more than 2.2 billion people do not have proper access to safely regulated potable water services (WHO and Unicef, 2000). The reasons for such problems are numerous, mostly due to weak governance, poor infrastructure, and fiscal woes as well as the absence of political will. However, even if these nontechnical problems are equally or often even larger than technical challenges, the issues and challenges can be overcome in certain regions by offering new and advanced water purification and remediation technologies. Recent developments strongly suggest that nanotechnology can be used to fix—and eventually overcome—many of today's water quality issues. Nanotechnology is already having a significant effect on water quality studies. Nanomembranes, zeolites, or nanoporous polymers for water purification and desalination, nanoscale sensor components for the detection of pollutants and infectious agents in water, filters and nanomembranes, and magnetic or catalytic nanoparticles for treating wastewater or water remediation are among some of the nanotechnology possibilities under exploration (Gehrke et al., 2015; Savage and Diallo, 2005; Theron et al., 2008).

16.3 CONCLUSION—A LOOKING AHEAD

Nanotechnology is no longer a theoretical concept; it is rapidly becoming a part of everyday life. Nanotechnology opens up a slew of new prospects for social and economic growth, both now and in the future. Enhanced environmental monitoring capabilities, increased energy usage, and reduced environmental effects of human activities are all significant positive outcomes of nanomaterial adaptation. The nanoproducts that are now on the market have ignited little public discussion and have gained little additional regulatory scrutiny aimed directly at their novel features. The world's largest major economic powers' dedication to nanoscale research reflects these potential applications. The applications of nanotechnology are rapidly expanding, and resources are still required, particularly to develop the essential infrastructure, which includes well-equipped research and prototyping facilities as well as a skilled workforce. Present scientific research aims to quickly leverage novel nanomaterial applications. Given the large-scale investments in product development, government authorities play a critical role in evaluating and resolving the dynamic consequences of widespread nanotechnology adoption, both short and long term. This is especially true in the case of nanoparticles that may be released into the atmosphere, whether deliberately or accidentally. Several complementary steps, ranging from meticulously performed laboratory studies and computer simulations to small-scale field trials, are mandated. It will also be important to establish specifications and instrumentation that will enable researchers to accurately characterize and track the effects of these new materials. Since several unintentional and unintended environmental impacts are long-term, current or new protocols for product traceability and life cycle analysis may be necessary. One important hurdle is that many nanotechnology products and processes are not commonly acknowledged. After several major obstacles are eliminated, real economic development through nanotechnology can be anticipated. The main underlying technology has been shown to function effectively, is remunerative, and is intended to last for longer period of time, viz., nanoprecision selective welding or glue technology, or a regulated surfaces procedure with a toolbox for incorporating functionalizations. Along with the advent of such vital technology, an additional big hurdle must be overcome: the absence of regulations and standards.

Simple evaluations of the financial benefit of green nanotechnologies would take into consideration the net costs of technological development and market entry relative to the value of outputs and outputs achieved, considering time and outlook factors (or standing). Public research and development, knowledge creation and infrastructure investment, private sector R&D costs, prototyping, assessment, advertising, and start-up costs are all included in the net costs. Contributions to research and information, standardized or specific new technologies developed, the development of intellectual property (including patents and licences), the development of standards, and company's new start-ups are all possible results of such investments. These outcomes can contribute to many developmental and community advantages, such as enhanced regional and national GDP, higher efficiency, and trade balance and ecological and other economic gains.

The economic growth implications of emerging technology, such as green nanotechnologies, are also of particular concern to policymakers, including the effects on employment and incomes. While it is difficult to predict new job opportunities, employment shall be created through research, production, distribution, utilization and maintenance of product and services, procedures, and associated industries and services of green nanotechnology. Governments and international organizations should partner with researchers and the private industry to develop scientifically and ethically sound risk-based criteria for emerging nanotechnology-based products, as well as encourage "industry standards" to avoid possible environmental and health problems. Standard nomenclature is also required to remove uncertainty when defining the distinctions between nanomaterials and bulk materials, as well as when disclosing regulatory processes. Today's globalized environment provides a once-in-a-lifetime opportunity to create, disseminate, and share the benefits of technological advancement with a greater number of people in a shorter time. As traditional products are phased out, existing employees can transition to green nanotechnology activities.

Environmental systems are becoming more scientifically known, as is public understanding of environmental concerns. Policymakers, industry, non-governmental organizations, and researchers, on the other hand, must cooperate to raise public awareness of nanotechnology's unique threats and challenges. They must also keep the public informed about the measures being taken to determine the possible effects of nanomaterials before they are published into the market. To maximize benefits while minimizing risks, a well-balanced approach is needed.

KEYWORDS

- **industrial boom**
- **nanotechnology**
- **technological advances**
- **sustainability**
- **economic growth**

REFERENCES

Abdin, Z.; Alim, M. A.; Saidur, R.; Islam, M. R.; Rashmi, W.; Mekhilef, S.; Wadi, A. Solar Energy Harvesting with the Application of Nanotechnology. *Renew Sust. Energ. Rev.* **2013,** *26*, 837–852.

Ahmed, W.; Jackson, M. J.; Ul Hassan, I. Nanotechnology to Nanomanufacturing. In *Emerging Nanotechnologies for Manufacturing, Micro and Nano Technologies*; Ahmed,W., Jackson, M. J., Eds.; William Andrew Publishing, 2010; pp 1–15.

Anastas, P. T.; Warner, J. C. *Green Chemistry: Theory and Practice*; Oxford University Press: New York, 2000.

Market Watch. *Global Nanotechnology Market Industry Analysis, Size, Share, Growth, Trends and Forecast 2021–2027* [Press release], January 9, 2021.

Barbier, E. The Policy Challenges for Green Economy and Sustainable Economic Development. *Nat. Resour. Forum* **2011,** *35* (3), 233–245.

Candelaria, S. L.; Shao, Y.; Zhou, W.; Li, X.; Xiao, J.; Zhang, J. G.; Wang, Y.; Liu, J.; Li, J.; Cao, G. Nanostructured Carbon for Energy Storage and Conversion. *Nano Energy* **2021,** *1* (2), 195–220.

Caprotti, F.; Bailey, I. Making Sense of the Green Economy. *Geografiska Annaler: Series B Human Geog.* **2014,** *6* (3), 195–200.

El Naschie, M. S. Nanotechnology for the Developing World. *Chaos, Solitons Fractals* **2006,** *30* (4), 769–773.

Fichtner, M. Nanotechnological Aspects in Materials for Hydrogen Storage. *Adv. Eng. Mater.* **2005,** *7* (6), 443–455. https://doi.org/10.1002/adem.200500022

Gaurav, N.; Sivasankari, S.; Kiran, G. S.; Ninawe, A.; Selvin, J. Utilization of Bioresources for Sustainable Biofuels: A Review. *Renew. Sust. Energ. Rev.* **2017,** *73*, 205–214. https://doi.org/10.1016/j.rser.2017.01.070

Gehrke I.; Geiser, A.; Somborn-Schulz A. Innovations in Nanotechnology for Water Treatment. *Nanotechnol. Sci. Appl.* **2015,** *8*, 1–17 https://doi.org/10.2147/NSA.S43773

Hamdouch, A.; Depret, M. Policy Integration Strategy and the Development of the "Green Economy": Foundations and Implementation Patterns. *J. Environ. Plan. Manag.* **2010,** *53* (4), 473–490. https://doi.org/10.1080/09640561003703889

Ritchie, H.; Roser, M. Access to Energy. Published online at OurWorldInData.org. 2019 https://ourworldindata.org/energy-access

Heath, J. R.; Nanotechnologies for Biomedical Science and Translational Medicine. *Proc. Natl. Acad. Sci. USA* **2015,** *112* (47), 14436–14443.

Hussein, A. K. Applications of Nanotechnology in Renewable Energies—A Comprehensive Overview and Understanding. *Renew. Sust. Energ. Rev.* **2015,** *42*, 460–476. https://doi.org/10.1016/j.rser.2014.10.027

Inshakova, E.; Inshakova, A. Nanomaterials and Nanotechnology: Prospects for Technological Re-Equipment in the Power Engineering Industry. *IOP Conf. Ser. Mater. Sci. Eng* **2020,** *709*, 033020.

Iqbal, P.; Preece, J. A.; Mendes, P. M. Nanotechnology: The "Top-Down" and "Bottom-Up" Approaches. In *Supramolecular Chemistry*; Gale, P. A., Steed, J. W., Ed.; John Wiley & Sons, Ltd., 2012. https://doi.org/10.1002/9780470661345.smc195

Loiseau, E.; Saikku, L.; Antikainen, R.; Droste, N.; Hansjürgens, B.; Pitkänen, K.; Leskinen, P.; Kuikman, P.; Thomsen, M. Green Economy and Related Concepts: An Overview. *J. Clean. Prod.* **2016,** *139*, 361–371. https://doi.org/10.1016/j.jclepro.2016.08.024

Lu, Y.; Ozcan, S. Green Nanomaterials: On Track for a Sustainable Future. *Nano Today* **2015,** *10* (4), 417–420. https://doi.org/10.1016/j.nantod.2015.04.010

Organisation for Economic Co-operation and Development. Global and Local Environmental Sustainability, Development and Growth. Element 4, Paper 1. OECD and Post -2015 Reflections. Published online at https://www.oecd.org/. 2015. FINAL POST-2015 global and local environmental sustainability.pdf (oecd.org)

Ottman, J. A. New Technologies and Environmental Innovation. *J. Prod. Innov. Manage.* **2005,** *22* (5), 456–457.

Paniagua-Michel, J. J.; Morales-Guerrero, E.; Olmos Soto J. Microalgal Biotechnology: Biofuels and Bioproducts. In *Springer Handbook of Marine Biotechnology*; Kim S. K., Ed.; Springer: Berlin, Heidelberg, 2015. https://doi.org/10.1007/978-3-642-53971-8_62

Patel, S.; Nanda, R. Nanotechnology in Healthcare: Applications and Challenges. *Med. Chem.* **2015,** *05* (12). https://doi.org/10.4172/2161-0444.1000312

Dickel, S.; Petschow, U. Green Economy. *Ecol. Econ.* **2013,** *28* (3), 14–16 https://doi.org/10.14512/oew.v29i3.1300

Savage, N.; Diallo, M. S. Nanomaterials and Water Purification: Opportunities and Challenges. *J. Nanopart. Res*, **2005,** *7* (4–5), 331–342. https://doi.org/10.1007/s11051-005-7523-5

Schulte, J. *Nanotechnology: Global Strategies, Industry Trends and Applications*; John Wiley & Sons, Ltd, 2005. https://doi.org/ 10.1002/0470021071

Science, N.; Scoreboard, I. *Leading Patenting Countries in Nanotechnology 2016.* http://statnano.com/news/57346

Seitz, R.; Moller, B. P.; Thielmann, A.; Sauer, A.; Meister, M.; Pero, M.; Kleine, O.; Rohde, C.; Bierwisch, A.; de Vries, M.; Kayser, V. Nanotechnology in the Sectors of Solar Energy and Energy Storage. *Technol. Rep.* **2013,** 1–102. http://www.iec.ch/about/brochures/pdf/technology/IEC_TR_Nanotechnology_LR.pdf

Serrano, E.; Rus, G.; García-Martínez, J. Nanotechnology for Sustainable Energy. *Renew. Sust. Energ. Rev.* **2009,** *3* (9), 2373–2384. https://doi.org/10.1016/j.rser.2009.06.003

Shafiei, S.; Salim, R. A. Non-Renewable and Renewable Energy Consumption and CO_2 Emissions in OECD Countries: A Comparative Analysis. *Energy Policy* **2014,** *66*, 547–556. https://doi.org/10.1016/j.enpol.2013.10.064

Theron, J.; Walker, J. A.; Cloete, T. E. Nanotechnology and Water Treatment: Applications and Emerging Opportunities. *Crit. Rev. Microbiol.* **2008,** *34* (1), 43–69. https://doi.org/10.1080/10408410701710442

US Energy Information Administration (EIA). International Energy Outlook 2017 Overview, 2017

WHO, Unicef. Global Water Supply and Sanitation Assessment 2000 Report. *Water Supply*, 2000

CHAPTER 17

Sustainability Aspects of Nano-Remediation and Nano-Phytoremediation

MISBAH NAZ[1], MUHAMMAD AMMAR RAZA[2], SARAH BOUZROUD[3], ESSA ALI[4], SYED ASAD HUSSAIN BUKHARI[5], MUHAMMAD TARIQ[6], and XIAORONG FAN[1,7*]

[1]State Key Laboratory of Crop Genetics and Germplasm Enhancement, Nanjing Agricultural University, Nanjing 210095, China

[2]College of Food Science and Biotechnology, Key Laboratory of Fruits and Vegetables Postharvest and Processing Technology Research of Zhejiang Province, Zhejiang Gongshang University, Hangzhou 310018, China

[3]Laboratoire de Biotechnologie et Physiologie Végétales, Centre de biotechnologie végétale et microbienne biodiversité et environnement, Faculté des Sciences, Université Mohammed V de Rabat, Rabat 1014, Morocco

[4]Institute of Plant Genetics and Developmental Biology, College of Chemistry and Life Sciences, Zhejiang Normal University, Jinhua 321000, China

[5]Department of Agronomy, Bahauddin Zakariya University, Multan 60800, Pakistan

[6]Faculty of Pharmacy and Alternative Medicine, The Islamia University Bahawalpur 6300, Pakistan

[7]Key Laboratory of Plant Nutrition and Fertilization in Lower-Middle Reaches of the Yangtze River, Ministry of Agriculture, and Nanjing Agricultural University, Nanjing 210095, China

[*]Corresponding author. E-mail: xiaorongfan@njau.edu.cn

Nanotechnology for Environmental Pollution Decontamination: Tools, Methods, and Approaches for Detection and Remediation. Fernanda Maria Policarpo Tonelli, Rouf Ahmad Bhat, & Gowhar Hamid Dar (Eds.)

ABSTRACT

Environmental pollution has an impact on the quality of the soil, water, and atmosphere. Major attempts have recently been made to reduce pollution sources and fix or restore degraded soil and water supplies. Compared with chemical and physical methods, phytoremediation and nanoremediation are cost-effective and have less adverse effects. They have become more and more important in academia as well as this area. Nanoparticles have been shown to have soil and water remediation ability in a variety of plant species. It is also to be anticipated that the most recent biotechnology research will play a significant role in the creation of new environmental remediation technologies. This chapter attempts to briefly introduce the latest developments in phytoremediation study and experimentation and nanoremediation of soil and water resources. According to the phytoremediation properties of plants, they have excellent development potential. These plants absorb pollutants from natural resources such as soil and water. Pollutants target different mechanisms, such as dividing enzyme activity into plant organelles. Heavy metals are pollutants that have undergone epidemiological research, and most of the pollutants are covered by industrial waste. Another type of waste is pharmaceutical waste that has a broad variety of adverse effects on human health. They come from mishandled antibiotics, hormones, and drugs. To understand the fate of nanomaterials in environment and environmentally friendly approaches to remove them from environment, it is interesting to explore the hazards of these pollutants in order to develop remedial measures against these pollutants.

17.1 INTRODUCTION

Green plants are used in phytoremediation to remove toxins from polluted sources such as soil, water, air, and sediment (Yan et al., 2021). In general, phytoremediation requires five purification processes, which are rhizofiltration, phytostabilization, phytoextraction, phytovolatilization phytodegradation, and phytoremediation (Tangahu et al., 2011). Phytoremediation techniques are not new to science (Lone et al., 2008). It is reported that about 300 years ago, plants were used to treat wastewater. The remedy, which improves the advantages of minimal site disruption, reduced erosion, and the elimination of the need to dispose of contaminated plant materials, the release of mercury into the environment may pass through precipitation cycles, making this technique the most contentious of all phytoremediation techniques and then re-deposit back into the ecosystem (Singh et al., 2012). Phytoremediation, the use of green plants to manage and regulate pollution in water, soil, and air, is an integral aspect of new fields of ecological engineering (McCutcheon and Schnoor, 2003). Site soil and water characteristics, as well as nutrient sustainability, control in situ and in-situ activities, meteorology, hydrology, pollutant attributes, and sustainable habitats (McCutcheon and Schnoor, 2003). Plant toxicity and large-scale transport restrictions or bioavailability are often critical in applications because of their dependence on sunlight and in-situ nutrient recovery, many applications are affordable. However, large treatment areas and longer treatment periods are typically restricted to root and shallow water areas wetlands, grasslands, crops, and tree plantations have been successfully linked to a number of wastes, often at low concentrations and without serious plant toxicity (Cruz et al., 2019). Metals and metal derivatives, certain heterogeneous contaminants and salt-soaked sources, sewage, sludge, and other typical wastes are examples of organic and inorganic wastes. Depending on the acute ness of toxicity, some redundancy or alternative treatment may be required to counteract the variability of the biological system (McIntyre, 2003). However, few phytoremediation techniques use the basic principles of eco-engineering to optimize sustainability. The application of monoculture of hybrids and sometimes alien species and simple ecosystems of plants and microorganisms is feasible, but in some cases difficult to apply. Self-engineering and self-design need to be explored and used to apply sustainable ecosystems to manage waste (Haq et al., 2020). This chapter provides valuable information about the consequences of nanomaterials released in the environment. Nanomaterials, especially AgNPs, can affect soil properties, microorganisms, and plants, and may be toxic to living organisms including humans. Plant extraction is considered to be a promising method for removing metal-based nanomaterials. However, there is still a lack of information in this regard. Further research is recommended to better understand the mechanism of plant uptake of AgNPs. Understanding the transporters and metabolites involved in removing AgNPs from the environment will help scientists to improve the success rate of phytoremediation of these materials by using plant biotechnology methods to change the expression of related genes.

17.2 CERTAIN PLANTS FOR PHYTOREMEDIATION IN ASIA

Heavy metals are a type of pollutant found in water and soil. A modern plant solar-driven plant technology in which plants are grown, becomes an eco-friendly and cost-calculating method to remove heavy metals from a variety of media to help hyper accumulating plant species (Sharma and Pandey, 2014). The purpose of this review paper is to give information on the mechanisms of compounding and metal removal of plants, with a special emphasis on cadmium (Cd) metals, and to highlight the role of various super-stored plants in the reuse of Cd metals in soil and water. It is consistent with a number of field case reports that are critical in understanding Cd removal in different factories. In addition, which refers to a variety of sources and consequences for Cd and other Cd-correction technology (DalCorso et al., 2008). This work is the latest advancement in Cd hyper accumulation mechanism for various plants to encourage further study in this area. Every year, new plants are identified, and known plants are used in experiments to observe the impact of pollution. *Brassica juncea* (Indian mustard) and *Eichhornia crassipes* (water hyacinth) have the highest trend of absorbing heavy metals from soil and water, respectively. Numerous studies have found that some plants improve the biodegradability of certain exogenous organic molecules in polluted soil. However, the information mechanisms, stages and the role of plants, directly or indirectly, in transforming these Compounds are scarce (Sarma, 2011). The purpose of this work is to put a plant one by one technology for cleaning contaminated soil is a useful alternative. General mechanism and plant use for specific stages, as well as complex interactions with plant native study of microorganisms-heteroclites in soil (Mejáre and Bülow, 2001). This knowledge will allow the proposed solve pollution problems and eventually recover. In recent decades, the annual increase in the pollution of synthetic organic compounds has become a significant human and environmental health problem. China, like other nations, has yet to find an effective solution to the issue of organic soil and water pollution (Schwarzenbach et al., 2010).

Over the past few years, China has been exposed to inorganic pollutants by billions of degrees, but as international cooperation has shown, Chinese scientists are increasingly polluting the environment than their different organic compounds. They are, like their foreign counterparts, are attempting to investigate cost-effective technology to solve this issue, as they are conscientious that traditional techniques for dealing with contaminated soil are inadequate. We concentrate on soil and water polluted with organic matter in China, as well as research activities on various plants in China, including plant intake of pollutants to conduct research on contaminated sites (Liu and Raven, 2010). Heavy metals, radionuclides, and organic contaminants are the major pollutants that lead to soil depletion in China. This chapter is a new technology for heavy metal and other inorganic pollutant soil contamination in China. In this section, we concentrate on organic matter-contaminated soil and water in China as well as research programs on various plants in China, such as plant intake of pollutants to conduct research on contaminated sites. The application of this method on bench and on a scale was briefly discussed (Wen et al., 2020). In India, urbanization, overuse of water bodies, and population increase are causing air, water, and soil pollution, as well as pollution. India's main environment is soil degradation (deforestation, overgrazing, overfarming, and irrigation failures), destruction of animals, destruction, erosion of habitats and genetic resources (crops and plants, as well as aquatic animals and fish), as well as pollution (air, water and soil pollution, toxic waste and other substances). The most pressing environmental issue in India is soil preservation and degradation (wasteland/marginal land) (Rajkumar et al., 2010). Soil erosion is a major issue in India from 130.5 million years old to 16.4 million stateless terrain deformations. Soil erosion was reported under different land-use options, with marginal loss noticed while planting trees, while in silicon field systems, trees, and grass coexist. For example, Shivaliks (one of the most vulnerable companies in the foothills of the Himalayas) includes a combination of poplar-baba grass; Acacia classification grass; Lukaena-Napier grass; teak-Lukaena-Babar; Yukaliptus-Lukaena-Turme; poplar-Lukana-Babar; and Sessamu-canola seeds. The soil of the Ganges alluvial plains in India has the characteristics of pH, poor permeability, higher exchangeable sodium and phosphorus, soil dispersion, a lack of organic matter as well as low fertility. Specific planting tools have been introduced for the cultivation of multipurpose tree species in Sodic and saline soils. A model of a silicon field consisting of Prosopijuli flora and Leputoclo Afosca has also been produced and the alkali soil seems to have been lyon. One major issue is land degradation; again, floods affect some 11.6 million hectares of land in India.

Environmental pollution has an effect on soil and water quality as a result of urbanization and makeup trends. For many years, including in Pakistan, the situation has been even more threatening, since no precautions have been taken to avoid the resolution of this issue. Although at present, many of the current technologies in Taiwan are used to carry out the environment, including one of the plants (Anjum et al., 2020). Plants are used to remove toxins from the atmosphere through this most environmental-friendly technology. Pakistan does have a diverse range of plants that can be used to clean up pollution in the environment. As far as we know, research on the use of plants rarely comes from Pakistan. According to recent research, 50 Pakistani plants have been studied. In this study, the ability of various plant species in Asian countries is addressed in the sections below (Garbisu and Alkorta, 2001) (Table 17.1).

TABLE 17.1 Phytoremediation Basic Natural Resource Best Plants for Phytoremediation in Asia.

Plant local name	Scientific name	Origin	Nature/function	References
Indian mustard/ mustard	*Brassica juncea L.*	India, Pakistan	Water-loving plants are useful for accumulating some metals as well as beautifying landscapes.	Qiu et al. (2021).
Willow	*Salix species*	India, China	The beneficial effect of poplar trees on soil and underwater has also been extensively researched.	Rockwood et al. (2004)
Poplar tree	*Populus deltoides*	India	The effect of the sunflower rhizosphere on the biodegradation of PAHs in soil shows that sunflowers reduce various PAH levels in soil.	Rockwood et al. (2004)
Indian grass	*Sorghastrum nutans*	India	Skilled of cleaning up petroleum hydrocarbons.	Sharma and Pandey (2014)
Sunflower	*Helianthus Annuus L.*	Pakistan, China, India	Pb, Zn (heavy metals extraction potential of sunflower) (*Helianthus annuus*) and Canola (Brassica napus)), N, P, K, Cd, Cu, or Mn are examples of heavy metals (capability of heavy metals absorption by corn, alfalfa and sunflower intercropping date palm)	Reza (2017)
Castor plant	*Ricinus communis,*	India, Pakistan, China	A plant of commercial importance, it has also been used as a biofuel crop. Commonly found growing in clusters, these plants can survive in stress and environmentally polluted areas.	Annapurna et al. (2016)
Jatropha	*Jatropha curcas*	India, China	Applied during the dry season and infertile and heavily contaminated soils. Accumulation of chromium (Cr) and lead (Pb).	Agamuthu et al. (2010)

17.3 PHYTOREMEDIATION FOR SOIL AND WATER RESOURCES

Heavy metals are among the most severe environmental contaminants. Several approaches have been used to eliminate these chemicals from the atmosphere, but the majority are costly and difficult to produce the best results (Tangahu et al., 2011). Currently, plant phytoremediation is an efficient process and cost-effective technological solution for removing or reusing active metals and metal pollutants from polluted soil and water. The technology is environmentally friendly and possibly less expensive. The goal of this paper is to gather some information about the sources, effects as well as the treatment of heavy metals like arsenic, lead, and mercury (ass, lead, and mercury) (Burakov et al., 2018). It also provided an in-depth review of plant restoration techniques, along with heavy metal acquisition mechanisms and a number of studies on these topics. In addition, it explains several sources of As, as well as the environmental impact of As., Pb, and Hg, the benefits of this technique in lowering them, and the mechanisms for obtaining heavy metals in plant restoration techniques and factors that affect the acceptance mechanism (Chibuike and Obiora, 2014). Several recommended plants that are frequently used in plant phytoremediation and their ability to reduce pollutants were also reported. Heavy metals are a type of environmental pollution. In addition to all those human activities, including natural activities, have the opportunity to enhance to heavy metal side effects (Jaishankar et al., 2014). Some of the case splendor severing events in ecosystems are the migration of these pollutants to uncontaminated areas like dust or seepage in soil, and heavy metals containing sewage sludge. To uphold the high quality of the soil and water from pollution, attempts are continually being made to build innovations that are simple to use and long-lasting and economically viable. To fix polluted soil and physical chemistry techniques have been commonly used and water, particularly on small levels (Keesstra et al., 2018). However, due to the high cost and side effects, they face greater large-scale remediation difficulties. Since the last decade, usage of plant species to mop up polluted soil and waters called plant phytoremediation (Mani and Kumar, 2014). It has attracted more and more attention as a new and cheap technology. Over the past two decades, many studies have been carried out in this area. A large number of plant species have been described and evaluated to detect and put their abilities to the test by ingesting and accumulating various heavy metals. The metal acquisition mechanism of the entire plant and cell level was studied. Progress in the mechanical and practical application of plant phytoremediation was studied by Morillo and Villaverde, 2017.

17.3.1 ROLE OF PHYTOREMEDIATION IN POLLUTED SOIL AND WATER

The importance of plants phytoremediation is that it is usually harmless, low cost, and uses plant materials or fixed elements. On the other hand, the process is long and allows some contaminants to escape (Sharma, 2012). In addition, it is not suitable for certain types of contamination, which is not possible if the toxicity level to plants is too high. Plant phytoremediation is a repair technique based on plants and their interactions with soil and microorganisms (Glick, 2003). In particular, the technology includes water purification and soil restoration. The efficiency of the technology in tackling air pollution is often controversial. Combining soil microorganisms, plants, and algae reduces the fluidity of some pollutants (plant stability), absorbs them (plant extraction), anchors them in tissues (plant stability), or metabolizes them, detoxifies, and eliminates them (plant degradation) (Liu et al., 2020). Heavy metals are a form of pollutant found in soil and water that is toxic. Plant restoration, a modern plant solar-powered technology, has become an eco-friendly and cost-effective way to extract heavy metals from different media, due to the super-accumulation of plant organisms (Chandra and Kumar, 2018; Shabbir et al., 2020). This chapter paper's aim is to give information on plant regeneration and heavy metal removal processes, especially cadmium (Cd) metals, and to identify the role of various super-stored plants in soil and water Cd metal abatement. It is consistent with a number of field case studies that are helpful in determining Cd removal in different factories (Yang et al., 2016). In addition, it identifies many sources and their consequences for Cd and other Cd-correction technology. The latest development of cd super-accumulation mechanism of different plants in order to promote further research in this field environmental pollution affects the quality of the biosphere, hydrosphere, atmosphere, lithosphere, and biosphere. Huge efforts have been made over the last couple of decades to reduce pollution sources and regulate polluted soil and water supplies. It was observed that pb levels in roots and shoots after 12 weeks indicated that more bioavailable pools of pb were transferred from the roots to seeds, leaves, and stems in this order. The plant has the ability of plant extraction and can be used to restore the soil contaminated by Pb. Heavy metal pollution has been a concern over the past few decades because of its health hazards to humans as well as other species that reside in biological systems (Lu et al., 2015). Obviously, plant restoration is good for restoring balance in a tense environment, but care must be taken. The potential of *glycine Max* L. to repair lead-contaminated soil. After 12 weeks of restoration, the plant's seeds usually contain the highest Pb, which means that the plant is in the late stages of its growth in cleaning contaminated soil (Huang et al., 1997). Therefore, this plant, when used to repair metal contaminated soil, should be harvested after 12 weeks and replanted seeds for another clean-up process. However, without thoroughly examining lead contamination in the soil, planting *G.max* L. in the soil poses a great risk to the population spending the harvested seeds, since the plant has been reported to be increased significant amounts of Pb in its seeds (Islam et al., 2007).

17.4 PHYTOREMEDIATION AND RHIZOFILTRATION

Rhizofiltration is a type of plant phytoremediation restoration in which polluted groundwater and surface water are filtered and wastewater through a large number of roots to extract harmful contaminants or excess nutrients (Dhir et al., 2009) (Fig. 17.1). Rhizofiltration, a technique for absorbing or precipitating contaminants from surrounding solutions to plant roots. Rhizofiltration is exactly the same as plant mining, Plants, rather than soil, are used in the process to clean polluted groundwater (e.g., plant mining) (Gunarathne et al., 2019). In the process, used plants grow in greenhouses, with their roots in water. Contaminated water is then stored from the waste site or plants are planted in polluted areas. When pollutants saturate the roots of plants, they are removed from contaminated areas (Etim, 2012). Sunflowers, for example, are used to clear radioactive pollutants from pond water in Chernobyl, Ukraine. Lead may be extracted from water by sunflowers, Indian mustard, tobacco, spinach, and corn. In particular, sunflowers decreased lead concentration dramatically, and an engineered rheumatic device (Akpor and Muchie, 2010). Plant phytoremediation is the use of plants to eliminate toxins from the atmosphere. The capacity to use terrestrial and marine plants for in situ or in-situ applications. Rhizo filtration is primarily used to eliminate heavy metals such as Pb, Cd, Cu, Ni, Zn, and Cr (Jadia and Fulekar, 2009). One important aspect of this procedure is that pollutants do not move to shoot buds (Jadia and Fulekar, 2009). Water filtration is an exciting technique to improve aquatic environments. Aside from being cost-effective, it is a more environmentally friendly choice. A few studies have shown that many macrophytes have an exceptional capacity to extract toxins from aquatic environments (Rai, 2009). These plants could be used to eliminate inorganic contaminants (heavy metals, ions, excess nutrients, and so on) as well as organic pollutants (dyes, pesticides, solvents, and so on) from surfaces and groundwater. *E. crassipes* (water hyacinth), *P. stratiotes* (water lettuce), *L.*

minor (duckweed), *T. natans* (water chestnut), and *Azolla pinnata* (water velvet) naturally grow in large animals living in aquatic environments and can effectively accumulate and pass pollutants to its airborne part (Bragato et al., 2006). It is needed to establish more appropriate and optimal conditions to further strengthen the rheumatic filtration process study and is also needed to develop a system for the proper disposal or actual use of harvested plants after removal. Flow pipes are the process of absorbing contaminants present in the nasal layer into the plant roots (Petrides et al., 2014).

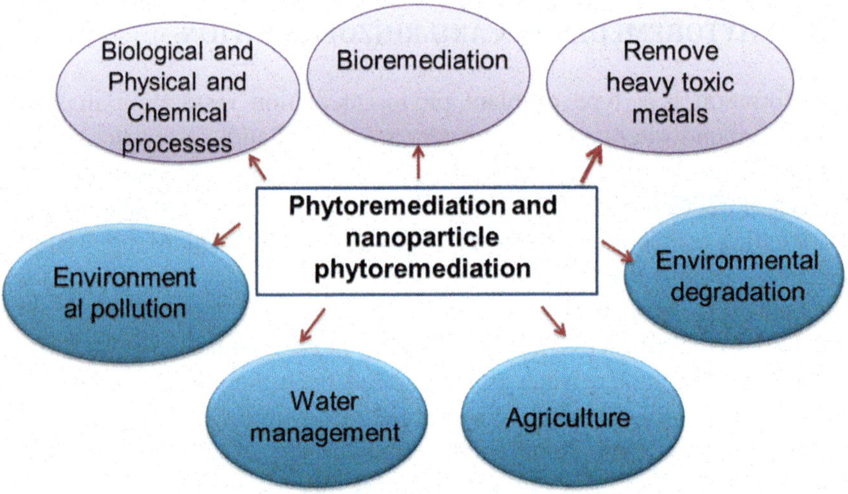

FIGURE 17.1 Role of phytoremediation and nanoparticle phytoremediation.

Rhizofiltration usually treats contaminated groundwater, while plant extraction targets contaminated soil. As plants usually grow in greenhouses and are exposed to contaminated water once they grow roots (Awa and Hadibarata, 2020) (Fig. 17.2). This adaptation allows the plant's endogenous process to adapt to chemicals in the water. After adaptation, plants are transplanted into contaminated aquifers or soil above the water level below the habitat, and when the plant's absorption is thought to have reached a threshold, they are harvested and biomass is treated in a similar way to the biomass extracted by the plant (Pivetz, 2001). Rhizofiltration is particularly effective in applications involving low concentrations and large amounts of water. Plants that are efficient in transferring metals to buds should not be used for soil filtration because more contaminated plant residues are produced. The depth of contamination, the type of heavy metals, and the level of contamination must be determined and monitored. Vegetation shall be aquatic, emergency, or water-affected (Nagajyoti et al., 2010). To design successfully, the hydraulic retention time and adsorption of the plant roots must be taken into account. Rhizofiltration is economical for large amounts of water with lower toxic elements and low (strict) standards. It is relatively cheap, but it's probably more effective than similar technologies (Toet et al., 2005).

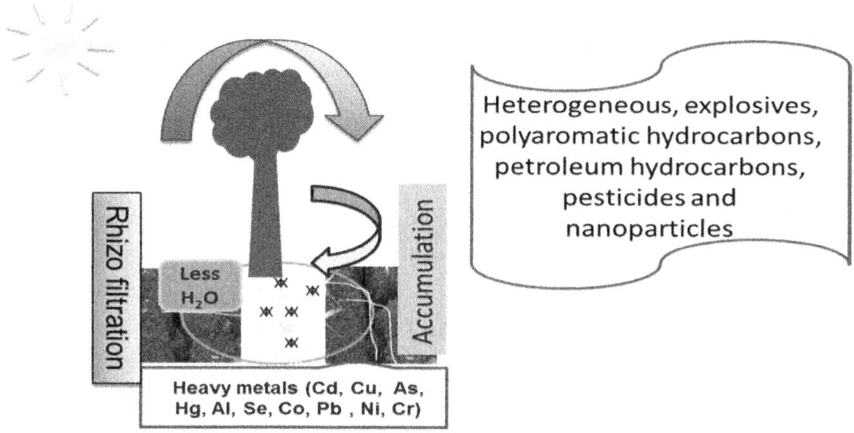

FIGURE 17.2 Phytoremediation of heavy metals rich soils by using plants that hyperaccumulate these metals in above-ground organs. The harvesting of the aerial part of the plants leads to the disposal of the huge amounts of toxic heavy metals removed from the soil or to the recovery of the valuable metals taken up.

17.5 NANOPARTICLES PHYTOREMEDIATION

Nanoparticles are particles that exist on the nanoscale (i.e., a minimum of one dimension is below 100 nm. They have physical properties such as uniformity, conductivity, or special optical properties, making them popular in materials science and biology. As global industrialization and urbanization intensify, so does concerns about emerging environmental pollutants (Ebrahimbabaie et al., 2020). Plant technologies are recommended as a viable option for maintaining environmental sustainability among the various options for repairing these pollutants. Current developments in phytoremediation, genetic/molecular/economic/metabolic engineering, and nanotechnology have provided new possibilities for the successful management of emerging organic/inorganic pollutants (Srivastav et al., 2018). For this context, the clarification of molecular processes and genetic modification of super-accumulated plants are required to improve environmental pollutant remediation. The aim of this analysis is to give useful information about the molecular mechanisms of plant regeneration and the prospects for genetically modified super-accumulators (Furukawa et al., 2020).

Heavy metals and metals, heterogeneous, explosives, polyaromatic hydrocarbons, petroleum hydrocarbons, pesticides, and nanoparticles are examples of such substances that are more stress-resistant to different pollutants. The role of gene restoration and nanoparticles in enhancing plant restoration techniques is also described in frameworks related to biotechnology prospects, such as plant molecular nano-agriculture (Krivoruchko et al., 2019). Finally, nanotechnology is the study and manipulation of matter at the nanoscale and the characteristics of nanomaterials are mainly derived as a result of their high relative surface area as well as quantum impacts. Any material intended to produce at the nanoscale with a specific characteristic or specific component is termed a manufacturing/engineering nanomaterial. Compared with conventional nanomaterials, this engineering nanomaterial has different properties (Zhang and Webster, 2009). Nanotechnology offers new agrochemical formulations Nanotechnology provides new agrochemical formulations and delivery pathways to improve crop yields, which is expected to minimize pesticide usage. Nanotechnology can promote agricultural production by using: (1) nanoformulations for agrochemicals for the use of pesticides and fertilizers to boost crop; precision farming techniques can be used to further increase crop productivity without damaging soil and water, reducing nitrogen losses from seepage and emissions, and increasing nutrients that soil microbes for a long term (Sekhon, 2014). Nanoparticles are used in crop production to achieve long-term sustainability as these particles are linked to the manipulation of essential plant life circumstances and are used for a number of purposes in agriculture, such as minimizing nutritional loss, reducing various types of environmental stress, including heavy metal stress, and increasing crop yields. Several forms of contaminants are released into the atmosphere, either inadvertently or on purpose. These contaminants must be repaired in order to avoid entry to various terrestrial, marine, and atmospheric habitats (Sarwar et al., 2010).

Various pollutants, including heavy metals, have been observed, characterized, and designed to enhance this capacity of these plants through the use of different types of formulations (e.g., plant hormones, nanoparticles, biochar, etc.). In fact, some plants that do not usually help remedy have been manipulating their physiology and biochemistry to repair contaminants through the chemical's exogenous applications (Margesin et al., 2011). However, in this regard, nanoparticles are remarkable because they can alleviate different forms of heavy metal stresses in plants, because they protect plants from induced oxidative stress and mimic the role of enzymes involved in oxidative metabolism, such as catalytic enzymes, peroxidases, and superoxide dislocations (Etesami and Jeong, 2018). Since olden history, silver metal, silver nitrate, and silver sulfonamide have been used to treat burns, wounds, and various bacterial infections. Nevertheless, prior to the advancement of many antibiotics, use of such silver compounds has decreased significantly. Nanotechnology has gained a huge boost this century because it has been able to regulate metals to nanosizes, dramatically modifying the chemical, physical, and optical properties of metals (Knetsch and Koole, 2011).

Metal silver, in the context of silver nanoparticles, has made a big comeback as a possible antibacterial agent. The need for silver nanoparticles is also important since many pathogenic bacteria have gained resistance to various antibiotics. Silver nanoparticles have appeared in a variety of medical applications, ranging from silver-based dressings to silver-coated medicinal devices such as nanogels. The promise of nanotechnology has been fulfilled and the greatest scientific and technological progress has been achieved in a number of areas (Rai et al., 2009).

The biological killing nature of metal nanoparticles in general and silver nanoparticles in particular (AgNPs) depends on the multiple morphology as well as the particles' chemical and physical properties. Many of the interactions between AgNPs and the human body remain difficult to understand, therefore, the development of nanoparticles with good control morphology and physical chemistry characteristics, the application in the human body is still an active advanced research (Durán et al., 2016). The discovery of nanostructure compounds appears to have a massive and all-encompassing effect on all scientific and technical fields. The security monitoring mechanism, on the other hand, is used and implemented. Nanotechnology has performed on its pledge to drive scientific and technological development in a range of fields. Nanostructure compounds have a wide range of engineering applications. The creation of biocompatible and environmentally friendly nanostructures will reduce ecosystem harm;, allow better use of resources in the synthesis of multiple processes, and save (Venugopal et al., 2008).

17.6 CONCLUSION

Metal phytoremediation in soil, sludge, wastewater, and water, the various techniques used, the biological, physical, and chemical processes involved, and the benefits and disadvantages of each strategy are discussed in this chapter. Special attention is given to the use of genetically modified species and the phytoremediation of metal nanoparticles. This also demonstrates the ability of phytoremediation to extract heavy toxic metals from polluted soil. As a result of the given debate, it is possible to conclude that, like its applications in other scientific fields, it has extensive application in bioremediation. Because of their high potential, their applications are expected to skyrocket in the foreseeable future, and plants play an important role in stable growth. Various plants are essential for toxic metal remediation.

KEYWORDS

- **phytoremediation**
- **polluted soil and water resources**
- **environmental remediation**
- **morphology**
- **physical chemical characteristics**

REFERENCES

Agamuthu, P.; Abioye, O. P.; Aziz, A. A. Phytoremediation of Soil Contaminated with Used Lubricating Oil Using *Jatropha curcas*. *J. Hazard. Mater.* **2010,** *179*, 891–894.

Akpor, O. B.; Muchie, M. Remediation of Heavy Metals in Drinking Water and Wastewater Treatment Systems: Processes and Applications. *Int. J. Phys. Sci.* **2010,** *5*, 1807–1817.

Anjum, M. S.; Ali, S. M.; Subhani, M. A.; Anwar, M. N.; Nizami, A.-S.; Ashraf, U. et al. An Emerged Challenge of Air Pollution and Ever-Increasing Particulate Matter in Pakistan; A Critical Review. *J. Hazard. Mater.;* **2020,** *402*, 123943.

Annapurna, D.; Rajkumar, M.;Prasad, M. N. V. Potential of Castor Bean (*Ricinus communis* L.) for Phytoremediation of Metalliferous Waste Assisted by Plant Growth-Promoting Bacteria: Possible Cogeneration of Economic Products. In *Bioremed. Bioecon.*; Elsevier, 2016; pp 149–175.

Awa, S. H.;Hadibarata, T. Removal of Heavy Metals in Contaminated Soil by Phytoremediation Mechanism: A Review. *Water Air Soil Pollut.* **2020,** *231*, 1–15.

Bragato, C.; Brix, H.; Malagoli, M. Accumulation of Nutrients and Heavy Metals in *Phragmites australis* (Cav.) Trin. ex Steudel and *Bolboschoenus maritimus* (L.) Palla in a Constructed Wetland of the Venice Lagoon Watershed. *Environ. Pollut.* **2006,** *144*, 967–975.

Burakov, A. E.; Galunin, E. V, Burakova, I. V, Kucherova, A. E.; Agarwal, S.; Tkachev, A. G. et al. Adsorption of Heavy Metals on Conventional and Nanostructured Materials for Wastewater Treatment Purposes: A Review. *Ecotoxicol. Environ. Saf.* **2018,** *148*, 702–712.

Chandra, R.; Kumar, V. *Phytoremediation: A Green Sustainable Technology for Industrial Waste Management*; CRC Press: Boca Raton, FL, 2018; pp 1–42.

Chibuike, G. U.; Obiora, S. C. Heavy Metal Polluted Soils: Effect on Plants and Bioremediation Methods. *Appl. Environ. Soil Sci.* **2014,** *2014.*

Cruz, N. C.; Silva, F. C.; Tarelho, L. A. C.; Rodrigues, S. M. Critical Review of Key Variables Affecting Potential Recycling Applications of Ash Produced at Large-Scale Biomass Combustion Plants. *Resour. Conserv. Recycl.* **2019,** *150*, 104427.

DalCorso, G.; Farinati, S.; Maistri, S.; Furini, A. How Plants Cope with Cadmium: Staking All on Metabolism and Gene Expression. *J. Integr. Plant Biol.* **2008,** *50*, 1268–1280.

Dhir, B.; Sharmila, P.; Saradhi, P. P. Potential of Aquatic Macrophytes for Removing Contaminants from the Environment. *Crit. Rev. Environ. Sci. Technol.* **2009,** *39*, 754–781.

Durán, N.; Durán, M.; De Jesus, M. B.; Seabra, A. B.; Fávaro, W. J.; Nakazato, G. Silver Nanoparticles: A New View on Mechanistic Aspects on Antimicrobial Activity. *Nanomed. Nanotechnol. Biol. Med.* **2016,** *12*, 789–799.

Ebrahimbabaie, P.; Meeinkuirt, W.; Pichtel, J. Phytoremediation of Engineered Nanoparticles Using Aquatic Plants: Mechanisms and Practical Feasibility. *J. Environ. Sci.* **2020,** *93*, 151–163.

Etesami, H.; Jeong, B. R. Silicon (Si): Review and Future Prospects on the Action Mechanisms in Alleviating Biotic and Abiotic Stresses in Plants. *Ecotoxicol. Environ. Saf.* **2018,** *147*, 881–896.

Etim, E. E. Phytoremediation and Its Mechanisms: A Review. *Int. J. Environ. Bioenergy* **2012,** *2*, 120–136.

Furukawa, K.; Ohmi, Y.; Yesmin, F.; Tajima, O.; Kondo, Y.; Zhang, P. et al. Novel Molecular Mechanisms of Gangliosides in the Nervous System Elucidated by Genetic Engineering. *Int. J. Mol. Sci.* **2020,** *21*, 1906.

Garbisu, C.; Alkorta, I. Phytoextraction: A Cost-Effective Plant-Based Technology for the Removal of Metals from the Environment. *Bioresour. Technol.* **2001,** *77*, 229–236.

Glick, B. R. Phytoremediation: Synergistic Use of Plants and Bacteria to Clean up the Environment. *Biotechnol. Adv.* **2003,** *21*, 383–393.

Gunarathne, V.; Mayakaduwa, S.; Ashiq, A.; Weerakoon, S. R.; Biswas, J. K.; Vithanage, M. Transgenic Plants: Benefits, Applications, and Potential Risks in Phytoremediation. In *Transgenic Plant Technology for Remediation of Toxic Metals and Metalloids*; Elsevier, 2019; pp 89–102.

Haq, S.; Bhatti, A. A.; Dar, Z. A.; Bhat, S. A. Phytoremediation of Heavy Metals: An Eco-Friendly and Sustainable Approach. In *Bioremediation and Biotechnology*; Springer, 2020; pp 215–231.

Huang, J. W.; Chen, J.; Berti, W. R.; Cunningham, S. D. Phytoremediation of Lead-Contaminated Soils: Role of Synthetic Chelates in Lead Phytoextraction. *Environ. Sci. Technol.* **1997,** *31*, 800–805.

Islam, E.; Yang, X.; Li, T.; Liu, D.; Jin, X.; Meng, F. Effect of Pb Toxicity on Root Morphology, Physiology and Ultrastructure in the Two Ecotypes of *Elsholtzia argyi*. *J. Hazard. Mater.* **2007,** *147*, 806–816.

Jadia, C. D.; Fulekar, M. H. Phytoremediation of Heavy Metals: Recent Techniques. *Afr. J. Biotechnol.* **2009,** *8*, 6.

Jaishankar, M.; Tseten, T.; Anbalagan, N.; Mathew, B. B.; Beeregowda, K. N. Toxicity, Mechanism and Health Effects of Some Heavy Metals. *Interdiscip. Toxicol.* **2014,** *7*, 60–72.

Keesstra, S.; Mol, G.; De Leeuw, J.; Okx, J.; De Cleen, M.; Visser, S. Soil-Related Sustainable Development Goals: Four Concepts to Make Land Degradation Neutrality and Restoration Work. *Land.* 2018, *7*, 133.

Knetsch, M. L. W.; Koole, L. H. New Strategies in the Development of Antimicrobial Coatings: The Example of Increasing Usage of Silver and Silver Nanoparticles. *Polymers (Basel).* **2011,** *3,* 340–366.

Krivoruchko, A.; Kuyukina, M.; Ivshina, I. Advanced Rhodococcus Biocatalysts for Environmental Biotechnologies. *Catalysts* **2019,** *9,* 236.

Liu, J.; Raven, P. H. China's Environmental Challenges and Implications for the World. *Crit. Rev. Environ. Sci. Technol.* **2010,** *40,* 823–851.

Liu, S.; Yang, B.; Liang, Y.; Xiao, Y.; Fang, J. Prospect of Phytoremediation Combined with Other Approaches for Remediation of Heavy Metal-Polluted Soils. *Environ. Sci. Pollut. Res.* **2020,** *27,* 16069–16085.

Lone, M. I.; He, Z.; Stoffella, P. J.; Yang, X. Phytoremediation of Heavy Metal Polluted Soils and Water: Progresses and Perspectives. *J. Zhejiang Univ. Sci. B.* **2008,** *9,* 210–220.

Lu, Y.; Song, S.; Wang, R.; Liu, Z.; Meng, J.; Sweetman, A. J. et al. Impacts of Soil and Water Pollution on Food Safety and Health Risks in China. *Environ. Int.* **2015,** *77,* 5–15.

Mani, D.; Kumar, C. Biotechnological Advances in Bioremediation of Heavy Metals Contaminated Ecosystems: An Overview with Special Reference to Phytoremediation. *Int. J. Environ. Sci. Technol.* **2014,** *11,* 843–872.

Margesin, R.; Płaza, G. A.; Kasenbacher, S. Characterization of Bacterial Communities at Heavy-Metal-Contaminated Sites. *Chemosphere.* **2011,** *82,* 1583–1588.

McCutcheon, S. C.; Schnoor, J. L. Overview of Phytotransformation and Control of Wastes. In *Phytoremediation: Transformations and Control of Contaminants*; Wiley: New York, 2003; pp 1–58.

McIntyre, T. Phytoremediation of Heavy Metals from Soils. *Phytoremediation* **2003,** 97–123.

Mejáre, M.; Bülow, L. Metal-Binding Proteins and Peptides in Bioremediation and Phytoremediation of Heavy Metals. *Trends Biotechnol.* **2001,** *19,* 67–73.

Morillo, E.; Villaverde, J. Advanced Technologies for the Remediation of Pesticide-Contaminated Soils. *Sci. Total Environ.* **2017,** *586,* 576–597.

Nagajyoti, P. C.; Lee, K. D.; Sreekanth, T. V. M. Heavy Metals, Occurrence and Toxicity for Plants: A Review. *Environ. Chem. Lett.* **2010,** *8,* 199–216.

Petrides, D.; Carmichael, D.; Siletti, C.; Koulouris, A. Biopharmaceutical Process Optimization with Simulation and Scheduling Tools. *Bioengineering* **2014,** *1,* 154–187.

Pivetz, B. E. Phytoremediation of Contaminated Soil and Ground Water at Hazardous Waste Sites. US Environmental Protection Agency, Office of Research and Development, 2001.

Qiu, Y.; Dixon, M.; Liu, G. Chinese Mustard Cultivation Guide for Florida. *EDIS,* **2021,** 1.

Rai, M.; Yadav, A.; Gade, A. Silver Nanoparticles as a New Generation of Antimicrobials. *Biotechnol. Adv.* **2009,** *27,* 76–83.

Rai, P. K. Heavy Metal Phytoremediation from Aquatic Ecosystems with Special Reference to Macrophytes. *Crit. Rev. Environ. Sci. Technol.* **2009,** *39,* 697–753.

Rajkumar, M.; Ae, N.; Prasad, M. N. V.; Freitas, H. Potential of Siderophore-Producing Bacteria for Improving Heavy Metal Phytoextraction. *Trends Biotechnol.* **2010,** *28,* 142–149.

Reza, M. Effects of Plant Growth-Promoting Bacteria on the Phytoremediation of Cadmium-Contaminated Soil by Sunflower. *Arch. fuì r Acker-und Pflanzenbau und Bodenkd.* **2017,** *6,* 807–816.

Rockwood, D. L.; Naidu, C. V, Carter, D. R.; Rahmani, M.; Spriggs, T. A.; Lin, C. et al. Short-Rotation Woody Crops and Phytoremediation: Opportunities for Agroforestry? In *New Vistas in Agroforestry*; Springer, 2004; pp 51–63.

Sarma, H. Metal hyperaccumulation in Plants: A Review Focusing on Phytoremediation Technology. *J. Environ. Sci. Technol.* **2011,** *4*, 118–138.

Sarwar, N.; Malhi, S. S.; Zia, M. H.; Naeem, A.; Bibi, S.; Farid, G. Role of Mineral Nutrition in Minimizing Cadmium Accumulation by Plants. *J. Sci. Food Agric.* **2010,** *90*, 925–937.

Schwarzenbach, R. P.; Egli, T.; Hofstetter, T. B.; Von Gunten, U.; Wehrli, B. Global Water Pollution and Human Health. *Annu. Rev. Environ. Resour.* **2010**, *35*, 109–136.

Sekhon, B. S. Nanotechnology in Agri-Food Production: An Overview. *Nanotechnol. Sci. Appl.* **2014,** *7*, 31.

Shabbir, Z.; Sardar, A.; Shabbir, A.; Abbas, G.; Shamshad, S.; Khalid, S. et al. Copper Uptake, Essentiality, Toxicity, Detoxification and Risk Assessment in Soil-Plant Environment. *Chemosphere*, **2020,** *259*, 127436.

Sharma, P.; Pandey, S. Status of Phytoremediation in World Scenario. *Int. J. Environ. Bioremed. Biodegrad.* **2014,** *2*, 178–191.

Sharma, S. Bioremediation: Features, Strategies and Applications. *Asian J. Pharm. Life Sci.* **2012,** *2231*, 4423.

Singh, D.; Tiwari, A.; Gupta, R. Phytoremediation of Lead from Wastewater Using Aquatic Plants. *J. Agric. Technol.* **2012,** *8*, 1–11.

Srivastav, A.; Yadav, K. K.; Yadav, S.; Gupta, N.; Singh, J. K.; Katiyar, R. et al. Nano-Phytoremediation of Pollutants from Contaminated Soil Environment: Current Scenario and Future Prospects. In *Phytoremediation*; Springer, 2018; pp 383–401.

Tangahu, B. V.; Sheikh Abdullah, S. R.; Basri, H.; Idris, M.; Anuar, N.; Mukhlisin, M. A Review on Heavy Metals (As, Pb, and Hg) Uptake by Plants through Phytoremediation. *Int. J. Chem. Eng.* **2011,** *2011*.

Toet, S.; Van Logtestijn, R. S. P.; Kampf, R.; Schreijer, M.; Verhoeven, J. T. A. The Effect of Hydraulic Retention Time on the Removal of Pollutants from Sewage Treatment Plant Effluent in a Surface-Flow Wetland System. *Wetlands* **2005,** *25*, 375–391.

Venugopal, J.; Prabhakaran, M. P.; Low, S.; Choon, A. T.; Zhang, Y. Z.; Deepika, G. et al. Nanotechnology for Nanomedicine and Delivery of Drugs. *Curr. Pharm. Des.* **2008,** *14*, 2184–2200.

Wen, D.; Fu, R.; Li, Q. Removal of Inorganic Contaminants in Soil by Electrokinetic Remediation Technologies: A Review. *J. Hazard. Mater.* **2020,** *401*, 123345.

Yan, A.; Wang, Y.; Tan, S. N.; Yusof, M. L. M.; Ghosh, S.; Chen, Z. Phytoremediation: A Promising Approach for Revegetation of Heavy Metal-Polluted Land. *Front. Plant Sci.* **2020,** *11*.

Yan, L.; Van Le, Q.; Sonne, C.; Yang, Y.; Yang, H.; Gu, H. et al. Phytoremediation of Radionuclides in Soil, Sediments and Water. *J. Hazard. Mater.* **2021,** *407*, 124771.

Yang, J.; Tang, C.; Wang, F.; Wu, Y. Co-Contamination of Cu and Cd in Paddy Fields: Using Periphyton to Entrap Heavy Metals. *J. Hazard. Mater.* **2016,** *304*, 150–158.

Zhang, L.; Webster, T. J. Nanotechnology and Nanomaterials: Promises for Improved Tissue Regeneration. *Nano Today* **2009,** *4*, 66–80.

Index

A

Acetylcholinesterase (AChE), 198
Aeroponic techniques
  yield greater biomass
    food crops, 474–478
    nonfood crops, 478–479
Air pollution management, 62, 409
  commercialized nanosenors, 428–430
  sensor devices, 414–416, 416–418
    0D nanomaterials, 418
    1D nanomaterials, 420–422
    2D nanostructured materials, 423
    3D nanostructures materials, 423–424
  toxic gases, nanotechnology adsorption
    adsorption of isopropyl alcohol, 427–428
    carbon dioxide capture, 426
    dioxins, 424–425
    NOX, 425
    volatile organic materials from air, elimination, 426–427
Alcanivorax borkumensis, 206
Applied nanotechnology, economic impact, 495
  commercialization of, 497–499
  enabling technology, 500
    in energy, 501–502
    in environment and agriculture, 503–504
    in healthcare, 500–501
    in water treatment and purification, 504
  green economy and nanotechnology, 499–500
Arbuscular mycorrhizal fungus (AMF)
  aeroponic techniques, yield greater biomass
    food crops, 474–478
    nonfood crops, 478–479
  radical mycelium network inoculum, 474
Artificial enzymes
  catalytic monolayers
    alkanethiol-protected AUNPS, 173–175
    moieties covered by alkanethiol, 176–177
    RNASE (DNASE), 175–176
    RNase, transphosphorylation, 177
    TACN ligand, 175
    thiolated biomolecules-protected, 176
  enzymatic catalytic properties, 158
  nanoenzymes applications
    air degradation, 178
    biofilm, formation, 180–181
    chemical contaminants, wastewater depletion, 178–179
    Chemical Warfare Agents (CWA), 180
    human factors, 177
    transphosphorylation, 178
  nanoenzymes catalytic pathways
    catalase, 164
    glucose oxidase, 161
    glutathione peroxidase, 163
    haloperoxidase, 163–164
    peroxidase, 162–163
    sulfite oxidase, 162
  nanoenzymes, classification
    enzyme mimics with applications, 161
    metabolic pathways, 159
    and natural enzymes, 160
  nanomaterials
    carbon nanotubes as, 171–172
    fullerene and derivatives, 169–170
    graphene and derivatives, 170–171
    hybrid enzymes, 159
    imitated by superoxide dismutase, 170
    imitators of peroxidase, 173
    peroxidase mimics, 171
  tuning of, 164
    composition, 166–168
    constructing hybrid nanoparticles, 167
    lighting, 169
    molecules or ions, 168–169
    morphology, 165
    pH and temperature, 168
    size, 165
    steric effect, 165
    surface modification, 166

B

Bioremediation
  types
    mycoremediation, 52
    phytodegradation, 54
    phytoextraction, 52–53
    phytostabilization, 53
    phytovolatization, 53–54
    rhizofiltration, 53
Biosensing
  carbon nanofibers (CNFs), 277
  carbon nanohorns, 278
  carbon nanotubes (CNTs), 273–275
  fullerene, 277
  Graphene (GR) structure, 275–276
  synthesis of
    carbon nanofibers (CNFs), 280
    carbon nanotubes (CNTs), 278–279
    fullerenes, 280–281
    graphene, 279–280
Bisphenol-A (BPA), 202–203

C

Carbon nanotube (CNTs)
  carbon quantum dots (CQDs), 108–109
  fullerenes, 107
  graphene sheet, 107–108
  nanohorns, 109
  types, 105–106
Carbon quantum dots (CQDs), 104
Carbonaceous materials, nanoremediation, 345
Carbon-based nanomaterials
  biological contaminants, sensing, 286
    pathogenic bacteria, 287–288
    pathogenic virus, 287
  biosensing
    carbon nanofibers (CNFs), 277
    carbon nanohorns, 278
    carbon nanotubes (CNTs), 273–275
    fullerene, 277
    Graphene (GR) structure, 275–276
  carbon nanotube (CNTs)
    carbon quantum dots (CQDs), 108–109
    fullerenes, 107
    graphene sheet, 107–108
    nanohorns, 109
    types, 105–106
  chemical contaminants, sensing
    antibiotics, 283
    endocrine-disrupting chemicals (EDCs), 284
    heavy metal pollutants, 284–285
    nitro-based explosives, 282
    noxious gases, 286
    pesticides, 285–286
    pharmaceutical pollutants, 283
    small organic molecules, 282
  classification of
    carbon quantum dots (CQDs), 104
  consumption, 264
  molecular level alterations, 103
  monitoring environmental pollution, 281
  nanobiosensors, 265
    components of, 266
  nanoparticle-based biosensors, 267
    acoustic wave biosensors, 268
    bionanomaterial-based biosensors, 270–271
    biorecognition or bioreceptor, 271
    electrochemical biosensors, 269, 272
    electrochemical transducer, 271–272
    magnetic biosensor, 268–269
    mechanical nanobiosensors, 272–273
    nanotubes-based biosensors, 269–270
    nanowires-based biosensors, 270
    optical nanobiosensors, 271
  properties of, 109
    bonding configuration, 110
    molecular interaction and sorption, 111–112
    new electronic properties, 110
  synthesis of biosensing
    carbon nanofibers (CNFs), 280
    carbon nanotubes (CNTs), 278–279
    fullerenes, 280–281
    graphene, 279–280
Catalytic monolayers
  alkanethiol-protected AUNPS, 173–175
  moieties covered by alkanethiol, 176–177
  RNASE (DNASE), 175–176
  RNase, transphosphorylation, 177
  TACN ligand, 175
  thiolated biomolecules-protected, 176
Chemical Warfare Agents (CWA), 180

D

Decabromodiphenyl ether (c-Deca-BDE), 17
Dichlorodiphenyltrichloroethane (DDT), 16–17
Dioxin-like compounds (DLCs), 17

E

ELISA (enzyme-linked immunosorbent assay), 208
Endocrine-disrupting chemicals (EDCs), 284
Engineered water nanostructures (EWNS), 212
Enrofloxacin (ENR), 198
Environmental pollutants
 detection of, 120
  C_{20} fullerene, 123
  carbon nanostructures, 126
  carbon-based sensors, 123
  density functional theory (DFT), 123
  drugs, 127
  ecological monitoring, 127
  electrochemical sensors, 127
  fullerene-based nanocomposites, 123
  GPN–quantum dots (QDs), 124
  Graphene–CNTs, 122
  Graphene-Nafion-based sensor, 125
  heavy metals, 124
  humidity, 125
  mechanical exfoliation (ME) method, 121
  MWCNTs, 122
  natural gas, 122
  peroxy-nanosensor (PNS), 128
  phenolic compounds, 123
  pressure sensors, 126
  QCM humidity sensor, 125
  UEPA (US Environmental Protection Agency), 121
  water eutrophication, 123
 and detections
  air pollution, 112
  examples, 113–114
  inorganic gases, 112
  monitoring, 116
  soil pollution, 115–116
  VOC (volatile organic compounds), 112
  water pollution, 115
 sensors based carbon nanomaterials, types
  biosensors, 117–118
  chemical sensors, 118–119
  experiment-based studies, 119
  flow sensors, 120
  mass sensing, 120
  pressure sensors, 119
  strain sensors, 119–120
  temperature sensors, 120
Environmental pollutions
 applications
  biosensing mechanism, 88
  heavy metal pollutants, 78–79, 82–84
  organic pollutants, 84–86
  paper-based SERS swab, 88–89
  PCBs determination, 87
  SPR nanobiosensor, 88
 chromatography technique, 63
 nanosensor, 63
  chemical nanosensor, 64
  classification, 65
  electrometer, 64–65
  materials, 65–68
  mechanism principles, 64
 sensing types, 68
  colorimetric nanosensor, 75–76
  electrochemical techniques, 69–72
  magnetic nanosensors, 76–78
  optical nanosensor, 72–75
 water pollution, 62
Environmentalist, 102
Enzymes, 158
European Food Protection Authority (EFSA), 205

F

Förster resonance energy transfer (FRET), 211

G

Gold nanoparticles (AuNPs), 202
Graphene quantum spots (GQDs), 209
Green algae *(Chlorella)*, 52

H

Hexabromocyclododecane (HBCDD), 20
Hexachlorobenzene (HCB), 21
Hexachlorocyclohexane (HCH), 21–22
Horseradish peroxidase (HRP), 236

I

Inorganic environmental pollutants, 49
biological approaches
green algae *(Chlorella),* 52
microbial fuel cell (MFC), 51–52
bioremediation, types
mycoremediation, 52
phytodegradation, 54
phytoextraction, 52–53
phytostabilization, 53
phytovolatization, 53–54
rhizofiltration, 53
chemical management
advertising material, 50
recurring biochar condition, 51
sorbents and adsorbents, use, 51
Inorganic pollutants
environmental pollutants
inorganic pollutants, 43–44
organic pollutants, 43
origin and types
human health risks, 48
Methane, 44
nutrient pollution, 48–49
radioactive substances, 47–48
waste collection, 44
water, 44
PTEs, 41–42

L

Lab-on-a-chip (LOC) devices, 143

M

Mercury-specific oligonucleotides (MSOs), 198
Metal organic frameworks (MOFs), 139
Microbial fuel cell (MFC), 51–52
MRSA (methicillin-resistant *Staphylococcus aureus*), 207

N

Nanobiosensors, 190
agricultural industry, 197
attributes and key elements
biologically sensitive material, 193
detection, 192
nanomaterials, 194
principal structure, 194
trans, 194
detection signal
colorimetric assay, 200
field-effect transistor (FET), 202
genetic mutation, 201
heavy metals, 200
nano-biomaterials, 199
peptide-based gold nanoparticle, 200
polycyclic aromatic hydrocarbons (PAHs), 201
water, 200
Xanthocerassorbifolia extract (XT-AuNP), 201
DNA nanosensors
ELISA method, 207
GMO, transgene-containing, 207
heavy metal pollutants, 208
MRSA (methicillin-resistant *Staphylococcus aureus*), 207
pathogen contamination, 206
pesticides, 206–207
Rayleigh scattering (RRS), 208
immuno-based nano-biosensor
electrochemical immunosensor, 209
ELISA (enzyme-linked immunosorbent assay), 208
graphene quantum spots (GQDs), 209
immobilization, 210
Pseudomonas aeruginosa, 209
Staphylococcus aureus, 209
surface plasmon reverberation (SPR), 209
two-photon Rayleigh scattering (TPRS) spectroscopy, 209
integrated with enzymes
minimal detection limit, 196
organophosphorus pesticide, 196
progression in, 195
microorganism-based
agriculture networks, 204
Alcanivorax borkumensis, 206
DOT-T1E biosensor, 205
European Food Protection Authority (EFSA), 205
monitoring environmental pollutants
acetylcholinesterase (AChE), 198
enrofloxacin (ENR), 198

fluorescence, 196
life-threatening heavy metal, 198
mercury-specific oligonucleotides (MSOs), 198
zymolytic-modified nanomaterials, 199
nanowire biosensors, 203
Boron-doped silicon nanowires (SiNWs), 204
CNT/Nafion complex, 204
phenolic compounds
amalgam of Nafion, 202
bisphenol-A (BPA), 202–203
detection of pollutants, 203
gold nanoparticles (AuNPs), 202
nanoporous gold film (NPGF), 203
tyrosinase enzyme, 202
pollution detection
array type, 210
engineered water nanostructures (EWNS), 212
Förster resonance energy transfer (FRET), 211
lead, 211
Ralstonia solanacearum, 211
rapid and sensitive detecting systems, 211
single-walled carbon nanotube field-effect-transistor (swCNT-FET), 212
Nanoenzymes
applications
air degradation, 178
biofilm, formation, 180–181
chemical contaminants, wastewater depletion, 178–179
Chemical Warfare Agents (CWA), 180
human factors, 177
transphosphorylation, 178
catalytic pathways
catalase, 164
glucose oxidase, 161
glutathione peroxidase, 163
haloperoxidase, 163–164
peroxidase, 162–163
sulfite oxidase, 162
classification
enzyme mimics with applications, 161
metabolic pathways, 159
and natural enzymes, 160
Nanomaterials
hazardous, safety, management, 487
environmental hazards, 490–491
exposure and handling, 491–492
health hazards, 488–490
management of risk, 492
Nano-phytoremediation, 445
nano zerovalent iron (NZVI OR FE0)
heavy metals, technologies, 455–456
materials, 452–454
pollutants and plants, 452–454
pollutants, technologies of, 448–449
remediation of environmental sites, 450–452
in soil and water
nanotechnology and organic pollutants, 456–461
Nanoporous gold film (NPGF), 203
Nano-remediation, 509
phytoremediation
in Asia, plants for, 511–514
nanoparticles, 519–521
polluted soil and water, role, 515
and rhizofiltration, 517–519
soil and water resources, 515
Nanozymes. *See* Artificial enzymes
Noncarbon-based nanomaterials, 136
advantagious and preferential choice
biosensing technology, 224
DLS processes, 223
factors, 223
metal oxides, 226
noncarbonbased, 223
pollutants, 226
challenges, 243–245
composite nanobiosensors
metal oxide-metal oxide composite biosensor, 239–240
metal-metal composite based biosensor, 237–238
metal-metal oxide based biosensor, 238
noncarbon/carbon composites, 241–243
contaminants mechanisms, detection
colorimetric sensors, 139–140
electrochemical techniques, 141–142
fluorescence emission, 142–143
lab-on-a-chip (LOC) devices, 143
surface enhanced Raman spectroscopy (SERS), 140–141

environmental application, 221
environmental contaminants, detection of, 143
heavy metals, 147–148
nanomaterials for, 145–146
organic contaminants and gases, 148–149
pathogenic microorganisms, 144
persistent organic pollutants (POPs), 148
quantum dots, 147
Raman reporter molecules, 146
receptors, 144
environmental monitoring and forensics
chitason, 236
horseradish peroxidase (HRP), 236
metal oxide based biosensor, 231
metal-based nanobiosensors, 227–231
SNO_2 based biosensor and environmental applications, 232–233
sol-gel method, 236
TIO_2-based biosensor and environmental applications, 231–232
WO_3-based biosensor and environmental applications, 234–235
ZNO-based biosensor and environmental applications, 233–234
metal organic frameworks (MOFs), 139
noble metal nanoparticles, 137
quantum dots, 137–138

P

Pentachlorobenzene (PeCB), 23–24
Pentachlorophenol (PCP), 24
Peroxy-nanosensor (PNS), 128
Persistent organic pollutants (POPs), 148
active pharmaceutical ingredients, 26–27
anthropogenic origin, 6
synthetic substances, 9
classification, 9
diversity of, 4
emerging contaminants, 26–27
hazardous
biological and chemical degradation, 5
exposure, health problems, 6
semivolatile and evaporate, 5
Stockholm Convention, 5
health problems, 7–8
indexes/obtaining fees
contaminated drinking water, 28
dietary exposure, 28
ingestion from soil, 27
inhalation, 27
skin absorption from soil, 27–28
molecules, 4
priority pollutants
aldrin, 10, 15
chlordane, 15
chlordecone, 15–16
decabromodiphenyl ether (c-Deca-BDE), 17
dichlorodiphenyltrichloroethane (DDT), 16–17
dioxin-like compounds (DLCs), 17
endosulfan, 18–19
furan and dibenzofurans, 18
heptachlor, 19
hexabromobiphenyl, 20
hexabromocyclododecane (HBCDD), 20
hexachlorobenzene (HCB), 21
hexachlorocyclohexane (HCH), 21–22
lindane, 22
mirex, 22–23
pentachlorobenzene (PeCB), 23–24
pentachlorophenol (PCP), 24
polybrominateddiphenyl ethers (PBDEs), 23
polychlorinated biphenyls (PCBs), 24–25
polychlorinated naphthalenes (PCNs), 25–26
Stockholm Convention, 10, 11–14
toxic components, 9
synthetic chemicals, 4
Polybrominateddiphenyl ethers (PBDEs), 23
Polychlorinated biphenyls (PCBs), 24–25
Polychlorinated naphthalenes (PCNs), 25–26
Polycyclic aromatic hydrocarbons (PAHs), 201
Pseudomonas aeruginosa, 209

R

Rayleigh scattering (RRS), 208
Remediate water pollution
degradation
nanomotors, 323–324
nanophotocatalysts, 324–327
domestic wastewater possesses, 302
industrial wastewater, 303

landfills, 302
nanomaterials (NMs), 304
nanotechnology field, 304
noncarbon-based nanomaterial, 305–307
Mentha piperita, 308
nZVI nanoparticles (nZVI-NPs), 308
silver nanoparticles, 309
water remediation, 310–314
water remediation, technologies, 314
filtration, 315–318
nanoadsorbents, 318–322
nanomotors, 322–323

S

Sensor devices, 414–416, 416–418
0D nanomaterials, 418
1D nanomaterials, 420–422
2D nanostructured materials, 423
3D nanostructures materials, 423–424
Single-walled carbon nanotube field-effect-transistor (swCNT-FET), 212
Staphylococcus aureus, 209
Surface enhanced Raman spectroscopy (SERS), 140–141
Surface plasmon reverberation (SPR), 209

T

Two-photon Rayleigh scattering (TPRS) spectroscopy, 209

U

UEPA (US Environmental Protection Agency), 121

V

VOC (volatile organic compounds), 112

W

Water pollution, 62

X

Xanthocerassorbifolia extract (XT-AuNP), 201